Smart Cities: Infrastructure, Innovation, Technology, Governance and Citizenship—Volume I

Smart Cities: Infrastructure, Innovation, Technology, Governance and Citizenship—Volume I

Topic Editors

Luis Hernández-Callejo
Sergio Nesmachnow
Pedro Moreno-Bernal

MDPI

Basel • Beijing • Wuhan • Barcelona • Belgrade • Novi Sad • Cluj • Manchester

Topic Editors

Luis Hernández-Callejo
Department of Agricultural and Forestry Engineering
University of Valladolid
Soria
Spain

Sergio Nesmachnow
Instituto de Computación
Universidad de la República
Montevideo
Uruguay

Pedro Moreno-Bernal
Faculty of Accounting, Administration, and Informatics, Autonomous University of Morelos State (UAEM)
Cuernavaca
Mexico

Editorial Office
MDPI AG
Grosspeteranlage 5
4052 Basel, Switzerland

This is a reprint of the Topic, published open access by the journals *Buildings* (ISSN 2075-5309), *Sensors* (ISSN 1424-8220), *Smart Cities* (ISSN 2624-6511), *Sustainability* (ISSN 2071-1050), *Urban Science* (ISSN 2413-8851) and *Applied Sciences* (ISSN 2076-3417), freely accessible at: https://www.mdpi.com/topics/1QQN8GMG0O.

For citation purposes, cite each article independently as indicated on the article page online and as indicated below:

Lastname, Firstname, Firstname Lastname, and Firstname Lastname. Article Title. *Journal Name* **Year**, *Volume Number*, Page Range.

Volume I
ISBN 978-3-7258-5013-6 (Hbk)
ISBN 978-3-7258-5014-3 (PDF)
https://doi.org/10.3390/books978-3-7258-5014-3

Set
ISBN 978-3-7258-4951-2 (Hbk)
ISBN 978-3-7258-4952-9 (PDF)

Contents

About the Editors

Luis Hernández-Callejo

Luis Hernández-Callejo is a full professor and researcher at Universidad de Valladolid, Spain. He has participated in numerous research projects and successfully supervised more than 10 doctoral theses. He is the editor-in-chief of several prestigious journals and a guest editor for numerous Special Issues in several other journals. His research interests include renewable energy, smart cities, microgrids, and artificial intelligence applied to engineering.

Sergio Nesmachnow

Sergio Nesmachnow is a full professor and researcher at Universidad de la República, Uruguay. He has more than 450 published articles across journals and international conferences. His main research areas are high performance computing, optimization, and smart cities.

Pedro Moreno-Bernal

Pedro Moreno-Bernal is a full-time Associate Professor at the Universidad Autónoma del Estado de Morelos and a National Research Associate at the Secretaría de Ciencia, Humanidades, Tecnología e Innovación in Mexico. His areas of interest include computational intelligence, metaheuristics, energy sustainability, and smart cities. He has co-authored multiple indexed scientific publications, 11 of which were indexed in Journal Citation Reports and 31 in Scopus, resulting in an h-index of 5. Furthermore, he has served as a guest editor for Special Issues in various journals.

Preface

Cities are constantly changing, and this is the main reason we are launching this Reprint of articles. The aim is to collect scientific articles on different topics that affect cities, and these topics mainly include energy, IoT, computing, and others.

The main purpose of this Reprint is to present some research advances across the above-mentioned topics, which can serve as a solid foundation for researchers who are starting to work on topics related to smart cities.

Luis Hernández-Callejo, Sergio Nesmachnow, and Pedro Moreno-Bernal
Topic Editors

smart cities

MDPI

Article

Open Data Insights from a Smart Bridge Datathon: A Multi-Stakeholder Observation of Smart City Open Data in Practice

Sage Cammers-Goodwin

Philosophy Section, Department of Behavioral, Management and Social Sciences, University of Twente, Drienerlolaan 5, 7522 NB Enschede, The Netherlands; sagecammersgoodwin@gmail.com

Abstract: "Open Data" efforts are growing, especially in Europe, where open data are seen as a possible ethical driver of innovation. As smart cities continue to develop, it is important to explore how open data will affect the stakeholders of smart public spaces. Making data open and accessible not only has a managerial and technical component but also creates opportunities to shift power dynamics by granting individuals (and entities) access to data they might not otherwise be able to obtain. The scope of those who could access these data is wide, including data-illiterate citizens, burgeoning startups, and foreign militaries. This paper details the process of making data "open" from the MX3D smart bridge in Amsterdam through a "datathon". The development and outcomes of opening the data and the event itself bring us closer to understanding the complexity of open data access and the extent to which it is useful or empowering for members of the public. While open data research continues to expand, there is still a dearth of studies that qualitatively detail the process and stakeholder concerns for a modern smart city project. This article serves to fill this gap.

Keywords: open data; datathon; citizen science; hackathon; smart bridge

1. Introduction

Data collection is everywhere, home utility measurement, our coddled smartphones, and increasingly in public spaces, where government and corporate actors use algorithmic decision making generated from sensor inputs for a multitude of purposes [1]. With so much data collected that either may be relevant to the public, partially funded by the public, or generated by the public, the power imbalance between the data owners and "the public" in smart public space is an underlying concern.

While tremendous effort has been put in place to increase consumer awareness of online data collection (primarily of data that is personal in nature), the standardization of data collection visibility in public space lags behind [2]. This means that the practitioners of public spaces rarely know to what extent data are collected around them, much less how and why it is used or how to access it. Moreover, the focus on personal data disappoints in the public arena. Any data collected from one's personal device or home that keeps track of the origin point is personal, thereby calling for increased consumer awareness, access, and control under the General Data Protection Regulation (GDPR) and similar laws. Public space is not inherently connected back to the individual, making the notification of collection and the right to access data not guaranteed.

The power imbalances between varying data stakeholders in the smart city and the rampant datafication of public spaces make the push towards "open data" potentially disruptive [3,4]. Open data can be viewed as the idea that data can be safely and ethically open to the public and should be made freely available. Andrew Schrock describes the history of open data in the United States as rooted in the early 1900s movement for freedom of information which naturally evolved with the internet, thereby shifting the focus of "openness" from information to data [5]. Schrock shares that "civic hackers" both instigated and transformed with the shift to open data. Depending on one's views, civic hackers, individuals who demand and use civic open data, can be geeky neoliberals or essential fixtures for democratic engagement and positive change [5].

In 2019, the European Union (EU) amended the Open Data Directive, which pushes for more datasets stemming from the public sector to be made openly accessible [6]. In the academic sector, the Open Science approach has been seen by some as the EU prioritizing innovation and "technical optimization" over ethics-first design principles like those fostered by Responsible Research and Innovation [7]. A literature review revealed that open data are "1.12-1: to 181:1" times more economically valuable than closed data [8] (p. 67).

While open data are gaining academic notice, there is still a dearth of studies that qualitatively detail the process and stakeholder concerns for a modern smart city project. This article serves to fill this gap by addressing the issue of open data in the smart city through the lens of business, academia, government, and the general public by documenting the trials of opening a small subset of data from a smart bridge in Amsterdam. In order to analyze what people think of open data and what they do with it, a datathon was hosted, where participants were interviewed about their experience. This analysis is useful because data sharing is more easily written about idealistically than put into practice. The often-hidden details that complicate the practice of open data are shared in this piece along with universal analysis.

The datathon described in this paper is by no means the first event exposing smart city data. Academia, government, and corporations alike have begun to embrace the concept of civic hackers as inspiration to host urban hackathons, usually posed with a design challenge, as a participatory means to urban development [9]. For example, in 2014, a hackathon hosted in Denmark, with 65 university students participating, was made possible via a "triple helix" collaboration between academia, corporations, and government. Notably, citizens were not included in the triple helix model [10]. The city of Mississauga in Canada hosted an open data challenge in 2020 with 18 days of competition that ended with a winning team [11,12]. India's Ministry of Housing and Public Affairs hosted an open data week from 17 to 22 January 2022 with participation from their 100 smart cities on their Smart Cities Open Data Portal [13]. On 5 March 2022, Open Data Day, an annual celebration beginning in 2019, drew 153 international events with people and datasets from all over the world [14].

Given that open data are posited to become more of a norm in the smart city [15], the time is ripe to analyze stakeholder interests and shifting power dynamics without the emphasis on civic hackers or specified design challenges. Civic hackers represent a minority of the population and specified design challenges do not necessarily translate to how a population might choose to use open data. This article proceeds in four parts. Section 2 outlines the parameters of this study. It gives context to the bridge, data, research consortium, and datathon centered in this work. Section 3 articulates the concerns and interests of public data sharing from within the research committee and Amsterdam municipality. This paper includes the process of opening data from the bridge and the organization of the datathon. Often these details are not included in academic research to the detriment of those wishing to duplicate similar projects or learn from the realistic challenges of smart city projects. Part Four describes and analyzes the outcomes of the datathon. Part Five discusses the question, "who wins" with open data based on the findings from this case study, which also foreshadows who loses. Finally, the paper concludes with predictions and recommendations for the future of infrastructural smart city open data.

2. Project Parameters

This paper draws upon a case study involving making a subset of data from the MX3D bridge open to the public through a datathon. Figure 1 shows the MX3D bridge, a 12-m-long 3D-printed stainless steel bridge that rests over Oudezijds Achterburgwal Canal in Amsterdam's Red Light district. It is the world's first 3D-printed stainless steel bridge.

Figure 1. MX3D Bridge in Amsterdam City Center, by Thea van den Heuvel.

The bridge has over 90 sensors including load cells, displacement sensors, strain gauges, thermistors, accelerometers, and inclinometers. Readings from the sensors are stored in a local server room and then uploaded to a private cloud space where they are paired with open data climate readings from nearby. The Smarter Bridges Research Consortium is a group composed of corporate and academic actors including Autodesk, the Alan Turing Institute, and the University of Twente. Most members are interested in studying the structure of the bridge. MX3D, the company that created the bridge, was commissioned by Amsterdam municipality to place the bridge for a period between two and ten years with an initial contract for two years. As part of MX3D's contract with the municipality, data from the bridge is owned by the city but shared with the Smarter Bridges Research Consortium through a contract. Consortium members have access to all the data from the bridge in almost real time.

The NWO-funded research group, "BRIdging Data in the built Environment" (BRIDE), is a member of the Smarter Bridges Research Consortium and has long been interested in opening data from the bridge to the public. Some of the goals of BRIDE are to study what people do with the data, foster discussion on smart infrastructure, and ideally make the data a shared resource. The Red Light District, where the bridge is placed, is a hotbed for data collection. According to Amsterdam's crowd monitoring map, as of summer 2022, there were four active crowd monitoring sensors within 100 m of the bridge and an additional three crowd monitoring sensors and four Wi-Fi trackers installed (but inactive) at the same distance from the bridge [16]. These can be seen in Figures 2 and 3. Currently, the public cannot access these data from the municipality, but the police may use these anonymous crowd data. These sensors should be taken as a low estimate because they do not include security cameras for businesses in the region.

Six months after the bridge was placed in Amsterdam, the project successfully applied for a grant to host a datathon to allow members of the public to access data from the bridge. The format of a datathon as opposed to a hackathon was to allow a "create-athon" option to not scare off non-technical participants and to emphasize that the event focused on data from the bridge. Traditionally, datathons are hackathons for data, but our event was divided into two categories, "hack" and "create," each offering a €40 gift card per team member for the winning team and €20 for second place. The panel of judges included a software engineer who works in the Smarter Bridges Research Consortium, an official from Amsterdam municipality, and a professor of philosophy of technology—none of the judges organized the event or had prior connections with the participants.

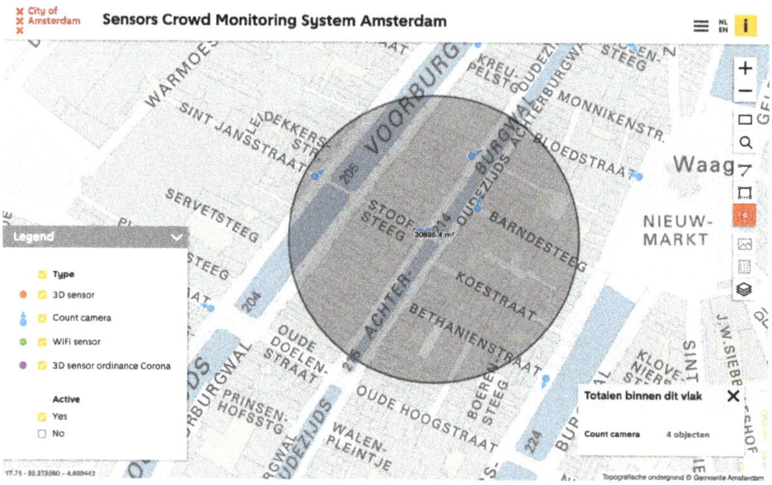

Figure 2. Active Crowd Monitoring Sensors within 100m of the Bridge.

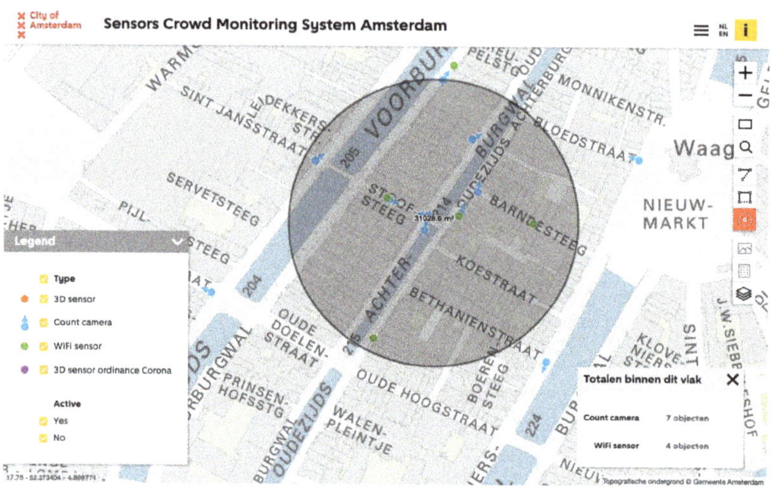

Figure 3. All Crowd Monitoring Sensors within 100m of the Bridge.

The datathon was advertised on social media, through the University of Twente Design Lab, through the University of Twente event page, on Eventbrite, at a local non-profit community space, and at student group meetings. In total, 19 participants joined from backgrounds including computer science, design, and civil engineering, 16 of which ended up actively participating in the majority of the event. Ethics Review Board permission for interviews and data sharing was obtained from the Behavioral, Management, and Social Sciences (BMS) ethics committee at the University of Twente. All participants signed a consent form and organizers informed them that the event was also being held for research purposes. The group that joined was mainly young (under 30) with at least some university education. There were three doctoral researchers as well. This study may be seen as a step towards answering questions regarding how smart city open data impacts the public in the smart city but cannot be generalized due to its size and specificity. In order to make predictions about any given population there needs to be a representative sample;

moreover, for a ±10% sampling error where P = 0.5 and the confidence rate is 95% for the Dutch population (approx. 17.5 million) we would need at least 100 participants [17]. This estimate assumes that the attributes are normally distributed [17]. Unfortunately, such a quantitatively rigorous study was outside the scope of the datathon.

3. Corporate Academic and Municipal Concerns

The first step in opening up data from the MX3D bridge to the public was to communicate with the Smarter Bridges Research Consortium. Although hosting an open data event seemed like a good test case for possibly opening up data for the future, it did not come without its concerns. One primary concern was that data from the bridge would be used for research that overlapped with that being done by group members. Given the current nature of academia, novel publications are often needed to justify and adhere to project funding. If a group at the hackathon was able to publish discoveries before the researchers, it would be considered a loss of investment.

Another concern was the possibility of people misinterpreting the data. Given the unique structure of the bridge, how the bridge was placed in the canal, and the possibility of sensors straying from the baseline, additional knowledge is necessary to have an accurate understanding of what sensor data means. The sensor data itself is "truthful" in that it does correctly share what the sensor registered and stored in the database. In actuality, there could be a power outage leading to a loss of data collection. The sensor could have "drifted," meaning that the relative changes in values are accurate but shifted from the true values. A sensor could also just be broken. The bridge is also not directly placed on the load cells, some of the weight from the bridge is distributed directly into the ground. Therefore, adding the data from all four load cells at a given time will not give an accurate reading for the weight on the bridge and the weight of the bridge itself. As shared in a research paper from the project, it also takes engineering expertise to analyze the sensor output to determine bridge safety [18].

The group also considered different options for sharing data at the event. Currently, an Autodesk API is used to share data from the bridge with the researchers. It was thought that participants at the datathon could use the API to access a separate instance of either the whole dataset or part of the dataset. The API route had been considered as a future option for researchers who want to join the project post embargo to access the full dataset. These researchers could register their research interests, name, and institution and sign a code of conduct to access the data from the bridge. This would have been a route for Autodesk to test their API and allow near real-time data access. Another option was to share .txt or .csv files from various sensors without granting access to the API. Sharing files as opposed to the API would reduce the likelihood of a collaborator hacking or breaking the database. It also would mean that the API would not have to be redesigned. For these reasons, the latter option was preferred.

When the group first considered opening data from the bridge, there was also a concern about whether the data would be fully open or only disclosed to event participants for a limited duration. The following question emerged: who can make the decision to share the data and what would be the license? Can individuals do whatever they wish with the data? Is the data then theirs to profit from? Who will act as overseer of the data if someone releases it into the wild on GitHub? It seemed that private data licensing would be difficult to control. The group decided that a public solution would be best and that the only entity that could legally make the decision would be the city of Amsterdam. That said, foreseeing the possibility that the route might not yield favorable results, the group considered using data from when the bridge was at the University of Twente or trying to use a non-disclosure agreement and have the event be part of an internal research project. Both of these latter options were not preferred.

It was eventually decided to preselect datasets that would reduce risk from all the listed concerns. Data would be shared from a select group of more functional sensors. This dataset would be from a time when no extreme values were recorded that could be severely misinterpreted and all the load cells were functioning properly. In a joint decision-making process, three datasets were chosen using eleven sensors: one hour of data at ten readings per second, twelve hours of data with a min, max, and mean reading each second, and two weeks of data with the min, max and mean reading per fifteen minutes. All units were changed to be in their most intelligible form, which meant transferring some units from voltage to what they actually measured using the conversion method for each sensor type. The sensors were also labeled on a two-dimensional image of the bridge so that participants could connect the file name to the sensor type, units, and location on the bridge. Table 1 and Figure 4 show what was finally agreed to be shared with participants.

Table 1. Public Dataset Sensors and Data Frequencies.

			Data Sets		
Sensor	**Name**	**Units**	**1 h, 10 Hz**	**12 h, 1 Hz**	**2 Weeks, 15 Min**
Accelerometers	A01y, A02y, A06y, A07y	Δg (acceleration change)	x	x	x
Load Cells	LC01, LC02, LC03, LC04	kilograms	x	x	x
Strain Gauges	SG21, SG23	Microstrain	x	x	x
Thermistors	T06	Celsius	x	x	x
Displacement Gauges	-	-	-	-	-
Inclinometers	-	-	-	-	-

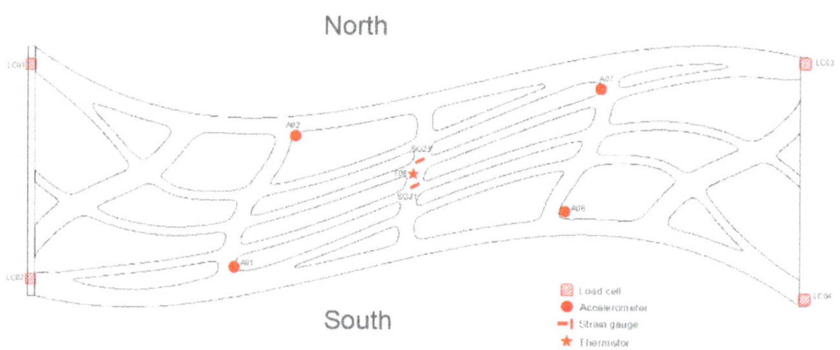

Figure 4. Public Dataset Sensor Locations.

The next step was to communicate with the municipality of Amsterdam. At the time of organizing the event, Amsterdam's privacy office was understaffed in comparison to the load of requests being made. Once contact was made, the author shared information with a privacy officer which was then forwarded to other agents in the municipality. First, they wanted to know what data exactly was going to be made available, where the data were stored, and how the data would be accessed. This is part of the city's initiative for "controlled transparency". This initiative requires the city to be aware of all the data that it opens to the public. For the event itself, the data were stored on the author's University of Twente Microsoft One Drive. Each participant got access to download the dataset through an email invitation. When asked if it would be easier to only open up the dataset for the event, the officer responded that data for some members of the public means that data should be made available for all.

Another issue that arose during discussions with one of the municipality's privacy officers is that the city also follows an assortment of regulations in addition to GDPR. At the time there was "Wet openbaarheid van bestuur" (WOB) which translates to "government (public access) to information act," which as of May 2022 was replaced by "Wet open overheid" (WOO) that translates to "the open government act." WOO promises more active disclosure of information than its predecessor but similarly has grounds for the privacy of information. For example, if the costs of sharing outweigh the benefits to the environment, then the data can remain closed [19]. Additionally, there is "Wet hergebruik overheidsinformatie" (WHO), the government information reuse act and "Wet maatschappelijke ondersteuning" (WMO), the market and government act. It was not enough that the dataset shared for the datathon would be anonymous. The officer explained that anonymous data can expose confidential data and pose security risks. For example, purportedly anonymous traffic light data might show a string of green lights. These light patterns are indicative of sensitive affairs such as large amounts of money being transported to banks or the Dutch Royal Family commuting.

The Dutch government created a document that outlines the various guidelines to help with opening up government data to the public [20]. The first guideline is that data should be proactively made open as long as it does not pose risks to privacy or other fundamental values. The second is that open data should be free of cost to reuse. The third is that the data be released under a CC-0 license, free of third-party rights. The fourth is that data can be accessed without registration. The fifth is that data are released in a standard format that is computer processable. The sixth is that metadata are included according to the DCAT-AP standard (see Appendix A). The seventh is that the data are made as complete and as raw as possible—additional processing can only occur to comply with the disclosure framework provided that it does not violate competition law. Lastly, open data should be registered on data.overheid.nl so that it can be more easily found.

The Open Data guidelines conflict with several of the initial ideas for open data posited by the research consortium. For the datathon, the research consortium decided on a dataset that was not in its most raw state. Output was transferred into the correct units and the collection frequency decreased for longer spans of time to protect the consortium's research advantage and to decrease the file sizes. Furthermore, a smaller subset of sensors was chosen. The full dataset of 10 Hz data from nearly a hundred sensors for several months would be a lot of information to store and possibly hold unforeseen information. Since the Netherlands is encouraging open data as a way to promote innovation, it makes sense for the license to be open for individuals to do what they want with the data and also profit from using it. For a corporation or institute that wants to publish their findings first, making data free from third-party rights takes away their advantage of being the data controllers. Another possible point of contention is that data can be accessed without registration. The Smarter Bridges Research Committee wanted to act like an ethics watchdog over the data, which is not possible without registration. It can be argued that people should have the right to use open data at their own discretion in private.

After a halting month of discussion, mainly due to competition with the many tasks Amsterdam's privacy office oversees, Amsterdam's Chief Technology Officer gave permission via email to release the chosen dataset to the public. In principle, the datathon dataset will be shared with the public beyond individual access via a link on the author's Microsoft One Drive account. However, the process of how and where to share these data is not explicit—data.overheid.nl is one obvious choice, but there is also data.amsterdam.nl. The privacy officer reached out to others in the municipality asking for assistance with these questions, but the public dataset remains in limbo. One relatively innocuous project justifiably is not the highest priority for the city.

Since the city commissioned the bridge and is the technical owner of the data, it would appear from the Open Data Act that the government maintains the responsibility of navigating the post-embargo data opening. Yet, it seems that the municipality does not have the bandwidth to manage all the smart city projects, especially when there is no one-size-fits-all option. The smart city is filled with private and public partnerships where data management and storage are handled by third parties. Denoting responsibility is already challenging to navigate for those building smart infrastructures. A lack of open data is not always due to a lack of interest, but a lack of clarity when in many instances there is not one individual tasked with carrying out the duty and navigating the risks.

4. The Datathon

The goal of the datathon was to observe and analyze what happens when people are given access to open data from the bridge without prompts. In order to attract a variety of experience levels, the event was divided into "Hack" and "Create" as shared in the Project Parameters section. In order to make the data as accessible for the average person as possible, a student assistant was tasked with making a Jupyter Notebook that would help people upload and graph the data. The author also provided MATLAB code to plot data from the bridge. Each participant got access to a folder with all the datasets and a README that explained what sensors were provided, their locations, and a link to a tool to make the high-frequency data play as though it was alive.

After registration, participants were given post-it notes to matchmake and form teams (Figure 5) and a mind map to help themselves brainstorm (Figure 6). Green indicated that they were looking for partners. Pink was for coding experience and orange was for design experience. Participants could put a 'checkmark' if they were experienced, an 'up arrow' if they had limited experience and want to learn more, or an 'X' if they had no experience and had no interest in doing any work in that domain. Next, they were introduced to the bridge, the dataset, and the judging criteria (Table 2). Three groups came in with already formed teams (one of which joined virtually). One group quickly formed of two people who already knew each other and someone sitting at the same table as them. This left the last group to coalesce due to circumstances. The groups were able to design and present their projects for 6.5 h between 10:30 and 17:00.

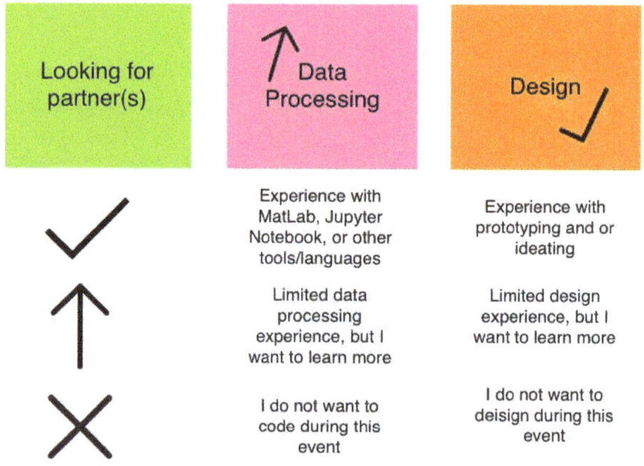

Figure 5. Match Making Post-it Notes.

Each group connected with the data differently based on their expertise. The first group of mainly computer science students and graduates had no trouble technically working with the data and programming with it. They did not bother reading the README file and independently found the GitHub code linked in the README that feeds high-frequency data as if it is alive. They also did not use any of the other tools provided such as the Jupyter Notebook and MATLAB code. As a fun challenge, they tried to think of something as unethical as possible, such as a drunk detection algorithm called "Stumble"—wordplay off of a dating app called "Bumble." Their final project ended up being a tool that would take live data and create music based on the bridge traffic. They were the only team that was able to demo a sample of their final product. Although they had a limited understanding of civil engineering, they were able to build something interesting within the span of a few hours without necessarily knowing how each sensor worked or the bridge mechanics. This group demonstrated that open data might be used in creative ways by hackers for personal projects and that those with coding experience are adept at creating programs with data they might not fully understand.

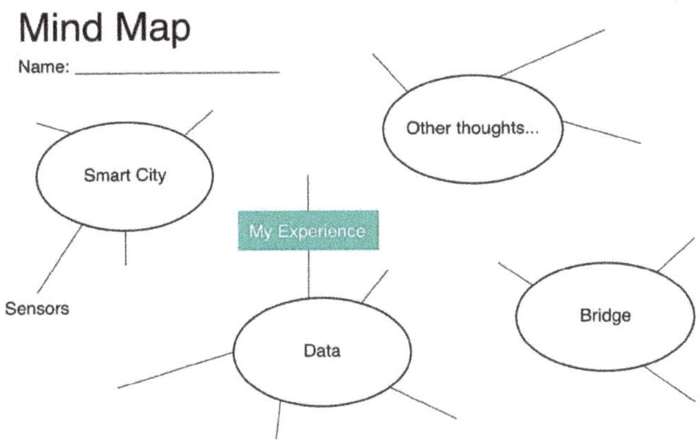

Figure 6. Mind Map (adapted from the event).

Table 2. Datathon Judging Criteria.

Presentation	How clearly explained is the product/solution? If code is shown, can you see how it might work? If a creative solution is made, is it visualized well enough to be understood?
Potential	Does it seem like, if followed through, this idea might work?
Value	A variety of values can be considered, groups should share in what value domains the project should be ranked (ex: usefulness, community, monetary, social, privacy).
Creativity/Innovation	Has the idea been done before? Is an aspect of the solution novel?
Technology	Is the bridge data used in some way (minimum)? Has something been built using code and the bridge data?
Judge Bonus	You bring your own expertise and can edit your ranking here. Please explain your thoughts in the "notes" section.

The second group possessed the most varied mix of backgrounds, reflecting the circumstances of how they came together—more informally than formally. They thought that it was "cool" that these data were shared for the event and that it was fine to share the bridge data with the public because the data were anonymous, but to some, the data still felt very abstract and hard to connect with and understand. Another person shared that they did not feel like they had a right to the data necessarily. The group decided to use the data from the bridge to create a light beam display of fireflies that would "die" when the bridge is too crowded. This would help discourage crowdedness in the area and subtly inform people that the bridge collects data. This group was unsure how the fireflies would be made, which datasets would be used, or how the system would work. The more technically able in their group did attempt to make a prototype using data from the bridge, but ran into issues during the short time frame. While the first group was able to use the data right away without any of the tools provided, for some members of the second group the additional resources were insufficient to connect with the dataset, at least within the timeframe of the datathon.

The third group was completely design focused and decided to not use or examine the bridge data besides a brief peak at some of the files. They shared data and as the data are anonymous, they thought that it was fine to share the data from the bridge and that it was "good" to be transparent. They pointed out that maybe when one individual was on the bridge, they could be tracked. Originally, they thought of creating a meditation system on the bridge that would change based on load so that people could meditate together. The handrails could vibrate to cue breaths. By the end, they changed to a fountain system that would spray water based on interactions with the bridge. This would bring people to the present moment and take them away from their distracting phones. They suggested that this system could be monetized and sold, likely influenced by their interpretation of "generating value" from the judging criteria. This team demonstrated that open data, even when not used, may inspire ideation on possible use cases. Theoretically, if live data from the bridge was available to the public, the second and third groups could apply for a grant and acquire permission from the municipality to create their designs with the assistance of hired engineers.

Group Four had an engineering student who had worked with the bridge before. This person found the dataset frustrating because it offered less information than had been previously accessible to them. It was more challenging to find out where someone was on the bridge, especially for the longer, lower-frequency datasets. They could also not do what they originally planned because the dataset only included two strain gauges in the middle of the bridge as opposed to more physically distanced strain gauges. This group envisioned using screens to visualize the bridge data that could help local businesses predict when they might have more customers. They also thought if everyone has access to the data, then people could work together to point out if there is a flaw in the bridge or in the data itself. This team confirmed that limiting open datasets may restrain innovation from the users of the data. Restraining innovation might be helpful if revealing the location of someone during an individual bridge crossing is deemed harmful. Group Four was also the first group to recognize that the data might not be accurate and that the public might also be able to help analyze the data to recognize if the bridge needs maintenance work. These insights were sparked by the student who had previously worked on the bridge.

The fifth group consisted of two engineering researchers who joined online and thought that working with the data was "challenging and inspiring." One of them thought that the data were an asset generated by the public, so there was no reason to not share the data. As engineers, they found the format of the data convenient but thought that it might be more challenging for the public to understand, giving the example of how the strain gauge orientations could be confusing. The other team member believed (1) people should be able to access raw data, (2) experts need to explain what they are doing, and (3) lots of interpretation should be given. They believed that "information is power" and pointed out that there is a freedom of information act (See Appendix A). This person thought data access was a complicated and not a simple "yes" or "no" question.

Group Five's final concept was to allow engineers to access geotagged social media data and use it to better understand sensor output on the bridge instead of more invasive techniques like cameras or time-consuming work such as standing and taking notes on site. They believed this would be ethical because people publicly share to social media, making it open data by choice, unlike being filmed or watched while crossing a bridge. Location-based social networks have already been recognized as useful in urban studies research despite their drawbacks of being unrepresentative of the total population [21]. Beyond research integrity, considering all public data as open data and all open data as free to use and profit from may have serious social consequences that are beyond the scope of this article. Most people do not post to social media expecting it to be used in a larger context. At the same time, not all personal data are confidential. Open data broadly might force refinement of what types of personal information can and cannot be repurposed. This group demonstrated that there is some likelihood that open data from the MX3D bridge would be used in combination with other datasets.

There were a few people who joined the datathon partially and never submitted a final project. One of these people had visited the bridge in the fall of 2021, knowing that the bridge was smart and walked back and forth on it, but thought others were unaware that the bridge was collecting data. Another person thought that sharing the data from the bridge is fine in its current location. However, they expressed concern that if the bridge were located in a rural area where it was mainly used by only a few households, the bridge and its data collection would pose a threat to privacy. They explained that there could be a situation where someone leaves the house at 3:00 in the early morning and comes back later in the morning after an affair. Data from the bridge could be used against them in a legal proceeding, such as a divorce court. These two semi-participants saw the value in data collection because many bridge repairs in Amsterdam are currently instigated by individuals reporting or recording damage as opposed to preemptive sensors.

The last solo participant who joined late and also did not present was concerned about military, government, and corporations gaining access to the data—noting that open data also means open internationally. They shared that entities have more power than individuals. If data collection becomes more universal and open data practices are in place for all datasets, then powerful institutions have the resources to deanonymize data and also create new sources of knowledge from generalizing and combining data. This concern involving open data allowing for deanonymization is called "the mosaic effect."

Mosaic theory refers to when data viewed independently would have no expectation of privacy, but in summation breaks societal expectations of discretion. Consider the 2012 United States Supreme Court case, United States v. Jones. In that case, the court reinterpreted the Fourth Amendment right to privacy, using mosaic theory. In that case, police officers violated their warrant requirements when installing a GPS tracker on Jones's automobile, which would collect data on where Jones traveled. The case centered on a principle in privacy law—that individuals have a diminished privacy right in public spaces. However, Jones argued that the lack of expectation of privacy in public does not extend to continued surveillance, which may reveal personal information. The court rejected the government's argument that there was no reasonable expectation of privacy in the continuous GPS tracking. The Jones case can be extrapolated and applied in other instances. For example, imagine a tracker revealing that a woman went to the gynecologist and later to a baby store [22].

As a data scientist, graphing and manipulating the data was easy for the final participant. They weighed the costs and benefits of open data. Opening the data could be good for accountability and innovation. On the other hand, for transparency, they believed that it was important to share (1) who is accessing and (2) what they are doing with it—keeping track of these two standards would be challenging with open data. Finally, they believed that there should be privacy between individuals and their government. As cities increase in "intelligence" it will become more essential to design for privacy and define the individual right to privacy in public spaces. As there is nothing inherently private about being in public, the permanence of sensor collection in public space constitutes a new urgency for anonymization and deletion of data. The push for open data only makes these regulations more needed.

The observations from the event are further condensed and summarized in Table 3 for quick referencing.

Table 3. Summary of Datathon Team Observations.

#	Size and Skillset	Thoughts on Open Data from Bridge	Final Project
1	3: All Computer Scientists	Open data can be used creatively. Dove into the data without using additional explanatory resources. Created a prototype in a few hours.	Music generated by bridge traffic patterns
2	4: Computer Science, Interaction Technology, Prototyping and Coding, Philosophy of Technology, and Public Administration	Mixed feelings within the group. One person found the data difficult to connect with even with the resources provided. Some thought that the data should be shared because it is anonymous. Others thought that they did not have a right to the data. They tried to produce something with the data but struggled during the timeframe.	Lights that represent fireflies and dim when the bridge is too crowded
3	4: Design and Human-Computer Interaction, Industrial Design, Applied Mathematics, Mechanical Engineering	Did not use the data and instead prototyped a system. Believed that data from the bridge is fine to share because it is anonymous and that it is "good" to be transparent. Nonetheless, they noted that maybe when one individual is on the bridge, they could be tracked.	Fountain that reacts to pedestrian interactions with the bridge
4	3: All Civil Engineers (one of which worked with the full dataset previously)	Data from the datathon was insufficient for their goals. Noted that bridge data could be inaccurate. Suggested that if data are open, then the public could assist in confirming data validity and alerting when the bridge needs maintenance.	Public screen that displays an analysis of bridge data for the public and local businesses
5	2: Both Civil Engineers	One believed that data from the bridge is a public asset that should be shared. Another thought is that people should be able to access raw data, experts need to explain what they are doing, and that data should be interpreted by the public. This person stated that "information is power" and there is a freedom of information act. Team was concerned with the public's ability to interpret the data in its current format and understand how the sensors work.	Data validity system that uses public social media data combined with bridge data to check sensor accuracy and establish ground truth
6	2: Computer Scientists	Most people on the bridge probably do not know that it is collecting data. Data sharing depends on location. The Red Light District is busy, so it is fine. These data could be helpful for bridge repair and safety.	None
7	1: Data Scientist	Concerned about data being open for everyone including foreign governments and militaries. Opening the dataset could be good for accountability and innovation. It is important to share 1) who is accessing it and 2) what they are doing with it, but keeping track of these two standards would be challenging with open data.	None

5. Discussion

Theoretically, the datathon provided the opportunity for up to 60 people ages sixteen and older to use data from the MX3D bridge, communicate with the chair of the Smarter Bridges Research Committee, and present their findings to a group with power relative to the project: an Amsterdam municipality officer, an engineer that builds a digital twin platform for the MX3D bridge and other infrastructures, and a professor of philosophy of technology. In practice, sixteen people actively participated in the majority of the event with an additional three hanging out or briefly manipulating the data. Although people could join online, the instructions were all in English and we would not be able to provide extensive virtual tech support. Five ideas emerged from the event: a sound system that creates music from bridge traffic, a firefly net that dies with crowdedness, an interactive fountain system, a commercialized crowd management system, and using geotagged crowdsourcing for ground truth data. As of now, there is no continuous, live, open dataset that would make most of the projects possible. Additionally, beyond gift cards for the project winners, no funding was made available to make the projects a reality.

Most datathon participants communicated that they enjoyed the event and were glad to participate. Of course, the event was self-selective—people independently signed up through advertising or were brought in by a friend. Software engineers could code using the data. Civil engineers were able to graph the data and determine how sensor behavior reflected the mechanics of the bridge. Designers were able to think creatively and share an artistic presentation, communicating the need and benefit of their design. Even as non-institutions, individuals brought their strengths to their interactions with the dataset. Those with greater expertise also had more concerns about open data sharing than those with less of a background in civil engineering or data science. These more "expert" views reflected some from the research committee, such as the concern over individuals misinterpreting data and that it might be helpful to prevent those with bad intentions from having access. Other views reflected the values and considerations of the Dutch government such as wanting data in its most raw form and that open data could generate value not possible by having the dataset closed. Most people thought that since the data are collected in a public space, they should be open to the public as long as they are anonymous (and can remain anonymous even after data fusion).

This case study revealed that in practice, "open data" has a different cost-benefit analysis for each stakeholder. For an individual academic entity that invests in a data collection system, opening data could mean losing the research advantage over other institutions. There might be a similar concern for corporate parties that choose to invest in the smart city in order to reap the greatest reward. Panning out to the market, though, competition drives innovation. Although open data may weaken the data producers' advantage, everyone else gets free access to a system they do not manage, build, or upkeep. From a modern government perspective, where success is often derived from economic growth, open data are advantageous, especially for data made possible by tax revenue. Even the Smarter Bridges Research Consortium uses open data to capture environmental readings and store them alongside sensor readings from the bridge. Using that external open dataset was much easier (and more sustainable) than setting up a separate environmental sensor system just for the bridge.

Nonetheless, competition is only one factor when discussing open data. Additionally, there are concerns about the data itself. Data are not necessarily accurate: sensors may break, systems can shut down, and information can be wrongly labeled. Open data has the potential to spread false information far from those who have the background knowledge to interpret or correct it. Imagine a dataset about regional city crime rates presented without valuable information such as the population density throughout the city, where officers are most often sent, or what types of crime happen where. The initial dataset would have the potential to prompt decisions that would not make sense with more contextual knowledge. The participants of the datathon were not given information about sensor drift, the weight distribution of the bridge on the canal, or how heat affects load and strain. This could give users of open data the impression that they know more about a system than they actually do and allow for misinterpretation.

Determining what open data are safe to share is also not trivial. As described above, open data may require more open data to provide context so that the data are interpreted correctly. The majority of participants in the hackathon seemed to assume that anonymous data are safe to share and that it is good to share data generated in public. However, as the Amsterdam privacy officer noted, even anonymous data can reveal things that might better serve the interests of the public if they remain hidden. Maybe open bridge data would cause individuals to test the limits of the bridge to see the sensor feedback. This might crowd the area, cause noise disturbances, and unnecessarily stress a safe infrastructure. Determining unreasonable risk becomes increasingly complex in an open data environment where a bad actor might have access to additional datasets and deanonymize innocuous anonymous data.

Participants of the datathon valued having access to data from the bridge, but also shared that it is important to have information about data collection—who is collecting it and how is it used. The request for transparency is repeatedly shared as a tenet of ethical smart cities, often in the context of open data [23]. However, when data are made open without the need for registration, transparency becomes impossible to manage. How do you inform the public about data usage if any entity can build something with the data without needing to explain their purpose? Moreover, what is the incentive for the data managers to share what they are doing with the data if others can also use the data for whatever purpose they wish, whether it be for business, research, crime, or a middle school science project? To the average person in the smart city, it may become a question of whether open data or transparency is more important.

When considering power dynamics between the public and the data controllers, it is important to note that open data are just one possible stage or avenue for public empowerment. There is the possibility for empowerment in the creation process, deciding what infrastructures are put into place and which should be avoided. There is also room for empowerment in the management phase through boards that include a diversity of citizens and responsive transparency with the public. Responsive transparency would involve making it publicly clear who is in charge, what decisions are being made with the data, what data are collected, and what are the justifications for collection. Responsiveness refers to the possibility for external actors to complain, praise, or edit the system. This concept is adapted from "contestablity," which refers to the ability of the public to contest the process and outcomes of AI decision-making algorithms, thereby changing the function of the AI system [24]. Responsive transparency, unlike open data, would allow the public to act like data controllers in terms of management and decision making but would not allow the freedom to use the dataset however they wish.

Open data from the MX3D bridge has a redistributive effect but not necessarily on the average person. It might have the highest relative impact for an individual or small institution without the resources to create the dataset but with the expertise to make use of it (as seen with the datathon and within the Smarter Bridges Research Consortium). When zooming out to the system at large, the most powerful companies, governments, and institutions have the highest capability to use multiple datasets and connect them to make new sources of knowledge that might be proprietary. This is not inherently negative—the Smarter Bridges Research Consortium makes use of an open dataset of environmental sensors, which is more sustainable than creating a new system from scratch. Additionally, it may encourage the installation of fewer systems to test, review, fix, and supply energy. Nonetheless, it shows that the direct benefits of open data largely bypass the average individuals in society. Useful applications may also be made available to the public through contextual open data access [2].

6. Conclusions

This case study of opening data from the MX3D bridge to the public revealed nuances that expose the positive and negative effects of open infrastructural data. Some of the findings were counterintuitive. Open data may limit the monopolization of smart city data and lead to products that benefit members of the public and generate more revenue in the economy. It also allows anyone to see exactly what data controllers collect and make sense of the data in their own way. However, most people are not data experts and it is currently recommended in the Netherlands that open for some means open for all, including companies, governments, and militaries. These entities might have more resources and motivation to process data in ways that may disempower the public. Furthermore, if data are open for all without registration, then that limits the power of transparency, because it is impossible to share all the groups that are using the data and what they are doing with it.

The best opportunity for open data would be if the movement forces data controllers to be more selective in the datasets that they collect so that data collecting systems are rigorously examined, limited, and not built to cause ecological or social harm. The worry remains that datasets, once released into the world, are not independent bodies, but can be combined with other data and create something unforeseen. We are still at the beginning of possibly a long future of heightened data collection and processing in public spaces. Due to GDPR and similar legislation, the public is predisposed to be wary of personal data collection, but there are other ways that data can be harmful, some of which might take more generations to recognize. Moreover, open data does cannot compensate for the main harm of smart city data sets being built based on values that do not reflect society at large [25].

When looking towards the future, and imagining a public with perhaps more data literacy than the current generation, open data seems both daunting and perhaps necessary. The large organizations capable of weaponizing data might be combining worldwide open data, but so too might activists, researchers, and new innovative businesses solving the pressing problems of the time. While the regular members of the public still might not be able (or interested) to use the multitudes of open datasets, they will benefit from others using data to improve the world on their behalf. The greatest challenge will be keeping lines of transparency open and allowing data systems and decision-making algorithms to be built and contested by those without data expertise and the political sway of the data controllers.

Data Availability Statement: Please contact the author for access to the dataset used for the datathon. It is available to the public.

Acknowledgments: This article would not have been possible without the assistance of members of Amsterdam municipality, the Smarter Bridges Research Consortium, or MX3D.

Conflicts of Interest: At the time of writing, the author was chair of the Smarter Bridges Research Committee. This allowed for additional insight and an insider perspective but is not as neutral as an external researcher. The funders had no role in the design of the study; in the collection, analyses, or interpretation of data; in the writing of the manuscript; or in the decision to publish the results.

Appendix A

Stand on the MX3D bridge using Google Street View here: https://goo.gl/maps/SoRHKjd2ZqLGp5Tv9, (accessed on 19 January 2023). Learn more about the DCAT standard here: https://data.overheid.nl/en/ondersteuning/open-data/dcat, (accessed on 19 January 2023). The current Dutch freedom of information act is WOO.

Funding: This research was funded by The Dutch Research Council (NWO) (project: Bridging Data in the built environment (BRIDE); project number: CISC.CC.018).

References

1. Dodge, M.; Kitchin, R. Code and the Transduction of Space. In *Machine Learning and the City*; Carta, S., Ed.; Wiley: Hoboken, NJ, USA, 2022; pp. 309–339. [CrossRef]
2. Cammers-Goodwin, S.; Van Stralen, N. Making Data Visible in Public Space. *McGill GLSA Res. Ser.* **2021**, *1*, 1–32. [CrossRef]
3. Milić, P.; Veljković, N.; Stoimenov, L. Using OpenGovB Transparency Indicator to Evaluate National Open Government Data. *Sustainability* **2022**, *14*, 1407. [CrossRef]
4. Douglas, S.; Meijer, A. Transparency and Public Value—Analyzing the Transparency Practices and Value Creation of Public Utilities. *Int. J. Public Adm.* **2016**, *39*, 940–951. [CrossRef]
5. Schrock, A.R. Civic Hacking as Data Activism and Advocacy: A History from Publicity to Open Government Data. *New Media Soc.* **2016**, *18*, 581–599. [CrossRef]
6. Official Journal of the European Union. *Directive (Eu) 2019/1024 of the European Parliament and of The Council of 20 June 2019 on Open Data and the Re-Use of Public Sector Information*; Official Journal of the European Union: Mercier, Luxembourg, 2019; Available online: http://data.europa.eu/eli/dir/2019/1024/oj (accessed on 20 February 2023).
7. Shelley-Egan, C.; Gjefsen, M.D.; Nydal, R. Consolidating RRI and Open Science: Understanding the Potential for Transformative Change. *Life Sci. Soc. Policy* **2020**, *16*, 7. [CrossRef] [PubMed]
8. Welle Donker, F. Funding Open Data. In *Open Data Exposed*; van Loenen, B., Vancauwenberghe, G., Crompvoets, J., Eds.; Information Technology and Law Series; T.M.C. Asser Press: The Hague, The Netherlands, 2018; Volume 30, pp. 55–78. [CrossRef]
9. Pogačar, K.; Žižek, A. Urban Hackathon—Alternative Information Based and Participatory Approach to Urban Development. *Procedia Eng.* **2016**, *161*, 1971–1976. [CrossRef]
10. Jetzek, T. Elements of a Successful Big Data Hackathon in a Smart City Context. *Geoforum Perspekt.* **2015**, *14*, 25. [CrossRef]
11. *2020 Smart City Annual Report for the City of Mississauga*. Mississauga. 2021. Available online: https://storymaps.arcgis.com/stories/3c0dc90e53864a91a44abf6780078ab1 (accessed on 19 January 2023).
12. *Smart City Open Data Challenge*. SMRTCTY Mississauga. Available online: https://smartcity.mississauga.ca/event/smart-city-open-data-challenge/ (accessed on 19 January 2023).
13. PIB Delhi. *MoHUA Launches 'Open Data Week'*; Press Release 1790466; Ministry of Housing & Urban Affairs; 2022. Available online: https://pib.gov.in/PressReleasePage.aspx?PRID=1790466 (accessed on 7 December 2022).
14. *Open Data Day 2022 Events*. Open Data Day. Available online: https://opendataday.org/events/2022/ (accessed on 7 December 2022).
15. Neves, F.T.; de Castro Neto, M.; Aparicio, M. The Impacts of Open Data Initiatives on Smart Cities: A Framework for Evaluation and Monitoring. *Cities* **2020**, *106*, 102860. [CrossRef]
16. *Sensoren Crowd Monitoring Systeem Amsterdam*. Gemente Amsterdam. Available online: https://maps.amsterdam.nl/cmsa/ (accessed on 30 June 2022).
17. Israel, G.D. *Determining Sample Size*; University of Florida Cooperative Extension Service, Institute of Food and Agriculture Sciences, EDIS: Gainesville, FL, USA, 1992.
18. Wynne, Z.; Buchanan, C.; Kyvelou, P.; Gardner, L.; Kromanis, R.; Stratford, T.; Reynolds, T.P.S. Dynamic Testing and Analysis of the World's First Metal 3D Printed Bridge. *Case Stud. Constr. Mater.* **2022**, *17*, e01541. [CrossRef]
19. van Nieuwaal, B.; van Genderen, D. *The End of the Wob and the Arrival of the Woo in a Nutshell*. lvh-Advocaten. Available online: https://lvh-advocaten.nl/en/the-end-of-the-wob-and-the-arrival-of-the-woo-in-a-nutshell/ (accessed on 1 December 2022).
20. *HANDREIKING 2.0*. Data.overheid.nl. 2020. Available online: https://www.digitaleoverheid.nl/document/handreiking-open-data-2-0/ (accessed on 19 January 2023).
21. Martí, P.; Serrano-Estrada, L.; Nolasco-Cirugeda, A. Social Media Data: Challenges, Opportunities and Limitations in Urban Studies. *Comput. Environ. Urban Syst.* **2019**, *74*, 161–174. [CrossRef]
22. Gray, D.; Citron, D.K. A Shattered Looking Glass: The Pitfalls and Potential of the Mosaic Theory of Fourth Amendment Privacy. *N. C. J. Law Technol.* **2013**, *14*, 381.
23. Lnenicka, M.; Nikiforova, A.; Luterek, M.; Azeroual, O.; Ukpabi, D.; Valtenbergs, V.; Machova, R. Transparency of Open Data Ecosystems in Smart Cities: Definition and Assessment of the Maturity of Transparency in 22 Smart Cities. *Sustain. Cities Soc.* **2022**, *82*, 103906. [CrossRef]
24. Alfrink, K.; Keller, I.; Kortuem, G.; Doorn, N. Contestable AI by Design: Towards a Framework. *Minds Mach.* **2022**. [CrossRef]
25. Cammers-Goodwin, S. Revisiting Smartness in the Smart City. In *The Oxford Handbook of Philosophy of Technology*; Vallor, S., Ed.; Oxford University Press: New York, NY, USA, 2022; pp. 168–188. [CrossRef]

sustainability

MDPI

Article

Development of an Intelligent Personal Assistant System Based on IoT for People with Disabilities

Abd-elmegeid Amin Ali [1], Mohamed Mashhour [1,*], Ahmed S. Salama [2,*], Rasha Shoitan [3] and Hassan Shaban [1]

1 Computer Science Department, Faculty of Computers and Information, Minia University, Minia 61519, Egypt
2 Electrical Engineering Department, Faculty of Engineering & Technology, Future University in Egypt, New Cairo 11835, Egypt
3 Computer and Systems Department, Electronic Research Institute, Cairo 12622, Egypt
* Correspondence: mohamedmashhor11pg@fci.s-mu.edu.eg (M.M.); a.salama@fue.edu.eg (A.S.S.)

Abstract: Approximately 15% of the world's population suffers from different types of disabilities. These people face many challenges when trying to interact with their home appliances. Various solutions are introduced to increase their quality of life, such as controlling their devices remotely through their voices. However, these solutions use command templates that fail to understand the unstructured or semi-structured command. Many authors have recently integrated intelligent personal assistant (IPA) systems, such as Google Assistant, Siri, and Alexa, with control circuits to exploit the advantages of the NLP of these IPAs to control traditional home appliances. However, this solution still struggles with understanding unstructured commands and requires the internet to be available for controlling the devices. This research proposes a new IPA system integrated with IoT, called IRON, for disabled people to use to control customizable devices with a structured and unstructured voice command. The proposed algorithm receives voice orders from the person in a structured or unstructured form and transforms them into text based on the Google Speech-to-Text API. The natural language processing technique splits the commands into tokens to determine the device name and the command type, whether it is a question about device status or a statement. Afterward, the logistic regression classifies the rest of the tokens as positive or negative to turn on or off the device, then sends the command to a Raspberry Pi to control the device. The proposed IRON system is implemented using logistic regression, naïve Bayes, and the support vector machine and is trained on a created dataset consisting of 3000 normal, negative, and unstructured commands. The simulation results show that the IRON system can determine 90% of the device's names for all commands. Moreover, the IRON correctly classifies 100% of the commands as positive or negative within approximately 30 s.

Keywords: NLP; intelligent personal assistant; IoT; logistic regression; text classification; smart home; Raspberry Pi

1. Introduction

More than one billion people worldwide—almost 15% of the population—live with one or more disabilities, according to the World Health Organization [1]. These disabilities may present early in childhood or develop later in life, such as impaired hand function as a result of stroke [2]. Every day, people with disabilities significantly struggle to control their home appliances. Therefore, traditional homes should be transformed into "smart homes" to improve the quality of life of those with disabilities. Over the last decades, the objective of IoT technology has been to allow communication between devices without the need for human involvement [3]. However, currently, IoT technology has been integrated with home devices to enable these devices to be controlled remotely through the internet [4]. The term IoT describes a network of physical objects, or "things", that have sensors, software, and other technologies built into them to enable the connecting and sharing of data with other systems and devices over the internet. These devices include light switches that respond to turn-on and off commands or thermostats that change the indoor temperature to reduce energy consumption. Different authors present several solutions for help the disabled person to control the devices remotely via IoT based on the user voice or the smartphone graphical user interface (GUI). Mittal et al. [5] propose a smart home automation system (SHAS) to control home devices based on user voice commands. The system is designed to control five groups: light, access, fan, utility, and safety. Voice Recognition Module V3 is used to recognize the voice commands, and then the Arduino microcontroller compares these commands with the previously stored command templates to perform the desired action. Abidi et al. [6] present a voice-control smart home automation system to control four home devices: a TV, LED, fan, and Compact Fluorescent Lamp (CFL). The system architecture consists of an Android phone, Arduino mega, GSM SIM900A, Bluetooth module, and ultrasonic sensor (HC-SR04). A mobile application transforms the voice command into text, and then a Bluetooth module (HC-05) transmits this text to the Arduino Mega to control the dedicated device. The ultrasonic sensor (HC-SR04) is also used to provide home security by detecting any movement and sending a message to the user's phone via GSM SIM900A to convey that there is movement in the home.

Repeatedly, Mayer et al. [7] suggest home automation architecture based on IoT technology to help people with disabilities who have trouble using a remote control or a smartphone to control their home devices. First, the voice command is recognized and processed using a speaker recognition system (ASR) on the ASR remote server. Afterwards, the ASR sends an XML file to the IoT gateway, which contains the command sent by Xbee to the appropriate device. Moreover, Saha et al. [8] have developed a home automation system that lets people use voice commands to control their electronic devices. The system uses an Android phone to capture the user's voice, which is then sent to an Arduino board via Bluetooth. The Arduino board processes the data and controls the devices accordingly through relays. The system also has an auto-mode that automatically utilizes sensors to automatically control lighting and temperature. Ismaeel et al. [9] present a home automation system that aims to control the electronic devices through voice or a webpage. If the user uses the voice application, the system receives it over microphone/mobile and employs a Raspberry Pi development board for processing. The user's voice is transcribed to text using the online speech-to-text interface (Wit). Suppose the user uses the web page to turn on or off a device. In that case, the content script text is sent to the Raspberry Pi via the Apache remote server to control the general-purpose input-output (GPIO) pins to activate or deactivate the relays connected to the devices.

Besides, Jat et al. [10] introduce the application of voice activity detection (VAD) in home automation systems to control home appliances through voice commands. VAD detects human speech using the microphone and activates a specific action. The system is implemented on a Raspberry Pi 3 Model B+, and the PocketSphinx speech recognition system captures the user's voice, suppresses the noise, and converts the analyzed speech to text. The transcribed text is then processed to determine if the text includes identified commands. According to the command, the system applies the action to the dedicated device through infrared, wireless LAN, or Bluetooth. Additionally, Alrumayh et al. [11] present a context awareness for voice assistants system (CANVAS) for use in smart homes with multiple occupants. The system consists of two major phases: a configuration webpage and a context-aware algorithm. The configuration webpage assists the owner of the home in configuring different rules to access the home devices by a drag-and-drop GUI interface. The context-aware algorithm receives the device commands from a human voices assistant to determine whether to discard or execute the user commands according to previous rules. Khan et al. [12] introduce a web application known as HAS (home automation system) to control many devices from their smartphones. This application uses Google Assistant to convert speech to text for controlling the appliances with the user's voice. It also monitors the devices' daily, weekly, and monthly power consumption. Mustafa A. Omran et al. [13] present a Wi-Fi-based home automation system for a smart home (SH) prototype. The system uses Raspberry Pi 3 Model B+ and Arduino Mega 2560 and is controlled by a smart IOS system via the Blynk app.

Most of the methods mentioned above design their systems based on command templates; if the user uses an unstructured command, the system will not understand it. Therefore, several authors add natural language processing (NLP) to the systems to process and understand various types of commands. Intelligent personal assistant (IPA) [14] systems have recently gained popularity as a way to help individuals with many of their everyday tasks. This personal assistant system analyzes human speech based on NLP and produces a set of direct instructions. Examples of well-known IPAs are Google Assistant from Google [15], Siri from Apple [16], and Alexa from Amazon [17]. These well-known IPAs set alarms, set ringtones, close applications, make phone calls, order food online, play music, etc., with differing accuracy, according to each company. However, these IPAs do not work with all home devices. These IPAs work with smart devices designed to be controlled by their applications. Many authors integrate these IPAs with a control circuit to control any device. Uma et al. [18] introduced a home application system to turn on or off home devices via voice or text based on Google Assistant. When the user is not physically present in the environment, he can schedule the state of the appliances and be given the option to turn them on for a predetermined period of time. The user uses the mobile application to send his voice command through Google Assistant. The voice command is fed to the firebase, which the NodeMCU can fetch to turn on or off the home devices. The application function is implemented using Node-RED technology, deployed in a Dialogflow account, and integrated with Google Assistant.

Furthermore, Kumer et al. [19] introduced a voice-based smart home automation system using IFTTT, Adafruit cloud, Ubidots IoT dashboard, and Google Assistant to efficiently control energy consumption. The voice commands are received by Google Assistant and sent to IFTTT. These commands are then sent to the microcontroller via Wi-Fi to turn the device on or off, according to the user's command. The microcontroller updates the device status on the Ubidots dashboard, and feedback is sent to the user concerning this status. Putthapipat et al. [20] develop a home automation system based on the PiFrame framework. The system goal is to create an open platform for researchers to extend the features and develop their designs, similar to products such as Google Home and Amazon Alexa, but with more flexibility. The system hardware consists of a Raspberry Pi, a microphone, a speaker, and relays to control the home devices using voice commands. Google Speech API and Wit.ai are used to understand the user's command and communicate with the home appliances. The system stores the home device's current status on the Google Cloud Platform Cloud SQL through the Node.js application on Raspberry Pi. The response to the user is generated by translating the text to speech using Google Translate. Tiwari et al. [21] proposed a home assistant system that can be controlled and monitored using voice commands. In this system, the user can control the home devices and sensors from any cloud-connected platform through speech, text, or a mobile application. The system uses advanced speech recognition techniques, such as Mel-Frequency Cepstrum Coefficients (MFCC), with other features to create the feature vector. Vector Quantization (VQ) and Principal Component Analysis (PCA) are used for feature vector dimensional reduction and a Gaussian Mixture Model (GMM) to classify the speakers. The system uses cloud services IBM's Bluemix and Google's cloud service for speech-to-text conversion. Mehrabani et al. [22] presented a personalized speech recognition system using dynamic hierarchical language modeling for controlling customizable devices in smart home applications.

The previous methods integrate the IPAs with control circuits to benefit from the NLP of IPAs to increase command understanding accuracy and control any device type. However, while these methods can understand structured and semi-structured voice commands, they still cannot understand unstructured commands. Moreover, these methods depend mainly on the internet to take actions because they save commands on firebase or use the mobile phone as a GUI to send voice commands.

This paper proposes a new IPA system integrated with IoT, called IRON, to help the elderly and disabled people control their home devices remotely through IoT, based on their voices. The IRON system is designed from three modules: a speech-to-text module, a text analysis and classification module, and a command execution module. Instead of mobile phones, the IRON detects the user's voice from different microphones spread around the home. This option helps disabled people operate the devices from any place inside the home, in case the phone is lost or the internet is disconnected. The speech-to-text module transforms the incoming voice into text. IRON possesses two ways to turn voice commands into text: if the internet is available, it uses the Google Cloud API, and if it is not, it uses a speech recognition library. Thus, IRON can work online or offline. In the text analysis and classification module, the transcribed text passes through two main steps for analysis. First, IRON splits the text into a list of tokens by applying NLP to the transcribed text to search for the device name, according to the list of predefined devices. If the device is not included in the devices list, the user can ask the IRON to add a new device to the devices list and configure it without recoding the IRON. Second, IRON takes the device's name out of the command and determines if the statement is a question or an order. If the statement is a question, the IRON answers regarding the device status and break; otherwise, IRON considers this statement as a command and sends the rest to the pre-trained machine learning model to classify whether this command is a positive or negative statement. According to the statement type, IRON executes the required action on the device through the GPIO pins. The IRON system is designed to overcome the cons of the previous methods of understanding unstructured commands by adding a machine learning algorithm with NLP to classify these commands as positive or negative for turning on or off a device. The main contribution of this research can be abstracted as follows:

1. An IRON system that helps the elderly and impairment to control their home devices through their voice is proposed.
2. The dataset consists of 3000 normal, negative, and unstructured commands to control 100 devices.
3. Natural language processing is applied to the text generated from the google speech API for splitting the text into tokens for further classification.
4. A machine learning algorithm is included in IRON for classifying the commands as positive or negative.
5. Multi-microphones are distributed in different locations in the home to ensure that the elderly and disabled can access IRON from their locations.
6. New devices can be added and reconfigured by the impaired person's voice without re-coding the IRON.
7. The IRON system is designed to work online or offline and to turn off, on, or adjust a device range as fan speed.
8. The IRON system is secured by requesting a password to start the controlling process.

2. System Architecture and Design

This research proposes an IPA system based on IoT, called IRON, to assist people with disabilities in controlling their home appliances based on their voices. Figure 1 shows the proposed IRON system architecture. This figure shows that the IRON hardware system consists of a microphone, speaker, Raspberry Pi, Wi-Fi module, and relays. When the user opens the IRON, it introduces itself and tells how to interact with it. IRON is usually in sleep mode. Once the user says "IRON", it will respond with a "hello" message and then ask the user about what he wants. The user speaks the command, which is then analyzed to control the dedicated device. Algorithm 1 presents the summarized steps for interacting with the IRON system.

Algorithm 1: The IRON system procedure

Input: *user voice command*
Output: *device control*
Initialization: *IRON introduces itself and informs the user instructions about how interact with it.*
While true
    Wait in sleep mode   *#microphone ready to receive voice.*
    If *the user says "IRON":*   *#user knows this from instruction.*
        IRON say: hello message and ask human what he need to do.
        Receive Order ()
        Speech to text ()
        Analysis Text ()
        Execute the command ()
        return feedback ()
    Else:
        Check if the user needs to end IRON or not.
    End
End

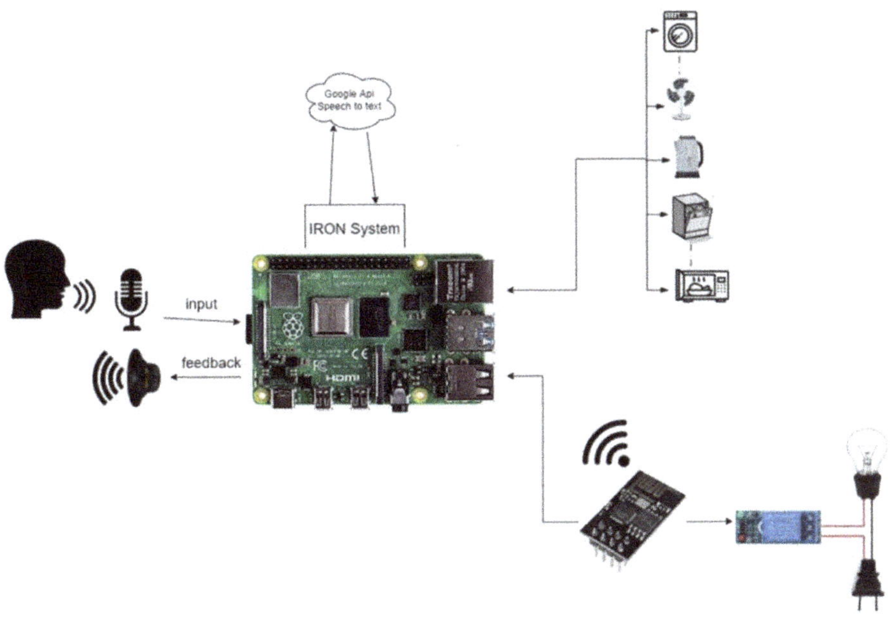

Figure 1. The proposed IRON system architecture.

Figure 2 introduces the IRON major modules and flowchart. The IRON system is designed from three modules: speech-to-text module, text analysis and classification module, and command execution module. In a speech-to-text module, Google API receives the user's voice command through the microphone and converts it to text. The text analysis and classification module receive the text and analyzes it to extract the device name and execute the required command. According to the device and the command, the command execution module applies an action on that device through Raspberry Pi. At the same time, the speaker responds and informs the user that the action is completed. The details of each module of the IRON are described in the following subsections.

- Module 1: Speech to text

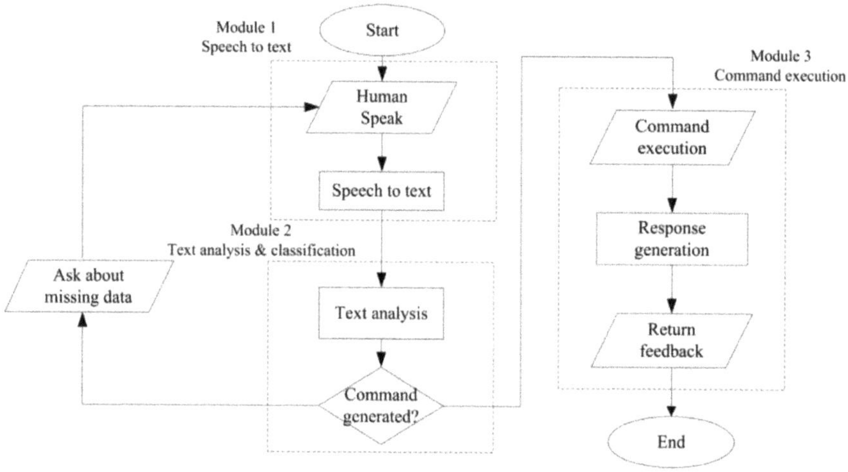

Figure 2. The IRON system flowchart.

In this module, the person's commands are converted from speech to text using the Google Speech to Text Cloud API. Google Speech to Text Cloud API transcribes the speech file using the most advanced deep learning neural network algorithms for automatic speech recognition (ASR) and returns the text statement [23]. Google Speech to Text Cloud API is one of the simplest methods for recognizing speech and can analyze up to 1 min of voice data.

- Module 2: Text analysis and classification

This module is responsible for understanding the correct command from the text generated by the Google API and then confirming it with a human to execute the desired action. Because of the uncertainties in human language, it is extremely challenging to create software that correctly ascertains the text's intended meaning, so NLP is used in this module for manipulating and recognizing the text. NLP deconstructs the text into small units to assist the computer in understanding the ingesting text. Different libraries and algorithms are proposed for NLP, such as the Natural Language Processing toolkit (NLTK) [24], Apache OpenNLP, Stanford CoreNLP, Pattern, GATE, and Spacy. Table 1 compares the most popular NLP techniques in terms of the programming language and the processes applied to the text statements. NLTK is considered the most robust and is the mother of all-natural language processing tools; thus, IRON utilizes it for analyzing the text. When IRON receives an input statement, tokenization, stemming, and stop word removal processes from the NLTK library are applied to the text.

Table 1. NLP toolkits Methods.

NLP Toolkit	Programming Language	Tokenization	Stemmer
NLTK	Python	Yes	Yes
Apache OpenNLP	Java	Yes	Yes
Stanford CoreNLP	Java	Yes	Yes
Pattern	Python	Yes	No
GATE	Java	Yes	No
Spacy	Python/Cython	Yes	No

Tokenization splits the text into smaller units called tokens to build a list representing all the vocabulary in the text. The tokens can be words, numbers, punctuation, stop words, and possessive markers. Moreover, the stemming step extracts the root word to reduce inflected words to their word base or root form [25]. It is a well-known fact that many words share morphological characteristics and semantic properties, making these words suitable for use in information retrieval applications. The stemmer removes the phrases to form the root term, such as "consider", which is the root term of consideration. To reduce processing time and supply more correct results from the enormous corpus of documents, several types of stemmers are employed to break down terms into their root terms. Searching is then performed on these terms rather than the real terms. In the stop words removal step, all the stop words encounter significant noise. These stop words, such as "the", "a", "an", and "in", do not add additional information, and removing them enhances the execution of the information retrieval system and text analytics [26].

After analyzing the text, splitting it into tokens, and removing the noise, IRON searches in the tokens list regarding the device name to know which device the user wants to control. When IRON determines the dedicated device, it should start to open or close the device according to the statement type. The statement type is deduced based on a trained classifier to recognize if the statement is positive or negative to switch on or off the device. The proposed IRON uses a logistic regression classification algorithm [27,28] to classify the rest of the tokens into positive and negative statements. The logistic regression classifier is an algorithm proposed for binary classification problems. It assesses the likelihood of the target variable, taking either 0 or 1. The classifier operates based on logistic regression, a statistical technique that makes predictions through the relationship between the dependent and independent variables. The logistic regression classifier model can be trained through maximum likelihood estimation or stochastic gradient descent methods. Additionally, the classifier includes regularization options L1 and L2 to counteract overfitting and enhance the model's generalization ability. The logistic regression equation is a mathematical formula that represents the relationship between the dependent and independent variables. It takes the form of a logistic function and can be written as:

$$p = \frac{1}{1 + \mathrm{e}^{-\mathrm{z}}} \tag{1}$$

where p is the predicted probability of the dependent variable taking the value 1, and z is a linear combination of the independent variables and their coefficients. The coefficients are estimated during the training phase through optimization algorithms, such as maximum likelihood estimation or gradient descent. The resulting equation is then used to make predictions on new data. In the case of classifying positive and negative statements, the dependent variable would represent the statement, with 1 representing positive and 0 representing negative. The independent variables would be features of the statement, such as words used, tone, or other relevant information. The logistic regression classifier would estimate the impact of each feature on the statement through the coefficients for each feature. The classifier would then use the estimated coefficients and features of a new statement to determine a predicted probability of the statement being positive or negative. The prediction can be a thresholder, with a value greater than 0.5 considered positive and less than 0.5 considered negative.

- Module 3: Command execution

In this module, IRON executes the desired action via Raspberry Pi to turn on or off the dedicated device. The devices are connected to the Raspberry Pi according to their location from the main circuit. If the devices are close to the main circuit, they should be wired together; otherwise, they should be wirelessly connected using the Wi-Fi module ESP8266, as shown in the Figure 3a,b, respectively. When IRON sends an activated signal to a device, this signal is transmitted through the ESP8266 to turn on or off the relays which are connected to the device. The ESP8266 is selected as a communication module because it is a low-cost Wi-Fi microcontroller, and it supports both TCP/IP and Wi-Fi protocols.

Furthermore, it covers a range of up to 400 m in the open air, but it can be reduced to a few meters when obstacles such as walls or furniture are present.

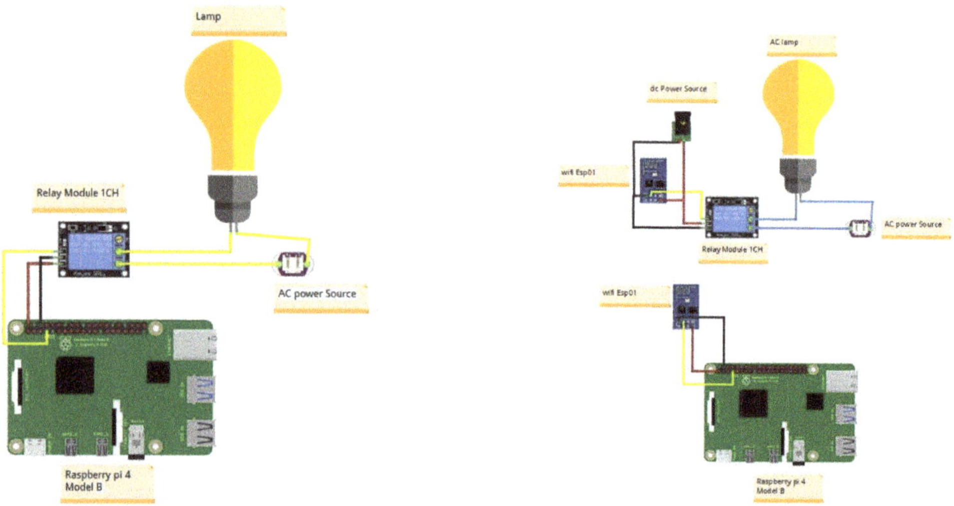

(**a**) (**b**)

Figure 3. The types of device connection in the IRON: (**a**) wired; (**b**) wireless.

Figure 4 shows the detailed system flow chart, which presents the detailed steps of the proposed IRON system as follows:

Step 1: The user asks IRON to switch on or off a device.

Step 2: IRON receives the user voice command through high-quality BY-MA microphones distributed in different locations in the home to assist the impaired person in reaching it quickly.

Step 3: IRON transcribes the user's speech-to-text using the Google Speech-to-Text API, if the internet connection is available, or a speech-to-text conversion library, if the internet is disconnected.

Step 4: The transcribed statement is then processed using NLTK, creating a list of tokens. IRON searches the tokens list for the device name to determine which device the user wants to control. If the device name exists, IRON will continue; otherwise, IRON asks the user to speak the device name again.

Step 5: IRON checks the tokens list to determine if the statement is a question about the device status or a command. If the statement is a question, IRON responds with a proper answer about the device status and break; otherwise, IRON treats this statement as a command.

Step 6: After considering the statement as a command, the IRON removes the device name and classifies the rest of the tokens to a positive or negative command using logistic regression. If it is a positive command, Raspberry Pi sends "on" to the device through the GPIO pin; otherwise, it sends "off" to the device to execute this action.

Step 7: IRON generates a response text that matches the user text and uses the offline text-to-speech conversion library to convert this response into voice. IRON plays the generated voice on the YST-1052 speaker, which is connected to Raspberry Pi via USB and an audio jack to inform the user that action is applied to the device, such as, "OK Sir, living room light is on now".

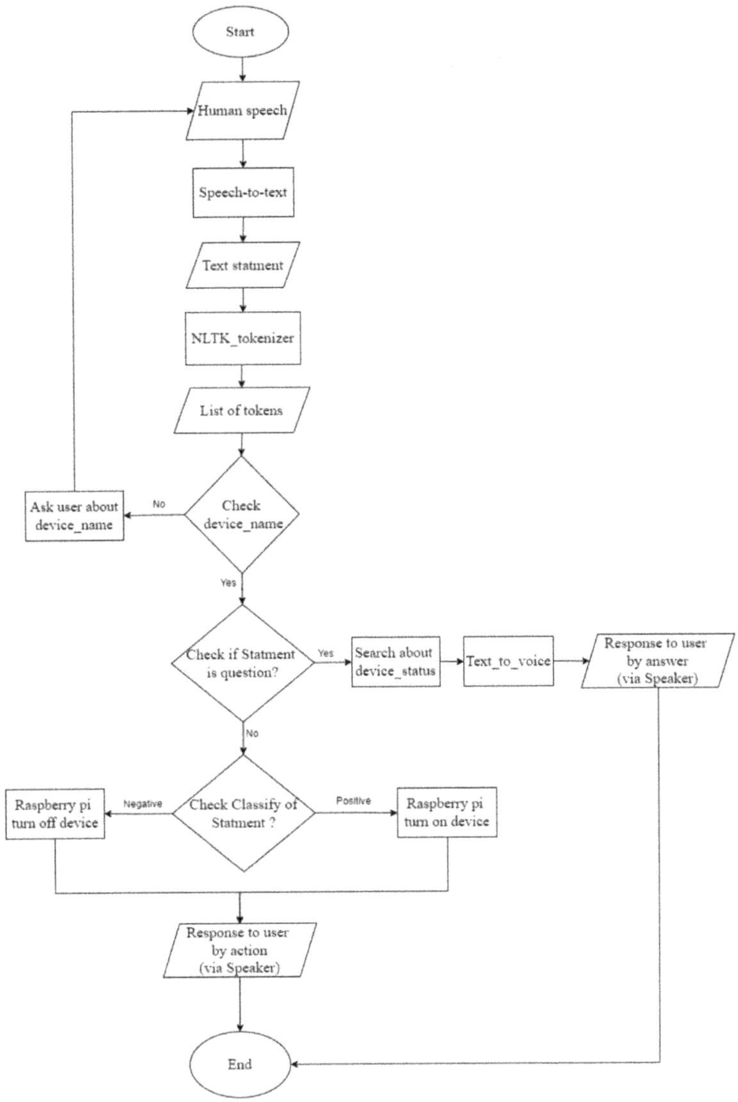

Figure 4. The detailed IRON system flowchart.

Over and above, IRON adds a feature for the user to add a new device via voice and configure it, as shown in Figure 5. First, the user connects the new device to the relay module unit. The user then asks IRON to turn on or off this new device. IRON checks whether this device is in the stored device list or not. If the device is not found in the devices list, IRON asks the user to confirm adding this device to the list. After that, IRON requests the device configuration from the user to add it to the file, as shown in Figure 6.

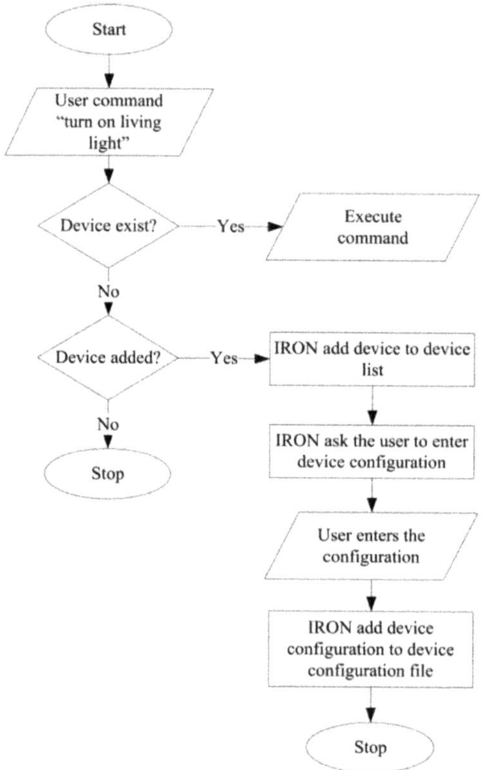

Figure 5. The new device's configuration.

```
{
        "devices": [
            {
                "device_name": "living room",
                "GPIO_pin": 2,
                "values": [0,1]
            },
            {
                "device_name": "air condition",
                "GPIO_pin": 4,
        "values": [18,19,20,21,22,23,24,25,26,27]
            }
        ]
}
```

Figure 6. Configuration file sample.

3. Evaluation Results

This section presents the evaluation metrics, dataset, and the simulation results of the proposed IRON system.

3.1. Evaluation Metrics

Three evaluation metrics, called accuracy, precision (AP), and recall (AR), are applied on the classification model output to evaluate the performance of the IRON system.

- Accuracy

Accuracy is a metric that measures the accuracy of the model in making the right prediction. It is computed as the percentage between the true predictions and the total number of predictions [29].

$$Accuracy = \frac{number\ of\ correct\ predictions}{total\ number\ of\ predictions} \tag{2}$$

$$Accuracy = \frac{TP + TN}{TP + TN + FP + FN} \tag{3}$$

where TP = true positives, TN = true negatives, FP = false positives, and FN = false negatives.

- Precision

Precision measures the percentage of correctly predicted positive samples compared to the total number of predicted positive samples.

$$precison = \frac{TP}{TP + FP} \tag{4}$$

- Recall

Recall measures the ratio between the numbers of positive samples correctly predicted as positive to the total number of actual positive samples.

$$Recall = \frac{TP}{TP + FN} \tag{5}$$

3.2. Dataset

A dataset of 3600 commands is created to control 100 devices. These commands are divided into negative commands, unstructured commands, and normal commands. Negative commands are the direct commands to turn off the device, such as "Don't turn on room1 light" and "I didn't want to open TV". Unstructured commands are the indirect commands to turn on or off the device as "Please make light off" and "We need to open Fan". Normal commands are direct commands to turn on the device, such as, "Turn on light".

3.3. Simulation Results

The proposed IRON system is trained on an Intel(R) Core (TM) i5-5300U CPU running at 2.30 GHz, 2.29 GHz, and 16 GB of RAM, as well as an NVIDIA GeForce 830 with 2 GB of memory. The trained model is run on a Raspberry Pi and has a Broadcom BCM2711, which includes a quad-core Cortex-A72 (ARM v8) 64-bit SoC @ 1.5GHz with 2 GB Ram, 4 USB ports, and communication modules such as Bluetooth, wireless LAN, and an Ethernet port.

As mentioned earlier, the first step of the IRON is to detect the device name, then analyze the commands to determine the action. The proposed dataset is divided into 66% training, 17% validation, and 17% testing. IRON is trained on the proposed dataset with different machine learning algorithms: logistic regression, naïve Bayes, and support vector machine (SVM) to classify the commands for determining the action. Table 2 presents the percentage of the devices and the correctly detected and classified commands. It can be noticed from the table that for all command types, IRON detects the device name with 90% accuracy. However, the percentage of classifying the commands correctly varies according to the command type. The higher classification percentage is for normal commands, while the lowest percentage is for unstructured commands.

Table 2. The percentage of correctly detected devices and classified commands.

Categories	Detected Devices	Correct Classified Commands
Normal commands	90%	93%
Unstructured commands	90%	80%
Negative commands	90%	83%

Table 3 introduces the performance of the IRON system for different machine learning algorithms. The IRON is trained on the smart home dataset [30] and the proposed dataset [31]. The smart home dataset consists of 663 commands. It can be observed from the results that the IRON based on the logistic regression and SVM achieves the best results, while naïve Bayes classify different commands as FP, which affects the precision values.

Table 3. The performance of the IRON system in terms of accuracy, precision, recall, and time.

Algorithm	Dataset	Accuracy	Precision	Recall	Time
Naive Bayes	Smart Home Commands Dataset	0.89	0.89	1.0	-
	The Proposed Dataset	0.63	0.59	0.64	0.0119
Logistic Regression	Smart Home Commands Dataset	0.969	0.967	1.0	-
	The Proposed Dataset	1.0	1.0	1.0	0.0112
SVM	Smart Home Commands Dataset	0.949	0.977	0.966	-
	The Proposed Dataset	1.0	1.0	1.0	0.020

For evaluating the processing time of each of the machine learning algorithms on the proposed dataset, the computational time is measured on the test set and presented in Table 3. The processing time of each technique is measured after running each technique 5 times, and the average of these 5 runs is calculated to obtain more reliable results. It can be demonstrated that the logistic regression consumes less time compared to naive Bayes and SVM.

The proposed IRON system is practically implemented, as presented in Figure 7. The hardware system consists of a lamp device connected with a relay module to Raspberry Pi, as shown in the Figure 7. The relay module transfers the signal from the Raspberry Pi to the lamp device. When the user says "IRON", the system is activated and gives instructions to guide the user on how to use IRON, as shown in Figure 8a. Afterward, the user starts to give the command to the system to control the device. If the user stops giving commands, the IRON goes into sleep mode and is deactivated until the user activates it again, as shown in Figure 8b. In the end, the user says, "Thank you", to turn off IRON. Additionally, the time required to execute a command in IRON is empirically calculated to be approximately 30 s.

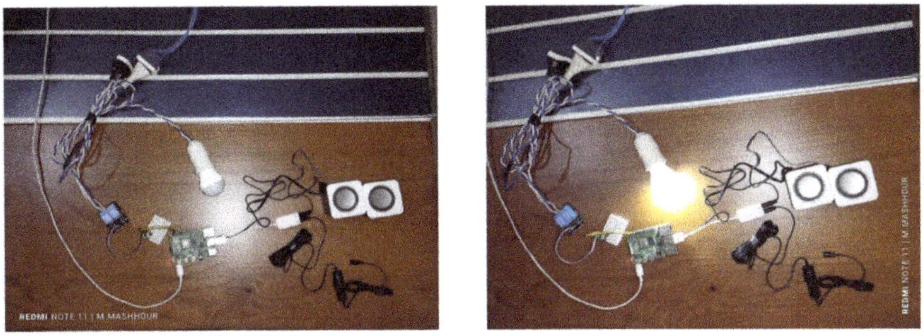

Figure 7. The IRON system Experimental setup.

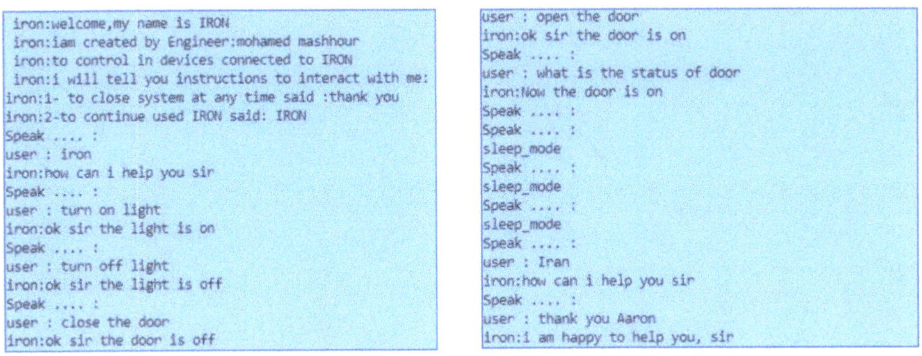

(**a**) (**b**)

Figure 8. (**a**) IRON instructions, (**b**) IRON sleep mode.

Figure 9 clarifies how the sensors and devices are distributed in the home. The distribution varies according to the necessary purpose and the location of the devices to be controlled.

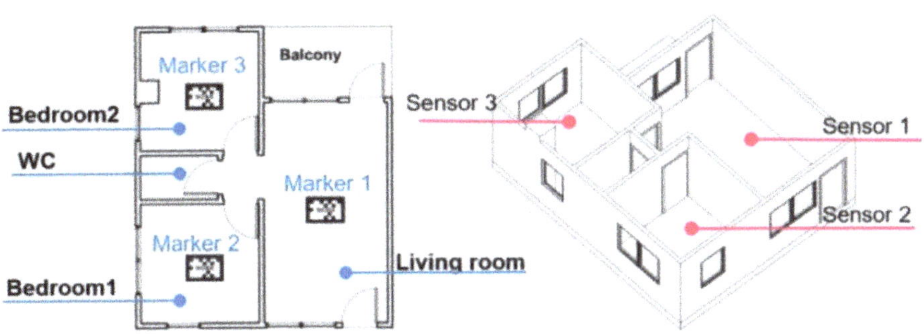

(**a**) (**b**)

Figure 9. Home sensors distribution: (**a**) floor plan and markers location setup, (**b**) 3D BIM model.

For evaluating the IRON system against the commercial intelligent personal assistant (IPA) systems, such as as Siri [16], Google Assistant [15], Alexa [17], Cortana [32], and Watson [33], Table 4 presents the comparison in terms of the speech recognition, text analysis, device detection, and negative and unstructured command understanding.

Table 4. Comparison between commercial IPA and IRON systems.

IPA	Speech Recognition	Text Analysis	Devices Detection	Negative and Unstructured Command Understanding
Apple Siri	Yes	Yes	No	Semi
Google Assistant	Yes	Yes	No	Semi
Amazon Alexa	Yes	Yes	Semi	Semi
Microsoft Cortana	Yes	Yes	No	Semi
IBM Watson	Yes	Yes	Yes	Semi
IRON system	Yes	Yes	Yes	Yes

4. Conclusions

The Internet of things, sensors, smartphones, smart appliances, and cloud computing have been used recently to transform ordinary homes into smart homes. These smart homes help persons with disabilities enhance their quality of life by remotely controlling the devices via mobile applications or their voices. This research proposes an intelligent personal assistant, called IRON, to help disabled people control their appliances with their voices. The IRON system architecture consists of three modules: a speech-to-text module, text analysis and classification module, and a command execution module, and is implemented on Raspberry Pi 3 model B. IRON receives the disabled person's voice commands through the microphone and transforms them into text using a speech to text module. The device name and the commands are extracted from the transformed text using NLP and logistic regression in the text analysis and classification module. According to the device name and the command, the IRON sends the required action to the dedicated device. IRON is built using different machine learning algorithms and is trained on 3600 commands to control 100 devices. The simulation results demonstrate that the logistic regression performs better than naïve Bayes regarding precision, recall, and accuracy. Furthermore, logistic regression requires less time to execute the commands than SVM and naive Bayes. The results also show that IRON can recognize different commands as structured and unstructured statements. IRON will be updated in future work to help people remember their notes and dates, in addition to controlling home appliances. Additionally, IRON can automatically regulate the devices according to some conditions (e.g., change the air condition temperature if the weather changes). Moreover, the user can schedule different commands that IRON can execute at a specific time under certain conditions.

Author Contributions: Conceptualization, M.M., and A.-e.A.A.; methodology, M.M. and H.S.; software, M.M.; validation, H.S. and A.S.S.; formal analysis, R.S.; investigation, M.M., R.S. and A.S.S.; resources, A.-e.A.A. and A.S.S.; data curation, M.M.; writing—original draft preparation, M.M., H.S. and R.S.; writing—review and editing, R.S. and M.M.; visualization, A.S.S., H.S. and A.-e.A.A.; supervision, A.-e.A.A. and H.S.; project administration, A.S.S., H.S. and R.S.; funding acquisition, A.S.S. All authors have read and agreed to the published version of the manuscript.

Funding: This research was funded by Future University in Egypt (FUE) research center.

Institutional Review Board Statement: Not applicable.

Informed Consent Statement: Not applicable.

Data Availability Statement: The datasets used in this investigation are available on request from the corresponding author.

Conflicts of Interest: The authors declare no conflict of interest.

Abbreviations

IPA	intelligent personal assistant
GUI	graphical user interface
SHAS	smart home automation system
CFL	compact fluorescent lamp
GPIO	general purpose input output
VAD	voice activity detection
CANVAS	context awareness for voice assistants
HAS	home automation system
MFCC	mel-frequency cepstral coefficients
VQ	vector quantization
PCA	principal component analysis
GMM	Gaussian mixture model
NLTK	Natural Language Toolkit
SVM	support vector machine

References

1. World Health Organization (n.d.). WHO. Available online: https://www.who.int/en/ (accessed on 9 March 2023).
2. Li, G.; Cheng, L.; Gao, Z.; Xia, X.; Jiang, J. Development of an untethered adaptive thumb exoskeleton for delicate rehabilitation assistance. *IEEE Trans. Robot.* **2022**, *38*, 3514–3529. [CrossRef]
3. Abdelhamid, S.; Aref, M.; Hegazy, I.; Roushdy, M. A survey on learning-based intrusion detection systems for IoT networks. In Proceedings of the 2021 Tenth International Conference on Intelligent Computing and Information Systems (ICICIS), Cairo, Egypt, 5–7 December 2021; pp. 278–288. [CrossRef]
4. Omran, M.A.; Saad, W.K.; Hamza, B.J.; Fahe, A. A survey of various intelligent home applications using IoT and intelligent controllers. *Indones. J. Electr. Eng. Comput. Sci.* **2021**, *23*, 490–499. [CrossRef]
5. Mittal, Y.; Toshniwal, P.; Sharma, S.; Singhal, D.; Gupta, R.; Mittal, V.K. A voice-controlled multi-functional smart home automation system. In Proceedings of the 2015 Annual IEEE India Conference (INDICON), New Delhi, India, 17–20 December 2015; pp. 1–6. [CrossRef]
6. Abidi, M.E.; Asnawi, A.L.; Azmin, N.F.; Jusoh, A.Z.; Ibrahim, S.N.; Mohd Ramli, H.A.; Abdul Malek, N. Development of voice control and home security for smart home automation. In Proceedings of the 2018 7th International Conference on Computer and Communication Engineering (ICCCE), Kuala Lumpur, Malaysia, 19–20 September 2018; pp. 1–6. [CrossRef]
7. Mayer, J. IoT architecture for home automation by speech control aimed to assist people with mobility restrictions. In *International Conference on Internet Computing (ICOMP)*; The Steering Committee of The World Congress in Computer Science, Computer Engineering and Applied Computing (WorldComp)—CSREA: Providence, RI, USA, 2017; pp. 92–97.
8. Saha, I.; Maheswari, S. Deep learning for voice control home automation with auto mode. In Proceedings of the International Conference on Computational Vision and Bio Inspired Computing, Coimbatore, India, 29–30 November 2018; pp. 1605–1614. [CrossRef]
9. Ismaeel, A.G. Home automation system support for persons with different types of disabilities using IoT. In Proceedings of the IMDC-SDSP 2020: 1st International Multi-Disciplinary Conference Theme: Sustainable Development and Smart Planning, Cyberspace, Ghent, Belgium, 28–30 June 2020; p. 18. [CrossRef]
10. Jat, D.S.; Limbo, A.S.; Singh, C. Voice activity detection-based home automation system for people with special needs. In *Intelligent Speech Signal Processing*; Academic Press: Cambridge, MA, USA, 2019; pp. 101–111. [CrossRef]
11. Alrumayh, A.S.; Lehman, S.M.; Tan, C.C. Context aware access control for home voice assistant in multi-occupant homes. *Pervasive Mob. Comput.* **2020**, *67*, 101196. [CrossRef]
12. Khan, M.A.; Ahmad, I.; Nordin, A.N.; Ahmed, A.E.-S.; Mewada, H.; Daradkeh, Y.I.; Rasheed, S.; Eldin, E.T.; Shafiq, M. Smart Android Based Home Automation System Using Internet of Things (IoT). *Sustainability* **2022**, *14*, 717. [CrossRef]
13. Omran, M.A.; Hamza, B.J.; Saad, W.K. The design and fulfillment of a Smart Home (SH) material powered by the IoT using the Blynk app. *Mater. Today Proc.* **2022**, *60*, 1199–1212. [CrossRef]
14. de Barcelos Silva, A.; Gomes, M.M.; da Costa, C.A.; da Rosa Righi, R.; Barbosa, J.L.V.; Pessin, G.; Federizzi, G. Intelligent personal assistants: A systematic literature review. *Expert Syst. Appl.* **2020**, *147*, 113193. [CrossRef]
15. Google Assistant (n.d.). Available online: https://assistant.google.com/ (accessed on 9 March 2023).
16. Apple Inc. (n.d.). *Siri*. Available online: https://www.apple.com/siri/ (accessed on 9 March 2023).
17. Amazon.com Inc. (n.d.). *Alexa Developer*. Available online: https://developer.amazon.com/en-US/alexa (accessed on 9 March 2023).
18. Uma, S.; Eswari, R.; Bhuvanya, R.; Kumar, G.S. IoT based voice/text controlled home appliances. *Procedia Comput. Sci.* **2019**, *165*, 232–238. [CrossRef]
19. Kumer, S.A.; Kanakaraja, P.; Teja, A.P.; Sree, T.H.; Tejaswni, T. Smart home automation using IFTTT and google assistant. *Mater. Today Proc.* **2021**, *46*, 4070–4076. [CrossRef]

20. Putthapipat, P.; Woralert, C.; Sirinimnuankul, P. Speech recognition gateway for home automation on open platform. In Proceedings of the 2018 International Conference on Electronics Information and Communication (ICEIC), Honolulu, HI, USA, 24–27 January 2018; pp. 1–4. [CrossRef]
21. Tiwari, V.; Hashmi, M.F.; Keskar, A.; Shivaprakash, N.C. Virtual home assistant for voice based controlling and scheduling with short speech speaker identification. *Multimed. Tools Appl.* **2020**, *79*, 5243–5268. [CrossRef]
22. Mehrabani, M.; Bangalore, S.; Stern, B. Personalized speech recognition for Internet of Things. In Proceedings of the 2015 Conference on Empirical Methods in Natural Language Processing (EMNLP), Lisbon, Portugal, 19–21 September 2015; pp. 1836–1841.
23. Google (n.d.). *Speech-to-Text API Requests*. Available online: https://cloud.google.com/speech-to-text/docs/speech-to-text-requests#overview (accessed on 9 March 2023).
24. Bird, S.; Loper, E.; Klein, E. Natural Language Toolkit (NLTK) (2009–2020). Available online: https://www.nltk.org/ (accessed on 9 March 2023).
25. Haroon, M. Comparative analysis of stemming algorithms for web text mining. *J. Adv. Inf. Technol.* **2018**, *9*, 20–25. [CrossRef]
26. Raulji, J.K.; Saini, J.R. Stop-word removal algorithm and its implementation for Sanskrit language. *Int. J. Comput. Appl.* **2016**, *150*, 15–17.
27. Pedregosa, F.; Varoquaux, G.; Gramfort, A.; Michel, V.; Thirion, B.; Grisel, O.; Vanderplas, J. Logistic Regression—Scikit-Learn 1.0 Documentation (2011–2021). Available online: https://scikit-learn.org/stable/modules/generated/sklearn.linear_model.LogisticRegression.html (accessed on 9 March 2023).
28. IBM (n.d.). *Logistic Regression—Overview and Use Cases*. Available online: https://www.ibm.com/topics/logistic-regression (accessed on 9 March 2023).
29. Google (n.d.). *Classification: Accuracy*. Available online: https://developers.google.com/machine-learning/crash-course/classification/accuracy (accessed on 9 March 2023).
30. Ceunen, B. Smart Home Commands Dataset. 2021. Available online: https://www.kaggle.com/datasets/bouweceunen/smart-home-commands-dataset (accessed on 9 March 2023).
31. Mashhour, M. Mashhour Dataset. 2021. Available online: https://www.kaggle.com/datasets/mohamedmashhour/mashhour-dataset (accessed on 9 March 2023).
32. Microsoft (n.d.). *Cortana*. Available online: https://www.microsoft.com/en-us/cortana (accessed on 9 March 2023).
33. IBM (n.d.). *Watson Assistant*. Available online: https://www.ibm.com/products/watson-assistant (accessed on 9 March 2023).

sustainability

MDPI

Article

A Review of the Theory and Practice of Smart City Construction in China

Zhengrong Liu [1] and Jianping Wu [1,2,*]

[1] School of Civil Engineering, Tsinghua University, Beijing 100084, China; liu-zr19@mails.tsinghua.edu.cn
[2] Autonomous Driving and Smart Transport Research Centre, Shenzhen Institute of Tsinghua University, Shenzhen 518057, China
* Correspondence: jianpingwu@tsinghua.edu.cn

Abstract: The construction of smart cities is an imperative trend for high-quality urban development and represents a crucial solution to various problems that afflict urbanization, such as population growth pressure, resource scarcity, environmental degradation, and economic frailty. To create a smart city, it is essential to gain a profound understanding of the evolution and experiences of urban development and have a clear grasp of the concept, emphasis, and developmental trajectory of smart cities. However, previous research and theoretical frameworks have mainly focused on the digital transformation of cities, the application of new generation information technologies, and the comprehensive intelligence of cities, neglecting the fundamental positioning and value of cities, which is the people's central position and pursuit of happiness. This has led to many situations where intelligence is pursued for the sake of intelligence, resulting in inefficient use and allocation of resources. This paper puts forth a framework for the three generations of smart city construction and their meanings at the conceptual level, conducts an in-depth analysis of the development process and challenges of smart cities in China, and provides suggestions for improvement.

Keywords: China; smart city; theoretical development; practical exploration; review

1. Introduction

Different cities have different cities' historical deposits, cultural traditions, social characteristics, economic bases, etc., making it challenging to propose a comprehensive and targeted theory or solution for smart city construction that considers the diverse construction situations in different locations. Moreover, research on the potential negative effects and issues associated with smart city construction remains relatively limited. China is the largest country and market for smart city construction in the world. Unlike developed countries, China's urban population continues to grow rapidly; the population's average education level is low; life is stressful, and many refined management initiatives that require popular participation or cooperation are difficult to implement, posing challenges for smart city construction. China's approach and progress in smart city construction presents a unique opportunity to observe and study the development of smart cities. It provides a natural laboratory for exploring the challenges and opportunities of smart city construction and can provide valuable insights for researchers and policy makers globally [1]. Current research on smart cities in China has focused on the policy implications [2], technological applications [3], construction effects [4], and development path [5], while insufficient consideration has been given to China's own unique urban–rural dual structure, large and concentrated urbanization, and rapid growth, among other factors [6]. Furthermore, significant public risks, such as the recent pandemic, have highlighted the need for enhanced emergency response and resilience in future smart cities. It is clear that the process of urbanization has a significant impact on the development of smart cities, and the strategies and solutions used to address the unique challenges faced by cities must evolve accordingly.

China has launched a smart city construction program over the past decade, and is at the forefront of smart city construction worldwide in terms of scale and speed of development [7,8]; however, there is a lack of theoretical research and support in the development of smart cities [9]. Despite the growth in smart city practice, there is a need to reflect on and summarize the progress of theoretical research and practical exploration. To address this, this article follows the logical progression underlying the theoretical and practical development of smart city construction in China. We delve further into the concept, elements, and structure of smart cities while improving the connotation of the concept of three generations of the smart city concept.

- Smart City 1.0: led by large technology-based companies, focused on technology development.
- Smart City 2.0: based on the Smart City 1.0, is led by the government and focuses on the role of technology as a tool, but the government's efforts are not always welcomed or accepted by the public. Most cities are built in the first or second generation.
- Smart City 3.0: based on the second generation, combines the "top-down" and "bottom-up" governance models, highlighting the people as the main body of smart city construction, and should promote participatory governance.

The Chinese context is also considered; the practical progress of smart cities is summarized and outlined; and the need for people-centric approaches that are adapted to local conditions and prioritized popular participation is emphasized. Ultimately, this article provides suggestions for the future development priorities of Chinese smart cities.

The structure of this paper is as follows. In Section 2, we introduce the theory, connotation, features, and evaluation system of the three generations of smart city construction. In Section 3, we summarize the global situation of smart city construction and its implications for China, providing important references for China's smart city construction. Section 4 starts from the characteristics of China's urban development, reviews the history and policy pilot of smart city construction in China. The guidance function of policies is crucial for urban development in China. In Section 5, we analyze the problems exposed in the current smart city construction in China and propose improvement suggestions. Finally, in Section 6, we conclude the paper.

2. The Theoretical Study of Smart Cities

2.1. The Connotation of the Concept of Smart City

The origin of the concept of a smart city can be traced back to the 1980s, when the New Urbanism movement emerged in the United States. This movement was accompanied by the introduction of information technology, and, in the 1990s, the term "smart city" was first used in the media [10]. In 2009, IBM CEO M.S. Peng proposed the concept of "Smart Earth", which aimed to realize a "sense, connect, and intelligence" core concept, leading to a significant increase in the popularity of smart cities worldwide [11,12]. The definition of smart cities varies across different countries and regions, due to varying economic, geographical, environmental, and cultural conditions, which has resulted in a lack of a unified understanding among the academic community. Although many scholars have attempted to outline the characteristics of smart cities, focusing on areas such as governance, technology, communication, sustainability, transportation, and environment, their definitions often lack universal recognition and have limitations. It is crucial to clarify the concept of smart cities in order to guide their actual construction, establish objective and feasible standards for evaluation, and build up a consensus. Achieving clarity is especially significant for addressing the challenges inherent to the global urbanization process. Scholars from various disciplines—including urban studies, computer and information technology, sociology, and public health—have proposed more than 30 different definitions, highlighting the need for further clarification [13]. Caragliu et al. offered a comprehensive and objective definition of smart cities as cities that invest in social and human capital, as well as traditional and modern communication infrastructure, and engage in participatory governance to promote sustainable economic and social development, improve the quality of life of their citizens, and achieve the intelligent management of natural resources [14]. After synthesizing the theoretical and practical development of smart cities, the authors concluded that the concept and connotation of smart cities have undergone two significant iterations, as shown in Figure 1.

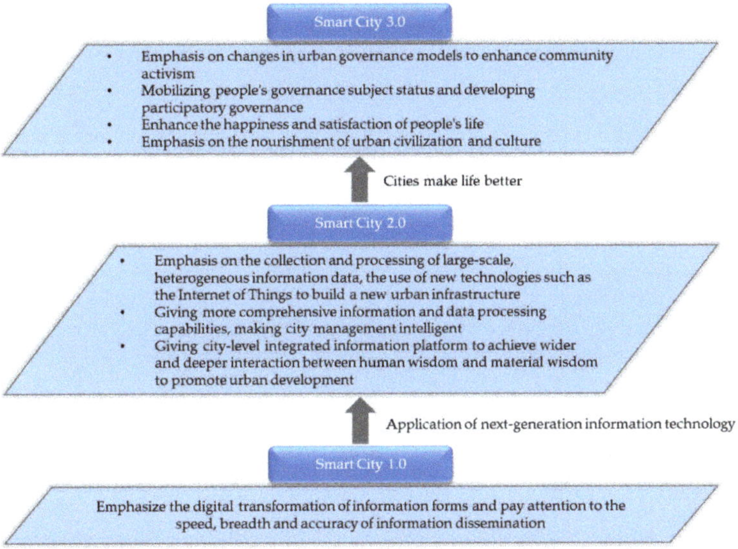

Figure 1. Comparison of the concepts of different generations of smart cities.

2.1.1. Smart City 1.0

The initial iteration of smart cities, centered on technology-centric approaches, was primarily aimed at enhancing urban development and operational models [15]. The first wave of smart cities was centered around the concept of utilizing new-generation information technologies, such as Internet of Things infrastructure, cloud computing, big data, and geospatial information integration, to promote intelligent urban planning, construction, management, and services. Such an approach primarily focuses on the role of technology in advancing urban development and construction [16]. Existing theories generally agree that large technology providers tend to play a leading role in the construction of smart cities at this stage, including IBM Corp [11], Songdo of the Republic of Korea [17–19], Portugal's PlanIT Valley [20,21], and China's Unicom [22,23]. The smart city operational model, which relies on IT support, data, and intelligent technologies, aims to upgrade industries, improve efficiency, and protect livelihoods. Regarding this stage of development, D.R. Lee [24] stated the representative view that a smart city is an intelligent digital city combined with the Internet of Things, which is essentially an intelligent urban management and operation mechanism that is visible, measurable, perceptible, analyzable, and controllable. The mechanism is established through comprehensive digitalization of the city's infrastructure, including sensors, networks, and computing resources, and utilizing a city information management and decision support platform that can analyze information and data in real-time [25].

With respect to this theoretical view, the development of smart cities has revealed several problems. For instance, Songdo, as a new technological city, has implemented various sustainable features such as green public spaces, efficient transportation, energy-saving buildings, sewage recycling, and waste management systems; however, despite these features, the quality of life of its residents remains unsatisfactory, the cost of living is high, and the city is far from achieving sustainable economic and social development. At present, the population of the city is only about 70,000, which is much lower than the original plan to accommodate nearly 300,000 people. Matsushima is another example demonstrating the shortcomings of the first-generation smart city concept, which focused solely on high-tech hardware without considering the humanistic aspect of urban planning. Currently, Matsushima, which was initially intended as a business district, now resembles a high-end international residential area that falls short of meeting the expected targets. Thus, while high-tech hardware plays a crucial role in urban planning, it is equally important to prioritize humanistic aspects in city planning for long-term sustainable development.

2.1.2. Smart City 2.0

The second generation of smart cities adopts a more holistic approach to urban development, focusing on concept and vision rather than just technology [15]. Governments are often the main driving force behind this approach. Unlike the first-generation smart cities, which emphasized technological development, the governments of second-generation smart cities use high-tech devices, software, and platforms to address urban problems and enhance the quality of life of citizens. They also regulate the use of technology in order to prevent it from dominating city life or decision making [26]. Scholars, such as Trencher, have highlighted specific aspects of second-generation smart cities—such as addressing social issues, improving citizen well-being, optimizing public services, and prioritizing citizen needs—that are not necessarily technology-related [27]. A typical case here is Barcelona, Spain [28–31], where development has focused on green, livable, and sustainable aspects. Smart city scholars believe that the primary objective of building a smart city is to enhance the quality of life of its residents by addressing the mismatch between supply and demand that regulates the various functions of a city [32]. For example, Chinese scholar Li Chunjia [33] argued that the intellectualization of city citizens, the intelligence of technology, and the wisdom of the environment are the basic means and contents of urban development. Meanwhile, Myeong [34], Vukovic [35], and others consider the smart city as a comprehensive urban development strategy that integrates urban operation and management, industrial development, public services, and administrative effectiveness, culminating as a high-end form of modern urban development. Moreover, scholars stress that smart cities should prioritize sustainable development [36]. For example, Souza et al. contended that contemporary smart cities should prioritize sustainability and efficient management of energy, transportation, healthcare, and governance to cater to the demands of urbanization [37]. Similarly, Xu Qingrui et al. put forth a smart city framework with a pyramid-star structure and asserted that the objective and strategic goal of smart city development in China should revolve around achieving comprehensive and sustainable economic, social, and ecological progress, ultimately fulfilling the public's needs for security and well-being in life [38].

The prevalent definition of a smart city revolves around the concept of next-generation development. According to the International Organization for Standardization, a smart city is "an integrated built environment that effectively blends physical, digital, and human systems to provide its citizens with a sustainable, inclusive, and prosperous future" [39]. The Telecommunication Standardization Sector of the International Telecommunication Union highlights sustainable development and describes a smart sustainable city as an innovative urban center that leverages information and communication technologies and other resources to improve quality of life, optimize urban operations and services, enhance the competitiveness of the city, and meet the economic, social, environmental, and cultural requirements of both current and future generations [40]. In the Chinese context, the official definition of a smart city by the central government is encapsulated in the New Type of Smart City Evaluation Index. This index defines a smart city as an innovative urban center that harnesses information and communication technology to seamlessly integrate diverse urban management systems, foster information resource sharing and business synergy, promote intelligent urban management and services, enhance public services and urban operations, boost the happiness and contentment of city residents, and achieve sustainable development [41]. In comparison to the Chinese government, this definition places a greater emphasis on clearly defining the objectives and tools of smart city development, prioritizing the application of new technologies such as ICT, emphasizing collaboration and resource sharing, and underscoring construction goals such as happiness and sustainable development for the benefit of the populace.

In summary, while there may be differences in the interpretation of smart cities among various scholars, institutions, enterprises, and cities, there are commonalities and overlaps in the underlying concept of a smart city system. The underlying concept is composed of people, data and information, digital technology (ICT), and physical systems [42]. The core principle of a smart city is the use of information technology to enhance wisdom in urban management and services. Smart infrastructure, public services, industrial systems, resource integration, security, and humanistic development comprise the primary content of a smart city, with the aim of achieving scientific development, efficient management, and a better quality of life for citizens [43]. Ultimately, the fundamental goal of smart city development is to address or mitigate the various issues confronting urban growth, optimize urban growth patterns, and enhance urban quality [44].

The Smart City 1.0 concept underscores the crucial role of emerging technologies in facilitating smart city development, which is an essential instrument for building smart cities. By contrast, Smart City 2.0 aims to differentiate itself from the previous iteration by placing greater emphasis on addressing urban challenges, enhancing public services, and improving the well-being of citizens. This new iteration recognizes the difference between the goals and tools of smart city construction, building upon the foundations of Smart City 1.0, and aims to promote a more holistic and comprehensive approach to smart city development.

2.1.3. Smart City 3.0

Based on the Smart City 2.0 concept, Smart City 3.0 pursues optimization of the way that urban groups are organized, highlighting the need for smart city construction paths to focus on participatory governance, to stimulate and bring the wisdom and power of the people into play, and to combine top–down and bottom–up approaches [15]. This iteration recognizes the limitations of Smart City 2.0, in terms of effective community feedback, and aims to promote more active citizen involvement in addressing urban challenges and shaping the future of their cities. Smart City 3.0 emphasizes the importance of inclusive decision-making processes, where citizens share their opinions and contribute to finding practical solutions to the social, environmental, and governance challenges of urban construction.

Central to the success of Smart City 3.0 is the concept of co-creation, which recognizes the vital roles of innovation and entrepreneurship in driving the construction and development of smart cities [15]. Smart City 3.0 construction requires the active participation of urban subjects to foster an innovative environment, promote inclusivity, and involve citizens in the problem-solving process for the city's development. This shift from passive service recipients to active co-creators of their own quality of life is essential. High-tech and popular innovation are the driving forces behind smart cities, and local demographic characteristics, geography, history, and culture should be considered in its development while adhering to the problem orientation.

In addition, there are many scholars who have sought to understand smart cities from the perspective of system theory. Scholars using system theory believe that modern cities are a class of open complex giant systems, such that their management presents the characteristics of dynamic non-linear complex giant systems with multiple dimensions, levels, structures, and intertwined and intricate sub-systems [45]. Some scholars in this field consider the smart city as a new urban ecosystem supported by new technologies covering citizens, enterprises, and governance, involving the digitally networked management of complex systems (e.g., urban geography, resources, ecology, environment, population, economy, and society) and the digitalization and informatization of all aspects related to the urban infrastructure and life development, according to the services and decision-making functions of the information system. Xia and Wang [44] proposed that modern cities are complex systems that combine cyber, physical, and social systems. They define the "wisdom" of smart cities in terms of the integration of urban informatization, knowledge, creative city, sustainable development, and ecological livability. This definition goes beyond the perspective of information and communication technology, highlighting the importance of the interactions between the different systems [46,47].

The concept of a smart city has evolved from being tool-driven to target vision and realization pathways. Scholars have deepened their understanding and grasp of smart cities by considering various perspectives, including partial system theory. As the concept has been continuously analyzed and improved, the characteristics of smart cities have become increasingly clear.

2.2. Analysis of the Characteristics of Smart Cities

Urban morphology refers to the physical and cultural evolution of a city's external appearance. Over time, factors such as resource availability, initial level of development, external environment, economic structure, technology level, lifestyle, national psychology, and cultural characteristics contribute to the unique shape of a city, giving it dynamic, diverse, hierarchical, multidimensional, and structural characteristics. The smart city is an advanced form of urban development, with its core features shown in Figure 2.

Figure 2. Smart City Core Feature Diagram.

Smart City 1.0 is primarily concerned with digitizing information data. The goal is to integrate various types of data throughout the city, making it easier to share and use. This generation places greater emphasis on the digital conversion of information forms, allowing information to be disseminated more quickly, widely, and accurately through the network.

Building on the foundation of Smart City 1.0, Smart City 2.0 focuses on changing how information is collected and processed. Utilizing new-generation information technology such as the Internet of Things, cloud computing, big data processing, and intelligent analysis, the City Information Model is constructed that integrates urban planning, construction, management, and operation. The result is a more comprehensive and robust information processing capability, enabling intelligent production and manufacturing, urban management, and service operation. In addition, this version gives due attention to environmental factors such as humanities and institutions, changes in urban governance models, and the expansion of smart technology applications in urban social fields, as well as the integration among urban systems.

Smart City 3.0 prioritizes people-centered cities as its core value. The concept of smart city construction must return to the root of urban services to increase residents' sense of access, happiness, and security. By expanding public participation, streamlining government scale, and empowering people's urban subject status, this generation aims to promote the innovation of the urban civilization paradigm. Future smart city construction should encourage the public's wide participation as it is an objective need for the economic development and social progress of the city. Additionally, this is an essential requirement for the city to improve its competitiveness and attractiveness.

2.3. Evaluation Index of Smart City Development

The advancement of smart city construction requires a set of authoritative evaluation index systems to guide the process. A unified evaluation index system lays a solid foundation for realizing massive multi-source heterogeneous data and achieving cross-sectoral integration applications. Technical elements must be constructed to meet evaluation criteria. Currently, many evaluation metric systems are proposed worldwide from different perspectives. For instance, IBM's smart city model proposes three main systems, including planning and management services, infrastructure services, and human services, each including three subsystems [48]. Similarly, Giffinger and Gudrun have identified 6 evaluation dimensions and 31 aspects, along with 74 indicators, in order to assess European smart cities in the economy, governance, environment, people, mobility, and life domains [49]. Marsal-Llacuna et al. argued that smart city construction progress requires real-time data-based monitoring indicators and proposed constructing a final synthetic index using a principal component analysis (PCA) approach [50]. In the Chinese context, the smart city criteria system includes 5 major categories of criteria divided into 4 levels, 16 technical areas, and 101 branches [51]. Some scholars have used specific city construction examples to develop city construction evaluation systems based on infrastructure fields, industry fields, government services, and cultural characteristics [52,53].

Overall, different evaluation index systems examine different directions, but indexes are selected based on the examination of digital transformation of data, integration processing and application of data, innovation, and application of new technologies in important areas of the city. These systems correspond to the construction characteristics of Smart City 1.0 and 2.0 stages but lack assessment of participatory governance and improvement of people's living standards. Recently, the Smart City Index evaluation system proposed by IMD has begun to focus on the feedback of the public on the construction situation, indicating a trend towards transition from Smart City 2.0 to 3.0.

Additionally, the construction of the evaluation index system should also consider the specific stage in which the city is developing. Some cities propose targeted evaluation systems based on their own development stages and characteristics. For instance, the European Smart Cities website introduces a European smart city model by analyzing indicators extracted from more than 90 smart city projects in Europe, and divides the model into sub-models reflecting small and medium-sized cities and large cities [54]. Furthermore, Cohen [55] put forward a comprehensive set of 18 measurement indicators and 46 sub-indicators based on 6 key elements. These categories cover all of the essential aspects of smart city indicators, making it one of the most comprehensive classification systems in the related literature.

As the study of smart cities in China has progressed, the evaluation system has undergone gradual improvements through the acquisition of relevant knowledge and exploration. In the current Chinese context, the smart city evaluation system consists of a set of scientific and systematic evaluation indicators that enable quantitative calculation and scientific evaluation of smart city construction results (see Figure 3).

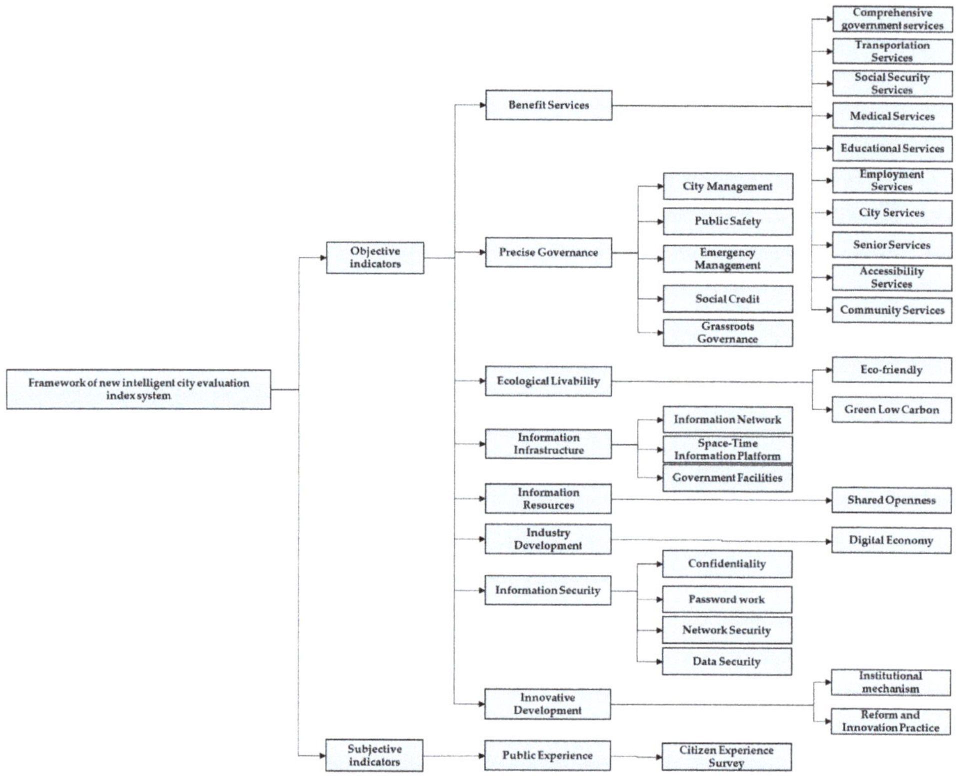

Figure 3. Framework of new smart city evaluation index system [41].

Since 2009, China has focused on the research and development of smart city construction standards; however, the country still lacks a third-party organization for standardization. In 2014, the Standardization Committee issued a notice regarding the establishment of the National Smart City Standardization Coordination and Promotion Group, along with the General Group and the Expert Advisory Group. The National Smart City Standardization General Group is the first official organization for standardization of smart cities in China, which is responsible for formulating the framework of China's smart city standardization system, standard development, strategy promotion, technical content, and application implementation of the standard at the national level. Meanwhile, local governments have also been exploring local standards that can guide their practices. For instance, Zhejiang Province, Shanghai City, and Guizhou Province have prepared the "Government Data Classification and Grading Guide (for trial implementation)" and the "Government Data Resources Catalogue" [56], which include metadata, core data elements, and compilation guidelines. These require city managers to systematically organize data types and business processes for standardization [57]. Despite some progress in standardizing smart city construction, it remains a complex project involving multiple industrial fields and city construction. Therefore, there is an urgent need to provide a collaborative platform for standardized knowledge services, project supervision, project management, standard testing, performance evaluation, and standard application demonstration.

3. The Construction and Inspiration of Smart Cities around the World

In order to have a comprehensive understanding of the current smart city construction, the keywords of "Smart Cities" and "Smart City" were searched through Web of Science,

Scopus, and Engineering Villages database, and the relevant documents and work reports were researched from the official government website. Based on the Web of Science database search, the number of published articles on the theme of smart cities from 2009 to 2022 is shown in Figure 4. We conducted a statistical analysis of cities ranked in the top 30 of the IMD Smart City Index from 2019 to 2023 [58], as shown in Table 1. The IMD Smart City Index focuses on analyzing the scope and impact of urban intelligence from a citizen's perspective. We believe that these cities, which were selected, can represent the advanced level of smart city development in the world today. We conducted a comparative analysis of the cities mentioned above and a few other classic city cases, focusing on economic development, citizen participation and experience, environment, mobility, governance, and living.

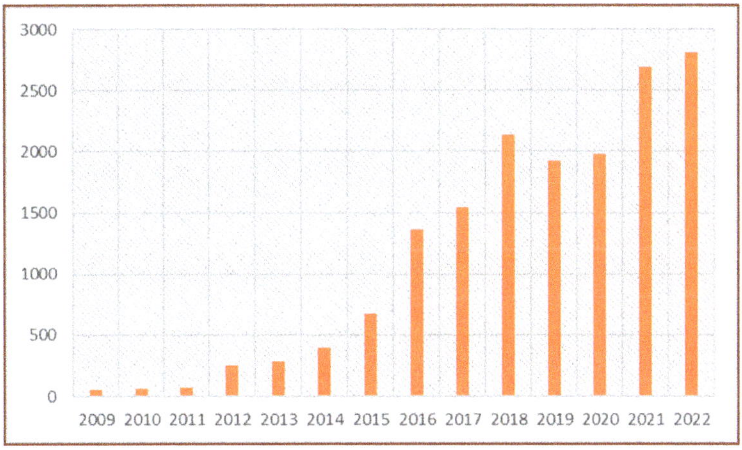

Figure 4. Number of published articles on the theme of smart cities from 2009 to 2022.

Table 1. Cities ranked in the top 30 of the IMD Smart City Index from 2019 to 2023.

City	City HDI	Rankings			
		2019	2020	2021	2023
Zurich	0.989	1	1	1	1
Oslo	0.98	2	2	2	2
Copenhagen	0.967	4	3	5	4
London	0.973	3	10	3	6
Singapore	0.939	10	7	7	7
Helsinki	0.96	6	5	9	8
Geneva	0.966	7	8	6	9
Stockholm	0.972	9	9	11	10
Beijing	0.907	30	22	17	12
Abu Dhabi	0.911	16	14	12	13
Prague	0.96	8	4	10	14
Amsterdam	0.962	11	11	13	15
Seoul	0.952	23	20	18	16
Dubai	0.911	13	19	14	17
Auckland	0.951	5	15	16	22
Bilbao	0.932	18	25	23	27
Vienna	0.942	12	18	20	28
Taipei City	0.916	24	23	25	29

Note: City HDI refers to the City Human Development Index and is an indicator used by the United Nations Development Program to measure the level of human development in cities. The higher the value is, the higher the level of human development is.

3.1. Smart City Construction Situation Worldwide

Many cities worldwide are facing infrastructure and service challenges due to population growth. As a result, traditional urban development models are no longer sufficient, and many cities have launched smart city development programs to enhance the quality of life of their citizens. Such cities include London, Stockholm, Dubai, New York, Barcelona, Hong Kong, Amsterdam, Singapore, Tokyo, Paris, and Copenhagen. However, the definitions of a smart city are inconsistent, due to distinct environmental, governance, and resource issues in each city. Some smart city definitions focus on sustainability or democracy; others emphasize data structure or organization, while some focus on smart infrastructure or transportation.

For example, smart city development in foreign countries typically focuses on smart living, smart economy, smart governance, smart environment, smart power, smart communications, and smart transportation. The US has prioritized smart city construction at a national strategic level, investing in infrastructure and smart grids. Europe, on the other hand, has focused more on improving energy use efficiency. Japan's i-Japan strategy aims to establish citizen-driven digital cities that foster trust and vitality, while Singapore's "Smart Nation 2015" plan has launched numerous projects to create a smarter city of the future.

Smart city exploration and construction are not limited to developed countries, as many developing countries are also actively involved in this field. Developing countries face faster urbanization rates and greater infrastructure challenges. For example, in 2014, India announced its intention to construct over 100 smart cities with high-tech communication capabilities. South Korea's National Strategic Smart City Program (NSSP) has established a new technological ecosystem and standards for smart cities. NSSP services leverage 5G telecommunication technologies and data structures, with the aim of expanding the industrial landscape of smart cities. Yang et al. have described the NSSP's smart city services and compared them with those offered by 15 smart cities in Europe, Asia, and North America [59].

Numerous scholars have analyzed the global development of smart cities from diverse perspectives. Angelidou compared and evaluated various smart city development strategies, identified their advantages and disadvantages, and provided recommendations [60]. Lee et al. conducted an empirical study on smart city construction in San Francisco and Seoul, revealing that effective and sustainable smart cities require dynamic coordination between managers and citizens on an open innovation platform. They also identified the need for complementary linkages among various participants, which should be adjusted according to the stage of urban development, social culture, and governance capacity [61].

Countries tend to adopt two primary models for smart city construction: Top–down technology-led models for new city construction, and a model that combines bottom–up and top–down approaches to encourage innovative services in established cities, based on relevant development needs. For instance, the Masdar Smart City project in the UAE, initiated in 2006, seeks to leverage solar technology to create a zero-carbon, sustainable, and smart habitat to house 50,000 people, 1500 clean technology firms, and employ 60,000 people daily. Despite some smart and traditional technological measures, such as smart wind towers, shade shelters, and shade connections, the city's economic and operational sustainability has not been well-addressed. The high cost of living associated with this urban construction model limits social integration and the heavily invested driverless bus system does not meet daily travel needs of residents. Additionally, local sand and dust issues severely impact the efficiency of solar panels. Thus, the lack of sustainability in Masdar's smart city project has become a significant challenge.

The Songdo smart city project in South Korea was awarded the 2008 Sustainability Award for being the most advanced new large-scale smart city construction project. However, it followed a top–down, supply-oriented construction model that focused more on physical hardware and industrial benefits, leading to huge financial investments and increased investment risk for the public sector, thus deviating from the knowledge culture starting point. The needs of local communities were ignored, public participation was lacking, and the ecological footprint increased, posing potential problems for future development.

On the other hand, the Amsterdam Smart City project, which began construction in 2009, is not technology-led but, rather, encourages the application of innovative technologies by multiple actors, promoting behavioral change among end-users. It emphasizes a quadruple helix partnership of government, business, academia, and community, engaging local communities in a bottom–up approach through the Smart City Living Lab to provide solutions for the city and its residents, ultimately combining environmental and social development goals with economics and technology. Similarly, Brisbane, Australia, has been successful in developing a knowledge corridor and upgrading public transport services to the city center.

Overall, the development of smart cities is still in its infancy, and technology alone cannot facilitate the complete construction and operation of a smart city. Smart decisions and policies that integrate smart city construction into a sustainable urban development framework with the quadruple bottom line of economic, social, environmental, and governance are necessary. The wisdom of citizens and communities is also critical in improving the efficiency of cities.

3.2. The World's Smart City Construction to China's Inspiration

The global construction of smart cities reflects the city's evolution in terms of material form and cultural connotation. This evolution has been shaped by a range of factors, including resource endowment, initial level, external environment, economic structure, technology level, lifestyle, national psychology, and cultural characteristics.

The development of new-generation information technology has had a direct impact on the operation of cities. The digitalization of various information data has significantly increased the speed, breadth, and accuracy of information transfer, people flow, and material exchange, thereby upgrading the city's operation system. The resultant improvement in operational efficiency has led to changes in the institutional and organizational systems that ensure the city's smooth functioning, creating a society in which everything is interconnected. This has led to the development of intelligent transportation, governance, medical care, energy, and other areas, thereby enhancing the city's wisdom. The ultimate goal of this wisdom is to serve the quality of life of people in the city by providing convenience and enhancing community activity, using smart technology to realize the fundamental purpose of the city, which is to make life better and innovate the paradigm of urban civilization. Figure 5 illustrates the logical progression model of the city's evolution. The model consists of two main parts, top–down (TD) and bottom–up (BU). TD means that the city's value system and relationship system guide the construction, development and operation of the city, while BU means that the city's institutional and organizational system will play a modifying role in influencing the city's value pursuit and relationships within the city.

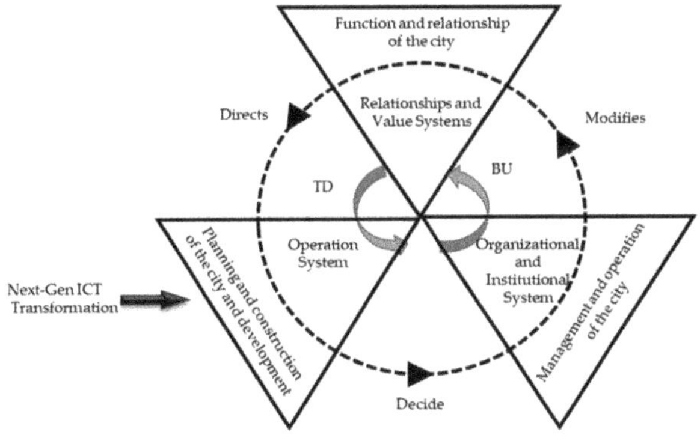

Figure 5. The logical progression model of the city's evolution.

The agglomeration of people is at the heart of urban development, and the relationship between individuals and groups and the relationship and interaction between people and nature constitute the deep structure of cities. Therefore, the deep development of cities requires the discovery and promotion of urban civilization and culture, the construction of both material and spiritual civilization, and the promotion of participatory governance.

Lan H. [62] identified three main challenges of urbanization: urban planning, construction, and development; urban management and operation; and urban functions. These challenges correspond to the three generations of smart city concepts, and the construction of smart cities worldwide provides valuable experience for China to address these problems. As cities experience rapid population growth, resource constraints, environmental pollution, and traffic congestion, smart cities with new-generation information technology provide a solution to these problems by promoting digital transformation, enhancing the speed and accuracy of information transmission, and improving infrastructure innovation.

The construction of Smart City 1.0 has realized the digital transformation and upgrading of the city, laying an essential foundation for the comprehensive transformation of urban governance, industry, development, and environment.

The development of Smart City 2.0 involves the integration and application of heterogeneous, large-scale data from various industries and intelligent technologies in different aspects and levels of urban governance. However, the ultimate goal of city construction is not only integrated and comprehensive intelligent development but also the agglomeration of people and the relationship and interaction between people and nature, which form the deep structure of cities. Therefore, promoting participatory governance and discovering and promoting urban civilization, culture, and characteristics are critical for the deep development of the city.

The development of Smart City 3.0 represents a new stage of urban civilization, which is to build a harmonious and livable city—that is, to create a city that is more compatible with the laws of nature, human society, and human development. The development of Smart City 3.0 is based on the integration and innovation of urban governance, industry, ecology, and civilization, and the construction of an organic and healthy symbiosis system that can adapt to the laws of nature, human society, and human development and realize the sustainable development of cities. Smart City 3.0 focuses on improving the overall quality of human life, emphasizing people-oriented, ecological harmony, and social governance, and promoting the harmony between man and nature, people and society, and people and people. Therefore, the construction of smart cities should take the people as the center and the harmonious development of human and nature as the goal, not only pursuing the intelligence of the city but also paying attention to the livability, cultural heritage, and ecological environment of the city.

4. Policy Development and Practice of Smart Cities in China

4.1. Characteristics of China's Urban Development

Since the promotion of the concept of smart cities in China, the development of Chinese cities has undergone many changes. The development concept of Chinese cities has shifted from emphasizing development as the top priority and achieving comprehensive, coordinated and sustainable development to building smart, livable and resilient cities. The people-oriented concept has always been upheld, with increasing attention paid to environmental protection, livability, and the quality of people's lives.

From the perspective of urban development strategy, Chinese urban development has evolved from the "point-driven" development model with special economic zones, new areas, and industrial parks as carriers to the "surface-driven" development model with comprehensive supporting reform pilot zones as carriers, and has now formed a trend of "19 + 2" urban clusters, transitioning from institutional reform to dual evolution of institutional and development modes. However, there is still uneven development among regions, and resource allocation needs to be tilted towards small and medium-sized cities.

In terms of urban governance, China has achieved breakthroughs in legal construction and policy-making at the level of city planning, urban land management, urban economic management, urban employment management, and urban environmental construction. A series of new policies and regulations have been introduced to solidify the foundation for the transformation of cities from extensive to intensive development.

In terms of urban development mode, the focus has shifted from solely pursuing economic efficiency to simultaneously paying attention to social development and improving people's quality of life. The development mode has gradually shifted from focusing on external expansion to improving the internal quality of cities. In particular, after the release of China's 14th Five-Year Plan, high-quality urban development has become a consensus. With the vigorous development of new-generation information technology, urban development is showing accelerated integration of industrialization, informatization, and urbanization. Concentrated urbanization, the construction of large urban clusters, and dispersed urbanization are all advancing at the same time, while urbanization and rural-urban integration are also advancing simultaneously. The coordinated development of large, medium, and small cities and small towns has become a new feature.

The current characteristics of smart city construction in China are significant differences in regional development. The market size of smart city construction is enormous, with over 900 cities at or above the prefecture level having smart city construction plans that have entered the implementation stage, and smart cities are extending upstream and downstream to smart city clusters and county-level smart city construction. Compared with the eastern region, the population size and density of Chinese cities in the central and western regions are vast. They hope to achieve common prosperity, material civilization, and spiritual civilization through smart city construction, and harmony between people and nature.

4.2. The Development Process of China's Smart Cities

The development of smart cities in China is unique due to the country's special national conditions, including a large population base, urban–rural dual structure, urbanization concentration, and continued rapid growth. China's smart city construction has gone through four stages of policy promotion.

The first phase of smart city development is the exploration and development period, marked by the Smart City 1.0 stage. From the end of 2008 to August 2014, the concept of "Smart Earth" proposed by IBM quickly spread worldwide and gained widespread recognition, resulting in a surge in smart city construction globally. China also began to explore smart city construction independently during this period. However, compared to other countries, China's smart city practice was relatively late. At that time, exploration lacked top-level design and coordination, and development showed a free trend, relatively scattered and disorderly. In 2010, relevant national ministries and commissions mainly guided pilot work in their own scope, while cities mainly promoted the idea of digital city construction. Cities such as Shenzhen, Ningbo, Foshan, Yangzhou, Nanjing, Shanghai, Beijing, and others joined the ranks of smart city construction. In early 2013, the Ministry of Housing and Construction announced the pilot list of the first 90 national smart cities and promulgated the Interim Management Measures of National Smart City Pilot and the National Smart City Pilot Index System, which kicked off the prelude to the comprehensive construction of smart cities in China.

The second stage of smart city development is characterized by standardization and integration and marks a transition from Smart City 1.0 to 2.0. The first phase of free development resulted in inconsistencies and challenges such as information silos, inadequate top-level design, and low-level duplication of construction. Between August 2014 and December 2015, the National Development and Reform Commission and seven other ministries and commissions issued the "Guidance on Promoting the Healthy Development of Smart Cities", which was the first systematic document approved by the State Council to comprehensively guide the construction of smart cities in China. This initiative resulted in the formation of an "Inter-ministerial Coordination Working Group for Promoting the Healthy Development of Smart Cities" at the national level, led by the National Development and Reform Commission and comprising 25 ministries. Various departments are now collaborating to guide the construction of local smart cities.

The third stage can be characterized as the strategic upgrading period that marks the transition to Smart City 2.0. During the period spanning from December 2015 to December 2017, the Central Urban Work Conference was held in 2015 [63], which represents a significant turning point. At the conference, smart cities were given greater attention and elevated to the national strategic level, emerging as a crucial aspect of the country's new urbanization plans. Following this, a new concept of smart city was proposed, with a greater emphasis on the comprehensive development and application of new-generation information technology in various fields of urban management, aimed at achieving sustainable urban development and enhancing the comprehensive competitiveness of cities. This new concept marks a shift towards Smart City 2.0.

The fourth stage is the period of deepening development, in which the exploration of Smart City 3.0 begins in terms of development concepts and planning. In October 2017, the 19th Congress of the Communist Party of China proposed the key construction of a smart society, after which the construction of new smart cities accelerated and formed the "19 + 2" smart city cluster development trend, gradually extending construction achievements to districts, counties, and rural areas. In this stage, relationships between people, nature, and cities are deeply explored, and the improvement of people's quality of life and participatory governance are emphasized as goals of urban development. The pursuit of urban construction values and innovation of civilization paradigms are highlighted, and the construction of smart cities is becoming increasingly mature.

Overall, the development of smart cities in China plays a significant role in achieving sustainable urban development, through leading the application of information technology, enhancing the comprehensive competitiveness of cities, and improving the quality of life of residents. As such, it is a key transformation direction for urban development in China.

4.3. Relevant Support Policies and Pilot Work

Policy support plays a crucial role in promoting the development of smart cities. In order to guide and ensure the healthy and sustainable development of smart cities, the State Council and various ministries have issued multiple documents and guidelines related to smart city construction in recent years (see Table 2).

Table 2. Key policies related to the construction of smart cities.

Release Time	Send a Message Unit	Policy Name	Core Content of the Document about Smart Cities
2012	State Department	The State Council on vigorously promote the development of information technology and effective protection of information security of a number of opinions	Promote the sharing of urban management information, promote the grid management model, accelerate the implementation of smart grid, intelligent transportation, and other pilot demonstrations to guide the healthy development of smart city construction
2012	Ministry of Housing and Construction	National Smart City Pilot Interim Management Measures" "National Smart City (District, Town) Pilot Index System (for trial implementation)	Regulating and promoting the healthy development of smart cities
2014	The Central Committee of the Communist Party of China and the State Council	National New Urbanization Plan (2014–2020)	Explicitly request to promote the construction of smart cities and put forward the main points to promote the construction of smart cities

Table 2. *Cont.*

Release Time	Send a Message Unit	Policy Name	Core Content of the Document about Smart Cities
2014	National Development and Reform Commission and other 8 ministries	Guidance on promoting the healthy development of smart cities	Clearly put forward the development ideas of China's smart cities, construction principles, major goals, and information security requirements, in order to strengthen guidance of the construction of smart cities around the practice
2015	National Standards Committee, Central Internet Information Office, National Development and Reform Commission	Guidance on the Construction and Implementation of Smart City Standard System and Evaluation Index System	Accelerate the development of relevant standards, ensure that the development of standardization of smart cities is officially on the national agenda
2016	State Department	"Thirteenth Five-Year Plan" National Informatization Plan	Proposed to build 100 new model smart cities by 2018; by 2020, the construction of new smart cities to achieve remarkable results
2016	The Central Committee of the Communist Party of China and the State Council	Further strengthen the management of urban planning and construction of a number of views	Strengthen the intelligent construction of urban management and service systems, promote the integration of modern information technology and urban management services, and enhance urban governance and services; strengthen the construction and functional integration of digital platforms for urban management
2017	National Bureau of Surveying, Mapping and Geographic·Information	Technical Outline for the Construction of Spatio-temporal Big Data and Cloud Platform for Smart Cities (2017 Edition)	Relying on the city cloud support environment, realize the promotion of space–time benchmark, big data, and information cloud platforms for smart cities, build urban space–time infrastructure and develop smart city thematic application systems
2017	Ministry of Transportation and Communications	Smart Transportation for Easier Travel Action Plan (2017–2020)	Accelerate the development of intelligent urban transportation travel
2017	Central Internet Information Office, etc.	Notice on the implementation of the national e-government comprehensive pilot	By the end of 2019, the pilot areas of e-government coordination capacity should be significantly enhanced, the level of infrastructure intensification significantly improved, the basic realization of the orderly sharing of government information resources on demand
2017	State Department	New Generation Artificial Intelligence Development Plan	Promote intelligent social governance
2017	Ministry of Industry and Information Technology	Notice on the comprehensive promotion of mobile Internet of Things (NB-IoT) construction and development	From 3 aspects, take 14 measures to comprehensively promote the development of mobile Internet of things construction
2018	General Administration of Market Supervision, etc.	Smart City Top Level Design Guide	Gives the general principles of the top-level design of the smart city, the basic process and the specific construction of demand analysis, overall design, architecture design, and implementation path planning
2019	National Development and Reform Commission	Key tasks of new urbanization construction in 2019; security assessment methods for cloud computing services	Clarify the key tasks of the current phase of urbanization, improve the security and controllability of the procurement and use of cloud computing services by party and government organs and operators of critical information infrastructure

Table 2. *Cont.*

Release Time	Send a Message Unit	Policy Name	Core Content of the Document about Smart Cities
2019	Ministry of Natural Resources	Smart City Spatio-temporal Big Data Platform Construction Technology Outline	Building a pilot space–time data platform for smart cities, guiding the construction of space–time data platforms and encouraging intelligent applications
2019	The Central Committee of the Communist Party of China and the State Council	Outline of the Development Plan of Guangdong, Hong Kong and Macao Greater Bay Area	Emphasis on promoting the pilot demonstration of a new type of smart city and the construction of a national big data comprehensive pilot area in the Pearl River Delta region
2020	National Development and Reform Commission	New urbanization construction and urban-rural integration development tasks in 2020	Implementation of a new type of smart city West East, improve the digital management platform and perception system, open up the community end, integrate information systems and data resources in various fields to support rapid response to emergencies
2021	National People's Congress	Outline of the Fourteenth Five-Year Plan and Vision 2035 for National Economic and Social Development of the People's Republic of China	Grading and classification to promote the construction of new intelligent cities
2022	Central Committee of the Communist Party of China	Report at the 20th National Congress of the Communist Party of China	Accelerate the transformation of the development of megacities, implement urban renewal actions, strengthen urban infrastructure construction, and build livable, resilient, and smart cities

Since 2010, the Development and Reform Commission, the Ministry of Housing and Construction, the Ministry of Industry and Information Technology, the Chinese Academy of Engineering, the Ministry of Science and Technology, the National Standards Committee, the National Bureau of Mapping and Geographic Information, and other relevant ministries and bureaus have conducted pilot smart city projects in areas where suitable conditions exist, in order to actively explore and accumulate valuable experience. As of 2022, a total of 14 provinces, municipalities, and autonomous regions in China, including Beijing, Shanghai, and Hebei, have released guidance or development plans to promote the construction of new smart cities. Additionally, 24 provincial, municipal, and autonomous regional governments, as well as several prefecture-level cities such as Qingdao, Shenzhen, and Ningbo, have issued guidance and development plans regarding the "digital economy". Liu, Peng et al. have categorized China's pilot smart city construction into three main application areas and their subfields: enhancing quality of life (e.g., family, community, healthcare, and education); improving public management and services (e.g., public safety regulation, food safety regulation, smart transportation, smart tourism, and environmental protection); and broad resource management (e.g., water, electricity, and agriculture) [64]. In a similar vein, Washburn et al. have identified seven key infrastructure components and services that constitute smart cities: city management, education, healthcare, public safety, real estate, transportation, and utilities [65]. A. Dang posited that China's new smart city development is propelled by two key factors: application scenarios and digital twins [66].

4.4. The Current Status of Smart City Construction in China

As of now, China is planning and constructing more than 900 smart cities. From a construction layout perspective, all vice-provincial cities, over 89.6% of cities at and above the prefecture level, and over 62.8% of counties have proposed building smart cities and forming a trend of coordinated development of smart city clusters [67], as is shown in

Figure 6. The figure displays two sets of numbers on the left side. The first set represents the number of smart cities planned to be built of each type, while the second set represents the total number of cities in each of those types.

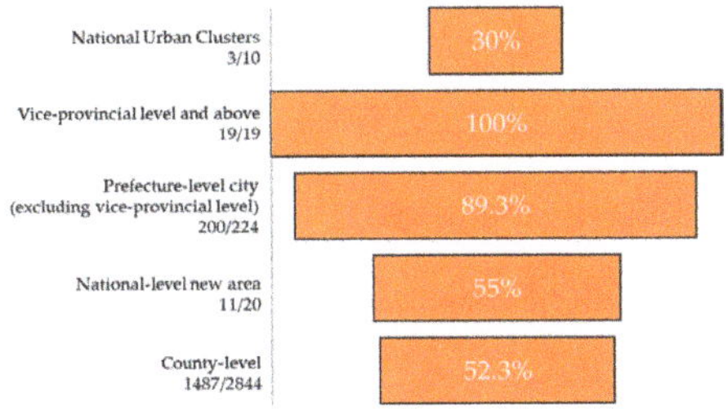

Figure 6. The situation of the top-level design of smart cities at all levels of cities in China.

From the perspective of smart city distribution density, the number of pilot projects is unevenly distributed among provinces and cities, mainly concentrated in the eastern and central regions, with the eastern coastal areas being the most concentrated. From a comprehensive construction situation perspective, first-tier cities such as Beijing, Shanghai, and Shenzhen continue to lead and perform well in smart city strategy, technology, domain, and innovation capabilities. According to the data from China Academy of Information and Communications Technology, the total investment in smart city projects in China in 2021 amounted to RMB 2.79 trillion, and the compound annual growth rate since 2015 is approximately 85%. In 2021, there were 94 smart city-related projects with a scale of RMB 100 million or more, with a total amount of RMB 12.196 billion. In terms of sub-fields, there were 16 smart transportation investment projects with an amount exceeding RMB 4.2 billion, which is the highest proportion [68]. The proportion of smart city construction enterprises in different regions and the proportion of per capita smart city construction enterprises are shown in the Figure 7 below. Figure 7D presents the total number of relevant enterprises and population in each region. To facilitate a more intuitive comparison of the difference in the number of enterprises per capita between regions, we have calculated the ratio of the total number of relevant enterprises and population in each region. We then normalized these ratios and reflected them in a pie chart. This allows for a clearer visualization of the relative proportion of enterprises per capita in each region. In summary, the uneven status of smart city construction between regions and different scale cities continue to exist.

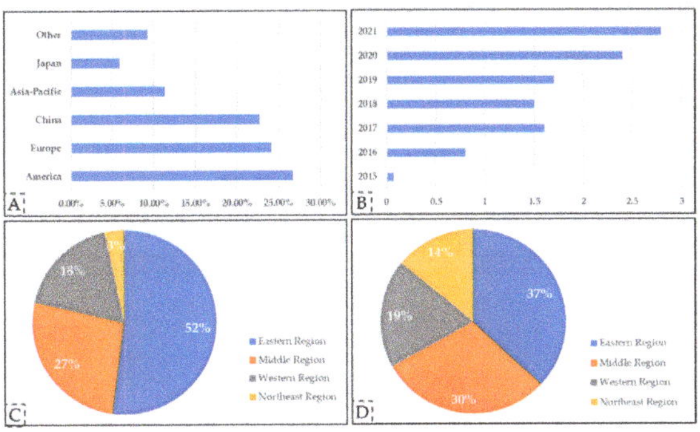

Figure 7. Investment and construction status of smart cities in China. (**A**) Proportion of the smart city market size in major regions worldwide in 2020; (**B**) Scale of central government investment in smart cities in China from 2015 to 2021 (unit: trillion yuan); (**C**) Regional distribution of the number of smart city construction enterprises in China in 2022; (**D**) Regional distribution of the per capita number of smart city construction enterprises in China in 2022.

5. Analysis of the Current Situation and Future Prospects of Smart Cities in China

5.1. Analysis of the Current Situation of China's Smart Cities

The smart city market in China has seen significant growth in recent years, with the China Smart City Working Committee reporting a projected market size of 25 trillion yuan by 2022. This investment scale will make China the second-largest smart city market globally, following the United States. China's push for smart city development has resulted in several rounds of pilot programs and the construction of over 900 smart cities, with coverage extending to all provinces, cities, and autonomous regions across the country. Despite an uneven distribution across regions, the formation of smart city groups has created a synergistic development trend. First-tier cities such as Beijing, Shanghai, and Shenzhen continue to lead the way in comprehensive construction, boasting impressive strategies, technologies, fields, and innovation capabilities.

Although China boasts a large number of smart city construction pilots, the smart city construction development process also reveals several shortcomings. According to our experience, the following major problems widely exist in the process of smart city construction throughout China.

First, too much emphasis has been placed on scale rather than effectiveness. Instead of combining urban construction and development goals and choosing reasonable smart city construction goals, scale, projects, and technical routes, many smart city construction projects pursue large and comprehensive development at the expense of expected results.

Second, the construction of smart cities is often not based on city-specific characteristics. Smart city construction fails to consider a city's unique advantages and main problems in terms of economic construction, the livelihood of residents, and ecological construction. This leads to a simple replication of the experience and methods of advanced cities and model cities, resulting in reduced social, economic, and environmental benefits.

Third, the construction is not deep enough. It does not pursue the quality of data perception, integrity, and data integration and sharing but, rather, aims for the superficial appearance of cameras and big-screen displays. This approach limits the collected data to large screen displays and does not allow experts and enterprises to conduct intelligent analysis, computation, and operation for effective smart city management.

Fourth, too much emphasis has been placed on hardware investment, at the expense of software development and maintenance. Smart city construction involves heavy investments into hardware, including the deployment of cameras, servers, and screens; meanwhile, less than one-third of the project funds are allocated for purchasing software or opening data for data analysis and intelligent computing, building model software, and/or effectively using the collected data for the intelligent operation and management of the city.

Finally, there is too much reliance on government-led initiatives. Most of China's smart city construction is dominated by government input, which has certain advantages in the urban information infrastructure construction stage; however, once the construction of urban information infrastructure is completed, a perfect data opening and management system must be established. In this context, a multi-party construction ecology with government guidance, universal participation, and government–enterprise cooperation should be formed to drive the city's data industry and digital economy. This also allows professional and technical personnel to effectively utilize the data and play key roles in smart city management.

5.2. Thoughts and Suggestions for Smart Cities in China

After considering the aforementioned issues, we propose that the future development and operation of China's smart cities should focus on the following aspects.

First, prioritizing top-level design is crucial, as smart cities are intricate and open systems that require a people-oriented approach. Comprehensive planning based on thorough research and consideration should strengthen the top-level design and fully integrate the concept of "people's city built by people, people's city for people" into the planning and design stage. This should be combined with future development needs and the situation of high-speed technological iteration, as well as setting reasonable targets, moderately planning ahead, and continuously enhancing the integrity, system, livability, inclusiveness, and growth of the city.

Second, the construction of smart cities should be tailored to the local conditions, including the unique history, geography, cultural background, and development stage of each location. Priorities should be distinguished, and coordination should promote the exploration of development paths and models that reflect the characteristics of the city.

Third, smart city construction should be problem-oriented, focused on solving real problems and addressing needs. It should be based on the demand for traction, focused on the key concerns of the masses, and aimed at solving problems to sustainably achieve urban development and promote the well-being of the people.

Fourth, smart city development should prioritize high-quality development. Data provide an essential lifeline for smart city development, and the effective use of data is critical for the operation of smart cities. Smart city construction should involve breaking down data barriers to effectively leverage data by combining hardware and software.

Fifth, promoting participatory governance is essential to attract the active participation and interaction of different actors, such as the government, researchers, enterprises, and the public, in order to stimulate social vitality. The government and enterprises should cooperate to clarify the boundaries of responsibility, and the public's importance as the demand-side, participant, and evaluator of the construction and operation performance should be considered, thus allowing the needs of the public for a better life to be met.

6. Conclusions

The concept of a smart city encompasses a range of fundamental elements, including subjects, objectives, means, foundations, environments, support systems, and safeguards. These elements interact with one another to form a complex urban system that exhibits shared characteristics across three intersecting layers: physical space, social space, and virtual space. These characteristics are also influenced by the developmental stages of smart cities. The advent of new information technologies has brought about revolutionary changes in information dissemination, which, in turn, has transformed urban operating systems, organizational systems, and institutional systems. Ultimately, these changes affect the relationships between individuals, between individuals and the city, between individuals and nature, and the value propositions of urban development.

This study focuses on the construction of smart cities in China. At the theoretical level, three generations of smart city evolution are proposed and explained. With the rapid development of new information technologies, Smart City 1.0 focuses on realizing the digitization of information, significantly improving the speed, breadth, and accuracy of information dissemination, and laying the foundation for achieving widespread perception, interconnectivity of all things, and comprehensive intelligence. Building on this foundation, Smart City 2.0 breaks down data barriers between departments, achieves the integration and widespread application of massive, multi-source, and heterogeneous data, and promotes the intelligent transformation and upgrading of various fields, providing inexhaustible power for economic development and industrial transformation. The development of intelligent cities promotes the innovation of urban civilization paradigms. Smart City 3.0, building on the achievements of 2.0, returns to the fundamental purpose of improving people's lives in cities, emphasizes public participation in urban governance, and truly achieves the goals of harmony, livability, and smart.

At the practical level, this study draws on the construction status of smart cities worldwide to identify and summarize experiences and lessons that are of great reference significance for the construction of smart cities in China, reducing exploration costs. Furthermore, the study conducts a systematic review of the history, current situation, and policy evolution of smart city construction in China, identifies existing problems, and provides improvement suggestions. The construction of smart cities in China still faces regional differences and should not be simply copied without a thorough understanding of the theoretical underpinnings of smart cities and taking into account factors such as geographic location, historical and cultural background, resource endowment, industrial characteristics, and population features. Instead, smart city construction plans should be developed based on these practical considerations.

Author Contributions: The conceptualization, methodology, formal analysis, investigation, data curation, writing, review, and editing were done by Z.L. and J.W. All authors have read and agreed to the published version of the manuscript.

Funding: This research received no external funding.

Institutional Review Board Statement: Not applicable.

Informed Consent Statement: This study did not involve humans.

Data Availability Statement: The data presented in this study are available on request from the corresponding author.

Conflicts of Interest: The authors declare no conflict of interest.

References

1. Wang, M.; Zhou, T.; Wang, D. Tracking the evolution processes of smart cities in China by assessing performance and efficiency. *Technol. Soc.* **2020**, *63*, 101353. [CrossRef]
2. Shen, L.; Huang, Z.; Wong, S.W.; Liao, S.; Lou, Y. A holistic evaluation of smart city performance in the context of China. *J. Clean. Prod.* **2018**, *200*, 667–679. [CrossRef]
3. Li, C.; Liu, X.; Dai, Z.; Zhao, Z. Smart city: A shareable framework and its applications in China. *Sustainability* **2019**, *11*, 4346. [CrossRef]
4. Li, C.; Dai, Z.; Liu, X.; Sun, W. Evaluation system: Evaluation of smart city shareable framework and its applications in China. *Sustainability* **2020**, *12*, 2957. [CrossRef]
5. Li, X.; Fong, P.S.; Dai, S.; Li, Y. Towards sustainable smart cities: An empirical comparative assessment and development pattern optimization in China. *J. Clean. Prod.* **2019**, *215*, 730–743. [CrossRef]
6. Wu, Y.; Zhang, W.; Shen, J.; Mo, Z.; Peng, Y. Smart city with Chinese characteristics against the background of big data: Idea, action and risk. *J. Clean. Prod.* **2018**, *173*, 60–66. [CrossRef]
7. Bouvier, L.F. Replacement migration: Is it a solution to declining and ageing population? *Popul. Environ.* **2001**, *22*, 377–381. [CrossRef]
8. Lim, T.W. Case study 3: The rise of smart cities in China—Domestic development and comparative perspectives. In *Industrial Revolution 4.0, Tech Giants, Digitized Societies*; Palgrave Macmillan: Singapore, 2019; pp. 117–125.
9. Zhao, F.; Fashola, O.I.; Olarewaju, T.I.; Onwumere, I. Smart city research: A holistic and state-of-the-art literature review. *Cities* **2021**, *119*, 103406. [CrossRef]
10. Bibri, S.E.; Krogstie, J. Smart sustainable cities of the future: An extensive interdisciplinary literature review. *Sustain. Citi. Soc.* **2017**, *31*, 183–212. [CrossRef]
11. Harrison, C.; Eckman, B.; Hamilton, R.; Hartswick, P.; Kalagnanam, J.; Paraszczak, J.; Williams, P. Foundations for smarter cities. *IBM J. Res. Dev. Policy Rev.* **2010**, *54*, 1–16. [CrossRef]
12. Palmisano, S. Smarter Planet. Available online: https://www.ibm.com/ibm/history/ibm100/us/en/icons/smarterplanet/words/ (accessed on 5 December 2022).
13. Ramaprasad, A.; Sánchez-Ortiz, A.; Syn, T. A unified definition of a smart city. In Proceedings of the Electronic Government: 16th IFIP WG 8.5 International Conference, EGOV 2017, St. Petersburg, Russia, 4–7 September 2017; pp. 13–24.
14. Caragliu, A.; Del Bo, C.; Nijkamp, P. Smart Cities in Europe. *J. Urban Technol.* **2011**, *18*, 65–82. [CrossRef]
15. Cohen, B. The 3 Generations of Smart Cities. Available online: https://www.fastcompany.com/3047795/the-3-generations-of-smart-cities (accessed on 10 December 2022).
16. Tahir, Z.; Malek, J.A. Main criteria in the development of smart cities determined using analytical method. *Plan. Malays.* **2016**, *14*, 1–14. [CrossRef]
17. Carrera, B.; Peyrard, S.; Kim, K. Meta-regression framework for energy consumption prediction in a smart city: A case study of Songdo in South Korea. *Sustain. Cities Soc.* **2021**, *72*, 103025. [CrossRef]
18. Patel, M.; Padhya, H. A Smart City Development Concept: The Songdo Experiences. *Int. J. Res. Eng. Sci.* **2021**, *9*, 7–10.
19. Kovalev, Y.; Burnasov, A.; Ilyushkina, M.; Stepanov, A. Political Models of Smart Cities and the Role of Network Actors in Their Implementation (the Case of Vienna, Lyon, and New Songdo in Seoul). In Proceedings of the AIP Conference Proceedings, I International Conference ASE-I-2021: Applied Science and Engineering: ASE-I, Grozny, Russia, 25 June 2021; AIP Publishing: Woodbury, NY, USA, 2021; p. 050002.
20. Thomas, R.M. Inside Smart Cities: Place, Politics and Urban Innovation, by Karvonen, Andrew, Federico Cugurullo, and Federico Caprotti. *J. Plan. Educ. Res.* **2022**, *42*, 497–498. [CrossRef]

21. Correia, D.; Teixeira, L.; Marques, J.L. Reviewing the State-of-the-Art of Smart Cities in Portugal: Evidence Based on Content Analysis of a Portuguese Magazine. *Publications* **2021**, *9*, 49. [CrossRef]
22. Shen, Z.; Guo, Z.; Shen, Q.; Guo, Z.; Sun, L. Promoting China's high-quality development with smart cities. *Inf. China* **2022**, *353*, 92–94.
23. Wang, Y. Rearch on the business development strategy of China Unicom's smart city. Master's Thesis, Shandong University, Jinan, China, 2020. [CrossRef]
24. Li, D.; Shao, Z. Research on Physical City, Digital City and Smart City. *Geospat. Inf.* **2018**, *16*, 1–4.
25. Li, D.; Yuan, Y.; Shao, Z. The Concept, Supporting Technologies and Applications of Smart City. *J. Eng. Stud.* **2012**, *4*, 313–323. [CrossRef]
26. Etezadzadeh, C. The City of the Future. In *Smart City—Future City? Smart City 2.0 as a Livable City and Future Market*; Springer Fachmedien Wiesbaden: Wiesbaden, Germany, 2016; pp. 47–55.
27. Trencher, G. Towards the smart city 2.0: Empirical evidence of using smartness as a tool for tackling social challenges. *Technol. Forecast. Soc. Change* **2019**, *142*, 117–128. [CrossRef]
28. Gascó-Hernandez, M. Building a smart city: Lessons from Barcelona. *Commun. ACM* **2018**, *61*, 50–57. [CrossRef]
29. Capdevila, I.; Zarlenga, M.I. Smart city or smart citizens? The Barcelona case. *J. Strategy Manag. Res. Pract.* **2015**, *8*, 266–282. [CrossRef]
30. Angelidou, M. Smart city policies: A spatial approach. *Cities* **2014**, *41*, S3–S11. [CrossRef]
31. Bakıcı, T.; Almirall, E.; Wareham, J. A smart city initiative: The case of Barcelona. *J. Knowl. Econ.* **2013**, *4*, 135–148. [CrossRef]
32. Brous, P.; Janssen, M.; Herder, P. The dual effects of the Internet of Things (IoT): A systematic review of the benefits and risks of IoT adoption by organizations. *Int. J. Inf. Manag.* **2020**, *51*, 101952.
33. Li, C. Research on the Connotation, Characteristic and Development of Smart City: A Case Study of Smart City Construction in Beijing. *Modern Urban Res.* **2015**, *5*, 79–83.
34. Myeong, S.; Jung, Y.; Lee, E. A study on determinant factors in smart city development: An analytic hierarchy process analysis. *Sustainability* **2018**, *10*, 2606. [CrossRef]
35. Vukovic, N.; Rzhavtsev, A.; Shmyrev, V. Smart city: The case study of Saint-Peterburg 2019. *Int. Rev.* **2019**, 15–20.
36. Pira, M. A novel taxonomy of smart sustainable city indicators. *Humanit. Soc. Sci. Commun.* **2021**, *8*, 197. [CrossRef]
37. Souza, J.T.d.; Francisco, A.C.d.; Piekarski, C.M.; Prado, G.F.d. Data mining and machine learning to promote smart cities: A systematic review from 2000 to 2018. *Sustainability* **2019**, *11*, 1077. [CrossRef]
38. Xu, Q.; Wu, Z.; Chen, Z. The Vision, Architecture and Research Models of Smart City. *J. Ind. Eng. Eng. Manag.* **2012**, *26*, 1–7.
39. *30182:2017*; Smart City Concept Model—Guidance for Establishing a Model for Data Interoperability. ISO/IEC: Geneva, Switzerland, 2017.
40. *Y.4900/L.1600(06/2016)*; Overview of Key Performance Indicators in Smart Sustainable Cities. ITU-T: Geneva, Switzerland, 2016.
41. *GB/T33356-2022*; Evaluation Indicators for New-Type Smart Cities. State Administration for Market Regulation: Beijing, China, 2022.
42. Song, Y.; An, X. Analysis on the Concepts and Conceptual Systems of Smart City in International Standards Based on Terminology Definitions of ISO, ITU-T and IEC. *Stand. Sci.* **2018**, *1*, 127–132.
43. Zhao, D. The Research on the Construction of the Smart City in China. Ph.D. Thesis, Jilin University, Changchun, China, 2013.
44. Xiao, H.; Wang, Z. Systemic Thinking on Smart Cities. *China Soft Sci. Mag.* **2017**, *7*, 66–80.
45. Gang, S.; Qiang, T. Modern Urban Management: An Open Complex Giant System. *Urban Stud.* **2007**, *14*, 66–70.
46. Sheth, A.; Anantharam, P.; Henson, C. Physical-cyber-social computing: An early 21st century approach. *IEEE Intell. Syst.* **2013**, *28*, 78–82. [CrossRef]
47. Wang, F. The emergence of intelligent enterprises: From CPS to CPSS. *IEEE Intell. Syst.* **2010**, *25*, 85–88. [CrossRef]
48. Rodrigues, M.; Monteiro, V.; Fernandes, B.; Silva, F.; Analide, C.; Santos, R. A gamification framework for getting residents closer to public institutions. *J. Ambient. Intell. Humaniz. Comput.* **2020**, *11*, 4569–4581. [CrossRef]
49. Giffinger, R.; Gudrun, H. Smart cities ranking: An effective instrument for the positioning of the cities? *ACE Archit. City Environ.* **2010**, *4*, 7–26. [CrossRef]
50. Marsal-Llacuna, M.-L.; Colomer-Llinàs, J.; Meléndez-Frigola, J. Lessons in urban monitoring taken from sustainable and livable cities to better address the Smart Cities initiative. *Technol. Forecast. Soc. Change* **2015**, *90*, 611–622. [CrossRef]
51. Guo, L.Q. *The Research Report on Standards System of China Smart City*; China Architecture & Building Press: Beijing, China, 2013.
52. Chen, M.; Wang, Q.C.; Zhang, X.H.; Zhang, X.W. Study on the System of Evaluation for Wisdom City Construction-Nanjing as the Case. *Urban Stud.* **2011**, *18*, 84–89.
53. Zhang, Z.Q.Y.; Zou, K.; Xiang, S.; Mao, T.T. Model Construction and Empirical Study on Smart City Construction Ability Assessment. *Sci. Technol. Manag. Res.* **2017**, *37*, 73–76.
54. Giffinger, R.; Kramar, H. Benchmarking, profiling, and ranking of cities: The "European smart cities" approach. In *Performance Metrics for Sustainable Cities*; Routledge: London, UK, 2021; pp. 35–52.
55. Shah, M.N.; Nagargoje, S.; Shah, C. Assessment of Ahmedabad (India) and Shanghai (China) on smart city parameters applying the Boyd Cohen smart city wheel. In Proceedings of the 20th International Symposium on Advancement of Construction Management and Real Estate, Melbourne, VIC, Australia, 20–23 November 2017; Springer: Singapore, 2017; pp. 111–127.

56. Zhejiang Released "Digital Reform Public Data Classification and Grading Guide" Provincial Local Standards. Available online: http://www.gov.cn/xinwen/2021-07/19/content_5625917.htm (accessed on 10 November 2022).
57. Wang, W. The Evaluation Model and Evaluation Standard of New Type Smart City Based on Ternary Space Feedback Control Structure. *J. Int. J. Syst. Sci.* **2023**, *3*, 58–64.
58. IMD Smart City Index 2023. Available online: https://www.imd.org/smart-city-observatory/home/ (accessed on 14 February 2023).
59. Yang, J.; Kwon, Y.; Kim, D. Regional smart city development focus: The South Korean national strategic smart city program. *IEEE Access* **2020**, *9*, 7193–7210. [CrossRef]
60. Angelidou, M. Smart city planning and development shortcomings. *TeMA J. Land Use Mobil. Environ.* **2017**, *10*, 77–94.
61. Lee, J.H.; Hancock, M.G.; Hu, M.-C. Towards an effective framework for building smart cities: Lessons from Seoul and San Francisco. *Technol. Forecast. Soc. Change* **2014**, *89*, 80–99. [CrossRef]
62. Lan, H. The Three Prominent Issues Faced in the Process of Urbanization Advancement. Available online: http://news.ccidnet.com/art/1032/20130810/5117077_1.html (accessed on 18 November 2022).
63. Central Urban Work Conference held in Beijin, Xi Jinping and Li Keqiang Made Important Speeches. Available online: http://www.gov.cn/xinwen/2015-12/22/content_5026592.htm (accessed on 2 February 2023).
64. Liu, P.; Peng, Z. China's smart city pilots: A progress report. *Computer* **2013**, *47*, 72–81. [CrossRef]
65. Washburn, D.; Sindhu, U.; Balaouras, S.; Dines, R.A.; Hayes, N.; Nelson, L.E. Helping CIOs understand "smart city" initiatives. *Growth* **2009**, *17*, 1–17.
66. Wu, Z.; Wang, J.; Li, D.; Ding, L.; Dang, A.; Zhen, F.; Yang, T.; Liu, H.; Yang, J.; Long, Y. Symposium of Calm Thinking amid the Smart City Fervor. *Urban Plan. Forum* **2022**, *2*, 1–11. [CrossRef]
67. Summary Speech of the 2023 China Development Forum. Available online: https://www.drc.gov.cn/Default.aspx (accessed on 19 April 2023).
68. Research Report on the Industrial Map of New Smart Cities in 2021. Available online: http://www.caict.ac.cn/ (accessed on 17 April 2023).

smart cities

MDPI

Hypothesis

Exploring the Challenges and Issues in Adopting Cybersecurity in Saudi Smart Cities: Conceptualization of the Cybersecurity-Based UTAUT Model

Nawaf Alhalafi * and Prakash Veeraraghavan

Department of Computer Science and Information Technology, La Trobe University, Bundoora, VIC 3086, Australia; p.veera@latrobe.edu.au
* Correspondence: 17814379@students.latrobe.edu.au

Abstract: This study aims to explore the challenges and issues in adopting cybersecurity practices in smart Saudi cities and to develop and validate a newly developed cybersecurity-based unified theory of acceptance and use of technology 3 (UTAUT3) model. The study has a twofold purpose. First, it identified the key challenges and issues in adopting smart cities in Saudi smart cities. Second, it developed a technology-based model to adopt cybersecurity practices in Saudi smart cities. Two surveys were conducted to achieve these objectives. The first survey identified challenges and gaps in adopting cybersecurity practices in smart cities, revealing concerns about weak cybersecurity platforms, privacy breaches, and the impact of IT infrastructure advancements on Saudi culture (N = 554: common public). The second survey focused on developing and validating a cybersecurity-based UTAUT3 model (N = 108: IT professionals), emphasizing nine factors: performance expectancy, effort expectancy, social influence, facilitating conditions, safety, resiliency, availability, confidentiality, and integrity of cybersecurity. The model's validity and reliability were assessed, demonstrating its potential for understanding user behavior and adoption patterns in smart cities. The study findings provide valuable insights into the factors influencing the adoption of cybersecurity measures in smart Saudi cities, highlighting the need for targeted strategies, effective awareness programs, and collaboration between stakeholders to promote a secure and resilient digital environment. Future research may focus on refining the model, extending its applicability to other regions or countries, and investigating the impact of emerging technologies and evolving cyber threats on user behavior and cybersecurity practices.

Keywords: cybersecurity-based UTAUT3 model; conceptual development; validation; smart Saudi cities

1. Introduction

Smart cities are emerging as an innovative solution to the challenges urban areas face in the 21st century. They integrate digital technology, data, and connectivity to improve the quality of life for citizens and foster sustainable development [1]. In recent years, various countries have strongly committed to implementing smart city initiatives, with ambitious projects to create intelligent urban environments [2]. Developing smart cities is becoming a top priority as these countries strive to diversify their economies and reduce dependency on traditional industries. However, implementing these innovative urban environments raises critical concerns related to cybersecurity, given the increased connectivity and reliance on digital infrastructure [3]. As the potential for future possibilities attracts people from various parts of the country and the world, the rapid population growth in these smart cities has highlighted significant security concerns and increased the informal economy [4]. Ensuring the safety and security of citizens, infrastructure, and data in smart cities remains a pressing issue that needs to be addressed to fully realize the potential benefits of these innovative urban environments.

Similarly, the smart cities in Saudi Arabia faced the same challenges. They reflected a 689% rise in residents compared to a 146% rise globally over the previous 56 years [5]. Identification of criminal activities, technological innovation, an excellent educational system, modern hospitals, and other social services are all required because of Saudi Arabia's growing population [6]. Smart facilities, networks, smart waste management framework, and smart management systems are all required because of Saudi Arabia's growing population. Technology today, such as cybersecurity systems, supports real-time data collection, recognizes, and analyzes possible issues to efficiently accomplish assets and resources [6]. In addition, Mohammad and Abdulqader [7] conducted a study of finding cybersecurity requirements in Saudi smart cities. They concluded that significant concerns include privacy, data security, and preventative measures. The proposed solutions in the research may be sufficient on their own, but it needs to be clarified what would occur if all of these solutions were combined into a single comprehensive system.

Smart cities worldwide, including those in development, still need to universally establish comprehensive action plans that detail their strategies to respond to potential cyberattacks targeting their facilities, infrastructure, and information and communication technology systems. Weaknesses in any one component can have far-reaching implications due to the inherent interconnectedness of all processes. As a result, recommendations should be made available to the organizations responsible for selecting and vetting the technologies that will be utilized in smart cities globally. As smart cities continue to grow and integrate advanced digital technologies, they become increasingly susceptible to cyber threats [8]. Cybersecurity is a crucial aspect of smart city development, as it aims to protect information, communication systems, and critical infrastructure from unauthorized access and attacks [9]. Despite the growing importance of cybersecurity, more research is needed on the cultural, social, and economic challenges and issues in adopting cybersecurity practices in smart cities worldwide [10]. Addressing these challenges and prioritizing cybersecurity will be essential to ensuring the safety and resilience of smart cities as they become more prevalent.

Therefore, this study seeks to address this knowledge gap by examining the challenges and issues that hinder the effective implementation of cybersecurity practices in the context of smart Saudi cities. Therefore, as smart cities continue to gain traction in Saudi Arabia, understanding the challenges and issues related to adopting cybersecurity practices becomes vital to ensuring the successful development and operation of these urban environments [10]. Additionally, analyzing the cultural, social, and economic factors that influence the implementation of cybersecurity measures can provide valuable insights into the unique context of Saudi Arabia, which can be used to develop targeted strategies and policies for improving cybersecurity in its smart cities [11]. Finally, the study offers the following research objectives:

1. To identify the challenges and issues in adopting cybersecurity and safety practices by IT professionals and the common public in smart Saudi cities.
2. To explore the economic, social, and cultural factors/challenges that make the cybersecurity framework practicable in smart Saudi cities.
3. To conceptualize a cybersecurity-based UTAUT3 model in smart Saudi cities.

2. Literature Review and Conceptualization

2.1. Challenges and Issues in the Implementation of ICT Technologies

2.1.1. Technological Challenges and Issues

Poor information and communication technology (ICT) infrastructure and security concerns hamper the adoption of e-government in Saudi Arabia. Hosam and Ahmad [12] assert that the lack of ICT infrastructure is a predominant issue. Additionally, there is a need for Saudi Arabia to develop standardized policies and regulations to oversee the use of ICT services. These changes will enable both the government and citizens in delivering services [12]. Privacy and security concerns are another major technical challenge affecting the implementation of cybersecurity in Saudi Arabia. A study by Alshehri and Drew [13] notes that most Saudi Arabians are concerned that using e-government platform to share personal information, such as names and ID numbers, could expose them to cybersecurity risks. The citizens are afraid that online platforms do not have adequate security to prevent hackers from accessing their data. Regrettably, the concern over the lack of security in e-government systems has created an unwillingness among Saudi Arabians to embrace e-services [13]. Hosam and Ahmad [12] claim that international privacy and security concerns are the leading issues among citizens when using e-government platforms. Thus, the government needs to enhance public awareness of how citizens should safeguard their data and privacy by building cybersecurity practices when using the Internet [12].

2.1.2. Organizational Challenges and Issues

Lack of qualified ICT experts, lack of training, resistance to change, inadequate ICT police, and lack of collaboration between Saudi Arabian agencies are also responsible for the slow adoption of cybersecurity. Alshehri and Drew [13] claim that in Saudi Arabia, most IT experts have moved from the public sector to the private sector since government jobs pay fewer wages compared to corporate jobs. The transition has deprived the public sector of experts who could have helped implement cybersecurity programs [13]. According to Hossam et al. [14], the new phenomenon experiences resistance because people fear the unknown. Since setting cybersecurity in smart cities is an emerging technological revolution in Saudi Arabia, it requires establishing an appropriate framework [12]. For any new government program to succeed, it should receive support from all agencies. Alshehri and Drew [13] claim that Saudi Arabia has experienced slow adoption of cybersecurity measures because of a lack of collaboration between government agencies in promoting their implementation. None of the Saudi Arabian agencies wishes to share their data with other organizations. The country could realize numerous benefits if all the agencies could share their information on cybersecurity practices [13]. These challenges and issues also include:

Lack of Trust

Certain distinctions exist between cybersecurity in Saudi Arabia and the United States. Alotaibi et al. [15] noted that perceived trustworthiness has a positive and major impact on behavioral intention to use m-government services [16]. This observation suggests that most participants in this research trust such applications and their merits since the government has deployed them. This result is consistent with the principle that when trust in the Internet and government is enhanced, the intention to use cybersecurity practices increases. In addition, security may not be a problem for most individuals who adopt security applications since they have been utilizing the Internet for a long time and are well familiar with security and privacy concerns. Privacy issues do not affect the purchase intentions of cloud computing services in the United States [15]. Moreover, they should deliver services through secure applications to motivate users to embrace these technologies [15].

In view of cybercrimes, related regulations in Saudi Arabia are influenced more by Islamic and social principles, as well as local cultural contexts [16]. These laws include the penalties for the different crimes and the fines that the religion considers consistent with the values it practices. During investigations, Saudi Arabia's Communication and IT Commission provides the required technical support to the established security body to assist with the examination of diverse cybercrimes. However, in the United States, the Federal Bureau of Investigation is responsible for investigating such offenses, in line with the Constitution.

Lack of Awareness and Training Developments

In the Middle East region, security awareness of information, particularly among undergraduates, scholarly researchers, and staff, has been examined to determine their degree of knowledge of information systems [17]. In a study, the researchers note that a lack of scholarship on information system principles is the main impediment to cybersecurity awareness. Several suggestions to minimize the severity of this situation were formulated, namely reinforcing awareness and training initiatives as well as embracing safety measures across learning institutions to promote data security.

Various factors necessitate an examination of cybersecurity in cybersecurity practices. In the view of Saudi Arabia, the e-government paradigm of administrative systems is a relatively new idea, originating in the 1980s. In government contexts, the execution of technology in providing diversified offerings is a costly plan that demands economic practicality. An effective model within which the electorate can access different services is necessary and must be protected to promote social welfare [15]. Given e-government implementation in the country, significant advantages—such as the adoption of digital technologies and a minimization in the cost of public service—have been evident. Saudi Arabia is a nation that maintains a positive outlook on the international stage with the goal of securing varied economic interests [17].

An assessment of Saudi Arabians indicates that their cybersecurity awareness level is considerably low. This situation can be attributed to the nature of the national culture. Additionally, while Saudi citizens exhibit a clear understanding of information technology (IT), their knowledge of cybersecurity risks and the government's role in facilitating information safety across the Internet is quite limited. Therefore, state agencies should create awareness of these issues through training initiatives in learning institutions.

Impacts of Culture

American social customs are broad, coupled with distinct values based on diverse populations and ethnic backgrounds [17]. This convergence means that one religion cannot form the basis for governing the penalties for people found guilty of cybercrime offenses; therefore, a challenge in managing the practice appears [18]. In addition, variations exist in the level of jail time for similar crimes in Saudi Arabia and the United States. Besides, phishing is the most dominant social engineering method in recent times. It entails stealing users' credit card numbers and login details to access their private data. This form of cyberattack accounted for 77% of all social engineering attacks in Saudi Arabia's educational sector in 2017. These attacks can be implemented through the Internet and social media.

Social harmony and human connections are major priorities among Asians. Efficiency and time management, on the other hand, are critical values for Westerners. Such cultural factors have a clear implication for cybersecurity.

Shortage of Local Expertise

In Saudi Arabia, a lack of IT professionals to guide the execution of e-government in their entities is clear [18]. A primary factor for such a lack is the transfer of IT expertise from the public sector to the private sector since government salaries are relatively low [17]. Further, a lack of IT personnel at all levels, including programmers and professional managers, is evident. Consequently, the training of existing employees, especially in the public sector, is vital to advance the adoption of any novel technology. On the other hand, a significant proportion of Saudi Arabia's population lacks the requisite IT competency needed to advance smart cities [17]. Therefore, this situation demands dependence on expatriates to assist in the management of distinct elements of the smart city initiative.

Finally, the study proposes the following assumptions to explore the challenges and issues of adopting cybersecurity practices in smart Saudi cities:

P1. The lack of trust related to cyberinfrastructure issues, services from governments or companies, cyber threat attacks, and a cyber-based economy limit cybersecurity framework adoption in Saudi Arabia.

P2. The lack of developments related to a shortage of cyber awareness programs and a lack of trained personnel limit cybersecurity framework adoption in Saudi Arabia.

P3. Cultural influences, including using social apps, influence Saudi Arabia's adoption of cybersecurity awareness methods.

P4. The lack of IT professionals in the public sector (a shortage of local expertise) limits the implementation of cybersecurity frameworks in Saudi Arabia.

After a careful literature review, the study identified the key challenges for citizens, governments, and organizations in services, mobility, and standards and protocols. Moreover, the study also identified the smart city factors (i.e., privacy, security, and risk) that need proper laws and regulations, wellbeing and quality of life, and governance. Therefore, the study divided all stakeholders (citizens, government, and organizations) into three main factors, including trust, operational and transitional, technological and sustainability challenges. Based on this information, the study also develops a conceptual framework (Figure 1) based on the following proposed propositions:

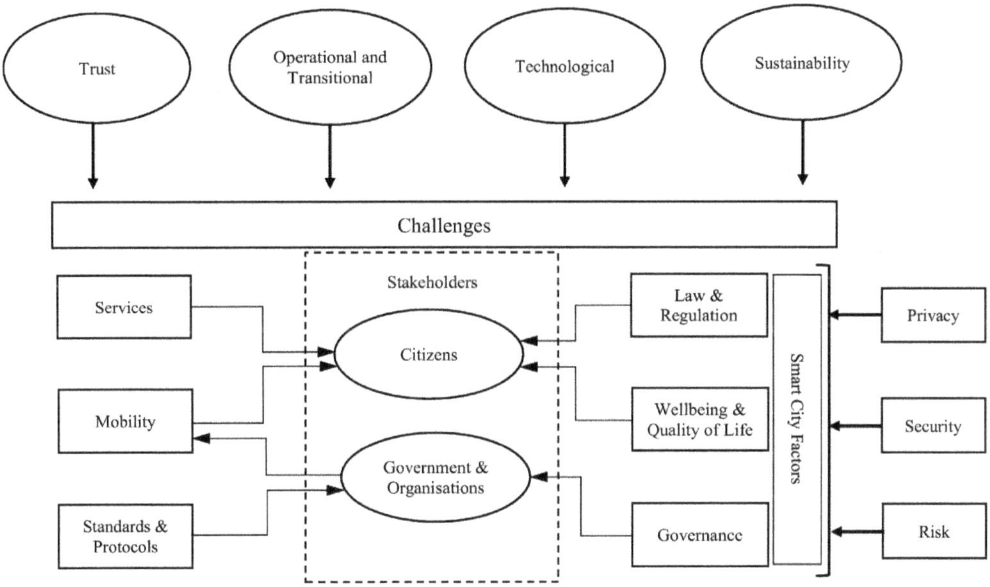

Figure 1. Conceptual framework.

3. Research Methodology

3.1. Research Design

This study adopts a quantitative research method to validate and test the survey items of cybersecurity practices in smart Saudi cities. Quantitative research is focused on collecting and analyzing statistical data, which is suitable for investigating factors in producing repeatable outcomes and generalizing findings [19,20]. Quantitative research is a predominant method due to its objectivity, generalizability, and replicability, as it relies on systematic procedures and numerical data for analysis [21]. Statistical techniques are employed to draw objective conclusions, make inferences about the population, and enhance the rigor and precision of the research. While quantitative research has limitations, such as the potential for overlooking qualitative nuances, its structured and statistical nature makes it accessible and efficient for data collection and analysis. According to the study of cybersecurity in smart cities, the quantitative method has been recommended in previous research [21] to test the assumptions. For this study, a survey-based cross-sectional approach was used to analyze relationships between various factors and the implications of the results for the research approach [19]. Therefore, the study used a survey questionnaire to develop a list of questions about exploring challenges and issues in adopting cybersecurity practices in smart Saudi cities.

3.2. Data Collection Procedure

The study adopted a survey questionnaire technique to collect data. The study collected data from the common public, targeting 710 citizens. The common public was the subject of a survey questionnaire that was distributed in order to collect data. The respondents using an online platform made available to Saudi citizens living in Saudi Arabia used a combined web-based online survey (Google Forms). In order to determine the cybersecurity challenges and issues in smart Saudi cities and the potential benefits they may provide, a literature review was conducted, and a survey questionnaire was created for this research. The survey questionnaire contained only questions that were straightforward, to the point, free of ambiguity, and simple to read. A list of research questions has been developed from an extensive literature review. When the questions were extracted, they were reviewed and proofread by two professors from universities and two IT experts in the same field (cybersecurity). The questionnaire comprised 36 questions identified from reviewing the literature (see Appendix A). A five-point Likert scale was used to measure the items to reduce the possibility of measurement errors and the mental strain placed on the participants [22].

The participants in this research were the public in information technology and cybersecurity (including web designers, software engineers, and programmers). Since Arabic is Saudi Arabia's primary language, but the questionnaire survey was initially written in English, it was necessary to have it translated into Arabic before it could be sent to the common public. Therefore, the study used both versions of the survey questionnaire to allow the common public to respond clearly. According to Sekaran [23], it is vitally important to select a language for the questionnaire that is clear, straightforward, and written at a level that respondents can comprehend. In this particular investigation, the researcher utilized the strategy of back translation. Back translation has recently become increasingly popular for academic translated versions and expert studies.

The designed questionnaire is improved through a process to reduce the number of errors caused by the creation of the questionnaire and ensure that the content is accurate [24]. The study surveyed the public (N = 710) in Saudi Arabia, but only 554 surveys were fully answered, so the study used only completed survey questionnaires. The study surveyed the public to determine the challenges and issues in adopting cybersecurity practices in smart Saudi cities. This strategy elaborates on the responses to the challenges, barriers, and needs of adopting cybersecurity in smart cities in Saudi Arabia. Finally, the study surveyed the public by taking 554 survey responses from two smart cities (for example, Neom and Riyadh) in Saudi Arabia. The researcher used the list of questions to test the assumptions/propositions of the research (see Appendix A).

3.3. Data Analysis

The study used a statistical package for social sciences (SPSS) 21 to analyze the pretesting survey data [25]. First of all, the study calculated descriptive statistics for the demographic characteristics of the respondents. The study also calculated the correlation coefficient [26] among the variables of the pre-testing survey questionnaire. As well, the study tests the survey questions in order to find the challenges and issues in adopting cybersecurity practices in smart Saudi cities.

4. Results

4.1. Descriptive Statistics (Public Survey)

In Table 1, the descriptive statistics showed that there are 62.8% (348) males and 37.2% (206) females. Moreover, 18.6% (103) of the respondents were from the public sector, 53.6% (297) of the respondents were from the private sector, 2.3% (13) of the respondents were unemployed, 9.2% (51) of the respondents were housewives, construction employees, teachers, medical doctors, or military men, 9.6% (53) of the respondents were students, and 6.7% (37) of the respondents were from free businesses. Additionally, 33% (183) of the respondents were using fast internet, 48.2% (267) of the respondents were using medium-speed internet, 15% (83) of the respondents were using low internet, and only 3.8% (21) of the respondents were using very fast speed internet.

Table 1. Frequency and percentage of the participants of the study (N = 554).

Gender	Frequency	Percent	Valid Percent	Cumulative Percent
Male	348	62.8	62.8	62.8
Female	206	37.2	37.2	100.0
Total	554	100.0	100.0	
	Occupation			
Public sector	103	18.6	18.6	18.6
Private sector	297	53.6	53.6	72.2
Unemployed	13	2.3	2.3	74.5
Others (housewife, construction, teacher, medical, military)	51	9.2	9.2	83.8
student	53	9.6	9.6	93.3
Free business	37	6.7	6.7	100.0
Total	554	100.0	100.0	
	How do you rate your current internet speed			
Fast	183	33.0	33.0	33.0
Medium	267	48.2	48.2	81.2
Slow	83	15.0	15.0	96.2
Very fast	21	3.8	3.8	100.0
Total	554	100.0	100.0	

4.2. Correlation Analysis

The strength and degree of the linear links between two sets of variables are evaluated using correlation coefficients. Therefore, the study used correlation analysis to check the relationship between the variables. Table 2 describes the correlation analysis of the factors related to the Internet and cybersecurity. The results showed that internet services are positively and strongly associated with the latest technologies (r = 0.406 **) and increased cybersecurity issues (r = 0.122 **). In addition, it was also observed that cybersecurity issues are significantly and positively related to the development of the latest technology (r = 0.092 *) and electronic services information (r = 0.099 *). At the same time, no significant relationship was observed between electronic services and cybersecurity scams (r = −0.071), and the relationship was negative. Finally, it was proven that there was the highest correlation coefficient between internet services and the latest internet technology. It happened due to the notion that if there is the latest internet technology, the use of internet services will increase automatically.

Table 2. Correlation coefficients (N = 554).

Correlations	*Internet Services*	*Latest Internet Technology*	*Cybersecurity Scam*	*Electronic Services Information*
Internet services	-	0.406 **	0.122 **	−0.028
Latest internet technology	-	-	0.092 *	0.099 *
Cybersecurity scam	-	-	-	−0.071
Electronic services information	-	-	-	-
	**. Correlation is significant at the 0.01 level (2-tailed).			
	*. Correlation is significant at the 0.05 level (2-tailed).			

Note: ** $p < 0.01$, * $p < 0.05$.

4.3. Testing Assumptions

4.3.1. Lack of Trust

The survey results indicate that there needs to be more trust in cyber infrastructure and the digital economy in Saudi Arabia (Table 3). A majority of respondents believe that the country has weak cybersecurity platforms, and they need more confidence in the nation's capacity to design and deploy smart cities. Despite this, many participants acknowledge the country's shift towards a digital economy. The respondents display awareness of various IT issues, such as social media scams, bank fraud, and fake advertisements. Concerns about external threats, fake accounts, information leakage, hacking, and improper cybersecurity systems are prevalent. Most participants have experienced medium or slow internet speeds, and many have faced security or privacy breaches. Furthermore, the majority of respondents admit to sharing their personal information online without reading the website's privacy policy. These findings suggest that cybersecurity and digital infrastructure improvements are crucial to increasing trust and confidence in Saudi Arabia's technological advancements.

Table 3. Lack of trust.

Lack of Trust (Cyber Infrastructure)		**Strongly Agree**	**Agree**	**Neutral**	**Disagree**	**Strongly Disagree**
Does Saudi Arabia have good cybersecurity platform?		1 (1%)	2 (1.9%)	20 (19%)	45 (42.9%)	35 (33.3%)
How strongly believe we have knowledge and capacity to design and deploy smart cities?		0	5 (4.8%)	31 (29.5%)	32 (30.5%)	35 (33.3%)
Lack of trust (digital economy)						
Rate how much Saudi Arabia is moving towards digital economy?		71 (67.6%)	25(23.8%)	4 (3.8%)	3 (2.9%)	
Awareness about various IT issues						
Email scam						2 (1.94%)
Bank fraud						5 (4.85%)
Social media scam						30(29.12%)
Viruses (that can lose the personal information from the workplace)						15 (14.56%)
Mobile data						14 (13.59%)
Fake advertisements (job advertisement, links on social media, advertisement messages)						20 (19.41%)
Trust and continuity competition						
External threats						8 (7.76%)
Fake accounts or hackers						9 (8.73%)
						Frequency
Information leakage						23 (22.33%)
Hacking bank accounts						22(21.35%)
Unprofessional work environment						27 (26.21%)
Improper cybersecurity system						31 (30.09%)
Use of smart devices based on internet speed			Are you aware of your internet-connected smart devices at home?			
				No		Yes
How do you rate your current internet speed?	Fast			5 (13.88%)		178 (34.36%)
	Medium			22 (61.11%)		245 (47.29%)
	Slow			9 (25%)		74 (14.28%)
	Very fast			0(0%)		21 (4.05%)
	I don't know			No		Yes
Have you had any problems with breaching security/privacy?	I haven't had problems breaching security/privacy, and I'm not worried about these issues.		5 (18.51%)	74 (16.26%)		5 (6.94%)
	I haven't had problems breaching security/privacy, but I worry about these issues.		18 (66.6%)	344 (75.60%)		40 (55.5%)
	I've recently had problems breaching security/breaching privacy, and I'm concerned about these issues.		4 (14.81%)	37 (8.13%)		27 (37.5%)
			Do you realize that you share your personal information with e-government services?			
			No	Yes		
Have you shared your phone number or personal information online without reading the website's privacy policy?	No		44 (17.39%)	209 (82.60%)		253
	Yes		48 (15.94%)	253 (84.05%)		301
Total			92	462		554

4.3.2. Lack of Awareness and Training Developments

The majority of Saudi citizens surveyed believe it is essential to be aware of cyber threats, with 74% strongly agreeing and 20.2% agreeing. A considerable percentage (40%) of respondents have already participated in a cybersecurity community awareness program, demonstrating an overall interest in the topic. Furthermore, an overwhelming majority

(97.09%) expressed interest in promoting a cyber-awareness program within the community, should the government provide incentives. Regarding participating in training, a higher proportion of respondents (62.68%) are willing to pay a small fee for the course, as opposed to 37.3% who would prefer not to pay. In conclusion, the findings indicate that Saudi citizens are highly aware of the importance of cyber threats and demonstrate a significant willingness to participate in and promote cybersecurity awareness programs, with a majority also being open to paying for relevant training courses.

Most participants recognize the importance of cyber threat awareness, indicating a successful initial stage of cybersecurity education and potential readiness for more in-depth knowledge. Despite the recognized importance of awareness, participation in cybersecurity community awareness programs could be much higher. This highlights a gap between understanding the importance of cyber awareness and taking active steps to increase it. Most participants would be interested in promoting cyber-awareness programs if the government provided incentives. This reveals a significant opportunity for government agencies to boost cybersecurity education through incentivized initiatives.

Participants' willingness to pay for a course indicates a market for paid cybersecurity education. However, only some interested people are willing to pay, suggesting that affordable or free training options might be necessary to reach a wider audience. While most interested participants are eager to invest financially in their cybersecurity education, many uninterested participants would not pay for such a course. This suggests that financial commitment may deter some individuals, necessitating additional strategies to engage this group, such as demonstrating the practical benefits of such training (Table 4).

Table 4. Lack of awareness and training developments.

Awareness of Cyber Threats	Strongly Disagree	Disagree	Neutral	Agree	Strongly Agree
How strongly you think it is good to have every Saudi citizen to be aware of Cyber threats?	0	1 (1%)	4 (3.8%)	21 (20.2%)	77 (74%)
Willingness to participate in cybersecurity awareness program		Yes		No	
Have you participated in cybersecurity community awareness?		42 (40%)		61 (58%)	
Variable		Yes		No	
Will you be interested in promoting cyber-awareness program to community, if the government provides some incentive for you?		100 (97.09%)		3 (2.91%)	
Interest and willingness to participate in trainings		No		Yes	Total
Are you ready to pay a small fee for the course?	No	159 (49.22%)		75 (37.3%)	234
	Yes	194 (54.95%)		126 (62.68%)	320
Total		353		201	554

4.3.3. Impact of Culture

The survey results indicate that a significant proportion of respondents believe that advancements in IT infrastructure will impact Saudi culture, with 14.3% strongly agreeing and 32.4% agreeing on this issue (Table 5). However, a notable percentage (40%) remain neutral, while 6.7% disagree and 4.8% strongly disagree. In terms of social apps affecting Saudi culture, Snapchat (51.4%) and Twitter (32.4%) are identified as the two most influential apps, with WhatsApp (14.3%) coming in third place and no respondents selecting Instagram or other apps. In conclusion, most respondents believe that advancements in IT infrastructure and the use of social apps, notably Snapchat and Twitter, have the potential to affect Saudi culture. However, a considerable portion remains neutral on the topic.

Table 5. Impacts of culture.

Impacts of Culture	Strongly Agree	Agree	Neutral	Disagree	Strongly Disagree
Do you believe advancement in IT infrastructure will affect the Saudi culture	15 (14.3%)	34 (32.4%)	42 (40%)	7 (6.7%)	5 (4.8%)
Social apps affecting Saudi culture	Twitter	Snapchat	WhatsApp	Instagram	Others
34. Provide two important technology apps (social apps) that affect Saudi culture.	34 (32.4%)	54 (51.4%)	15 (14.3%)	0	0

4.3.4. Shortage of Local Expertise

The survey results reveal that only a small percentage (11.4%) of respondents have the technical ability to perform network-wide deep-packet inspections (Table 6). In comparison, a more significant portion (41%) do not have this capability, and 45% are uncertain. There is a generally positive outlook regarding job satisfaction and the potential of public and governmental agencies to match private sectors shortly, with 36.2% strongly agreeing and 36.2% agreeing with the statement. However, 19% of respondents remain neutral, while 3.8% disagree, and 2.9% strongly disagree. In conclusion, although the technical ability to perform deep-packet inspections is limited among the respondents, there is optimism that public and governmental agencies can match private sectors shortly.

Table 6. Shortage of local expertise.

Technical Ability of IT Specialists			Yes	No	I Don't Know
Do you have the technical ability to perform network-wide deep-packet inspections?			12 (11.4%)	43 (41%)	48 (45%)
Job satisfaction	Strongly agree	Agree	Neutral	Disagree	Strongly disagree
How strongly you agree with the following statement: Public sector and governmental agencies can match private sectors in the new future:	(36.2%)	(36.2%)	(19%)	(3.8%)	(2.9%)

5. Conceptualization of the Cybersecurity-Based UTAUT Model

After identifying the challenges and gaps in the survey, it is shown that other economic, cultural, and social factors are related to behavioral intention and actual use of behavior. The literature review showed that economic, cultural, and social factors significantly contribute to behavioral intention and the actual behavior of using smart city technologies. The study identified several factors from the literature studies and tested the results. The factors for each antecedent are termed economic factors, i.e., privacy design (safety), cyber threat intelligence and analysis platform (resiliency); social factors, i.e., digital trust (confidentiality), cyber responses and resilience (availability); and cultural factors, i.e., cyber competencies and awareness program (integrity).

The study used the unified theory of acceptance and use of technology (UTAUT) as the basis to develop a cybersecurity-based UTAUT3 model in smart Saudi cities. Popova and Zagulova [27] in the smart city of Riga, Latvia, use UTAUT models in predicting cybersecurity behaviors and intentions, such as in a study. The study findings showed that all factors of the UTAUT model (i.e., performance expectancy, effort expectancy, social influence, and facilitating conditions) significantly contributed to the behavioral intention and actual use of behavior in adopting smart city technologies. However, they claimed that UTAUT factors might not predict adopting smart city technologies due to the underdeveloped culture. Kuberkar and Singhal [28] conducted a study by testing the behaviors of passengers toward smart city transport technologies. The study stated that performance expectancy, effort expectancy, social influence, facilitating conditions, trust, and anthropomorphism affect behavioral intention and actual use of behavior in adopting smart city transport technologies. Meanwhile, Van Zoonen [29] recognized security and privacy as contributing factors in the smart city awareness campaign model, in which privacy and security variables are connected to applying the appropriate realm of the smart city.

In this context, some of the issues include developing resilience in information security; increasing awareness of privacy risks; growing the security and integrity of online content, thereby encouraging greater use of information systems [30]; generating policy standards for cybersecurity based on global best practice; constructing resilient information systems; and increasing concern regarding security risks [17]. Alzahrani [31] conducted a study claiming that cybersecurity's availability, confidentiality, and integrity are linked to cybersecurity trust and culture. It is helpful to increase the safety of critical systems and cyber resilience capability, conduct research and training, and market cybersecurity remedies, products, or services when government organizations and commercial sectors engage. Therefore, Alzahrani [31] stated that data availability, confidentiality, and integrity should be enhanced to cover the lack of system development and IT experts. On the other hand, Alhalafi and Veeraraghavan [32] proposed recommendations for Saudi Arabia to implement a cybersecurity framework for smart cities. They claimed that smart cities require a model of cybersecurity that will safeguard their integrity and guide the reaction to possible assaults and hazards [32]. Based on the study's findings and evidence in the literature, the study develops a cybersecurity-based UTAUT3 model in smart Saudi cities (see Figure 2):

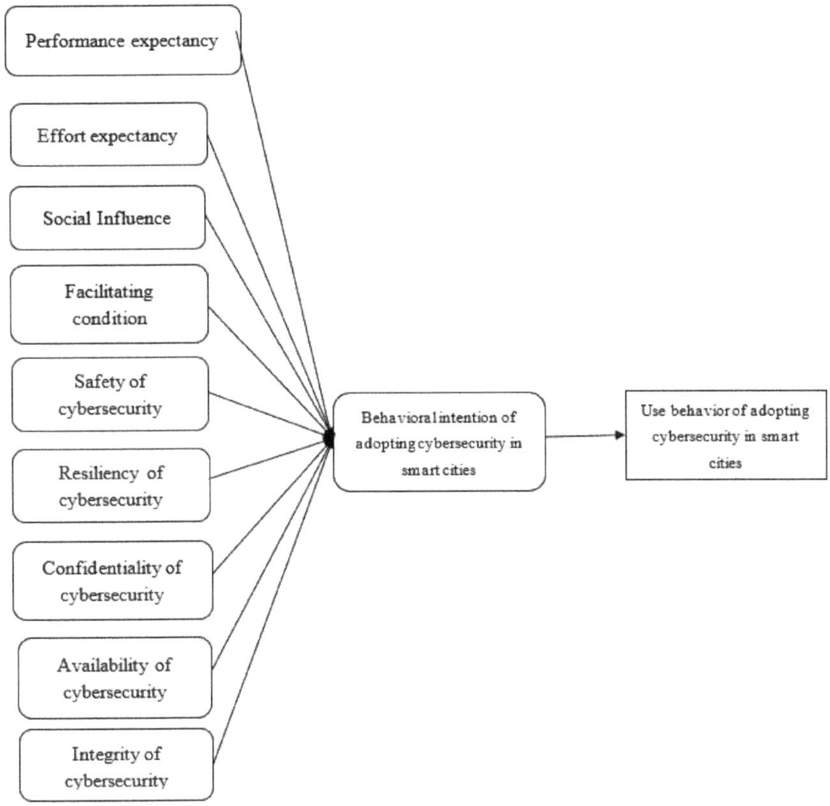

Figure 2. Conceptual framework 2.

6. Research Methodology

6.1. Data Collection Procedure

The questionnaire was developed based on the findings of the previous survey conducted in smart Saudi cities. The configuration of the research questionnaire uses an online platform, which details the study's objectives and provides the researcher and the supervisors' team with their respective contact information. On the cover page, which was explicitly designed to inform the respondents of the objectives and significance of this research, the researcher provided an explanation of the objectives of the survey and directions for completing the questionnaire at the start of the study. The questionnaire was designed to validate the conceptual framework in smart Saudi cities using nine factors. An emphasis was placed on the nine factors of the UTAUT3 model (performance expectancy, effort expectancy, social influence, facilitating conditions, safety, resiliency, availability, confidentiality, and integrity of cybersecurity) that had been taken. The questionnaire is divided into two distinct sections. Respondents' demographic data were gathered in the first part of the survey. The second section of the questionnaire consisted of UTAUT3 factors and items to validate the cybersecurity-based UTAUT3 model. The survey questionnaire validity and reliability testing involved the researcher sending the questionnaire to five scientists who are currently pursuing their doctoral degrees and have substantial expertise in cybersecurity and e-applications, in addition to having a solid understanding of Arabic, which is also their native language. The questionnaire was written in English and Arabic. The feedback from the doctoral students recommended minor alterations to the wording; furthermore, the feedback suggested splitting up a few questions and evolving the order in which they were asked. Their feedback was incorporated into the revision of the draft questionnaire, and then the completed survey instrument was presented to them and received their approval. Finally, a second survey consisting of 108 IT professionals was carried out for the purpose of validating the conceptual model.

6.2. Measurement Scales

This study's assessment constructs and items were derived from previous research (see Appendix B). Four performance expectancy items, three effort expectancy items, three social influence items, and three facilitating conditions items were gathered from earlier studies and modified for this investigation [33,34]. Based on Arpaci and Sevinc's [35] research, the study adapted the following cybersecurity factors: four safety items, three resilience items, three secrecy items, three availability items, and four integrity items. Additionally, the research adjusted three items of behavioral intention and three items of actual use behavior from past studies to facilitate data comparison [33,34]. A Likert scale was employed for the measurement scales, with levels ranging from 1 (strongly disagree) to 5 (strongly agree). McDaniel and Gates [36] argue that a Likert scale is practical when the research measures participants' attitudes towards various factors.

6.3. Data Analysis

This stage of the survey questionnaire was conducted to develop and validate the factors of the UTAUT3 cybersecurity framework in smart Saudi cities. The study used the Smart PLS 3.3.3 version to develop and validate the new UTAUT3 cybersecurity framework in smart Saudi cities. The study used the first step of the algorithm technique to assess the measurement model. The research-based guidelines of previous researchers in the literature have been followed in this research [37,38] to assess and test the research-developed conceptual framework.

7. Validation and Development of Cybersecurity-Based UTAUT3

7.1. Demographic Information

The demographic information of the pre-test shows that 50% (54) of the respondents were from the IT sector, 19.4% (21) of the respondents from the education sector, only 7.4% (8) from the construction sector, and 23.1% (25) of the respondents from the food sector. The study also reported that 71.3% (77) of the respondents were male and 28.7% (28.7) were female. Meanwhile, the study reported that 15.7% (17) of the respondents have 1–2 years of IT experience, 13% (14) of the respondents have 2–5 years of IT experience, 32.4% (35) of the respondents have 5–8 years of experience, 15.7% (17) of the respondents have 8–10 years of experience, and 23.1% (25) of the respondents have 10+ years of experience. On the other hand, the study finds that only 8.3% (9) of the respondents have matriculation education, 54.6% (59) of the respondents have intermediate education, 34.3% (37) of the respondents have a bachelor's degree, 1.9 (2) of the respondents have a master's degree, and 0.9% (1) of the respondents have a Ph.D. degree.

7.2. Convergent Validity and Reliability

The convergent validity and reliability of the measurement scales can be assessed using the threshold values established in the literature [39–41]. Convergent validity is demonstrated when all factor loadings are greater than 0.7 [41] and the average variance extracted (AVE) is greater than 0.5 [42]. Reliability is established when both Cronbach's alpha and composite reliability (CR) values are greater than 0.7 [39,40]. The study ran an algorithm with 5000 sub-samples and found that two items of facilitating conditions (FC1 = 0.446, FC2 = 0.508), one item of safety (SAFE4 = 0.626), one item of confidentiality (CON1 = 0.638), and one item of availability (AVAI2 = 0.655) have lower factor loading than 0.7, so they have been deleted from the model. In this study, all measurement scales show satisfactory convergent validity and reliability, as all factor loadings are above the 0.7 threshold, and AVE values for each scale exceed 0.5 (Table 7). Moreover, Cronbach's alpha and CR values are greater than 0.7 for all scales, indicating high internal consistency and reliability (Table 7).

Table 7. Convergent validity and reliability.

Measurement Scales	Items	Factor Loadings	AVE	Cronbach Alpha	Composite Reliability
Actual use of behavior	AUB1	0.837	0.704	0.790	0.877
	AUB2	0.822			
	AUB3	0.858			
Availability of cybersecurity	AVAI1	0.746	0.641	0.719	0.842
	AVAI3	0.794			
	AVAI4	0.858			
Behavioral intention	BI1	0.780	0.651	0.735	0.848
	BI2	0.840			
	BI3	0.799			
Confidentiality of cybersecurity	CON2	0.833	0.748	0.831	0.899
	CON3	0.887			
	CON4	0.874			
Effort expectancy	EE1	0.852	0.670	0.753	0.859
	EE2	0.803			
	EE3	0.799			
Facilitating conditions	FC3	0.753	0.644	0.725	0.844
	FC4	0.797			
	FC5	0.854			
Integrity of cybersecurity	INT1	0.798	0.643	0.815	0.878
	INT2	0.852			
	INT3	0.752			
	INT4	0.803			
Performance expectancy	PE1	0.824	0.628	0.803	0.871
	PE2	0.769			
	PE3	0.785			
	PE4	0.792			
Resiliency of cybersecurity	RESI1	0.811	0.662	0.746	0.855
	RESI2	0.810			
	RESI3	0.821			
Safety of cybersecurity	SAFE1	0.888	0.679	0.841	0.894
	SAFE2	0.810			
	SAFE3	0.837			
	SAFE5	0.756			
Social influence	SI1	0.789	0.717	0.803	0.884
	SI2	0.862			
	SI3	0.887			

7.3. Discriminant Validity

Cross-loadings are the correlation coefficients between the items and the factors, and they are used to evaluate the discriminant validity of the measurement scales [39]. Discriminant validity is established when the items load higher on their respective factors than on others [43]. A common rule of thumb is that the difference between the item's loading on its own factor and its highest loading on other factors should be at least 0.1 [43]. The findings showed that most items have a higher loading on their respective factors (diagonal elements in bold) than on other factors (off-diagonal elements). This suggests a reasonable level of discriminant validity for most of the items in this study (see Appendix C).

Another part of discriminant validity is testing the Fornell–Larcker [44] criterion. The Fornell–Larcker criterion is an approach used to assess discriminant validity, which is the degree to which a construct is distinct from other constructs [44]. This criterion is fulfilled when the square root of a construct's average variance extracted (AVE) is greater than the correlations between that construct and all other constructs in the model. In Table 8, the diagonal values (in bold) represent the square root of AVEs for each construct, while the off-diagonal values represent the correlations between the constructs. Based on the Fornell–Larcker criterion, it can be concluded that discriminant validity is established for most of the constructs as the square root of the AVE for each construct is higher than its correlations with other constructs.

Table 8. Fornell–Larcker criterion.

Factors	1	2	3	4	5	6	7	8	9	10	11
Availability of cybersecurity	**0.801**										
Behavioral intention of adopting cybersecurity in smart cities	0.548	**0.807**									
Confidentiality of cybersecurity	0.631	0.428	**0.865**								
Effort expectancy	0.373	0.564	0.373	**0.819**							
Facilitating conditions	0.542	0.609	0.462	0.640	**0.802**						
Integrity of cybersecurity	0.474	0.574	0.342	0.560	0.555	**0.802**					
Performance expectancy	0.538	0.510	0.317	0.493	0.680	0.440	**0.793**				
Resiliency of cybersecurity	0.534	0.651	0.501	0.559	0.606	0.619	0.523	**0.814**			
Safety of cybersecurity	0.505	0.644	0.470	0.581	0.638	0.621	0.501	0.725	**0.824**		
Social influence	0.376	0.531	0.340	0.693	0.551	0.546	0.459	0.531	0.501	**0.847**	
Use behavior of adopting cybersecurity in smart cities	0.455	0.733	0.407	0.508	0.589	0.654	0.450	0.677	0.762	0.551	**0.839**

7.4. Model Fitness (R-Square)

R-square (R^2) measures the proportion of variance in the dependent variable that the independent variables can explain in a multiple regression model. It is a commonly used statistic to evaluate the goodness of fit of a model. R-square values range from 0 to 1, with higher values indicating a better fit between the model and the data [39]. In the given study, the R^2 values for the two dependent variables, behavioral intention of adopting cybersecurity in smart cities and use behavior of adopting cybersecurity in smart cities, are 0.567 and 0.537, respectively. These values suggest that the nine factors in the model explained 56.7% of the variance in behavioral intention and 53.7% of the variance in user behavior.

8. Discussion and Conclusions

The existing literature and studies underscore the significant influence of economic, cultural, and social factors on the behavioral intention and actual use of smart city technologies. Applying the unified theory of acceptance and use of technology (UTAUT) to cybersecurity contexts, particularly in smart cities, has yielded significant findings concerning performance expectancy, effort expectancy, social influence, and facilitating conditions. However, gaps in cultural development and specific factors such as privacy, security, trust, and anthropomorphism were also identified as critical aspects of technology adoption. Emphasizing the need for resilient information systems and raising awareness of privacy and security risks are pivotal steps toward strengthening smart city infrastructure. Therefore, factors including data availability, confidentiality, and integrity are addressed to mitigate the dearth of system development and IT expertise. Consequently, developing and implementing a comprehensive cybersecurity framework is paramount to safeguarding the integrity of smart cities, providing guidance for respoding to potential threats, and promoting a robust and secure digital environment. After that, this study examines the widespread phenomenon of testing the behavioral intentions in adopting cybersecurity awareness programs in two cities in Saudi Arabia (i.e., Riyadh and Neom). The study conducted two surveys. The first survey was carried out to explore the challenges and issues in adopting cybersecurity practices in smart Saudi cities. The second survey was carried out to develop and validate the conceptual cybersecurity-based UTAUT3 model. In the first phase, the study targeted 554 respondents (common public). In the second phase, the study targeted 108 IT professionals because they provided the most authentic responses to validate a new model in smart Saudi cities. The survey results reveal several key insights about trust in cyberinfrastructure and the digital economy in Saudi Arabia. First, there is a widespread perception of weak cybersecurity platforms and a need for more confidence in the country's ability to design and deploy smart cities. Second, respondents are highly aware of various IT issues and have experienced security or privacy breaches. Third, Saudi citizens understand the importance of cyber threats and express a strong interest in participating in cybersecurity awareness programs. Furthermore, most respondents believe that advancements in IT infrastructure and social apps, notably Snapchat and Twitter, have the potential to impact Saudi culture. However, a considerable portion of respondents remains neutral on this issue. While technical abilities such as deep-packet inspections are limited among the respondents, there is optimism that public and governmental agencies can match the private sector shortly.

The study results emphasize the importance of economic, cultural, and social factors in shaping behavioral intention and actual use of smart city technologies. Key factors identified include privacy design, cyber threat intelligence, digital trust, cyber responses, resilience, and cyber competencies and awareness programs. Moreover, the R^2 values of 0.567 and 0.537 for behavioral intention and user behavior, respectively, indicate that the nine factors (performance expectancy, effort expectancy, social influence, facilitating conditions, safety, resiliency, confidentiality, availability, and integrity of cybersecurity) in the cybersecurity-based UTAUT3 model explain a significant proportion of the variance in these two dependent variables. This suggests that a comprehensive approach addressing economic, social, and cultural factors is crucial for increasing trust in cyberinfrastructure and promoting the adoption of cybersecurity measures in smart cities within Saudi Arabia.

Combining the two surveys' findings, the study concludes by emphasizing the interconnectedness of public perception, technical abilities, and the role of economic, social, and cultural factors in influencing the adoption of cybersecurity measures in Saudi Arabia's transition to a digital economy and smart cities. Additionally, the results highlight the need for effective government policies, incentives, and awareness programs to enhance the overall trust and confidence in the country's cyber infrastructure and digital advancements. The findings of the study presented cybersecurity aspects and factors, including the actual use of behavior, availability, behavioral intention, confidentiality, effort expectancy, facilitating conditions, integrity, performance expectancy, resiliency, safety, and social influence in the context of smart cities. These were each measured with several items, which resulted in factor loadings ranging from 0.746 to 0.888, indicating a high level of validity and reliability. Moreover, all constructs scored above 0.6 in average variance extracted (AVE), implying good convergent validity. Furthermore, all constructs' Cronbach alpha and composite reliability values were above the acceptable 0.7 thresholds, indicating high reliability and internal consistency. The correlation matrix noted that all the constructs were significantly correlated with each other, ranging from 0.317 to 0.733, with the diagonal showing the square root of AVE, illustrating adequate discriminant validity. This indicates that the constructs are related yet distinct from each other, thereby supporting the validity and reliability of the scales used in measuring these constructs. These measurements could be instrumental in shaping cybersecurity policy and best practices in the context of smart cities.

8.1. Practical Implications

The practical implications of this study, which focuses on developing and applying a cybersecurity-based UTAUT3 model in smart Saudi cities, are manifold. Implementing this model can lead to several changes in cybersecurity practices, including the following:

1. **Enhanced understanding of user adoption factors:** The UTAUT3 model provides a comprehensive framework for understanding the factors influencing the adoption of cybersecurity measures in smart cities. By identifying key economic, social, and cultural factors, policymakers and city planners can develop targeted strategies to increase trust and confidence in cyberinfrastructure and promote the adoption of advanced security measures.
2. **Improved cybersecurity awareness programs:** The UTAUT3 model emphasizes the importance of cyber competencies and awareness programs in shaping users' behavioral intentions and actual use of cybersecurity measures. By designing and implementing effective cybersecurity awareness campaigns, government agencies can equip citizens with the necessary knowledge and skills to identify, prevent, and respond to cyber threats.
3. **Informed policy development:** Understanding the factors influencing the adoption of cybersecurity measures in smart cities can help policymakers develop informed strategies that address economic, social, and cultural barriers. This may include providing incentives for participating in cyber-awareness programs, supporting training and education initiatives, and fostering public–private partnerships to improve cyberinfrastructure.
4. **Strengthened cybersecurity practices:** By addressing the concerns and needs of users, as identified in the UTAUT3 model, smart cities can design and deploy robust cybersecurity platforms that address privacy, resiliency, confidentiality, availability, and integrity. This will result in a more secure digital environment, fostering trust and confidence among citizens.
5. **Promoting digital trust and collaboration:** By addressing the concerns and needs of users in smart cities, the UTAUT3 model can help build digital trust among citizens, businesses, and government agencies. This will encourage greater collaboration in developing and maintaining cybersecurity practices, ultimately leading to more secure and resilient smart cities.
6. **Preserving cultural values:** The UTAUT3 model considers the potential impact of IT infrastructure advancements on Saudi culture. By considering these cultural factors, policymakers and city planners can ensure that cybersecurity practices align with societal values and contribute positively to the overall development of smart cities.

8.2. Limitations and Future Directions

While the study provides valuable insights into the factors influencing cybersecurity practices in smart Saudi cities, it has some limitations. The sample size and demographic distribution of respondents may only represent part of the population, potentially limiting the generalizability of the findings. The most significant limitation of the study is that it developed and validated the cybersecurity-based UTAUT3 model but did not examine the effect of all nine factors on behavioral intentions and actual use of behavior in adopting cybersecurity practices; a future study should consider this aspect to investigate the impacts. Furthermore, the study focuses solely on Saudi Arabia, and the results may not directly apply to other countries or regions with different cultural, economic, and social contexts. Future research could address these limitations by employing more extensive and diverse samples and conducting comparative studies across multiple regions or countries. Additionally, researchers could explore the long-term effects of implementing the UTAUT3 model on the overall security and resilience of smart cities, examine the role of emerging technologies such as artificial intelligence and blockchain in enhancing cybersecurity, and investigate the impact of evolving cyber threats on user behavior and adoption patterns.

Author Contributions: N.A. and P.V. contribute to saudi smart cities to identify the challenges and issues in adopting cybersecurity practices. All authors have read and agreed to the published version of the manuscript.

Funding: This research received no external funding.

Data Availability Statement: Study data cannot be shared due to ethical considerations.

Conflicts of Interest: The authors declare no conflict of interest.

Appendix A. Survey Items

Demographic Information					
Gender					
Male			Female		
Occupation					
Public sector	Private sector	Unemployed	Others (housewife, construction, teacher, medical, military)	Student	Free business
How do you rate your current internet speed?					
Slow	Medium	Fast	Very Fast		
Proposed questions based on assumptions					
Lack of Trust					
1. Does Saudi Arabia have a good cybersecurity platform?					
2. Does Saudi Arabia have a good cybersecurity platform?					
3. How strongly believe we have the knowledge and capacity to design and deploy smart cities?					
4. Rate how much Saudi Arabia is moving towards digital economy?					
5. What is the most important concern that you may face while using the Internet?					
6. The major parameter of lack of trust on cyber infrastructure					
7. How do you rate your current internet speed?					
8. Are you aware of your internet-connected smart devices at home?					
9. Do you know about the digital economy?					
10. Have you ever made any transactions online (e.g., paying an online bill, shopping, etc.)?					
11. Have you had any problems with breaching security/privacy?					
12. Has your card been subjected to any fraud?					
13. Have you shared your phone number or personal information online without reading the website's privacy policy?					
14. Do you realize that you share your personal information with e-government services?					
15. Have you clicked on any electronic links you have received via SMS, anonymous WhatsApp messages, email, or any other anonymous media?					
16. What is Saudi Arabia's shift towards e-commerce?					
17. Would you like to participate in an e-commerce training course?					
18. Do you think e-commerce will be a good alternative to traditional trade?					
19. Do you expect e-commerce to have a successful future in Saudi Arabia over the next five years?					

Lack of awareness and training developments	
1.	How strongly do you think it is good to have every Saudi citizen to be aware of Cyber threats?
2.	Have you participated in cybersecurity community awareness?
3.	Will you be interested in promoting cyber-awareness program to the community, if the government provides some incentive for you?
4.	How strongly do you agree with this statement: Keeping personal and identifiable information safe and secure is at most priority.
5.	How satisfied are you if we give you benefits and incentives when you serve the community?
6.	How satisfied are you if given a chance in a competitive program with special privileges?
7.	Do you know anything about cybersecurity?
8.	Do you think it's important to learn the basics of the concept of cyberattacks?
9.	Are you interested in attending an awareness course on cybersecurity or any course on security?
10.	Do you know the basics of using technology?
11.	Are you ready to pay a small fee for the course?
12.	Have you ever participated in a technology training course?
Impacts of culture	
1.	Do the citizens of Saudi Arabia want to live in a Smart digital city?
2.	To your knowledge, what percentage of Saudi citizens are not aware of current cyber threats?
3.	Do you believe advancement in IT infrastructure will affect the Saudi culture?
4.	Provide two important technology apps (social apps) that affect Saudi culture.
5.	Would you like to live in a smart digital city?
6.	Are you aware of the terms of smart cities?
Shortage of Local Expertise	
1.	Do you have the technical ability to perform network-wide deep-packet inspections?
2.	Preference for working in the private sector?
3.	Reasons for working in the private sector
4.	Reasons to avoid working in the public sector.
5.	How strongly do you agree with the following statement: Public sector and governmental agencies can match private sectors in the new future

Appendix B. Survey Items

Measurement Scales
Actual use of behavior in adopting cybersecurity in smart cities
I use cybersecurity in smart cities frequently during my job period
I use many functions of cybersecurity in smart cities
I depend on cybersecurity in smart cities.
Availability of cybersecurity
I use an up-to-date antivirus program on my devices.
I keep the firewall installed on my devices turned on.
I do not open the files I downloaded from the Internet without scanning with an anti-virus program.
Behavioral intention in adopting cybersecurity in smart cities
I intend to continue using cybersecurity in smart cities.
For my smart city, I would use cybersecurity.
I will continue to use cybersecurity on a regular basis.
Confidentiality of cybersecurity
I do not share information and documents in smart city cyberspace that I do not want to share with third parties in real life.
I ensure that the necessary people can only view the data I share in smart city cyberspace.
I do not share my contact information in smart city cyberspace.
Effort expectancy
I find cybersecurity technologies clear and easy to use.
I have the skills I need to use the cybersecurity technologies in my city.
Learning to operate and use new cyber technologies is easy for me.
Facilitating conditions
My technical support team are experts in their fields and they support cybersecurity very well in my city
Technical support is important for cybersecurity usage and it facilitates my work
A specific person (or group) is available for assistance with cybersecurity in smart cities.
Integrity of cybersecurity
It is safe to store data in smart city cyberspace.
Information and documents I have stored in smart city cyberspace are not lost or deleted.
Sharing data in smart city cyberspace does not involve any risk.
Information and documents stored in smart city cyberspace cannot be accessed by third parties.
Performance expectancy
I find cybersecurity technologies useful for smart cities.
Using cybersecurity technologies enables me to accomplish tasks more quickly.
Cybersecurity increases communication between IT professionals and citizens.
Using cybersecurity technologies makes it easier to protect smart cities
Resiliency of cybersecurity
I tend to bounce back quickly after hard times in cybersecurity
It does not take me long to recover smart protection from a stressful event
I usually come through difficult times with little trouble due to cybersecurity
Safety of cybersecurity
I use the correct cybersecurity safety for carrying out smart city projects
I ensure the highest levels of cybersecurity safety when I carry out smart city projects
I voluntarily carry out tasks or activities that help to improve cybersecurity safety
I help my coworkers when they are working under risky or hazardous Conditions on small city projects
Social influence
The supervisor thinks that I should use cybersecurity for smart cities.
My colleagues have helped me to use cybersecurity technologies in smart cities.
Most staff in my IT department think cybersecurity for smart cities is important

Appendix C. Cross-Loadings

	1	2	3	4	5	6	7	8	9	10	11
AUB1	0.406	0.645	0.470	0.427	0.556	0.429	0.412	0.541	0.632	0.485	0.837
AUB2	0.366	0.595	0.210	0.366	0.380	0.606	0.293	0.530	0.677	0.352	0.822
AUB3	0.372	0.601	0.332	0.485	0.543	0.620	0.423	0.635	0.610	0.548	0.858
AVAI1	0.746	0.426	0.695	0.438	0.511	0.459	0.456	0.545	0.497	0.365	0.422
AVAI3	0.794	0.404	0.464	0.241	0.376	0.313	0.364	0.283	0.402	0.212	0.328
AVAI4	0.858	0.481	0.375	0.224	0.416	0.368	0.467	0.447	0.326	0.320	0.347
BI1	0.307	0.780	0.154	0.428	0.373	0.420	0.274	0.422	0.433	0.349	0.536
BI2	0.464	0.840	0.421	0.548	0.588	0.485	0.523	0.599	0.663	0.579	0.690
BI3	0.545	0.799	0.432	0.369	0.486	0.481	0.402	0.535	0.426	0.318	0.524
CON2	0.478	0.387	0.833	0.378	0.484	0.348	0.387	0.384	0.454	0.247	0.346
CON3	0.653	0.377	0.887	0.310	0.382	0.244	0.230	0.490	0.357	0.310	0.352
CON4	0.504	0.344	0.874	0.274	0.321	0.294	0.195	0.425	0.407	0.328	0.357
EE1	0.275	0.455	0.269	0.852	0.564	0.398	0.453	0.396	0.463	0.554	0.345
EE2	0.286	0.473	0.247	0.803	0.494	0.402	0.322	0.426	0.427	0.554	0.418
EE3	0.355	0.455	0.402	0.799	0.515	0.577	0.438	0.550	0.538	0.594	0.483
FC3	0.421	0.401	0.312	0.461	0.753	0.405	0.541	0.380	0.317	0.359	0.374
FC4	0.355	0.480	0.255	0.590	0.797	0.392	0.550	0.455	0.546	0.447	0.444
FC5	0.519	0.566	0.515	0.494	0.854	0.526	0.553	0.594	0.628	0.502	0.575
INT1	0.471	0.522	0.394	0.471	0.458	0.798	0.356	0.509	0.540	0.502	0.520
INT2	0.346	0.415	0.255	0.506	0.417	0.852	0.362	0.529	0.456	0.498	0.427
INT3	0.372	0.437	0.208	0.360	0.483	0.752	0.310	0.491	0.472	0.347	0.508
INT4	0.312	0.450	0.217	0.454	0.415	0.803	0.379	0.451	0.510	0.393	0.632
PE1	0.582	0.444	0.364	0.340	0.550	0.433	0.824	0.505	0.483	0.281	0.452
PE2	0.352	0.373	0.257	0.426	0.551	0.381	0.769	0.428	0.374	0.388	0.357
PE3	0.445	0.393	0.219	0.380	0.536	0.226	0.785	0.352	0.323	0.334	0.282
PE4	0.307	0.401	0.155	0.427	0.522	0.348	0.792	0.368	0.398	0.464	0.325
RESI1	0.434	0.505	0.453	0.436	0.502	0.433	0.459	0.811	0.568	0.424	0.521
RESI2	0.444	0.515	0.333	0.558	0.520	0.554	0.503	0.810	0.590	0.529	0.526
RESI3	0.427	0.567	0.435	0.378	0.461	0.521	0.327	0.821	0.609	0.352	0.601
SAFE1	0.502	0.587	0.510	0.492	0.539	0.497	0.460	0.658	0.888	0.402	0.631
SAFE2	0.327	0.478	0.319	0.459	0.472	0.423	0.259	0.620	0.810	0.470	0.573
SAFE3	0.449	0.532	0.423	0.460	0.565	0.529	0.482	0.618	0.837	0.383	0.677
SAFE5	0.371	0.518	0.278	0.503	0.523	0.597	0.433	0.489	0.756	0.406	0.628
SI1	0.353	0.389	0.238	0.485	0.385	0.367	0.417	0.445	0.374	0.789	0.353
SI2	0.326	0.464	0.334	0.625	0.491	0.506	0.382	0.403	0.450	0.862	0.505
SI3	0.287	0.490	0.286	0.638	0.513	0.502	0.377	0.502	0.444	0.887	0.525

Note: 1 = availability of cybersecurity, 2 = behavioral intention of adopting cybersecurity, 3 = confidentiality of cybersecurity, 4 = effort expectancy, 5 = facilitating conditions, 6 = integrity of cybersecurity, 7 = performance expectancy, 8 = resilience of cybersecurity, 9 = safety of cybersecurity, 10 = social influence, 11 = use behavior of adopting cybersecurity.

References

1. Hollands, R.G. Critical interventions into the corporate smart city. *Camb. J. Reg. Econ. Soc.* **2015**, *8*, 61–77. [CrossRef]
2. Albino, V.; Berardi, U.; Dangelico, R.M. Smart cities: Definitions, dimensions, performance, and initiatives. *J. Urban Technol.* **2015**, *22*, 3–21. [CrossRef]
3. Kitchin, R. Making sense of smart cities: Addressing present shortcomings. *Camb. J. Reg. Econ. Soc.* **2015**, *8*, 131–136. [CrossRef]
4. Abou-Korin, A.A.; Al-Shihri, F.S. Rapid urbanization and sustainability in Saudi Arabia: The case of Dammam metropolitan area. *J. Sustain. Dev.* **2015**, *8*, 52. [CrossRef]
5. Lacinák, M.; Ristvej, J. Smart city, safety and security. *Procedia Eng.* **2017**, *192*, 522–527. [CrossRef]
6. Alharbe, M.A. Cyber Security, Forensics, and Its Impact on Future Challenges in Saudi Arabia Smart Cities. *Int. J. Adv. Trends Comput. Sci. Eng.* **2020**, *9*, 2464–2470. [CrossRef]
7. Mohammad, R.M.A.; Abdulqader, M.M. Exploring cyber security measures in smart cities. In Proceedings of the 2020 21st International Arab Conference on Information Technology (ACIT), Giza, Egypt, 28–30 November 2020; IEEE: New York, NY, USA, 2020; pp. 1–7.
8. Townsend, A.M. *Smart Cities: Big Data, Civic Hackers, and the Quest for a New Utopia*; WW Norton & Company: New York, NY, USA, 2013.

9. Chen, Q.; Mislove, A.; Wilson, C. Peeking beneath the hood of Uber. In Proceedings of the 2015 ACM Conference on Internet Measurement Conference, Tokyo, Japan, 28–30 October 2015; pp. 495–508.
10. Alkhater, N.; Wills, G.; Walters, R. Factors influencing an organisation's intention to adopt cloud computing in Saudi Arabia. *Int. J. Cloud Comput.* **2018**, *7*, 248–282.
11. Alghazzawi, D.M. A framework to implement cloud computing in e-government in Saudi Arabia. *Int. J. Comput. Sci. Issues* **2012**, *9*, 64–74.
12. Hosam, O.; Ahmad, M.H. Hybrid design for cloud data security using combination of AES, ECC and LSB steganography. *Int. J. Comput. Sci. Eng.* **2019**, *19*, 153–161. [CrossRef]
13. Alshehri, M.; Drew, S.; AlGhamdi, R. Analysis of citizens acceptance for e-government services: Applying the UTAUT model. *arXiv* **2013**, arXiv:1304.3157.
14. Hossam-Eldin, A.A.; Negm, E.; Elgamal, M.S.; AboRas, K.M. Operation of grid-connected DFIG using SPWM-and THIPWM-based diode-clamped multilevel inverters. *IET Gener. Transm. Distrib.* **2020**, *14*, 1412–1419. [CrossRef]
15. AlOtaibi, S.; Aljohani, N.R.; Hoque, M.R.; Alotaibi, F.S. The satisfaction of Saudi customers toward mobile banking in Saudi Arabia and the United Kingdom. *J. Glob. Inf. Manag.* **2018**, *26*, 85–103. [CrossRef]
16. Alrubaiq, A.; Alharbi, T. Developing a cybersecurity framework for e-government project in the Kingdom of Saudi Arabia. *J. Cybersecur. Priv.* **2021**, *1*, 302–318. [CrossRef]
17. Talib, A.M.; Alomary, F.O.; Alwadi, H.F.; Albusayli, R.R. Ontology-based cyber security policy implementation in Saudi Arabia. *J. Inf. Secur.* **2018**, *9*, 315. [CrossRef]
18. Basahel, A.; Yamin, M. Measuring success of e-government of Saudi Arabia. *Int. J. Inf. Technol.* **2017**, *9*, 287–293. [CrossRef]
19. Bryman, A.; Bell, E. *Business Research Methods*; Oxford University Press: New York, NY, USA, 2015.
20. Yilmaz, K. Comparison of quantitative and qualitative research traditions: Epistemological, theoretical, and methodological differences. *Eur. J. Educ.* **2013**, *48*, 311–325. [CrossRef]
21. Weerakkody, V.; Irani, Z.; Lee, H.; Hindi, N.; Osman, I. A review of the factors affecting user satisfaction in electronic government services. *Int. J. Electron. Gov. Res.* **2014**, *10*, 21–56. [CrossRef]
22. Dillman, D.A.; Smyth, J.D.; Christian, L.M. *Internet, Phone, Mail, and Mixed-Mode Surveys: The Tailored Design Method*; John Wiley & Sons: Hoboken, NJ, USA, 2014.
23. Sekaran, U. *Research Methods for Business: A Skill-Building Approach*, 4th ed.; John Wiley & Sons, Inc.: Hoboken, NJ, USA, 2003.
24. Saunders, M.; Lewis, P.; Thornhill, A. *Research Methods for Business Students*; Pearson Education: Upper Saddle River, NJ, USA, 2009.
25. Walliman, N. *Your Research Project. A Step-by-Step Guide for the First-Time Researcher*; Sage Publications: London, UK, 2000.
26. Tabachnick, B.G.; Fidell, L.S. *Experimental Designs Using ANOVA*; Thomson/Brooks/Cole: Belmont, CA, USA, 2007; Volume 724.
27. Popova, Y.; Zagulova, D. UTAUT Model for Smart City Concept Implementation: Use of Web Applications by Residents for Everyday Operations. *Informatics* **2022**, *9*, 27. [CrossRef]
28. Kuberkar, S.; Singhal, T.K. Factors influencing adoption intention of AI powered chatbot for public transport services within a smart city. *Int. J. Emerg. Technol. Learn.* **2020**, *11*, 948–958.
29. Van Zoonen, L. Privacy concerns in smart cities. *Gov. Inf. Q.* **2016**, *33*, 472–480. [CrossRef]
30. Mijwil, M.; Doshi, R.; Hiran, K.K.; Al-Mistarehi, A.H.; Gök, M. Cybersecurity Challenges in Smart Cities: An Overview and Future Prospects. *Mesop. J. Cybersecur.* **2022**, *2022*, 1–4.
31. Alzahrani, L. Statistical Analysis of Cybersecurity Awareness Issues in Higher Education Institutes. *Int. J. of Adv. Comput. Sci. and Appl.* **2021**, *12*, 630–637. [CrossRef]
32. Alhalafi, N.; Veeraraghavan, P. Cybersecurity Policy Framework in Saudi Arabia: Literature Review. *Front. Comput. Sci.* **2021**, *3*, 89. [CrossRef]
33. Alalwan, A.A.; Dwivedi, Y.K.; Rana, N.P. Factors influencing adoption of mobile banking by Jordanian bank customers: Extending UTAUT2 with trust. *Int. J. Inf. Manag.* **2017**, *37*, 99–110. [CrossRef]
34. Venkatesh, V.; Thong, J.Y.; Xu, X. Unified theory of acceptance and use of technology: A synthesis and the road ahead. *J. Assoc. Inf. Syst.* **2016**, *17*, 328–376. [CrossRef]
35. Arpaci, I.; Sevinc, K. Development of the cybersecurity scale (CS-S): Evidence of validity and reliability. *Inf. Dev.* **2022**, *38*, 218–226. [CrossRef]
36. McDaniel, C., Jr.; Gates, R. *Marketing Research*; Wiley: Hoboken, NJ, USA, 2006.
37. Hair, J.F., Jr.; Sarstedt, M.; Ringle, C.M.; Gudergan, S.P. *Advanced Issues in Partial Least Squares Structural Equation Modeling*; Sage Publications: Thousand Oaks, CA, USA, 2017.
38. Henseler, J.; Hubona, G.; Ray, P.A. Using PLS Path Modeling in New Technology Research: Updated Guidelines. *Ind. Manag. Data Syst.* **2016**, *116*, 2–20. [CrossRef]
39. Hair, J.F., Jr.; Hult, G.T.M.; Ringle, C.M.; Sarstedt, M.; Danks, N.P.; Ray, S. *Partial Least Squares Structural Equation Modeling (PLS-SEM) Using R: A Workbook*; Springer Nature: London, UK, 2021; p. 197.
40. Ali, F.; Rasoolimanesh, S.M.; Sarstedt, M.; Ringle, C.M.; Ryu, K. An Assessment of the Use of Partial Least Squares Structural Equation Modeling (PLS-SEM) in Hospitality Research. *Int. J. Contemp. Hosp. Manag.* **2018**, *30*, 514–538. [CrossRef]
41. Hult, G.T.M.; Hair, J.F., Jr.; Proksch, D.; Sarstedt, M.; Pinkwart, A.; Ringle, C.M. Addressing endogeneity in international marketing applications of partial least squares structural equation modeling. *J. Int. Mark.* **2018**, *26*, 1–21. [CrossRef]

42. Shiau, W.-L.; Sarstedt, M.; Hair, J.F. Internet research using partial least squares structural equation modeling (PLS-SEM). *Internet Res.* **2019**, *29*, 398–406. [CrossRef]
43. Richter, N.F.; Hauff, S.; Ringle, C.M.; Gudergan, S.P. The use of partial least squares structural equation modeling and complementary methods in international management research. *Manag. Int. Rev.* **2022**, *62*, 449–470. [CrossRef]
44. Fornell, C.; Larcker, D.F. *Structural Equation Models with Unobservable Variables and Measurement Error: Algebra and Statistics*; Sage Publications Sage CA: Los Angeles, CA, USA, 1981.

buildings

MDPI

Article

Evaluating the Impact of Public Participation Processes on Participants in Smart City Development: A Scoping Review

Coline Senior *, Alenka Temeljotov Salaj, Agnar Johansen and Jardar Lohne

Department of Civil and Environmental Engineering, Faculty of Engineering, Norwegian University of Science and Technology (NTNU), 7491 Trondheim, Norway; alenka.temeljotov-salaj@ntnu.no (A.T.S.); a.johansen@ntnu.no (A.J.); jardar.lohne@ntnu.no (J.L.)

* Correspondence: coline.senior@ntnu.no

Abstract: This paper provides an overview of the current state of research on the evaluation methods of participatory processes in smart cities. Specifically, it aims to identify and analyze existing evaluation methods and frameworks for public participation (PP) in smart city development. The study focuses on the evaluation of participatory processes to find key indicators and enable an assessment of PP from multiple perspectives. A scoping literature review was conducted to analyze the past ten years of scientific literature on the topic. Relevant literature was retrieved from Scopus, Web of Science, and Google Scholar, and articles were selected based on a set of criteria to ensure quality and relevance. We found that digital participation, also known as "e-participation", is the most used method either exclusively or mixed with in-person methods. The level of participation achieved was mostly limited to the first two degrees of PP (i.e., information and consultation), and only a few papers addressed the highest degrees (i.e., agenda-setting and co-management). The impact on participants was mostly related to knowledge and skills, awareness raising, and satisfaction with the process and method. This paper highlights the potential to upskill citizens and enhance their understanding of sustainable urban development, fostering their commitment to achieving the United Nations' sustainability goals for climate change mitigation in the urban context.

Keywords: public participation; smart cities; evaluation; upskilling citizens

1. Introduction

There has been a substantial amount of literature produced on the concept of smart city since the term was first introduced in 1994, with a significant increase in attention following its adoption as a cornerstone of the European Union's development strategy in 2010 [1,2]. The European Union continues to intensify its efforts and funding capacity for smart city projects. In total, the funding program Horizon Europe will invest 360 million euros in research and innovation actions linked to the "Climate-neutral and Smart cities" mission [3]. Smart cities rely on the deployment and use of information and communication technologies, not as an end in itself but as a means to foster sustainable urban development and enhance the quality of life for their citizens [4]. Thus, smart cities play a critical role in the pursuit of sustainable development; by leveraging technology, smart cities can enhance energy efficiency, reduce greenhouse gas emissions, improve public transportation, and promote sustainable economic growth and social development [5].

The development of smart cities is a complex process that requires the involvement and engagement of citizens to ensure sustainability and success. The United Nations Sustainable Development Goal (SDG) 11 emphasizes the importance of sustainable urbanization, which includes making cities inclusive, safe, resilient, and sustainable. To achieve this goal, cities must adopt participatory approaches that engage citizens in decision-making processes, such as those related to urban planning and transportation infrastructure. Similarly, SDG 13 focuses on climate action and the need to reduce greenhouse gas emissions. Citizen engagement and upskilling are essential in achieving this goal as well, as individuals need to be aware of their impact on the environment and to understand how they can reduce their carbon footprint [6,7]. Citizens' lack of motivation to engage in participatory processes makes it challenging to recruit representative samples [8–10]. In addition, the choice of participation method determines which demographics are capable or not of engaging. At the end of the process, the results of public participation can be difficult to implement in the plans, especially if essential knowledge about the issue at stake is not sufficiently transferred to participants [11]. A change toward more sustainable behaviors is also necessary to achieve the goals of sustainable urban development [12]. Therefore, evaluating the effectiveness of participatory processes is crucial for understanding how to improve citizen participation in smart city projects. As the goals of public participation in smart cities go beyond the mere recruitment of a high number of participants and the collection of individual opinions, it is important to identify what citizens can get out of these processes rather than only what they contribute to them in order to create a common project that caters for the needs of all citizens today and in the future, ensuring the uptake of technologies and inspiring change toward sustainable behaviors.

This paper proposes a scoping literature review of the current state of research covering evaluation methods of public participation in smart cities. This review aims to identify existing frameworks and indicators used in the scientific literature to analyze not only the outcomes but also the extent to which these consider what citizens gain in the process. We will address the following research questions:

RQ1. What research has been carried out on the evaluation of public participation methods in smart cities?

RQ2. In which ways do the public participation processes impact the participants?

The findings from this review will be used to develop an approach for upskilling citizens in smart city projects, specifically in topics related to sustainability and smart cities. This is directly connected to achieving the United Nations' sustainability goals for climate change mitigation in the built environment.

The study uses a scoping literature review analysis of specific elements in the evaluation methods of public participation, focusing on the impact on participants. The past ten years of scientific literature on the topic are reviewed, and methods and sets of indicators are analyzed to identify possible research gaps and potential for further research. Google Scholar, Web of Science, and Scopus were used to search the literature, and the selection was based on a set of criteria to ensure the scientific quality of each source. Only papers written in English and published in peer-reviewed journals were further analyzed.

We first provide an overview of the importance of participatory processes in the development of smart cities. We then proceed to describe the methodology used in the scoping literature review, including the search strategy and selection criteria. The findings are presented, including an analysis of the concentration of the literature on European smart cities, the level of public participation, and the impact on participants covered. The paper concludes by discussing the implications of the findings, including the potential for upskilling citizens in sustainability topics, and proposing avenues for further research.

The study provides a valuable contribution to the literature by identifying research gaps and the potential for further research in the evaluation of public participation in the specific context of smart city development. The originality of the study is its focus on the participants themselves rather than the outcomes of participatory processes. Additionally, the study offers practical implications for policymakers and stakeholders involved in the development of smart cities by demonstrating the importance of citizen engagement and the potential for upskilling citizens in sustainability topics.

2. Public Participation in Smart Cities

Public participation is a broad and multi-dimensional concept that has been discussed extensively in various fields of research, including urban planning, public policy, and environmental studies [13–16]. According to Arnstein (1969) [17], public participation is a process that allows citizens to actively engage in decision-making processes and to have a genuine impact on the outcomes of those processes. This definition emphasizes the importance of power dynamics between the government and citizens, with public participation serving as a means of empowering citizens and ensuring that their interests are taken into account. Several scholars have proposed typologies or models to conceptualize the different levels and types of public participation. For instance, Pretty (1995) [18] proposed a ladder of citizen participation which ranges from "manipulative participation" to "interactive participation and self-mobilization", with each rung representing a different level of citizen involvement and influence in the decision-making processes. Rowe and Frewer (2000) identified three levels of public participation: passive, where citizens receive information but do not provide feedback; consultative, where citizens provide feedback but do not have decision-making power; and collaborative, where citizens and decision-makers work together to create a mutually beneficial solution. More recently, Ringholm, Nyseth, and Sandkjær Hanssen (2018) [19] presented a model in which the legal requirements are the starting point for assessing the level of participation, thus catering for the specific context in which the participatory process is implemented. In the context of urban planning and development, public participation is increasingly being recognized as an essential component of sustainable urban governance [20–22]. In recent years, there has also been growing interest in public participation in smart cities, in which information and communication technologies are used to enhance the quality of life for citizens [23–25]. This emerging literature on smart cities emphasizes a citizen-centric approach rather than solely focusing on technology [22,24,26]. In this context, a smart city can be defined as a space where local governments, citizens, and various stakeholders collaborate on initiatives aimed at improving overall living conditions [27,28]. By recognizing the importance of involving citizens in decision-making processes, public participation processes can be initiated as a means to minimize conflicts and objections to planning proposals. Given that citizens are the primary beneficiaries of smart city services and environments, it is argued that their involvement should be integral to the development process [23,25,29]. In smart cities, public participation requires a new approach that takes into account their unique features, such as the use of technology and data-driven decision-making [23,30]. This new governance approach should also be more flexible and adaptable than traditional urban governance models to make room for integrating knowledge-sharing perspectives and developing organizational learning capabilities [10,31].

In smart cities, the potential for technology to increase citizen engagement and participation in decision-making processes is often referred to as "e-participation". E-participation refers to the use of digital technologies to facilitate citizen participation in government decision-making processes. Smart cities are often seen as ideal contexts for e-participation due to their advanced technological infrastructure and the availability of data [32].

By engaging citizens in decision-making processes, governments and other stakeholders can ensure that their policies and projects are aligned with citizens' needs and values, which can improve the effectiveness and legitimacy of governance [33]. However, the effectiveness of public participation can be limited by various factors, including power imbalances, lack of resources and skills among citizens, and inadequate communication between stakeholders [13,34].

Therefore, it is essential to understand the various dimensions and levels of how public participation is evaluated as well as the factors that enable or hinder its effectiveness in order to design and implement participatory processes that are inclusive, transparent, and effective.

3. Methods

3.1. Research Design and Search Process

This study uses the scoping review methodology to give an overview of a relatively recent field, namely, the evaluation of public participation in smart cities. The flexibility of this method compared with a systematic literature review makes it especially relevant to conducting an exploratory study [35]. The scoping review method was preferred to a systematic one as it helps clarify the context, inquiries, and the extent of the available evidence. By pinpointing areas where evidence is insufficient, scoping reviews can assist sponsors and researchers in prioritizing their limited resources to address the most pressing questions to be further researched and represent a stepping stone toward a systematic review or an empirical study [36]. This scoping review followed the five stages of the methodological framework for scoping studies [37]. The Preferred Reporting Items for Systematic Reviews and Meta-Analyses extension for Scoping Reviews (PRISMA-ScR) flow chart (Figure 1) illustrates the preliminary steps to arrive at the final sample to be examined in the study [35]. PRISMA-ScR is an adapted version of the original twelve points of the PRISMA method [38,39]. After defining the research question (as described in Section 1), we identified relevant studies using the following databases and search engines: Scopus, Web of Science, and Google scholar. As the field is rather narrow, the latter was used as an additional tool to expand the sample to include studies that might not have appeared in Scopus and Web of Science due to the search on "Title-Abstract-Keywords" in these databases. Then we selected the studies, extracted the data, and compiled and reported the results. The following section describes each step of the scoping review process. The initial search string for "Evaluation of public participation" AND "Smart cities" was extended to a parallel search on "Evaluation of citizen engagement" AND "Smart cities" to retrieve publications that focused on evaluating the methods used to engage citizens and not only the public participation process itself.

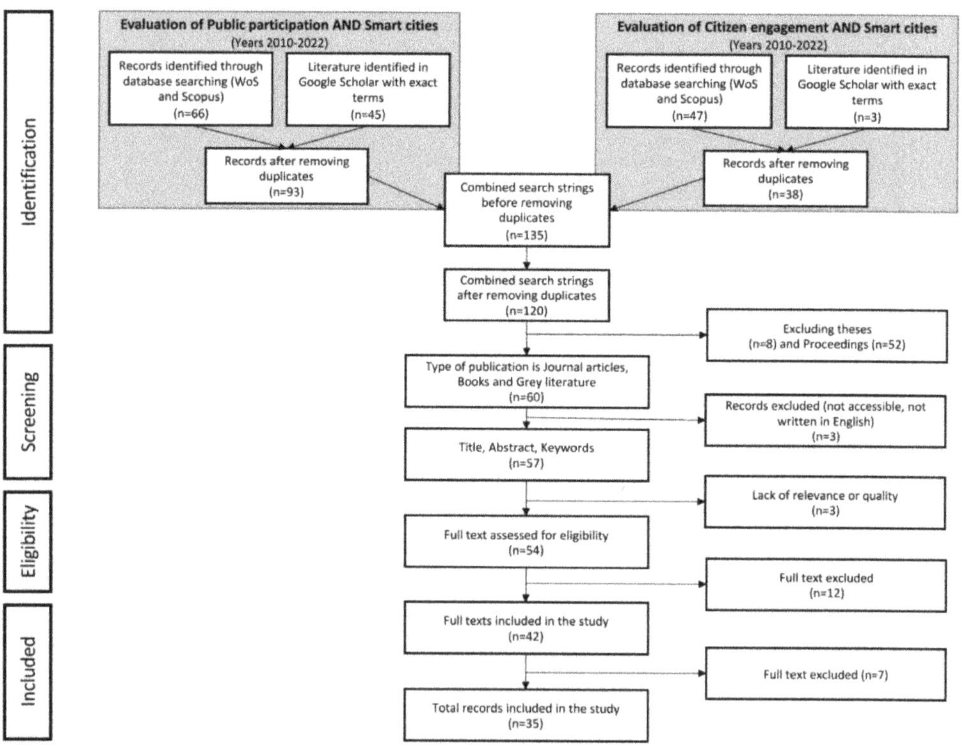

Figure 1. Preferred Reporting Items for Systematic Reviews and Meta-Analyses and Meta Analyses extension for Scoping Reviews (PRISMA-ScR) flow chart. Source: authors.

The PRISMA-ScR guidelines were followed in the current study for conducting the scoping literature review, as shown in Figure 1. The following 12 items (Table 1) for conducting the review process were addressed, with the exception of items 9, 11, and 12 as they are not applicable for scoping reviews [35].

Table 1. PRISMA checklist.

Item	PRISMA Checklist
1	The review is based on keyword-based articles retrieved from Scopus, WoS, and Google Scholar repositories published in the last 10 years.
2	The keywords must be present in one of the main sections of the articles to be eligible for inclusion in the study.
3	Scopus, WoS, and Google Scholar repositories can be respectively accessed at scopus.com/search/form.uri?display=basic, https://www.webofscience.com/wos/woscc/basic-search and https://scholar.google.com/.
4	Search strings are used for conducting the literature search process, as listed above in Section 3.1.
5	The selection process consists of keyword searching, screening, and removing duplicates. See a detailed description in Section 3.2.
6	The selected articles are analyzed by charting the data in Excel as described in Section 3.3.
7	The categories for charting the data consist of general information, core data, and definitions (see Section 3.3).
8	The included sources of evidence are typically not critically appraised for scoping reviews [35]. Therefore, the present study does not report on the appraisal.
9	Summary measures are not applicable for scoping reviews [35].
10	The synthesis of the results is presented as a "map" of the data in the form of diagrams and tables and in a descriptive format aligning with the review's objectives (see Section 4).
11	Risk of bias across studies is not applicable for scoping reviews [35].
12	Additional analyses are not applicable for scoping reviews [35].

3.2. Selecting Studies

The inclusion criteria were as follows:

- The topic is Smart cities and Public participation and/or Citizen engagement, and it mentions its evaluation or a dimension thereof.
- The study should include criteria and/or indicators to evaluate public participation or citizen engagement in smart cities.
- Articles are written in English and accessible to authors.

Exclusion criteria:

- Editorials and perspectives, conference proceedings, theses, project reports.
- Articles written in languages other than English.

Some studies were kept in the next step (i.e., charting the data) despite not primarily aiming to evaluate public participation or citizen engagement in smart cities, as some of their data addressed it as one of the dimensions to consider in the overall evaluation of smart cities.

3.3. Charting the Data

Three types of categories were defined:

- General information
- Core data
- Definitions

The "general information" included the sub-categories Authors, First author's country, Title of article or book chapter, Publication year, Journal or Book, Location of the study, Authors' keywords, Purpose of the study, Methods, Results, and Implications. Information regarding the first author's country and location of the study was collected to map the concentration of scientific production by world region and to identify the "hot spots". Purpose, Method, Results, and Implications were directly extracted from the abstracts and were used to assess the relevance of the study. This category was used to obtain a general overview and identify characteristics of the scientific literature connected to the research questions. At this stage, the full texts were read, and twelve articles and seven books were excluded from the review due to poor quality of writing, not meeting the inclusion criteria, or falling outside of the scope of this review.

The "core data" included "Participation method", "Purpose of the participation method/process", "Evaluation method", "What is measured", "How it is measured", and "What is missing". These sub-categories were defined to address the scope of the research questions (i.e., public participation in smart cities and how it is evaluated).

The category "Definitions" collected each paper's definition of smart city, public participation, and/or citizen engagement. The collected data were used to build a common understanding of the research field among authors and to define the theoretical framework of this scoping literature review.

3.4. Data Analysis

The extracted data were then analyzed by the authors, following the aforementioned categories, and visualized with the help of selected graphic charts. A refined coding step was introduced to ease the further analysis of the data, and the level of participation achieved was introduced during this phase.

Ringholm, Nyseth and Sandkjær Hanssen (2018) [19] defined five levels of participation to assess participatory processes (Table 2). These levels are adapted from Sherry Arnstein's ladder of participation [17].

Table 2. Levels of citizen participation defined by Ringholm, Nyseth and Sandkjær Hanssen (2018).

Levels	Definitions by Ringholm et al. (2018) [19]
Co-management	"citizens' initiatives, which can be used to initiate new planning processes or revise old ones"
Agenda setting	"have the opportunity to influence the agenda by putting forward themes that should be included in the planning strategy"
Dialogue	"two-way communication—a discussion, debate or deliberation regarding an issue between the municipalities and interested parties"
Consultation	"the plan proposal is made publicly available so that everyone has access to it and can provide feedback"
Information	"advertising the establishment of the planning process"

In this scoping literature review, the levels were used to categorize the studies based on the degree of participation achieved in the participatory processes. The main advantage of using the adapted model is that it is a simplified version with practical implications for assessing the level of participation according to national regulations.

When more "mixed" or "combined approaches" were identified in the literature which could not be classified within those levels, the studies were marked with a "*". These included multipurpose studies that did not focus on a specific process or method, but proposed evaluation models based on theoretical evidence. These will be discussed as another cluster of papers in the next section.

Methodological Limitations

One of the limitations of this scoping review is that the categories for data collection and further analysis were defined by the authors themselves and do not follow a specific framework. This approach, called grounded theory, is based on the idea that theory should emerge from data, rather than being imposed on it. Glaser and Strauss (1967) [40] argue that the researcher should immerse themselves in the data, constantly comparing and contrasting different pieces of information to identify patterns, relationships, and concepts. Through this process, the researcher can generate theories that are grounded in the data and are thus more likely to be relevant and applicable to the real world.

Another methodological limitation is the exclusion of gray literature and conference proceedings in the examined sample, limiting the scope of this literature review to peer-reviewed articles and books.

4. Results

4.1. Overview of the Sampled Literature

In total, 35 papers were identified, and data were collected into the categories previously described in Section 3.3. Even though the initial search included the period 2010–2022, relevant literature was only found in the period 2015–2022 (Figure 2). The first authors are primarily affiliated with the Netherlands (n = 6), Italy (n = 4), and Germany (n = 4) (Figure 3). European cities are over-represented (78%) in the examined papers compared with other world regions. The sampled literature is rather evenly spread across journals with only three journals counting more than one article (Figure 4).

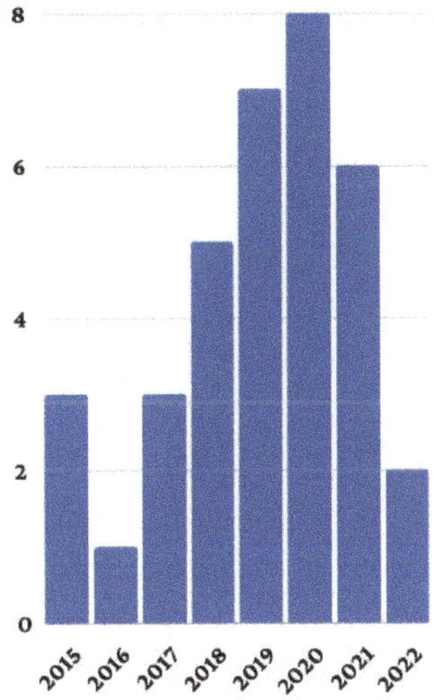

Figure 2. Sampled literature clustered by number of articles published per year. Source: authors.

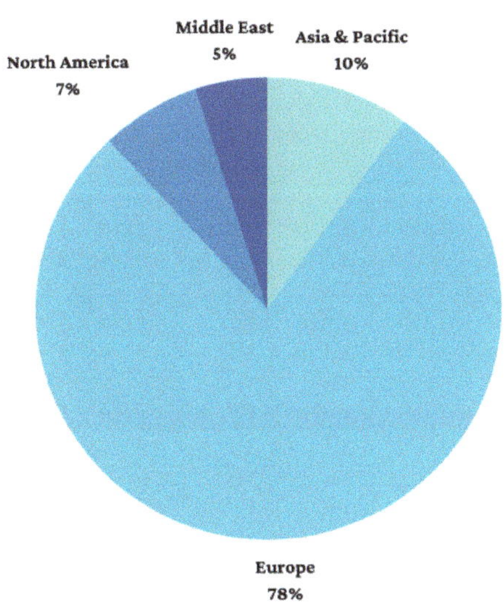

Figure 3. Geographic distribution of the location of the studies by world region. Source: authors.

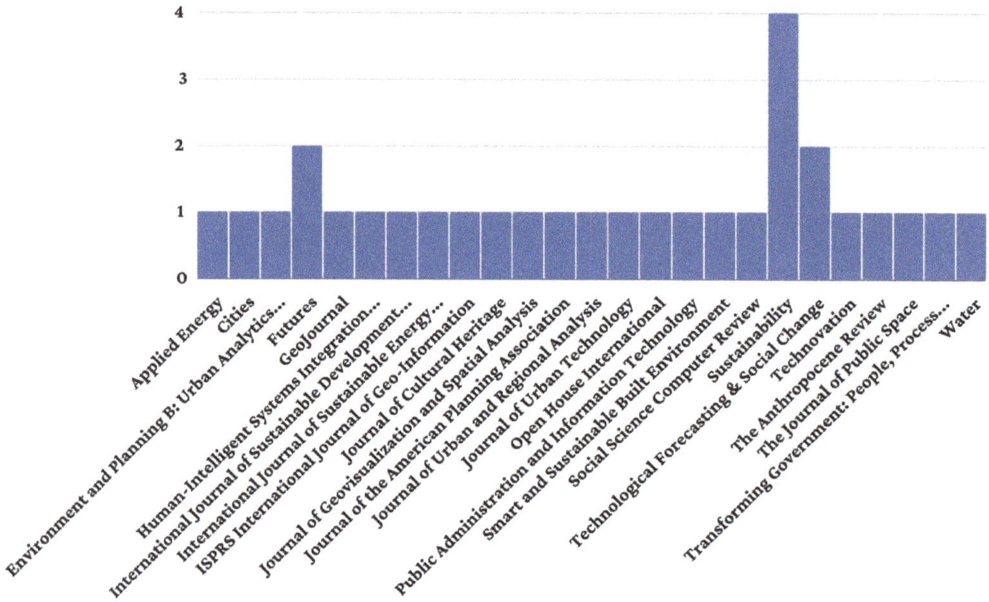

Figure 4. Journal distribution in the sampled literature.

4.2. Choice of Research Methods

Qualitative research methods dominate in the sampled literature (68%), followed by mixed-methods studies (25%). Only a few studies used quantitative methods (9%), while the book included in the sample [41] is unclassified as the majority of the chapters do not present a clear method. The sampled literature presents mostly case studies (54%), followed by literature reviews (31%).

The qualitative studies ranged from 2 [42] to 50 [43] participants, with workshops and interviews being the most-used methods. Studies using survey questionnaires in combination with other methods presented samples ranging between 16 [44] and 328 [45] respondents. The only study using a solely quantitative method analyzed a sample of 309 city policies [46]. Two pilot studies were also identified in the sampled literature. In the pilot study by Ambrose (2020) [47], 15 self-selected citizens took part in a one-time event. Paskaleva and Cooper (2018) [48] presented a framework first developed together with 25 participants and then pilot-tested in 6 cities without documenting the total number of participants in each of them.

Literature reviews range from 16 [49] to 114 [50] analyzed papers. In addition, four studies presented a literature review as part of a conceptual paper [51–54].

4.3. Participation Method and Evaluation

This section aims to classify the papers according to the participation method used in the studies, the level of participation, the type of impact on participants, and the specific attributes associated with it.

This section describes the sampled literature to address the research question of how public participation is evaluated in smart cities. First, we give an overview of the type of participation methods used in the studies (Figure 5), then the level of participation achieved, the impacts on participants that were in focus, and the evaluation indicators.

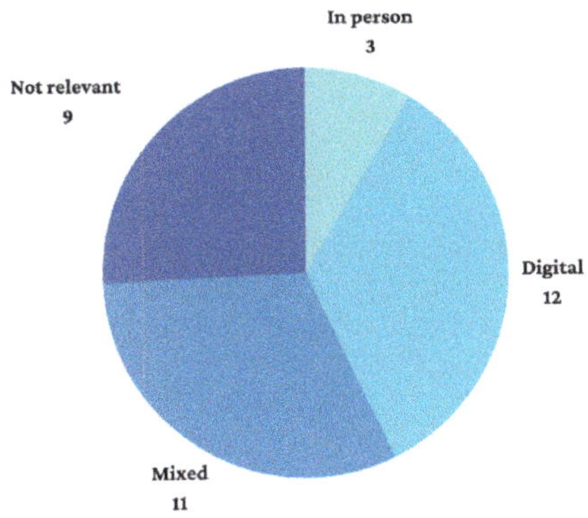

Figure 5. Type of public participation method identified in the sampled literature.

Digital participation, also known as "e-participation", was the most commonly represented method in the sampled literature, either exclusively (n = 12) or mixed with in-person methods (n = 11). Twelve of these studies that focused on e-participation reported on the use of visualization tools and methods (e.g., planning support systems, 3D models, data visualization apps).

4.3.1. Level of Participation

The majority of the papers (n = 18) included in this scoping review addressed the first two degrees of citizen participation, namely, "information" and "consultation". Only a few (n = 5) addressed the highest degree of citizen participation (i.e., "co-management"); these will be presented in more detail at the end of this section (Figure 6). Some of the studies did not refer to a specific case or participatory process but proposed new methodologies for the evaluation of public participation (n = 8). The latter are not included in the figure below but are described at the end of the section.

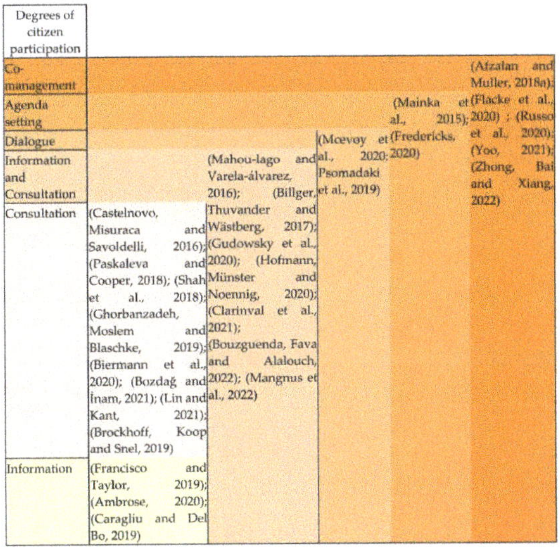

Figure 6. Level of participation achieved in the studies. The colors are set according to the degree of participation, from the lowest (lightest shade) to the highest (darkest shade) [1,42–48,50,53–70]. Source: authors.

Information Only

Three studies referred solely to the "information" level of public participation [44,46,47]. In these papers, the goal of participation was to inform people and ultimately induce a change in behavior.

Consultation Only

The primary focus of four studies addressing only "consultation" was on collecting input from citizens to inform the planning process [42,54,57,58]. The others aimed to induce a change of a behavior [56], to initiate a social-ecological transformation [55], and to co-create public value for and with citizens [48,54].

Information and Consultation

Eight papers addressed both "information" and "consultation" [1,43,50,53,59,61,62]. Public participation in these studies had a primary goal of informing the urban-planning process. In addition, [60] also included a goal of improving public service delivery.

Dialogue

One of the studies referring to "dialogue" [45] describes the design and evaluation of a prototype computational model that fuels audience engagement and collaboration of cultural organizations. The authors of this paper consulted cultural stakeholder groups (e.g., journalists, artists, art lovers, and representatives of art institutions). Another study by Mcevoy et al. (2020) [63] presents a research approach to evaluate the role of a collaborative planning support tool in an urban-planning process. The authors used a case-study approach to examine the role of the tool through a conceptual framework of analysis that focused on the context, use, and impacts of the tool. They found that the tool supported dialogue and improved communication.

Agenda Setting

Two papers addressed "agenda setting" in addition to "information" and "consultation" [64,65]. In these papers, the authors focused on processes that involved citizens in their co-design in order to improve smart city services [64] and to increase their sense of ownership and to improve their quality of life and social cohesion [65].

Co-Management

Only five studies focused on the highest degree of citizen participation— "Co-management" —and included dimensions of training in different domains. The common goal of these processes was to upskill citizens to empower them to co-manage the planning process.

Russo et al. (2020) [68] aimed to construct collective knowledge, taking a further step into co-management by training citizens to develop new skills and raising aspiration and inspiration, thereby creating a collaborative circle among all partners. Afzalan and Muller (2018) [66] proposed building community consensus by cultivating trust and mobilizing citizen action and generating support. Zhong (2022) [70] describes a collaborative process in which different groups (district, neighborhood, community) and citizens were not only informed, consulted, and trained but also empowered to undertake parts of the reconstruction work themselves. Flacke et al. (2020) [67] introduced a process of public dialogue, social learning, and citizen training using an interactive collaborative tool to facilitate active participation and co-management to achieve policy goals.

Yoo (2021) [69] presents a case study of the Seoul smart city portal, whose main goals are participation, communication, cooperation, and urban problem-solving. Training citizens is discussed in terms of promoting creativity to foster technological vocations in children, actively training people in digital skills for future employment, and providing social, intergenerational, territorial, and gender-inclusion training. Citizens make suggestions regarding projects, attend events, and receive training on new technologies in smart cities. The portal also includes a 'Smart Citizen Lab' section to recruit citizens as experts to participate in finding solutions to urban problems. In addition, citizens can be integrated into evaluation teams for new technology product alternatives and participate in tutoring programs.

Multipurpose Studies

In addition to the above, eight multipurpose studies were identified. These were not focused on a specific participatory process or method but proposed evaluation models based on theoretical evidence [41,49,51,52,71–74].

In their studies, Foth (2016) [52] and Foth (2017) [51] discuss the concept of participation, co-creation, and public space in urban interaction design. The author introduces five possible ways to think about the "urban user" and the implications that follow, including the user as a city resident; the user as a consumer of city services; the user as a participant in the city's community consultations; the user as co-creator in a collaborative approach to city-making; and the user re-thought as part of a much larger and more complex ecosystem of more-than-human worlds and cohabitation. The article by Clemente, Civierob and Celluralea (2019) [71] focuses on the transition of urban districts to Positive Energy Districts (PEDs). The authors developed a framework to facilitate the evaluation of the positioning and improvement of each smart city solution considered in the study. The framework identified strategies and stakeholders' commitment to promote Smart Urban District or PED transitions, and identified implementation domains for PED transition and products and solutions. Ben (2020) [72] focuses on understanding how Internet of Things (IoT) technologies used to deliver services affect the working and living logistics of citizens. She conceptualizes a participatory approach to the design of smart city services, in which citizens are engaged from the start to implementation. Flacke, Shrestha, and Aguilar (2020) [49] reviewed the literature to discuss the use of interactive Participatory Sensing Systems (PSS) applications, tools, and participatory processes in order to analyze the impacts of these systems on participation. The results of the study showed that these systems can be used to increase community engagement, create a shared understanding of the issues under investigation, and allow participants to work through their interests. The article by Mohseni (2021) [73] focuses on the concept of smart cities and how they can potentially empower citizens. Based on a categorization of the available smart city definitions, the author presents a model which classifies smart cities based on their potential to empower citizens. The key factors influencing the improvement of the quality of smart urban life are explored by Reza, Bigdeli and Moinifar (2022) [74]. The results of the study indicate that the integration of dimensions such as health and hygiene, public education, security, social coherence and participation, consumption of resources and the environment, culture, and housing and buildings with the development of smart information and communications technology is the most significant contributory factor in improving the quality of smart urban life. The book written by Chilvers (2015) [41] explores how participation can be reconfigured and remade, particularly from a co-productionist standpoint. It examines the situated pragmatics of participation, considering how a co-productionist standpoint offers a different way of imagining participation and how it alters notions of reflexivity, learning, and critique.

4.3.2. Evaluation Method

The majority of the studies used a qualitative approach to evaluate public participation (n = 16). These are followed by mixed methods (n = 9) and quantitative methods (n = 7) (Figure 7). Three of the papers did not include a clear evaluation method (N/A) as these were providing reflections on the purpose of public participation.

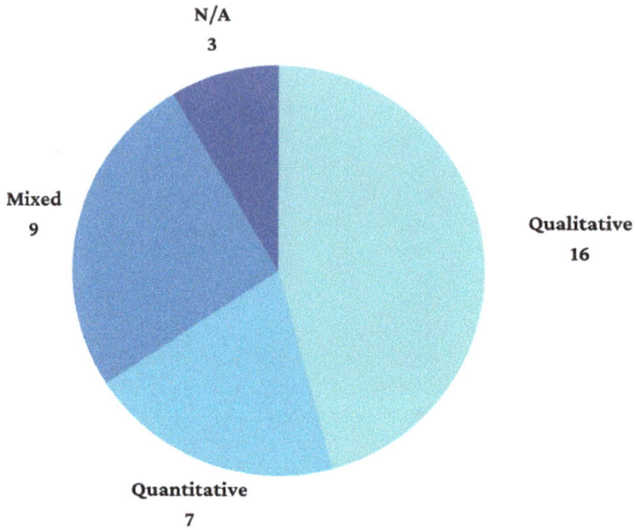

Figure 7. Evaluation method. Source: authors.

4.3.3. Impact on Participants

This section focuses on the purpose of evaluation, looking for indicators to assess the impact of public participation processes on the participants in the studies. "Impact on participants" encompasses what participants get out of the process (e.g., gained awareness, behavioral change, likeliness of future engagement). The main dimensions identified in these studies were "awareness", "change of behavior", "knowledge and skills", "participants' satisfaction", and "participants' attitudes" (Table 3).

Table 3. Impact on participants identified in the papers.

Type of Impact	Definition	Evaluation Indicators Identified in the Studies
Awareness (n = 6)	Participants and the general public are informed about the issue and realize what is at stake in this particular matter [34].	Level of awareness—pre- and post-experience [47]; community knowledge, local sense of urgency [59]; participants are informed and increase their knowledge about an issue [49,67]; raising awareness through community networks [65]; raising awareness on the concept of urban renewal and its implementation [58].
Behavioral change (n = 4)	Process that involves modifying or altering one's behavior or habits in response to internal or external cues, stimuli, or incentives. Behavioral change can be intentional or unintentional and can occur at individual, group, or societal levels.	Likeliness to change behavior based on the tour of the facility [47]; likeliness to change behavior based on the feedback provided by the tool [44]; participants change behavior to address a challenging situation [49,67].
Future engagement (n = 4)	The likelihood or willingness of individuals to participate in urban planning processes again in the future after their initial experience [75].	Actions to be incorporated in future engagement practices [1]; subjective assessments of deliberation quality and likelihood of participating in future participation processes [53]; community building: development of new collaborations, improved social cohesion [67]; integrating new approaches for further collaboration—casting a wider net across the top and bottom stakeholders and collectively bringing people together [65].

Table 3. *Cont.*

Type of Impact	Definition	Evaluation Indicators Identified in the Studies
Knowledge, skills, learning (n = 7)	In single-loop learning, participants seek to achieve their intended goals by making incremental improvements to their strategies and actions. In double-loop learning, participants question and potentially change their underlying assumptions, which leads to a deeper understanding of the causes of the problems and enables them to challenge and change established ways of thinking and operating [76].	Allows those who had minimal knowledge of city logistics to go up and carry out quality projects [68]; improved understanding of the problem [49,67]; preconditions for a successful process include a good learning situation [50]; actual and perceived knowledge gains regarding the process' subject matter [53]; appear to have contributed to learning through the provision of content and improved communication and interaction [63]; digital competency enhancement education: citizens can register and access online education computer resources, workshops, etc. and can review the history of the training they have received [69].
Satisfaction (n = 6)	Individuals' positive or negative impression of their experience with a particular initiative or participatory process [77].	Satisfaction of different groups concerning the public transportation system [56]; levels of satisfaction with the participation method [62]; participants' satisfaction with process and outcomes [49]; overall satisfaction of participating in post-disaster reconstruction [70]; satisfaction with the process [53]; user-satisfaction measures [54].
Attitudes (n = 4)	Attitudes toward public participation are shaped by factors such as trust in government, perceived risks and benefits, and individual values and beliefs [78].	Participants' attitudes toward technology [57]; no dominating person or group [49,67]; negative attitudes between NGOs and local administration [58].

The other papers discussed different dimensions of public participation, for example, the level of participation, the usability of digital platforms, expectations toward the process, outcomes of the process, and the level of interactivity between municipalities and citizens.

"Satisfaction" is mentioned in six of the studies. Here there is a difference between those examining the participants' satisfaction with the participatory process [49,53,70], the choice of method for participation [62], and the general public service delivery [56,79].

"Awareness" is also mentioned in six articles; five of them consider it as a dimension to monitor dynamically throughout the participation process [47,49,58,65,67], while the last one takes it as a starting point to assess the pre-existing awareness of the population on the risks of flood in the city [59]. This is justified by the focus of the study, which aims to evaluate a city's preparedness to face environmental risks related to climate change.

Out of 35 papers, only few studies were classified at the highest level of participation, namely, "Dialogue", "Agenda setting", and "Co-management" (Figure 6). The majority of the papers (65%) cover the lowest levels of participation (i.e., "Information" and "Consultation"). We find that awareness and behavioral change are focused upon at the information level; awareness, satisfaction, and attitudes at the consultation level; future engagement, knowledge, skills, and learning and satisfaction for those combining information and consultation levels. We also observe that within studies mentioning the highest level of participation, "Co-management", four of five papers discuss the impact on participants, but only one, the study by Flacke et al. (2020), mentions more than one dimension: awareness, future engagement, knowledge, skills, and learning and satisfaction.

5. Discussion

In this paper we set out to address the following two research questions: RQ1—What research has been carried out on the evaluation of public participation methods in smart cities? RQ2—In which ways do the public participation processes impact the participants? With these questions, we aim to identify evaluation frameworks that could provide indicators to assess whether the public participation processes succeed in empowering citizens and inspire a shift toward more sustainable behaviors, upskilling them and enhancing their understanding of sustainable urban development and fostering their commitment to achieving the United Nations' sustainability goals for climate-change mitigation in the urban context.

Most of the research that has been carried out regarding the evaluation of public participation in smart cities is concentrated on European cases, which can be understood by the intensified funding efforts of the European Union into research on smart cities. The predominant research methodology to evaluate public participation is qualitative approaches. Our review of the scientific literature on the evaluation of public participation in smart cities identified several limitations. Most of the studies did not actually measure the learning progress or outcomes achieved through public participation, and even though citizens were engaged, they were more informed and consulted than empowered. There is a lack of clear evaluation criteria and indicators, with evaluations mostly conducted through discussions and interviews with researchers without providing evidence of the implementation of the proposed evaluation frameworks in practice. Only one article presented how they measured learning outcomes for the citizens [67]. They showed that improved learning led to the results of the participation process being easily implemented as they were directly related to the needs and challenges of the urban area. Some studies only considered specific public participation methods, such as public displays, and planning support systems (PSS), limiting the applicability of their findings to other methods. Lastly, some studies proposed frameworks for the co-evaluation of smart city services or governance capacity in general, but with public participation being only one of the dimensions and not the main focus. Some of the studies broadened the traditional sampled population to involve artists, activists, older citizens, and NGOs, but we were not able to identify studies involving the next generation of smart citizens, i.e., children and youth.

This section is structured in a way that compares the level of participation achieved (from information to co-management) with the impact on the participants considered in the studies (awareness, behavioral change, future engagement, knowledge, skills and learning, satisfaction and attitudes). We seek to identify whether the level of participation that can be achieved relates to different impacts on participants considered.

Knowledge, skills, and learning are found at various levels of participation, both at the basic level of information and consultation [1,50,53], at the intermediate level of dialogue [63], and at the highest level of co-management [67–69]. Citizen empowerment is the common thread among the studies at the co-management level, as they utilize digital tools to not only empower citizens during the participatory process but also to enable them to acquire new knowledge and skills for future use. In the case of PSS tools and game simulations, these are not presented as "stand-alone" solutions as they are used during supervised sessions. Those tools allowed the improvement of the citizens' understanding of the issue at stake but were not developed to be intuitively used by citizens alone as they required the supervision of planners or other professionals. The digital platform presented by Yoo (2021) [69] supports the education and training of citizens in topics directly relevant to the implementation of smart city projects. In this case, citizens are not mere recipients of new technologies but take an active role in developing and implementing new solutions using the resources made available by the municipality.

Focusing solely on giving sufficient knowledge about the project and the process to participants will only allow them to reach the level of information and consultation. To reach the "Dialogue" level, the use of digital tools can help to facilitate the communication and collaboration between different stakeholders [63]. They found that the PSS tool played an important role in facilitating learning through the improved input and process. Flacke et al. (2020) [67] takes a step forward toward co-management by using a PSS tool for public participation that enables participants to understand the complexity of the planning process and to take an active role in it. They presented the benefits of communicating complex matters through visualization in order for citizens to understand the context and challenges related to a specific urban area. The tool enabled citizens to share their local knowledge and helped the planners to identify the most relevant locations for the projects. In the case of Russo et al. (2020) [68], the knowledge acquired by participants through gaming simulations allowed those with minimal knowledge to "go up and carry out quality projects" in city logistics.

The "knowledge and skills" dimension is the most represented one in the sampled literature. This indicates its importance in public participation in smart cities. However, most studies provide only recommendations and only a few give actual measurements or concrete methods for assessing increased or gained knowledge. The study by Yoo (2021) presents a case in which citizens can self-monitor their progress by keeping track of the training they have received on the digital platform. In their study, Hofmann et al. (2020) propose comparing perceived with actual knowledge gains in the process, which also relies on the participants' self-assessment and reflection. Overall, the knowledge and skills that participants gain during the participatory processes seem to be difficult to assess and insufficiently documented, despite being one of the dimensions to be accounted for in the effectiveness of public participation [34,80].

The scope of this review was limited to peer-reviewed articles, and gray literature and conference proceedings were excluded from the final sample. More recent studies, especially those conducted during the COVID-19 pandemic, are not represented due to the time-period limitation and the longer publishing time in certain journals. In addition, within the sampled literature, emerging countries are not sufficiently represented, which limits the practical implications of our scoping review mostly to developed countries. This is especially important to point out as the uptake of technologies is slower and the culture for public participation in planning is generally lower in these countries, and these can be, to some extent, tokenized by the political institutions [81,82]. These aspects should be factored in when looking to evaluate public participation processes in smart cities, especially those involving the use of digital technologies to encourage a shift toward more sustainable behaviors. In the context of emerging countries, the sole implementation of ICTs will not be sufficient if it does not go hand-in-hand with improved quality of life and access to satisfactory infrastructures and services [82].

6. Conclusions

This scoping literature review revealed the scarcity of scientific production on the evaluation of public participation in smart cities in general and a high representation of European smart cities. There is an underrepresentation of other world regions in the identified studies.

Within this body of literature, only a few studies presented a citizen-centric evaluation. Despite having identified some common criteria focused on the impact on participants, only a few studies provided clear indicators to be measured. More tools for actively involving citizens in decision-making processes in smart city projects are needed.

The main methodological limitation of the present scoping literature review is the exclusion of gray literature and conference proceedings. One of the resulting shortcomings is the absence of studies conducted during the COVID-19 pandemic period, during which e-services and e-participation methods developed rapidly in response to the global restrictions on physical interactions. More work is needed to understand how public participation is evaluated in practice beyond the conceptual frameworks developed in the scientific literature. Gamification and the use of serious games are gaining momentum in the literature on public participation, with studies demonstrating their potential to improve processes by breaking down the complexity of smart city development projects and engaging the younger public that tend to be underrepresented in traditional processes [83–86].

Furthermore, the influence of the COVID-19 pandemic on public participation processes should be investigated to reveal how the communication and knowledge transfer operated during this period.

This scoping review provides a valuable contribution to the literature by identifying research gaps and potential for further research in the evaluation of public participation in the specific context of smart city development. The predominant research methodology to evaluate public participation is based on qualitative approaches, and most studies do not provide a method for measuring the learning progress or outcomes achieved during the participatory process.

To address these limitations, further research is needed to investigate other world regions and to measure the learning progress and outcomes of public participation. Additionally, more empirical studies need to be conducted to reveal the most effective methods and tools to foster a positive culture for participation in communities and to improve their knowledge and understanding of sustainability issues in the urban context.

Author Contributions: Conceptualization, C.S. and A.T.S.; methodology, J.L.; validation, C.S., A.T.S., A.J. and J.L.; formal analysis, C.S. and A.T.S.; investigation C.S.; resources, C.S.; writing—original draft preparation, C.S. and A.T.S.; writing—review and editing, C.S., A.T.S., A.J. and J.L.; visualization, C.S.; supervision, A.T.S. and A.J.; funding acquisition, A.J. All authors have read and agreed to the published version of the manuscript.

Funding: This research was funded by Nordforsk, research project "CaPs-Citizens as Pilots of smart cities", grant number 95576.

Data Availability Statement: Data sharing is not applicable.

Acknowledgments: This study is supported by the Department of Civil and Environmental Engineering, Faculty of Engineering, Norwegian University of Science and Technology (NTNU).

Conflicts of Interest: The authors declare no conflict of interest.

References

1. Mangnus, A.C.; Vervoort, J.M.; Renger, W.; Nakic, V.; Rebel, K.T.; Driessen, P.P.J.; Hajer, M. Envisioning Alternatives in Pre-Structured Urban Sustainability Transformations: Too Late to Change the Future? *Cities* **2022**, *120*, 103466. [CrossRef]
2. Collins, D.; Johansen, A.; Kalsaas, B.T.; Temeljotov-Salaj, A.; Hamdy, M. Brought by Degrees: A Focus on the Current Indicators of Lean 'Smartness' in Smart Cities. In Proceedings of the 29th Annual Conference of the International Group for Lean Construction (IGLC), Lima, Peru, 14–17 July 2021; pp. 167–176. [CrossRef]
3. European Commission Climate-Neutral and Smart Cities. Available online: https://research-and-innovation.ec.europa.eu/funding/funding-opportunities/funding-programmes-and-open-calls/horizon-europe/eu-missions-horizon-europe/climate-neutral-and-smart-cities_en (accessed on 14 April 2023).
4. Caragliu, A.; del Bo, C.; Nijkamp, P. Smart Cities in Europe. *J. Urban Technol.* **2011**, *18*, 65–82. [CrossRef]
5. Schaffers, H.; Komninos, N.; Pallot, M.; Aguas, M.; Almirall, E.; Bakici, T.; Barroca, J.; Carter, D.; Corriou, M.; Fernadez, J. *Smart Cities as Innovation Ecosystems Sustained by the Future Internet*; HAL: Lyon, France, 2012.
6. Preston, S.; Mazhar, M.U.; Bull, R. Citizen Engagement for Co-Creating Low Carbon Smart Cities: Practical Lessons from Nottingham City Council in the UK. *Energies* **2020**, *13*, 6615. [CrossRef]
7. Schleicher, K.; Schmidt, C. Citizen Science in Germany as Research and Sustainability Education: Analysis of the Main Forms and Foci and Its Relation to the Sustainable Development Goals. *Sustainability* **2020**, *12*, 6044. [CrossRef]
8. Temeljotov Salaj, A.; Gohari, S.; Senior, C.; Xue, Y.; Lindkvist, C. An Interactive Tool for Citizens' Involvement in the Sustainable Regeneration. *Facilities* **2020**, *38*, 859–870. [CrossRef]
9. Jowkar, M.; Temeljotov-Salaj, A.; Lindkvist, C.M.; Støre-Valen, M. Sustainable Building Renovation in Residential Buildings: Barriers and Potential Motivations in Norwegian Culture. *Constr. Manag. Econ.* **2022**, *40*, 161–172. [CrossRef]
10. Lim, S.B.; Yigitcanlar, T. Participatory Governance of Smart Cities: Insights from e-Participation of Putrajaya and Petaling Jaya, Malaysia. *Smart Cities* **2022**, *5*, 71–89. [CrossRef]
11. Chen, F.-H. Assessing Sustainable Development Initiatives in Central Taiwan Science Park: A Study of Residents’ Opinions and the Impact on the Urban Ecosystem. *Buildings* **2023**, *13*, 1202. [CrossRef]
12. Khansari, N.; Mostashari, A.; Mansouri, M. Impacting Sustainable Behavior and Planning in Smart City. *Int. J. Sustain. Land Use Urban Plan.* **2014**, *1*, 46–61. [CrossRef]
13. Innes, J.E.; Booher, D.E. Reframing Public Participation: Strategies for the 21st Century. *Plan. Theory Pract.* **2007**, *5*, 419–436. [CrossRef]
14. Rowe, G.; Frewer, L.J. Public Participation Methods: A Framework for Evaluation. *Sci. Technol. Hum. Values* **2000**, *25*, 3–29. [CrossRef]
15. Reed, M.S. Stakeholder Participation for Environmental Management: A Literature Review. *Biol. Conserv.* **2008**, *141*, 2417–2431. [CrossRef]
16. Bovaird, T. Beyond Engagement and Participation: User and Community Coproduction of Public Services. *Public Adm. Rev.* **2007**, *67*, 846–860. [CrossRef]
17. Arnstein, S.R. A Ladder Of Citizen Participation. *J. Am. Inst. Plan.* **1969**, *35*, 216–224. [CrossRef]
18. Pretty, J.N. Participatory Learning for Sustainable Agriculture. *World Dev.* **1995**, *23*, 1247–1263. [CrossRef]

19. Ringholm, T.; Nyseth, T.; Sandkjær Hanssen, G. Participation According to the Law? *Eur. J. Spat. Dev.* **2018**, *67*, 1–20. [CrossRef]
20. Gordon, E.; Schirra, S.; Hollander, J. Immersive Planning: A Conceptual Model for Designing Public Participation with New Technologies. *Environ. Plan. B Urban Anal. City Sci.* **2011**, *38*, 505–520. [CrossRef]
21. Degbelo, A.; Granell, C.; Trilles, S.; Bhattacharya, D.; Casteleyn, S.; Kray, C. Opening up Smart Cities: Citizen-Centric Challenges and Opportunities from GIScience. *ISPRS Int. J. Geo-Inf.* **2016**, *5*, 16. [CrossRef]
22. Gohari, S.; Baer, D.; Nielsen, B.F.; Gilcher, E.; Situmorang, W.Z. Prevailing Approaches and Practices of Citizen Participation in Smart City Projects: Lessons from Trondheim, Norway. *Infrastructures* **2020**, *5*, 36. [CrossRef]
23. Berntzen, L.; Johannessen, M.R. The Role of Citizen Participation in Municipal Smart City Projects: Lessons Learned from Norway. In *Public Administration and Information Technology*; Gil-Garcia, J.R., Pardo, T.A., Nam, T., Eds.; Springer: Berlin/Heidelberg, Germany, 2016; Volume 11, pp. 299–314.
24. Simonofski, A.; Vallé, T.; Serral, E.; Wautelet, Y. Investigating Context Factors in Citizen Participation Strategies: A Comparative Analysis of Swedish and Belgian Smart Cities. *Int. J. Inf. Manag.* **2019**, *56*, 102011. [CrossRef]
25. Goodman, N.; Zwick, A.; Spicer, Z.; Carlsen, N. Public Engagement in Smart City Development: Lessons from Communities in Canada's Smart City Challenge. *Can. Geogr.* **2020**, *64*, 416–432. [CrossRef]
26. Hollands, R.G. Will the Real Smart City Please Stand up? Intelligent, Progressive or Entrepreneurial? *City* **2008**, *12*, 303–320. [CrossRef]
27. Gil-Garcia, J.R.; Pardo, T.A.; Nam, T. A Comprehensive View of the 21st Century City: Smartness as Technologies and Innovation in Urban Contexts. In *Smarter as the New Urban Agenda: A Comprehensive View of the 21st Century City*; Gil-Garcia, J.R., Pardo, T.A., Nam, T., Eds.; Public Administration and Informaiton Technology; Springer: Berlin/Heidelberg, Germany, 2016; Volume 11, pp. 1–19, ISBN 978-3-319-17620-8.
28. Biloria, N. From Smart to Empathic Cities. *Front. Archit. Res.* **2020**, *10*, 3–16. [CrossRef]
29. Nam, T.; Pardo, T.A. Smart City as Urban Innovation: Focusing on Management, Policy, and Context. In Proceedings of the 5th International Conference on Theory and Practice of Electronic Governance, Tallinn, Estonia, 26–29 September 2011; pp. 185–194. [CrossRef]
30. Yue, A.; Mao, C.; Chen, L.; Liu, Z.; Zhang, C.; Li, Z. Detecting Changes in Perceptions towards Smart City on Chinese Social Media: A Text Mining and Sentiment Analysis. *Buildings* **2022**, *12*, 1182. [CrossRef]
31. Israilidis, J.; Odusanya, K.; Mazhar, M.U. Exploring Knowledge Management Perspectives in Smart City Research: A Review and Future Research Agenda. *Int. J. Inf. Manag.* **2021**, *56*, 101989. [CrossRef]
32. Kopackova, H.; Komarkova, J.; Horak, O. Enhancing the Diffusion of E-Participation Tools in Smart Cities. *Cities* **2022**, *125*, 103640. [CrossRef]
33. Johansen, A.; Collins, D.; Temeljotov-Salaj, A.; Hagehaugen, G. Down by the Fjord: Successful Public and Private Collaboration in a Neighbourhood Redevelopment Project. *Manag. Procure. Law* **2022**, *40*, 1–10. [CrossRef]
34. Laurian, L.; Shaw, M.M. Evaluation of Public Participation: The Practices of Certified Planners. *J. Plan. Educ. Res.* **2009**, *28*, 293–309. [CrossRef]
35. Tricco, A.C.; Lillie, E.; Zarin, W.; O'Brien, K.K.; Colquhoun, H.; Levac, D.; Moher, D.; Peters, M.D.J.; Horsley, T.; Weeks, L.; et al. PRISMA Extension for Scoping Reviews (PRISMA-ScR): Checklist and Explanation. *Ann. Intern. Med.* **2018**, *169*, 467–473. [CrossRef]
36. Munn, Z.; Peters, M.D.J.; Stern, C.; Tufanaru, C.; McArthur, A.; Aromataris, E. Systematic Review or Scoping Review? Guidance for Authors When Choosing between a Systematic or Scoping Review Approach. *BMC Med. Res. Methodol.* **2018**, *18*, 1–7. [CrossRef]
37. Arksey, H.; O'Malley, L. Scoping Studies: Towards a Methodological Framework. *Int. J. Soc. Res. Methodol. Theory Pract.* **2005**, *8*, 19–32. [CrossRef]
38. Ullah, F.; Al-Turjman, F. A Conceptual Framework for Blockchain Smart Contract Adoption to Manage Real Estate Deals in Smart Cities. *Mach. Learn. Appl. Secur.* **2023**, *35*, 5033–5054. [CrossRef]
39. Moher, D.; Liberati, A.; Tetzlaff, J.; Altman, D.G.; Altman, D.; Antes, G.; Atkins, D.; Barbour, V.; Barrowman, N.; Berlin, J.A.; et al. Preferred Reporting Items for Systematic Reviews and Meta-Analyses: The PRISMA Statement. *PLoS Med.* **2009**, *6*, e1000097. [CrossRef] [PubMed]
40. Glaser, B.; Strauss, A. Applying Grounded Theory. In *The Discovery of Grounded Theory: Strategies of Qualitative Research*; Aldine de Gruyter: Berlin, Germany, 1967; p. 13.
41. Chilvers, J. *Remaking Participation*; Kearnes, M., Ed.; Routledge: Abingdon-on-Thames, UK, 2015; ISBN 9780415857390.
42. Lin, Y.; Kant, S. Using Social Media for Citizen Participation: Contexts, Empowerment, and Inclusion. *Sustainability* **2021**, *13*, 6635. [CrossRef]
43. Gudowsky, N.; Sotoudeh, M.; Capari, L.; Wil, H. Transdisciplinary Forward-Looking Agenda Setting for Age-Friendly, Human Centered Cities. *Futures* **2020**, *90*, 16–30. [CrossRef]
44. Francisco, A.; Taylor, J.E. Understanding Citizen Perspectives on Open Urban Energy Data through the Development and Testing of a Community Energy Feedback System. *Appl. Energy* **2019**, *256*, 113804. [CrossRef]
45. Psomadaki, O.I.; Dimoulas, C.A.; Kalliris, G.M.; Paschalidis, G. Digital Storytelling and Audience Engagement in Cultural Heritage Management: A Collaborative Model Based on the Digital City of Thessaloniki. *J. Cult. Herit.* **2019**, *36*, 12–22. [CrossRef]

46. Caragliu, A.; Del Bo, C.F. Smart Innovative Cities: The Impact of Smart City Policies on Urban Innovation. *Technol. Forecast. Soc. Change* **2019**, *142*, 373–383. [CrossRef]
47. Ambrose, A. Walking with Energy: Challenging Energy Invisibility and Connecting Citizens with Energy Futures through Participatory Research. *Futures* **2020**, *117*, 102528. [CrossRef]
48. Paskaleva, K.; Cooper, I. Technovation Open Innovation and the Evaluation of Internet-Enabled Public Services in Smart Cities. *Technovation* **2018**, *78*, 4–14. [CrossRef]
49. Flacke, J.; Shrestha, R.; Aguilar, R. Strengthening Participation Using Interactive Planning Support Systems: A Systematic Review. *ISPRS Int. J. Geo-Inf.* **2020**, *9*, 49. [CrossRef]
50. Billger, M.; Thuvander, L.; Wästberg, B.S. In Search of Visualization Challenges: The Development and Implementation of Visualization Tools for Supporting Dialogue in Urban Planning Processes. *Environ. Plan. B Urban Anal. City Sci.* **2017**, *44*, 1012–1035. [CrossRef]
51. Foth, M. Participation, Co-Creation, and Public Space. *J. Public Space* **2017**, *2*, 21–36. [CrossRef]
52. Foth, M. Participatory Urban Informatics: Towards Citizen-Ability. *Smart Sustain. Built Environ.* **2016**, *7*, 4–19. [CrossRef]
53. Hofmann, M.; Münster, S.; Noennig, J.R. A Theoretical Framework for the Evaluation of Massive Digital Participation Systems in Urban Planning. *J. Geovisualization Spat. Anal.* **2020**, *4*, 3. [CrossRef]
54. Castelnovo, W.; Misuraca, G.; Savoldelli, A. Smart Cities Governance: The Need for a Holistic Approach to Assessing Urban Participatory Policy Making. *Soc. Sci. Comput. Rev.* **2016**, *34*, 724–739. [CrossRef]
55. Shah, S.H.; Rodina, L.; Burt, J.M.; Gregr, E.J.; Chapman, M.; Williams, S.; Wilson, N.J.; Mcdowell, G. Unpacking Social-Ecological Transformations: Conceptual, Ethical and Methodological Insights. *Anthr. Rev.* **2018**, *5*, 250–265. [CrossRef]
56. Ghorbanzadeh, O.; Moslem, S.; Blaschke, T. Sustainable Urban Transport Planning Considering Different Stakeholder Groups by an Interval-AHP Decision Support Model. *Sustainability* **2019**, *11*, 9. [CrossRef]
57. Biermann, H.; Philipsen, R.; Brell, T.; Ziefle, M. User Perspectives, Expectations, and Challenges of Data and Information Distribution in Autonomous Driving. *Hum. Intell. Syst. Integr.* **2020**, *1*, 53–70. [CrossRef]
58. Bozdağ, A.; İnam, Ş. Collaborative Land Use Planning in Urban Renewal. *J. Urban Reg. Anal.* **2021**, *13*, 323–342. [CrossRef]
59. Brockhoff, R.C.; Koop, H.A.S.; Snel, A.W.K. Pluvial Flooding in Utrecht: On Its Way to a Floof-Proof City. *Water* **2019**, *11*, 1501. [CrossRef]
60. Mahou-lago, X.M.; Varela-álvarez, E.J. Innovation and Opportunities for Citizen Participation in Spanish Smart Cities. In *Smarter as the New Urban Agenda: A Comprehensive View of the 21st Century City*; Springer: Berlin/Heidelberg, Germany, 2016; pp. 367–392, ISBN 9783319176208.
61. Clarinval, A.; Simonofski, A.; Vanderose, B.; Dumas, B. Public Displays and Citizen Participation: A Systematic Literature Review and Research Agenda. *Transform. Gov. People Process Policy* **2021**, *15*, 1–35. [CrossRef]
62. Bouzguenda, I.; Fava, N.; Alalouch, C. Would 3D Digital Participatory Planning Improve Social Sustainability in Smart Cities? An Empirical Evaluation Study in Less-Advantaged Areas. *J. Urban Technol.* **2022**, *29*, 41–71. [CrossRef]
63. Mcevoy, S.; van de Ven, F.H.M.; Brolsma, R.; Slinger, J.H. Evaluating a Planning Support System's Use and E Ff Ects in Urban Adaptation: An Exploratory Case Study from Berlin, Germany. *Sustainability* **2020**, *12*, 173. [CrossRef]
64. Mainka, A.; Hartmann, S.; Meschede, C.; Stock, W.G. Open Government: Transforming Data into Value-Added City Services. In *Citizen's Right to the Digital City*; Springer: Berlin/Heidelberg, Germany, 2015. [CrossRef]
65. Fredericks, J. From Smart City to Smart Engagement: Exploring Digital and Physical Interactions for Playful City-Making. In *Making Smart Cities More Playable*; Springer: Berlin/Heidelberg, Germany, 2020. [CrossRef]
66. Afzalan, N.; Muller, B. Online Participatory Technologies: Opportunities and Challenges for Enriching Participatory Planning. *J. Am. Plan. Assoc.* **2018**, *84*, 162–177. [CrossRef]
67. Flacke, J.; de Boer, C.; Van den Bosch, H.; Pfeffer, K. *Handbook of Support Science*; Geertman, S., Stillwell, J., Eds.; Edward Elgar Publishing Inc.: Cheltenham, UK, 2020; ISBN 9781788971089.
68. Russo, F.; Calabrò, T.; Iiritano, G.; Pellicanò, D.S.; Petrungaro, G.; Sostenibile, E.; Studi, U. International Journal of Sustainable Development and Planning City Logistics Between International Vision and Local Knowledge to Sustainable Development: The Regional Role on Planning and on Public Engagement. *Int. J. Sustain. Dev. Plan.* **2020**, *15*, 619–629. [CrossRef]
69. Yoo, Y. Toward Sustainable Governance: Strategic Analysis of the Smart City Seoul Portal in Korea. *Sustainability* **2021**, *13*, 5886. [CrossRef]
70. Zhong, L.; Bai, L.; Xiang, M. Evaluation Index System for Public Participation in Post-Disaster Housing Reconstruction in Dujiangyan. *Open House Int.* **2022**, *47*, 107–121. [CrossRef]
71. Clemente, C.; Civierob, P.; Celluralea, M. Solutions and Services for Smart Sustainable Districts: Innovative Key Performance Indicators to Support Transition. *Int. J. Sustain. Energy Plan. Manag.* **2019**, *24*, 95–106.
72. Ben, E.R. Methodologies for a Participatory Design of IoT to Deliver Sustainable Public Services in "Smart Cities". In *Beyond Smart and Connected Governments: Sensors and the Internet of Things in the Public Sector*; Springer: Berlin/Heidelberg, Germany, 2020; pp. 49–68. [CrossRef]
73. Mohseni, H. Public Engagement and Smart City Definitions: A Classifying Model for the Evaluation of Citizen Power in 2025 Tehran. *GeoJournal* **2021**, *86*, 1261–1274. [CrossRef]
74. Reza, M.; Bigdeli, V.; Moinifar, M. Technological Forecasting & Social Change The Structural Model of Indicators for Evaluating the Quality of Urban Smart Living. *Technol. Forecast. Soc. Chang.* **2022**, *176*, 121427. [CrossRef]

75. Mannarini, T.; Talò, C. Evaluating Public Participation: Instruments and Implications for Citizen Involvement. *Community Dev.* **2013**, *44*, 239–256. [CrossRef]
76. Argyris, C. Single-Loop and Double-Loop Models in Research on Decision Making. *Adm. Sci. Q.* **1976**, *21*, 363–375. [CrossRef]
77. Boyer, E.J.; Van Slyke, D.M.; Rogers, J.D. An Empirical Examination of Public Involvement in Public-Private Partnerships: Qualifying the Benefits of Public Involvement in PPPs. *J. Public Adm. Res. Theory* **2016**, *26*, 45–61. [CrossRef]
78. Frewer, L. Risk Perception, Social Trust, and Public Participation in Strategic Decision Making: Implications for Emerging Technologies. *Ambio* **1999**, *28*, 569–574.
79. Castelnovo, W. Co-Production Makes Cities Smarter: Citizens' Participation in Smart City Initiatives. In *Co-Production in the Public Sector*; Fugini, M., Bracci, E., Sicilia, M., Eds.; Springer International Publishing: Berlin/Heidelberg, Germany, 2016; pp. 97–117, ISBN 9783319305585.
80. Innes, J.E.; Booher, D.E. Collaborative Rationality as a Strategy for Working with Wicked Problems. *Landsc. Urban Plan.* **2016**, *154*, 8–10. [CrossRef]
81. Malek, J.A.; Lim, S.B.; Yigitcanlar, T. Social Inclusion Indicators for Building Citizen-Centric Smart Cities: A Systematic Literature Review. *Sustainability* **2021**, *13*, 376. [CrossRef]
82. Alonso, F.; Faus, M.; Tormo, M.T.; Useche, S.A. Could Technology and Intelligent Transport Systems Help Improve Mobility in an Emerging Country? Challenges, Opportunities, Gaps and Other Evidence from the Caribbean. *Appl. Sci.* **2022**, *12*, 4759. [CrossRef]
83. Cavada, M.; Rogers, C.D.F. Serious Gaming as a Means of Facilitating Truly Smart Cities: A Narrative Review. *Behav. Inf. Technol.* **2019**, *39*, 695–710. [CrossRef]
84. Gabriel, S.; Schmölzer, B. Climate4Kids: A Gamified App Teaching about Climate Change. In Proceedings of the 16th European Conference on Game-Based Learning, Lisbon Portugal, 6–7 October 2022; pp. 236–243.
85. West, M.; Yildirim, O.; Harte, A.E.; Ramram, A.; Fleury, N.W.; Carabias, V. Enhancing Citizen Participation through Serious Games in Virtual Reality. In Proceedings of the 24th International Conference on Urban Planning, Regional Development and Information Society, Karlsruche, Germany, 2–4 April 2019; Volume 4, pp. 881–888.
86. Latifi, G.R.; Monfared, M.P.; Khojasteh, H.A. Gamification and Citizen Motivation and Vitality in Smart Cities: A Qualitative Meta-Analysis Study. *GeoJournal* **2022**, *87*, 1217–1230. [CrossRef]

urban science

MDPI

Communication

Planning on the Verge of AI, or AI on the Verge of Planning

Thomas W. Sanchez

Urban Affairs and Planning, Virginia Tech, Arlington, VA 22203, USA; tom.sanchez@vt.edu

Abstract: The urban planning process is complex, involving social, economic, environmental, and political systems. Knowledge of how these systems interact is the domain of professional planners. Advances in artificial intelligence (AI) present planners with a ripe opportunity to critically assess their approaches and explore how new data collection, analysis, and methods can augment the understanding of places as they seek to anticipate futures with improved quality of life. AI can offer access to more and better information about travel patterns, energy consumption, land utilization, and environmental impacts, while also helping to better integrate these systems, which is what planners do. The adoption process will likely be gradual and involve significant time and resources. This article highlights several topics and issues that should be considered during this process. It is argued that planners will be well-served by approaching AI tools in a strategic manner that involves the topics discussed here.

Keywords: urban planning; artificial intelligence; expert systems

1. Introduction

Research into artificial intelligence (AI) for urban planning began in the 1960s, influenced by the development of computer technologies with a broad range of scientific and industrial applications [1]. The development and implementation of planning applications, such as transportation and land use forecasting systems, however, were significantly limited by the lack of large-scale datasets and computing capabilities. Over time, academic and industrial research using advanced quantitative and spatial analyses commonly associated with AI has increased steadily; however, few of these have been adopted by the planning profession. While much of the current planning-related discussion concerning AI is related to "smart city" technologies that are being used to capture and analyze data for optimization processes, relatively less attention is being paid to urban planning and associated decision-making activities. Examples of this include scenario planning and generative designs. As global populations become increasingly urban, planning and management of these places are essential for sustainability, resilience, and equity in both the short and long term.

Just as urban planning gradually adopted computer technologies during the 1980s and GIS during the 1990s, the profession is now on the inevitable progression toward AI-augmented systems. The development and adoption of these technologies will likely need to be encouraged by universities where faculty are researching AI and where AI-related coursework will become integrated into planning curricula. Planning educators and professional organizations are on the verge of another wave of technological change. Yet, how will this current phase differ from the 1970s and 1980s when AI-related tools were being developed but were then not welcomed into planning practice? The primary differences today are changes in data availability, the pervasiveness of computer technology, and the general adoption of AI in many facets of daily life. The reluctance of the profession to accept technological change and the lack of appropriate education and training contributes to what Geertman referred to as the "implementation gap" [2] (p. 70).

This is an important time to assess current planning practices and how (or whether) they can benefit from emerging technologies such as artificial intelligence (AI). Methods focused on pattern recognition (e.g., machine learning) are a particular area where urban planning can greatly benefit, especially with enormous quantities of urban data being collected and analyzed [3]. Procedures used by planners should be examined to identify opportunities for improvement and steps required for adoption. A challenge involved with this is eliciting expert knowledge, particularly for issues including quality of life, design, engineering, health, and law [4,5]. Planners are not the sole experts on urban processes and will also rely on technical expertise such as computer science and information technologies to generate solutions. An emphasis on technical expertise alone is insufficient in a planning context because planners also facilitate collective decision making and navigate political landscapes. Understanding the mix of urban expertise, politics, and power plays an important role in the further adoption of technology by the planning profession where expertise can support hierarchical structures that avoid transparency and inclusiveness [6]. Add to this the different types of planning expertise and experiential knowledge that are difficult to document and operationalize [7], and leveraging AI technologies within the planning profession becomes challenging.

This article begins by introducing AI and how it is entering the public sphere. It then goes on to briefly describe the complexity of urban planning and future cities, which can be seen as challenges for the adoption of AI tools. The role of expertise is then discussed as it relates to the planning process—particularly because expert systems have been tried in the past and may offer some useful insight into how we consider this next attempt at building intelligence into planning systems. Finally, the article highlights key challenges that land use planners will be confronted with as they move toward AI adoption. The objective is to call attention to some strategic matters that need to be considered along the way.

2. What Is AI?

AI can be defined as using computers to mimic or improve upon human intelligence, such as reasoning and experience-based learning. Although it has been used by computer scientists for some time, AI is now associated with a wide range of consumer goods and services that we use every day, from product design to marketing to customer feedback. To do this, many disciplines contribute to the process. AI employs methods from probability theory, economics, and algorithm design. In addition, computer science, mathematics, psychology, and linguistics are all used in AI development [8].

AI is also starting to impact governance and policymaking. Governments all over the world are utilizing AI technology to influence their responses to some of the most pressing problems of our time, such as the coronavirus pandemic [9], climate change [10], and the implementation of new data laws and governance structures [11]. At the same time, data laws concerning individual privacy are of significant concern prompting citizen privacy protection laws like the General Data Protection Regulation (GDPR) in the European Union (EU) [12]. Organizations at different levels of government are using AI-related tools to address the many challenges they face. Public sector organizations contribute to the delivery of services to citizens, such as healthcare and education and the creation of smarter and greener communities. However, challenges persist for the public sector in the face of a wave of new technologies (and their acronyms). Cities and counties are outlining their future information technology (including AI) initiatives through plans and strategies that highlight the costs and benefits of implementation (see, for example, [13–16]).

While AI refers to a very broad range of analytical methods, the primary elements involve a sequence of inputs, analysis, and outputs (Figure 1). An AI program takes inputs of a large quantity of data, analyzes it according to a set of rules, and produces outputs, which can take a variety of forms depending on the types of inputs and types of questions being addressed. What makes AI "intelligent" is its ability to use results from the analysis and outputs to update or improve the overall process: to learn. Planners might note that this logic is not unlike the basic process of planning practice—gathering information about an issue or a challenge, analyzing it to answer important questions, and producing plans or policies that improve the situation or address that challenge.

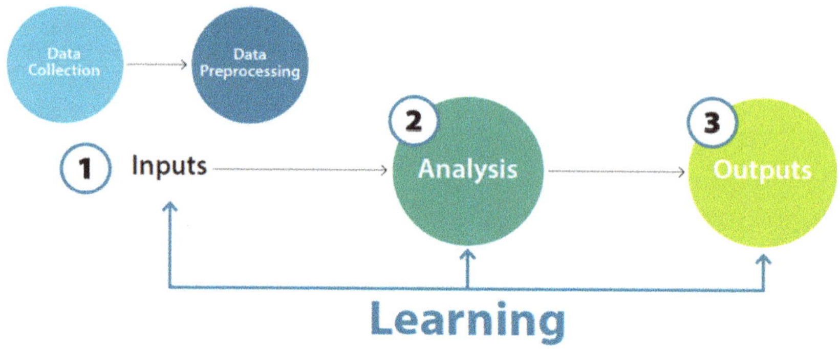

Figure 1. The basic analytical process for AI (source: Author).

Decades of AI research by computer scientists have focused on understanding the complexities of human thought processes and logic. The objective has been not only to understand how humans think but also to develop the means to improve upon it in a way that could be applied to nearly anything that involves intelligence and decision making [1]. AI is now associated with a wide range of consumer goods and services that we use every day, from product design to marketing to customer feedback.

Planning the cities of the future will require appropriate expertise and proper planning methods. There is general agreement that urbanization will continue to increase as a function of economic opportunity, while at the same time equity, environmental pressures, and infrastructure needs will continue to be major challenges. In addition, technology, connectedness, and digitalization are poised to accelerate at a rapid pace [17–20]. AI is expected to play an important role in future development activities as well as in the planning and management of cities for sustainability, resilience, and equity in both the short and long term. Planners must be prepared for the changes this means in how the cities of the future are planned, designed, and managed.

In addition to learning some of these AI-specific skills, planners will increasingly need to be conversant in other urban technology applications, such as those being used to create smart cities. Technology investments in infrastructure, the internet of things (IoT), and other technologies will transform how urban places function and therefore will be factored into how cities are planned and managed [21]. To stay relevant in a changing world, planners will need to know about these technologies and be prepared to engage with other disciplines involved in shaping the urban landscape, such as civil engineers, architects, landscape architects, and public administrators.

3. The Complexity of Urban Planning

The complex nature of human settlement has been long recognized. In discussing the future of cities, Harris [22] highlighted the fundamental challenges of urban planning that intermingle design, science, and policy—especially as they relate to understanding and optimizing urban systems. The mix of social, economic, environmental, and political aspects means that planning problems are sometimes considered wicked [23]. In addition, Wildavsky's [24] perspective on planning's expansiveness and perhaps overly ambitious reach are important to contemplate relative to planning's difficult-to-measure benefits [24]. It is useful to consider these pre-technological (i.e., pre-computerization) views of the complexity of urban planning (like those of Harris, Wildavsky, Rittel, Weber, and others) because they focus on the fundamental objectives and roles of planning and planners.

Dealing with a complex and diverse problem set requires solutions drawn from multiple domains that likely encompass various methodologies and sources of expertise. Given this caveat, one question is whether the urban planner is indeed the expert to be relied upon to understand how planning should occur [25]. Hendler [26] and Vigar [27] argued that planners possess a unique knowledge base and a definable set of expertise.

Others argue that planners' areas of expertise are represented by their skills as "network managers" to coordinate fragmented and assorted disciplines involved with managing the urban development process [28] (p. 245). Planning's venture into AI will require collaboration with domain experts, as did the moves into computerization (with computer scientists) and GIS (with geographers). The unique location of planning at the nexus of technical and political problems and the potentially fragmented nature of relevant expertise poses two major challenges to AI adoption in urban planning practice.

It is also important to consider what future cities will be like as we think about new approaches to planning these cities. The issues to be confronted will require appropriate expertise and proper planning methods. Well, we can only speculate about the condition and needs of cities in the next 30, 50, or 100 years, but there is general agreement that urbanization will continue to increase as a function of economic opportunity, while at the same time equity, environmental pressures, and infrastructure needs will continue to be major challenges. In addition, it is expected that technology, connectedness, and digitalization will continue to rapidly increase in importance [12,16–20]. AI is expected to play a role in future development activities as well as in the planning future.

4. Role of Expertise

Considering the multi-dimensional nature of urban planning, it is important to also ask, "What is 'to plan'? What do we need to plan for? Whose knowledge is necessary to plan?-which leads to "What do (professional) planners need to know?" [29] (p. 92). While these seem like large philosophical questions, they lay a necessary foundation for selecting planning tasks and operations that have the greatest potential for improvement through the application of AI technologies. Just as for any expert system, this requires an inventory of planning processes, knowledge, and expertise. Alexander [29] suggests that the kinds of knowledge needed for urban planning include (a) theory, (b) methods and skills, (c) judgment and good sense, (d) normative, and (e) substantive. It is easy to recognize from this mix of knowledge types that only some urban planning tasks fit into these categories. Hayes-Roth and Hayes-Roth [30] describe planning as the "predetermination of a course of action aimed at achieving some goals" (p. 275). While their definition is not specific to urban planning, it implies that designing processes (i.e., courses of action) and problem solving are principal aspects of planning.

The reality is that planning is not purely knowledge-driven but requires imagination in approaching challenging situations that have not been previously encountered and, therefore, involves significant uncertainty. In terms of connecting planning with AI, this highlights an issue stemming from the limited imagination of AI. For instance, machine learning (ML) relies on historical data, but the future may include innovations that fundamentally change problem spaces and cannot be predicted from even perfect knowledge of the past. Along with imagination, the element of common sense is another problem area for AI in planning, being a nearly boundless type of knowledge. As Mitchell notes, a prerequisite to trustworthy moral reasoning is generally common sense: what is harm, what are cause and effect, and counterfactuals [31]. Perhaps "common sense" is not the proper terminology in this case because it implies subjectivity that is strongly influenced by culture and socioeconomic status. Then again, perhaps it is appropriate because it suggests that an AI would need to have some awareness of cultural and socioeconomic factors to perform many of the tasks currently performed by urban planners.

From a knowledge acquisition perspective, common sense and tacit knowledge are closely aligned and are frequently difficult to capture and operationalize. Replicating or augmenting the network of urban planning tasks first requires the identification of relevant knowledge sources, both expert and non-expert, which assumes that needed information and experience can be explicated to create courses of action to solve urban planning problems. Along with the expert question is the need to incorporate stakeholder feedback. Because urban planning has a strong public orientation, stakeholders such as constituents or residents are commonly involved to provide feedback on objectives and outcomes but rarely in a process that involves highly technical expert knowledge. In cases of "technocratic models of practice", the stakeholder cannot participate meaningfully and is effectively excluded [32] (p. 578). Sorting out the expert and stakeholder balance of power remains a challenge that planners must contend with, as there are and will be questions of transparency involved [33].

It is unlikely that the whole of planning will be replaced by AI-based rules and automation. So where will AI be most appropriate and how will the right planning tasks be selected? Batty [34] characterized "routine and nonroutine" (p. 1) aspects of planning to distinguish the variety of planning tasks in terms of frequency, scale, and intensity. This has implications for the selection of tasks, application of technologies, and anticipated benefits. Regularly executed tasks (routine planning) potentially generate more opportunities for knowledge accumulation, so long as adequate acquisition and evaluation mechanisms are in place. With repetition, experts develop "a skill of recognition, of 'seeing' old patterns in the new problem" [35] (p. 152). Therefore, developing an artificial expert represents a potential savings of time, effort, and cost. Under the current model, each generation of human planning experts needs to be trained and accumulate knowledge and experience over time. This is not to say that nonroutine functions cannot benefit from AI, but in general, they provide fewer learning opportunities and are operationalized with longer-term anticipated efficiency gains. The combination of task frequency and scale is also associated with increased data volume. This is important because machine learning methods tend to need substantial amounts of data before they can achieve acceptable levels of accuracy and reliability.

5. Expert Systems in Urban Planning

Expert systems in urban planning are computer applications that apply AI to narrowly defined tasks or functions [36]. As previously mentioned, the expected benefits of expert systems include efficiency gains through time savings and improved decision making. An enhanced understanding of tasks through the examination of procedural steps can help to codify expertise for transparency and knowledge management [37]. Significant attention was given to this form of AI during earlier periods of urban modeling and expert systems from the 1960s to the late 1980s. On one hand, the planner designs the modeling process based on an understanding of real-world dynamics, while drawing upon historical and cross-sectional data that explains interactions for predictive purposes. A missing element is the ongoing knowledge capture that provides dynamic sources of feedback to these planning processes, particularly as planners enter the profession and then retire, as alluded to previously. Academic research can be seen as playing a role but usually lacks the specificity of planning tasks, particularly related to experiential expertise. Sources of information about urban dynamics come from the past or represent current conditions and interrelationships as understood by experts. Both natural and artificial (i.e., urban) systems are complex, with interrelated actors and processes. However, modelers of urban systems cannot rely on the same long histories of scientific exploration that underpin their natural counterparts [38]. The application of these techniques to urban processes is still relatively new.

Identifying the parts or tasks of planning that can benefit most from AI involves understanding the intentions and knowledge of planners with direct experience. This process has been discussed in the context of expert systems, both inside and outside of planning. To identify planning tasks that are appropriately handled by expert systems, Waterman, Silverman, and Goodall [39–41] identified seven characteristics to guide the selection process. They suggest the following:

1. Genuine experts exist who can articulate their (problem-solving) methods;
2. Experts agree on solutions;
3. The task is not poorly understood;
4. The problem typically takes a few minutes to a few hours to solve;
5. No controversy over problem domain rules exists;

6. The problem is clearly specifiable and well bounded;
7. The problem solving should be judgmental, not numerical.

Han and Kim [42] argue that few urban planning tasks meet these criteria, particularly because of the complexity of urban places and human behavior mentioned earlier. In addition, referring to general decision making and cybernetics, Thomas [43] identified four criticisms that also challenged the appropriateness of expert systems for planning:

1. It is impossible to evaluate all the alternatives;
2. Goals are not always agreed upon and fixed in advance;
3. Real decision making was not like this;
4. Claims for objective neutrality were just excuses for methods without context that sidestepped political issues in planning (p. 383).

The first objection is not as relevant as compared to years ago because of improved AI technologies. For example, there is now little brute force on decision trees in AI chess programs or Go engines. Instead, they use Monte Carlo simulation and other methods to approximate probabilities of actions [44]. The same approach holds for urban planning actions. Goals and problem solving are often a question of optimization than arriving at a solution, especially because urban systems rarely reach equilibrium. However, references to real decision making and objectivity are perhaps less relevant in the context of more current technological applications. Along with the challenge of selecting tasks is also the definition and selection of planning problems to be addressed [45]. Prioritizing problems and tasks may likely be an iterative process as each task is better understood not only individually, but also relative to the planning process. To design expert systems or other AI applications, the process of knowledge acquisition or knowledge engineering is an essential first step [34,46,47]. This is where planning practice can benefit from collaborating with experts to understand these processes and seek to develop ontologies to inform the design of applications.

6. Connecting Research and Practice

Urban planning is no exception to the tension seen in many professions between theory and practice. Researchers are frequently more interested in characterizing problems, gathering information, and developing theories rather than finding actionable solutions [48]. Planning literature is rife with instances of how academic work either fails to influence day-to-day planning decisions or ignores political realities [49]. Planning academics are encouraged to conduct more action-oriented research, to ground their findings in real planning problems, to strive to understand what planners do, and to learn from practice to close this gap between research and practice-oriented problem solving [50,51]. At the same time, planning professionals are encouraged to use academic research as a source for new perspectives and approaches. The point is that academics and planning professionals should reevaluate the potential role that research can play—more specifically, the function of evidence and its place in planning practice. There is a sizable amount of research on AI applications and represents an opportunity for practice and research to connect.

7. Potential Impacts of AI on the Planning Process

Concerns have been expressed that separating planning tasks can result in specialties that inhibit "articulating a synthetic understanding of urban society" [52] (p. 211). This can force planners to take a reductionist view of urban development by performing analysis on specific questions that can more easily involve technological solutions such as computer modeling. These pieces then need to be reassembled as part of the overall plan-making process. On the one hand, it seems that a planner's focus on technical expertise can place them in a more objective position to advise decision making [53]. On the other hand, this assumes that planning tasks are modular and that each can be integrated to represent parts or the whole of urban systems. This is where the technical planner takes on a scientific role to reach a type of certainty while creating "computer simulation[s] of relations whose consequences are imperfectly understood" [54] (p. 42). Should a planner be knowledgeable across these domains or are we asking too much?

However, the adoption of new methods and technologies is usually less disruptive than might be anticipated, because it is usually a gradual process that occurs incrementally. Consider geographic information systems (GIS), possibly the most significant technological advancement for planners of the past few decades. The use of GIS has progressed from the creation and updating of maps by specialized GIS technicians to the relatively sophisticated spatial analysis now available to anyone with access to a smartphone and GIS applications. This has taken place incrementally over many years and has positively transformed the practice of planning with minimal negative disruptive impacts. The adoption of AI by planners could follow a similar path.

AI is a set of many tools, and it is already integrated into many types of software, including word processing, email, spreadsheets, and GIS. While there will likely be new standalone AI applications or platforms developed for planners, most will be layered onto existing software or easy-to-access web-based platforms. However, like statistical methods used for making predictions, planners will need to know the underlying assumptions and be able to communicate the reliability of results. In some cases, new approaches to planning problems may be very different from those used in the past. These methods will involve new training and educational efforts.

8. Three Challenges to Implementation

Three challenges will impact how quickly, and efficiently urban planners will implement AI. Each of these will be briefly discussed to highlight how they play a role in the adoption of AI technology. These include (1) the need for new skills, (2) changing data needs, and (3) incorporating transparency.

8.1. The Need for New Skills

The digitalization of urban planning requires that planners learn new skills. However, AI is not the only driver of this change. Today's workplace is in a constant state of change with new technologies, changes in workplace culture, and evolving business practices. For many professions, the COVID-19 pandemic required the rapid adoption of video conferencing technologies and protocols to accommodate remote work, and electronic alternatives quickly replaced in-person contact. By necessity, we managed to learn new technologies quickly, some of which replaced long-standing previous practices. Updating skill sets to keep up with such changes is becoming a given.

The adoption of AI will differ among planners depending on their roles and responsibilities. It may include learning a new vocabulary of AI-related terminology, new software packages, programming languages, advanced statistical methods, and other AI-related concepts as discussed in this report. Upskilling to learn and maintain these skills will be an important element of adopting these new methods [55]. In addition to learning some of these AI-specific skills, planners will increasingly need to be conversant in other urban technology applications, such as those being used to create smart cities and "urban digital twins" [56]. Technology investments in infrastructure, the internet of things (IoT), and other technologies will transform how urban places function.

In addition, though most planners will not be directly involved with developing new AI applications, they may be engaging with technologists, such as computer scientists and application developers, to help them do so. As new urban technologies are being implemented, planners need to have seats at the table to represent the interests of community stakeholders as well as promote connections to better understand urban futures.

The readiness of planners to adapt and best use these new technologies is a crucial factor at this point. There are at least three factors involved in this: (1) the ability of AI now and in the future to have a significant impact on urban social and constructed environments, (2) the ability of AI now and in the future to meet planners' needs, and (3) the readiness and ability of planners to use AI technologies. The first of these is supported by research, and the second will emerge as we see planners embracing new tools. Last, but not least, whether planners employ AI in their work will be greatly influenced by their existing degree of awareness and expertise in the field.

The successful implementation of these new methodologies will require new expertise in data analytic techniques and information systems, which has immediate implications for planning practice. Both internal and external elements, such as the knowledge and skill set of new employees (i.e., young planners), will play a role in the adoption process for planning organizations. Upskilling might take place through training exercises for experienced planners, perhaps led by newly hired employees who bring these abilities into an organization.

The adoption of new digital technology also entails investments in new computer equipment and systems, as well as human resources. It is important to evaluate the potential benefit, which will also depend on how many current procedures will be disrupted or altered.

8.2. Changing Data Needs

Urban planning has a tradition of being data-hungry for the many types of analysis that are used in the plan-making process [57], and data are critical for effective AI implementation. Three primary factors are data volume, quality, and management.

One of the biggest obstacles to AI implementation is the lack of sufficient high-quality data [58]. Complicated algorithms can require thousands or millions of observations. A model risks poor performance when additional data are added if it is initially trained on inadequate amounts of data. Planning often relies on information about people, households, and drivers, as well as their behaviors, which usually come from several sources. Planners must often combine federal, state, and local data. In some cases, the data will need to be anonymized, which is standard practice for data collected by the U.S. Census. This includes approaches such as "data swapping", which transfers data points from one observation to another, increasing security but potentially decreasing the value of the data for analysis to draw conclusions [59].

Potentially using more finely grained disaggregated data for AI models means running the risk of infringing on personal privacy rules that will constrain the availability of data. In the United States, the General Data Protection Regulation (GDPR), Payment Card Industry Data Security Standard (PCI DSS), Health Insurance Portability and Accountability Act (HIPAA), Federal Information Security Management Act of 2002 (FISMA), Family Educational Rights and Privacy Act (FERPA), and Gramm–Leach–Bliley Act are just a few of the laws that forbid the use of confidential or sensitive information (i.e., personally identifiable information, or PII) as input for machine-learning models [60]. While urban planners have not often used a credit card, healthcare, or educational data in the past, the increasing ubiquity of data collection and its commercialization will require ethical decisions about the appropriateness of certain data that may reveal personal identities.

Data quality and integrity can be a challenge as well. Any data that a machine-learning model reads will be used to inform its assumptions. If the data are inaccurate or incomplete, the rules produced will be fundamentally flawed. Training data must accurately reflect the population by including data sets that represent every demographic category. According to several studies, dark-skinned females are mistakenly identified by facial recognition 40 percent more frequently than white males [60]. Nonrepresentative training data sets are mostly to blame for this. Even if the data are inclusive of all demographic groups, it may still suffer from being incomplete or out of date [61].

Compounding this challenge is that many planners are not effectively trained in the basic elements of data management. Without effective IT management practices, planning organizations can find it difficult to answer basic questions, such as how many databases an organization has, which database contains a particular piece of information, or how data were originally collected. Within organizations, siloed functional groups and poor communication create challenges for sharing data resources with coworkers, particularly for policymakers and administrators. Given that AI techniques rely heavily on data, these are serious issues. Organizations that lack the tools to understand and manage their data will find it more difficult to benefit from AI.

Public agencies struggle to deploy and manage AI systems because their staff lack fundamental data management skills [62]. Data governance and management practices should extend beyond IT departments to planning staff so they can better use and protect data assets. Though the learning curve for data administration is manageable, it is far more difficult to acquire the necessary abilities to create AI solutions. Planning organizations may need to prioritize the recruitment of planners with AI expertise.

8.3. Transparency and Explainability

Transparency in the public realm can be a challenge, and it has several facets. For AI, the main priority for planners should be being open about the use of AI for making predictions, recommendations, or decisions [63]. Transparency also involves enabling users to understand the elements of an AI system. The ability to deliver clear, relevant information regarding the outputs of an AI system and the justification for its use is another example of transparency, as is the facilitation of open, multistakeholder conversations and the creation of specialized organizations, where required, to promote public understanding and acceptance of AI systems.

Because AI and machine-learning models may be too technically challenging to be practical or helpful for understanding an output, transparency generally does not include disclosing source code or sharing proprietary datasets. Source code and datasets could also be considered intellectual property, which has specific legal protections.

Enabling those who will be affected by the output from an AI-based decision to understand how it was reached is referred to as explainability. This requires giving stakeholders simple-to-understand information that will allow those who are negatively impacted to contest outcomes, and, when possible, the causes and logic that led to those outcomes. For some AI systems, requiring explainability may harm accuracy, performance, privacy, and security, as it may necessitate condensing the solution variables to a set small enough for humans to understand, which may not be optimal in complex, high-dimensional problems. However, this will not likely be the case for many planning-related analyses.

The main factors in a decision; the data, logic, or algorithm behind the specific outcome; or an explanation of why similar-looking circumstances generated different outcomes may all be included when AI actors explain an outcome in clear and simple terms, as appropriate to the context. If applicable, personal data protection standards should be respected while allowing people to understand and contest the conclusion.

Related to transparency, AI can assist in identifying and minimizing the effects of human biases, but it also has the potential to exacerbate the issue by systematically introducing biases into sensitive application domains. Algorithms are susceptible to bias in several ways, even when sensitive factors like gender, ethnicity, or sexual orientation are accounted for [64]. Even large amounts of accumulated training data can be incomplete or inaccurate, reflecting previous poor decisions or biased analyses resulting from historical conditions.

We are all susceptible to and responsible for combating bias. Not only does bias harm those discriminated against, but it also harms everyone else by limiting people's participation in the economy and society. As a result of fostering distrust and delivering skewed results, bias lowers AI's potential for use in government, business, and society in general.

As complex as deep learning models can be, however, it should be noted that the human brain is the ultimate "black box". Humans can exaggerate or not even be aware of the reasons why they made a particular decision or selected a specific alternative. It may be nearly impossible to detect the source or type of bias in a human brain, while the logic and data used by computers can be carefully analyzed.

Solutions to combating bias involve expanding academic study regarding AI prejudice in urban planning and public policy and carefully considering the many ways that AI may supplement our current methods of making decisions. We can improve AI decision making by using models that are meticulously constructed to avoid past biases. It is also important to immediately resolve any instances we witness of bias in AI. There may be no easy fixes in these cases, however. Defining and assessing "fairness" is exceedingly difficult. Researchers have devised technical definitions of fairness, such as mandating that models have similar outcome values across socioeconomic groups [65]. Different fairness criteria typically cannot be satisfied at the same time, which is a considerable challenge.

It is also necessary to determine when a system is deemed fair for use by deciding under what circumstances automated decision making can be permitted [66]. In some cases, human-in-the-loop algorithms or AI system responses, which include human intervention or review, will be needed to maintain control or oversight, especially in unusual circumstances that may not have significant machine intelligence to draw upon. These issues call for interdisciplinary approaches from planners, engineers, designers, ethicists, and social scientists. This can be especially tricky in the case of planning, where political forces play a role in decision making.

Some basic steps are recommended to address potential ethical concerns relating to bias. Planners should keep up to date on the ethical dimensions of rapidly evolving urban technologies, including AI, which is being continuously uncovered as more and more such applications are being implemented. In addition, planners and planning organizations can create accountable procedures that help reduce prejudice when they use AI, including the use of technical tools or operational techniques such as oversight committees or external evaluations.

9. Conclusions

AI is an emerging technology that is here to stay. As such, though planners and planning organizations will face challenges in adopting this new technology, the benefits of AI may be worth the risk. As AI becomes more commonly used and useful over time, the challenges of adoption will lessen, just as they have in other cases of innovation adoption. AI will play a significant role in the future operation and planning of cities. As more and more data are being generated all around us, urban planners will increasingly use AI to analyze and process this data to detect patterns, make predictions, and have a better understanding of urban dynamics. However, AI will not take over all aspects of planning, and neither will it replace planners. Rather, it will comprise a useful set of tools for a subset of planning tasks. The most suitable tasks for AI will be those that are repetitive, have quantifiable dimensions, and involve data that can be captured and maintained. While there may be concerns about the loss of entry-level positions that typically handle these tasks, shifting needs will be accommodated by changes in skills that young planners have already acquired.

Organizations such as the American Planning Association have acknowledged the accelerating pace of AI application development, particularly as it relates to the planning profession [8]. There are both substantial opportunities and challenges presented by these technologies, with more being revealed each day. AI can help planners in their work, enhance current planning procedures, increase efficiency, and allow planners to refocus their work on the human components of planning. The task at hand is to increase our knowledge and skills with appropriate AI methods to enhance planning practice in the service of creating more livable, resilient, and sustainable communities.

The promises and potential of AI are both intriguing and absorbing, but organizations must carefully prepare before investing resources in AI. It is important to prepare for change, and some interruption is to be expected. We are accustomed to our routines; therefore, getting staff to accept AI may be challenging. As described in this article, it is important to develop a thorough understanding of what AI offers an organization and be strategic in thinking through the potential challenges. Once an organization adopts AI, it is also important to track performance and evaluate the value and advantages of AI deployment to document a comprehensive understanding of how AI benefits the organization, staff, and stakeholders. In addition, throughout the process of adopting AI, planners should be aware of the potential for ethical concerns related to its use as new types of data and analysis emerge for AI applications.

Finally, the future of AI holds transformative potential for urban planners, introducing tools that can significantly enhance decision making, optimize resource allocation, and improve overall livability. The integration of AI can lead to smart, responsive, and sustainable cities by predicting urban growth, analyzing traffic patterns, and assisting in planning sustainable infrastructure. Key considerations include the ethical use of data, maintaining privacy, and ensuring equitable access to the benefits of AI. As we move towards this AI-assisted future, it is crucial for urban planners to continuously engage in interdisciplinary dialogues, collaborate with AI experts, and involve the public to ensure that AI-based urban planning is transparent, accountable and, above all, serves the needs of all.

Funding: This research was partially supported by the National Science Foundation, Grant No. 2125259.

Acknowledgments: The author thanks the three reviewers for their very insightful comments and suggestions. He would also like to thank Trey Gordner and Theo Lim for their comments on previous versions of the article.

Conflicts of Interest: The author declares no conflict of interest.

References

1. Batty, M. The emergence and evolution of urban AI. *AI Soc.* **2022**, *38*, 1045–1048. [CrossRef]
2. Geertman, S. PSS: Beyond the implementation gap. *Transp. Res. Part A Policy Pract.* **2017**, *104*, 70–76. [CrossRef]
3. Sanchez, T.W.; Shumway, H.; Gordner, T.; Lim, T. The prospects of artificial intelligence in urban planning. *Int. J. Urban Sci.* **2022**, *27*, 179–194. [CrossRef]
4. Wang, M.; Zhou, T. Does smart city implementation improve the subjective quality of life? Evidence from China. *Technol. Soc.* **2023**, *72*, 102161. [CrossRef]
5. Han, Z.; Xia, T.; Xi, Y.; Li, Y. Healthy Cities, A comprehensive dataset for environmental determinants of health in England cities. *Sci. Data* **2023**, *10*, 165. [CrossRef]
6. Robin, E. The Politics of Urban Expertise. Ph.D. Thesis, University College London, London, UK, 2019.
7. Khakee, A.; Barbanente, A.; Borri, D. Expert and experiential knowledge in planning. *J. Oper. Res. Soc.* **2000**, *51*, 776–788. [CrossRef]
8. Sanchez, T.W. *Planning with Artificial Intelligence*, Planning Advisory Service Report 604; American Planning Association: Chicago, IL, USA, 2023.
9. Organisation for Economic Co-Operation and Development (OECD). Using Artificial Intelligence to Help Combat COVID-19. 2020. Available online: https://www.oecd.org/coronavirus/policy-responses/using-artificial-intelligence-to-help-combat-covid-19-ae4c5c21/ (accessed on 11 May 2023).
10. Dimock, W.C. What AI Can Do for Climate Change, and What Climate Change Can Do for AI. *Scientific American*, 5 April 2022. Available online: https://www.scientificamerican.com/article/what-ai-can-do-for-climate-change-and-what-climate-change-can-do-for-ai/ (accessed on 11 May 2023).
11. Rahnama, H.; Pentland, A. The New Rules of Data Privacy. Harvard Business Review. 2022. Available online: https://hbr.org/2022/02/the-new-rules-of-data-privacy (accessed on 24 December 2022).
12. European Commission. 2018 Reform of EU Data Protection Rules. 2018. Available online: https://commission.europa.eu/law/law-topic/data-protection/reform_en (accessed on 29 October 2022).
13. Barcelona City Council. Government Measures for Municipal Algorithms and Data Strategy for an Ethical Promotion of Artificial Intelligence. 2021. Available online: https://ajuntament.barcelona.cat/digital/sites/default/files/mesura_de_govern_intel_ligencia_artificial_eng.pdf (accessed on 1 November 2022).
14. The City of Los Angeles. SmartLA 2028: Technology for a Better Los Angeles. 2020. Available online: https://ita.lacity.org/smartla2028 (accessed on 15 October 2022).
15. Parikh, N.; Hohman, A. *NYC Artificial Intelligence Strategy*; The City of New York, Mayor's Office of the Chief Technology Officer: New York, NY, USA, 2021. Available online: https://nparikh.org/assets/pdf/nyc/nyc_ai_strategy.pdf (accessed on 13 February 2023).
16. County of Santa Clara, Technology Services and Solutions. FY22–24 Strategic Plan. 2021. Available online: https://it.sccgov.org/home (accessed on 11 May 2023).
17. Woetzel, J.; Rajadhyaksha, V.; Frem, J. Thriving Amid Turbulence: Imagining the Cities of the Future. McKinsey & Co. 2018. Available online: https://www.mckinsey.com/industries/public-and-social-sector/our-insights/thriving-amid-turbulence-imagining-the-cities-of-the-future (accessed on 2 October 2022).

18. National League of Cities. Future of Cities. 2022. Available online: https://www.nlc.org/initiative/future-of-cities/ (accessed on 5 October 2022).
19. Antunes, M.E.; Barroca, J.G.; de Oliveira, D.G. Urban Future with a Purpose: 12 Trends Shaping the Future of Cities by 2030. 2021. Available online: https://www2.deloitte.com/global/en/pages/public-sector/articles/urban-future-with-a-purpose.html (accessed on 11 October 2022).
20. U.N. Habitat. *World Cities Report 2022: Envisaging the Future of Cities*; United Nations Human Settlements Programme: Nairobi, Kenya, 2022. Available online: https://unhabitat.org/world-cities-report-2022-envisaging-the-future-of-cities (accessed on 2 October 2022).
21. Son, T.H.; Weedon, Z.; Yigitcanlar, T.; Sanchez, T.; Corchado, J.M.; Mehmood, R. Algorithmic urban planning for smart and sustainable development: Systematic review of the literature. *Sustain. Cities Soc.* **2023**, *94*, 104562. [CrossRef]
22. Harris, B. The city of the future: The problem of optimal design. In *Papers of the Regional Science Association*; Springer: Berlin/Heidelberg, Germany, 1976; Volume 19, pp. 184–195.
23. Rittel, H.W.; Webber, M.M. Dilemmas in a general theory of planning. *Policy Sci.* **1973**, *4*, 155–169. [CrossRef]
24. Wildavsky, A. If planning is everything, maybe it's nothing. *Policy Sci.* **1973**, *4*, 127–153. [CrossRef]
25. Pinson, D. Urban planning: An 'undisciplined discipline'? *Futures* **2004**, *36*, 503–513. [CrossRef]
26. Hendler, S. Do professional codes legitimate planners' values? In *Dilemmas of Planning Practice*; Thomas, H., Healey, P., Eds.; Aldershot: Avebury, UK, 1991.
27. Vigar, G. Planning and professionalism: Knowledge, judgment and expertise in English planning. *Plan. Theory* **2012**, *11*, 361–378. [CrossRef]
28. Sehested, K. Urban planners as network managers and metagovernors. *Plan. Theory Pract.* **2009**, *10*, 245–263. [CrossRef]
29. Alexander, E.R. What do planners need to know? Identifying needed competencies, methods, and skills. *J. Archit. Plan. Res.* **2005**, 91–106.
30. Hayes-Roth, B.; Hayes-Roth, F. A cognitive model of planning. *Cogn. Sci.* **1979**, *3*, 275–310. [CrossRef]
31. Mitchell, M. *Artificial Intelligence: A Guide for Thinking Humans*; Farrar, Straus and Giroux: New York, NY, USA, 2019; p. 127.
32. Goodspeed, R. Digital knowledge technologies in planning practice: From black boxes to media for collaborative inquiry. *Plan. Theory Pract.* **2016**, *17*, 577–600. [CrossRef]
33. Fischer, F. *Citizens, Experts and the Environment: The Politics of Local Knowledge*; Duke University Press: Durham, NC, USA, 2000.
34. Batty, M. Planning support systems and the new logic of computation. *Reg. Dev. Dialogue* **1995**, *16*, 1–17.
35. Olson, J.R.; Rueter, H.H. Extracting expertise from experts: Methods for knowledge acquisition. *Expert Syst.* **1987**, *4*, 152–168. [CrossRef]
36. Ortolano, L.; Perman, C.D. A planner's introduction to expert systems. *J. Am. Plan. Assoc.* **1987**, *53*, 98–103. [CrossRef]
37. Feigenbaum, E.; McCorduck, P.; Nii, H.P. *The Rise of the Expert Company*; Times Books: New York, NY, USA, 1988.
38. Dueker, K.J. Urban planning uses of computing. *Comput. Environ. Urban Syst.* **1982**, *7*, 59–64. [CrossRef]
39. Waterman, D.A. *A Guide to Expert Systems*; Addison-Wesley Longman Publishing Co., Inc.: Boston, MA, USA, 1985.
40. Silverman, B.G. Should a manager "hire" an expert system? In *Expert Systems for Business*; Boyd & Fraser Publishing Company: San Francisco, CA, USA, 1987; pp. 1–4.
41. Goodall, A. *Guide to Expert Systems*; Learned Information (Europe) Ltd.: London, UK, 1985.
42. Han, S.Y.; Kim, T.J. Can expert systems help with planning? *J. Am. Plan. Assoc.* **1989**, *55*, 296–308. [CrossRef]
43. Thomas, M.J. The procedural planning theory of A Faludi. *Plan. Outlook* **1979**, *22*, 72–77. [CrossRef]
44. Silver, D.; Hubert, T.; Schrittwieser, J.; Antonoglou, I.; Lai, M.; Guez, A.; Lanctot, M.; Sifre, L.; Kumaran, D.; Graepel, T.; et al. A general reinforcement learning algorithm that masters chess, shogi, and Go through self-play. *Science* **2018**, *362*, 1140–1144. [CrossRef] [PubMed]
45. Prerau, D.S. Knowledge acquisition in the development of a large expert system. *AI Mag.* **1987**, *8*, 43.
46. Hua, J. Study on knowledge acquisition techniques. In Proceedings of the 2008 Second International Symposium on Intelligent Information Technology Application, Shanghai, China, 20–22 December 2008; Volume 1, pp. 181–185.
47. Cooke, N.J. Varieties of knowledge elicitation techniques. *Int. J. Hum. Comput. Stud.* **1994**, *41*, 801–849. [CrossRef]
48. Baum, H.S. Teaching Practice. *J. Plan. Educ. Res.* **1997**, *17*, 21–29. [CrossRef]
49. Sager, T.; Ravlum, I.A. The political relevance of planners' analysis: The case of a parliamentary standing committee. *Plan. Theory* **2005**, *4*, 33–65. [CrossRef]
50. Forester, J. Reflections on the future understanding of planning practice. *Int. Plan. Stud.* **1999**, *4*, 175–193. [CrossRef]
51. Ozawa, C.P.; Seltzer, E.P. Taking our bearings: Mapping a relationship among planning practice, theory, and education. *J. Plan. Educ. Res.* **1999**, *18*, 257–266. [CrossRef]
52. Beauregard, R.A. Bringing the city back in. American Planning Association. *J. Am. Plan. Assoc.* **1990**, *56*, 210.
53. Checkoway, B. Paul Davidoff and advocacy planning in retrospect. *J. Am. Plan. Assoc.* **1994**, *60*, 139–143. [CrossRef]
54. Dyckman, J.W. The scientific world of the city planners. *Am. Behav. Sci.* **1963**, *6*, 40–44. [CrossRef]
55. Clint, A.; Cooke, K.; Gomez, A.; Hurtado, P.; Thomas, W.; Sanchez, S.S.; Wright, N. *AI in Planning Opportunities and Challenges and How to Prepare Conclusions and Recommendations from APA's "AI in Planning" Foresight Community*; American Planning Association: Chicago, IL, USA, 2022.

56. Ye, X.; Newman, G.; Lee, C.; Van Zandt, S.; Jourdan, D. Toward Urban Artificial Intelligence for Developing Justice-Oriented Smart Cities. *J. Plan. Educ. Res.* **2023**, *43*, 6–7. [CrossRef]
57. Boeing, G.; Batty, M.; Jiang, S.; Schweitzer, L. Urban analytics: History, trajectory, and critique. *arXiv* **2021**, arXiv:2105.07020. [CrossRef]
58. Roh, Y.; Heo, G.; Whang, S.E. A survey on data collection for machine learning: A big data-ai integration perspective. *IEEE Trans. Knowl. Data Eng.* **2019**, *33*, 1328–1347. [CrossRef]
59. U.S. Bureau of the Census. Comparing Differential Privacy with Older Disclosure Avoidance Methods. 2021. Available online: https://www.census.gov/content/dam/Census/library/factsheets/2021/comparing-differential-privacy-with-older-disclosure-avoidance-methods.pdf (accessed on 27 February 2023).
60. Moallem, A. (Ed.) *Human-Computer Interaction and Cybersecurity Handbook*; CRC Press: Boca Raton, FL, USA, 2018.
61. U.S. Government Accountability Office. *Facial Recognition Technology: Privacy and Accuracy Issues Related to Commercial Uses*; GAO-20–522; U.S. Government Accountability Office: Washington, DC, USA, 2020.
62. Sarker, M.N.I.; Wu, M.; Hossin, M.A. Smart Governance Through Big Data: Digital Transformation of Public Agencies. In Proceedings of the 2018 International Conference On Artificial Intelligence And Big Data (ICAIBD), Chengdu, China, 26–28 May 2018; pp. 62–70.
63. Gaur, L.; Sahoo, B.M. Introduction to Explainable AI and Intelligent Transportation. In *Explainable Artificial Intelligence for Intelligent Transportation Systems*; Springer: Cham, Switzerland, 2022. [CrossRef]
64. Roselli, D.; Matthews, J.; Talagala, N. Managing bias in AI. In Proceedings of the Companion 2019 World Wide Web Conference, Montreal, QC, Canada, 25–31 May 2019; pp. 539–544.
65. Corbett-Davies, S.; Goel, S. The measure and mismeasure of fairness: A critical review of fair machine learning. *arXiv* **2018**, arXiv:1808.00023.
66. Araujo, T.; Helberger, N.; Kruikemeier, S.; De Vreese, C.H. In AI we trust? Perceptions about automated decision-making by artificial intelligence. *AI Soc.* **2020**, *35*, 611–623. [CrossRef]

smart cities

MDPI

Article

Strengthening Urban Resilience: Understanding the Interdependencies of Outer Space and Strategic Planning for Sustainable Smart Environments

Ulpia-Elena Botezatu [1,2,*], Olga Bucovetchi [2,3,*], Adrian V. Gheorghe [4] and Radu D. Stanciu [3]

1 Romanian Space Agency, 010362 Bucharest, Romania
2 National Institute for Research & Development in Informatics—ICI Bucharest, 011555 Bucharest, Romania
3 Faculty of Entrepreneurship, Business Engineering and Management, National University of Science and Technology "Politehnica" Bucharest, 060042 Bucharest, Romania; radu.stanciu@upb.ro
4 Department of Engineering Management and Systems Engineering, Old Dominion University, Norfolk, VA 23529, USA; agheorgh@odu.edu
* Correspondence: ulpia.botezatu@rosa.ro (U.-E.B.); olga.bucovetchi@upb.ro (O.B.)

Abstract: The conventional approach to urban planning has predominantly focused on horizontal dimensions, disregarding the potential risks originating from outer space. This paper aims to initiate a discourse on the vertical dimension of cities, which is influenced by outer space, as an essential element of strategic urban planning. Through an examination of a highly disruptive incident in outer space involving a collision between the Iridium 33 and Cosmos 2251 satellites, this article elucidates the intricate interdependencies between urban areas and outer space infrastructure and services. Leveraging the principles of critical infrastructure protection, which bridge the urban and outer space domains, and employing simulation methods and software, this study articulates the intricate governance complexities of urban security and presents viable solutions for its enhancement. Consequently, the study contributes to the ongoing deliberations regarding the spatial integration of security practices by providing scholarly discourse on urban governance with potential strategies for cultivating sustainable smart cities. In essence, the intrinsic resilience of urban areas heavily relies on the interconnections between cities and outer space, necessitating urban strategists to acknowledge and comprehend these intricate interdependencies. To ensure sustainable urban development, it is imperative to fortify smart cities' resilience against space debris through the implementation of more stringent regulations.

Keywords: strategic urban planning; smart city; ICT advancements; critical outer space infrastructure; governance of security; complex systems

1. Introduction: Strategic Urban Planning and Development

Have we passed the "point of no return" regarding the development of our (smart) cities? Damocles sword is increasingly hanging over humanity, underscored not just by accelerating urbanization and environmental degradation but, as this paper argues, also by vulnerabilities originating from outer space. While information and communication technology (ICT) advancements have transformed urban landscapes, they have also created a web of dependencies tied to outer space infrastructure, such as satellites. These space assets are vulnerable to natural and human-made disturbances, from solar storms to satellite collisions, which can have cascading effects on essential services on Earth, such as communication, transportation, and even national security.

The necessity to integrate outer space considerations into strategic urban planning is no longer optional; it is imperative. Failing to address this vertical dimension can put at risk the sustainable and smart development that urban planners aim to achieve. This paper argues that understanding the intricate interdependencies between Earth-bound urban environments and outer space elements is critical for shaping resilient and sustainable smart cities.

As technological advancements continue to progress, the emergence of "smart cities" and the transition towards more efficient and sustainable urban living have become evident. Yet, distinguishing between today's cities and their historical predecessors is essential. How has the notion of a "smart city" evolved over the course of history, and what factors have contributed to its evolution? Additionally, how do present-day cities compare to ancient cities in terms of their infrastructure, technology, and urban design?

Throughout history, the significance of well-thought-out urban planning has been evident, reaching as far back as ancient civilizations. Numerous age-old metropolises featured painstakingly planned architecture, focusing on spatial layout, essential infrastructure, and effective governance. The primary goal for such historic urban planning was to meet community needs, with aspects such as livability, safety, and protection valued more than aesthetic design principles. Consequently, the evolution of ancient urban centers was often directed by environmental factors such as their proximity to clean water, food provisions, and important trade routes. However, even in those distant times, individuals emerged who displayed remarkable foresight and ingenuity in the realm of urban planning.

The concept of organized street grids can be traced back to antiquity, with notable examples having been documented by *The Economist* [1] (p. 21). Cities such as Mohenjo-daro in Pakistan, dating back to 2600 BC, and the visionary plans of Hippodamus of Miletus in the 5th century BC exemplify early instances of grid-planned urban development. These designs featured square blocks arranged in a rectangular grid surrounding a fortified core, which offered defensive capabilities and facilitated orderly urban life. Greek and Roman cities, including Athens, also showcased systematic street networks and public spaces, emphasizing the importance of functional and sustainable urban environments. Similarly, other ancient civilizations, such as the Mayans, Egyptians, and Mesopotamians, embraced strategic urban planning, placing a high value on thoughtful design.

During the Middle Ages, European urban planners adopted radial city designs characterized by roads spreading out from central plazas and fortifications of solid walls and towers. Filarete, a Florentine architect and urban planner from the 15th century, is renowned for his design of an ideal fortress city with a star-shaped layout, converging roads leading to a central square dominated by a church, and fortified walls [2]. These symmetrical and orderly designs incorporated central plazas, wide avenues, and grand public buildings.

In more modern times, urban grid designs have seen a resurgence, primarily driven by economic considerations [1] (p. 21). This revival can be traced back to historical factors such as the Law of the Indies [3], which mandated grid-based planning resembling European models in Spanish colonies across South America and Asia. Shortly after, in North America, Thomas Jefferson's land ordinance of 1785 further influenced the development of new American cities, advocating for structured grids in unexplored western lands. The grid pattern was not only effective for efficient land parcel sales, thereby helping to generate government revenue, but also advantageous for transportation. Specifically, the design facilitated the implementation of tram networks, enabling swift movement between emerging developments and the city center with just two tram lines running on orthogonal axes [1] (pp. 21–22). Grid designs also offer benefits such as effective traffic segregation [4], allowing for the separation of pedestrians and vehicles, and accommodating varying speeds and traffic volumes, thereby improving traffic flow and reducing pollution.

However, a debate between grid and radial city designs emerged in the 20th century. Advocates such as Ebenezer Howard [5] proposed the concept of small radial "garden cities" with curved streets and a clear separation between industrial and residential areas. Renowned architects such as Le Corbusier [6] applied these principles in the design of new cities, exemplified by, for instance, Chandigarh, India. Cul-de-sacs became prevalent in many new settlements worldwide, offering the advantage of limiting through traffic but in turn disadvantaging non-residents [7,8]. This debate exposes the larger issue faced by urban planners: the trade-off between comfort and cost, or quality of life and economic considerations, which is ultimately a false dilemma, as quality of life always comes at a price.

Urban areas have also experienced significant population growth, not only in developed countries but also in less urbanized regions such as African countries [9]. Critics such as Jane Jacobs [10] argue that outdated city infrastructure, including street grids, poses limitations, with short blocks and narrow roads contributing to congestion. Building a new city from scratch, as seen in the case of Brasilia, Brazil's capital inaugurated in 1960 and designed by architect Oscar Niemeyer, presents an opportunity to explore alternative strategic approaches to urban planning that go beyond traditional grid or radial designs. Cul-de-sacs alone are insufficient for effective city planning and strategy development [11], prompting the need for innovative thinking and design in shaping the cities of the future.

Addressing the Gap, Strengthening Urban Resilience

The notion of urban resilience is a multifaceted subject that has garnered significant scholarly interest, often concentrated on terrestrial factors such as climatic fluctuations, extreme meteorological occurrences, and issues of cybersecurity [12–21]. Nevertheless, the field has generally overlooked the intricate dimensions of extraterrestrial vulnerabilities, specifically those related to space debris collisions, that bear upon smart cities [22].

Existing investigations into urban resilience provide limited insights into the cascading effects within urban infrastructure and services [23,24], especially the second- and third-order repercussions. This study aims to rectify this shortcoming by creating an interdisciplinary dialogue between research into the sustainability of extraterrestrial activities and the resilience and operational efficacy of technologically advanced urban settings.

The concept of resilience is gaining increasing relevance in the discourse on smart cities [25,26], particularly concerning urban systems' capacity to recover from both intentional and inadvertent types of disruptions. This manuscript aims to augment the existing corpus by introducing an under-explored facet: the role of space-based technological networks. Specifically, we scrutinize smart cities, interlinked through space technologies and services, as highly complex and adaptive entities susceptible to extraterrestrial disturbances such as space debris collisions.

Building on extant scholarship [27,28], we propose a nuanced definition of urban resilience that is tailored to smart cities under threat from space debris and satellite mishaps. Here, urban resilience refers to the adaptability of a smart city's socio-technical and orbital networks in maintaining or swiftly recuperating optimal functionality when faced with disruptions emanating from space. This expanded conceptualization allows for the incorporation of multiple adaptive strategies, such as contingency planning and system redundancy. It also underscores the complex interplay between urban centers and orbital activities, advocating for resilient governance structures involving various stakeholders, ranging from urban planners to space agencies.

As for policy implications, much of the existing research [29,30] falls short of addressing the unique challenges that extraterrestrial threats pose to smart cities. Although emerging studies have started to shed light on the significance of space debris as a factor in urban planning [22,28], comprehensive assessments detailing their precise impacts on smart cities are still lacking. This study aims to bridge this academic void by investigating the effects of space debris, whether resulting from satellite collisions or other means, on the resilience and efficient operation of smart cities.

2. ICT Advancements and Smart Cities

The rapid growth of information and communication technology (ICT) infrastructure has provided new ways to manage the challenges of urbanization, particularly in developing countries. As cities transform into more efficient and livable spaces, ICT will continue to play a crucial role. Artificial intelligence and complex algorithms are now used to optimize networks for transportation, communication, services, and utilities. These technologies specifically address issues such as improper waste management, water and sewage problems, pollution, poor traffic control, and energy grid failures—all issues that are exacerbated by the world's rising population. Consequently, city-level decision making has become increasingly guided by key performance indicators that focus on these technological interconnections [31,32].

On the other hand, emergent concepts such as "sustainable development" and "smart growth" have gained momentum in the international arena, with governments needing to adhere to sets of rules aimed at finding localized solutions for eco-friendly urban places. In 1987, the Brundtland Report defined "sustainable development" as development that "meets the needs of the present without compromising the ability of future generations to meet their own needs" [33]. The study promoted the idea of "sustainable development" and helped mainstream "sustainability" in policy discourse. These two concepts have been used interchangeably ever since; nevertheless, to distinguish between them, UNESCO formulated a distinction as follows [33,34]: "Sustainability is often thought of as a long-term goal... while sustainable development refers to the many processes and pathways to achieve it". Later, Agenda 21 claimed that the economic, social, and environmental dimensions are inseparable when discussing sustainability [34]. Finally, the 2030 Agenda for Sustainable Development put forth a list of 17 goals and 169 targets to balance "the three dimensions of sustainable development, the economic, social and environmental" [34]. Regardless, strategic thinking in these areas is enriched with concepts and methods based on agility, flexibility, and the integration of various technologies, such as artificial intelligence, machine learning, and other algorithmic ways of reducing uncertainty and calculating managerial efficiency.

From this, it is clear that a sustainable smart city is one that is designed to address its social, environmental, and economic impacts through urban planning and city management, incorporating in the process various forms of ICT [35]. However, whereas the literature on smart cities focuses, on the one hand, on framing dimensions (i.e., smart mobility, smart environments, smart people, smart living, smart governance, and smart economies [36]) and, on the other hand, on parameters of technological intelligence, the domain of outer space and its connection with smart cities has been left untouched. Moreover, the implications of such a novel manner of thinking have not at all been considered.

While there has been some scholarly attention on the role of extensive data analytics in ensuring the effective governance of smart urban cities [37], the literature has, overall, neglected to consider the connections between events originating in outer space and the performance metrics of tech-savvy municipalities. However, a few recent studies are notable exceptions [22,28]. Furthermore, the influence of outer space disruptive events on the technological scaffolding that modern smart cities rely upon is an area that is virtually untouched in the current literature. This manuscript aims to contribute towards filling this void by emphasizing the vital importance of outer space sustainability due to its direct repercussions on earth-bound activities, and particularly within the context of smart cities. By employing an analytical lens that starts by focusing on distinct, space-born disruptions, this investigation aims to forge an integrative dialogue between technology and sustainability, offering nuanced insights into strategic urban planning.

2.1. Related Work on Technology and Urbanization

The transition from traditional cities to smart cities involves the integration of technology and data to enhance the efficiency, sustainability, and quality of life in urban areas. This transition is driven by the need to address the challenges facing modern cities, such as urbanization, climate change, and resource depletion, as well as to improve the delivery of public services and infrastructure.

In recent scholarly discourse, the evolution from conventional urban centers to technology-enabled smart cities has been a focal point of investigation. A recurrent theme in academic debates centers on the pivotal role of community involvement and civic participation in shaping the trajectory of smart urban initiatives. As articulated in Anthony M.

Townsend's groundbreaking work [38], the journey towards smart urbanism is not solely reliant on technological advancements; it also demands the reconfiguration of governance structures, civic responsibilities, and planning methodologies. His work underscores the capability of smart urban environments to democratize citizen agency, thereby encouraging a more interactive form of democratic engagement. Some scholars [39–41] contend that participatory funding methods, such as civic crowdfunding, can serve as robust mechanisms for actively involving the populace in both the financing and practical execution of smart city ventures. Their studies elucidate the capacity of civic crowdfunding to foster transparent, accountable, and community-driven processes.

In the field of smart cities, numerous authors have made seminal contributions that span various dimensions of this evolving paradigm. For instance, urban planner Carlo Ratti has been instrumental in advocating for citizen-centric smart city models [42,43] and has contributed to significant initiatives, such as MIT's Sensable City Lab.

The existing literature offers profound insights into the role technology can play in elevating urban efficiency and sustainability. Researchers such as Zanella [44] have studied a range of technologies, including, but not limited to, wireless sensing technologies, cloud infrastructure, and big data analytics. Such authors have underlined these technologies' ability to optimize public services, curb energy usage, and heighten civic participation.

Governance and policy frameworks also form a crucial axis of discussion in existing research. Various scholars [45–49] have dissected the governance structures involved in smart city initiatives, stressing the need for multi-stakeholder collaboration, explicit policy directives, and definitive regulatory guidelines.

Additional key figures in this field include Martin Dodge and Rob Kitchin, who have focused on the intersections between technology, data, and the urban environment [50,51]. Further, sociologist Saskia Sassen [52] has offered nuanced views on the interplay between globalization, urbanization, and how smart cities could mitigate socio-economic disparities. Adam Greenfield [53] has highlighted the potential pitfalls and ethical considerations surrounding smart cities, particularly regarding data privacy and surveillance issues. Geographer Jennifer Clark [54] has explored the nexus between innovation, urbanization, and economic stimulation, while sociologist Manuel Castells [55] has investigated the broader societal impacts of global digital transformations on urban ecosystems.

The transition to intelligent urban ecosystems represents a complex and multi-layered endeavor that requires the proactive engagement of a broad variety of participants—ranging from governmental agencies and industrial actors to community members and civic organizations. Achieving the promise of smart cities mandates an integrative methodology, blending expertise from fields as varied as urban development, engineering, informatics, and social sciences. Current scholarly contributions to the study of smart cities indicate a rising fascination with technology's capacity to enhance urban sustainability, invigorate civic participation, and reframe governance structures. This heightened focus highlights the pivotal role technology will play in guiding the urban future towards more adaptive and efficient landscapes.

Nonetheless, this emphasis also flags the necessity for a meticulous scrutiny of the socio-economic and political ramifications of technologically driven urban initiatives. A critical gap in the extant literature is an examination of smart cities' relationship with outer space—a theme this article aims to address.

Beyond conventional urban challenges such as traffic management, earthquake resilience, energy-efficient architecture, intelligent power distribution, eco-friendly infrastructure, and reducing carbon footprints, it is vital to consider an often-overlooked facet: the spatial component of urban design. This element transcends the act of merely building taller edifices for residential, commercial, or public functions, and necessitates a transition from a planar to a volumetric framework, facilitated by geospatial tools such as Geographic Information Systems (GIS) [56,57].

The third spatial dimension of smart cities will also incorporate new aerial vehicles, such as drones, airborne taxis, and even flying automobiles, each assigned specific operational altitudes [54]. This integration will further elevate the significance of geospatial instruments in the conceptualization of future smart cities, synergizing with technologies such as Building Information Modelling (BIM) and three-dimensional city planning. This naturally extends our attention to outer space—a realm already facing issues of congestion. A critical question thus emerges: Does the cosmos exert any measurable influence on terrestrial urban living? If it does, what are the implications of this?

This study aims to furnish nuanced answers to these queries by examining the symbiotic interactions between urban and extra-atmospheric domains, seeking in turn to illuminate the prospective impacts of this interplay.

2.2. Sustainable Smart Cities and Outer Space Technologies

Outer space begins right in our cities. Physically, every place or individual on Earth is just 100 km away from it. Operationally, space technology significantly enhances the facets of our lives, with about 5500 operational satellites [58] providing a wide range of services that augment our environment in a variety of ways—from the GPS systems in our automobiles to daily weather predictions. In the event of an unintentional or natural interruption, these critical services, or rather the lack thereof, also serve as a representation of our collective fragility. Thus, how do we reconcile technological progress with technological stagnation? More importantly, how should we strategically address and integrate this dichotomy into city design?

Cities can benefit greatly from space technologies during this transformation from analogue to digital (e.g., the monitoring of water and air quality, planning, traffic management, green areas mapping, et cetera). One example is the EU's Galileo system, a European Space Agency (ESA) satellite constellation that supports services such as public transportation optimization, smart parking, autonomous electric transportation, emergency tracking, and more. Space-related technologies that have terrestrial uses provide a different kind of contribution to cities. By using their digital infrastructure to integrate space assets both on Earth and in space, as well as to connect with user segments in the cloud, digital giants such as Amazon, Microsoft, and Google have recently joined the space business. The space sector is embracing new business models such as Ground-Station-as-a-Service and Satellite-as-a-Service, which adhere to the pay-as-you-go paradigm and allow users to only pay for the services they utilize.

Outer space activities and phenomena, such as satellite launches, space exploration, and space debris, can have a significant impact on urban management. Interruptions in satellite services can affect a wide range of industries, including telecommunications, transportation, and emergency services. Meanwhile, satellite imagery and remote sensing provide valuable data for urban planning and management. For example, satellite data can be used to monitor land-use changes, urban sprawl, and environmental conditions, which can inform decisions on urban development, disaster management, and climate change adaptation.

Moreover, the increasing amount of space debris resulting from decades of space activity poses a threat to urban areas. As this debris re-enters Earth, it can cause damage to urban infrastructure and pose a risk to human safety. Urban managers must consider the potential impacts of space debris and ensure that their cities are resilient to such impacts. In the event of a natural disaster or other emergency, outer space activities can play a critical role in urban management. Satellites can provide real-time information on the extent of damage and help emergency responders to plan and execute their response efforts.

These elements represent an exponentially growing market that comes with great benefits but also with uncertainties, turbulence, instabilities, and potential errors. If urban sustainability as a desirable process can be enhanced with the help of outer space technologies, this implies that urban decision making should first be kept in mind and then integrate events in outer space into urban management at all scales.

In summary, outer space activities have a significant impact on urban management, and urban managers must consider their potential impact. By understanding the risks and opportunities of space activities, urban managers can ensure that their cities remain resilient and adaptable to the changing landscape of space activities.

2.3. Sustainable Outer Space and Smart Cities: Definitions and Scope

To provide clarity and context, several key terms and areas of focus need to be defined:

In the context of this paper, "outer space technology" refers to any technology designed to function in the extraterrestrial environment, such as satellites, space stations, and lunar habitats, which can have both direct and indirect effects on sustainability in smart cities. By adhering to these boundaries, we aim to elucidate the tangible links between sustainable smart cities and sustainable outer space activities.

By "influence of outer space on smart cities", this paper refers to the direct and indirect impacts of outer space activities on the sustainable development and functioning of smart cities, and in turn to their impacts on human activities such as communication, navigation, positioning and timing, and others.

While acknowledging catastrophic space-related events such as planetary collisions, the study's primary focus is on the intersection of technologies and policies that influence both smart cities and outer space sustainability. Specifically, this includes the use of space-based services such as satellite imagery and GNSS in smart cities, the adaptation of sustainable technologies such as renewable energy in both domains, and the policy frameworks that govern these activities. While catastrophic events such as planetary collisions are a consideration in space sustainability, they are not the focus of this paper unless they have direct implications for current space-based services such as early warning systems.

The concepts of sustainable outer space and smart cities encompass the utilization of advanced technologies to foster sustainable development both on Earth and in space. There are several interconnected ways in which these concepts can be linked:

1. Space-based services for sustainable cities: Smart cities can leverage space-based services, including satellite imagery, remote sensing, and global navigation satellite systems (GNSS), to monitor environmental conditions and inform decisions that promote sustainable development. For instance, satellite imagery enables the monitoring of land-use changes, facilitating informed urban planning choices aimed at sustainable growth.
2. Smart cities as a testing ground for space technologies: Earth's smart cities can function as testing grounds for sustainable technologies that may eventually find applications in outer space. Renewable energy sources such as solar and wind power can be tested and refined in smart cities before their deployment in space habitats.
3. The sustainable development of space infrastructure: The development of space infrastructure, such as space stations and lunar habitats, necessitates sustainability considerations to prevent resource depletion on celestial bodies. The long-term sustainability of space exploration efforts can be promoted by implementing sustainable practices during the design and construction of space infrastructure.
4. Space debris management: As previously mentioned, space debris poses a threat to both space infrastructure and urban areas on Earth. Implementing sustainable practices is imperative for managing space debris and minimizing the risk of collisions.
5. The remote monitoring and management of space infrastructure: Advanced technologies such as artificial intelligence and robotics enable the remote monitoring and management of space infrastructure. These technologies reduce the need for human intervention in space, thereby promoting sustainability by minimizing the resource requirements for supporting human presence.

While the sustainability of outer space has garnered attention from political decision makers, the focus of this attention tends to align with international affairs rather than local urban management. Given its direct association with peaceful uses of outer space, space sustainability represents a focal point of global politics, with governments actively involved in decision making in this domain. In this context, there is limited intersectionality with urban settlements, particularly concerning the broad notion of "quality of life". Furthermore, within the framework of the United Nations' sustainable development concept, the operationalization of sustainability in global space politics has relied on documents serving as moral guidelines rather than legally binding directives. The notion of sustainability in outer space usage aims to secure socioeconomic benefits in the present and in the long-term future. Nevertheless, these guiding documents play a crucial role in facilitating long-term strategic thinking in decision-making processes.

In more recent times, there has been a growing realization that Earth's orbital space environment is a limited resource that is being utilized by an increasing number of space actors. Consequently, the need for effective measures to protect this environment has become urgent. States, commercial entities, and other non-governmental organizations contribute to the proliferation of orbital debris through the deployment of spacecraft, which poses risks to space operations by elevating the probability of collisions and interference with satellite functionality. These threats, particularly space congestion, have a global scope and necessitate global resolutions in forums such as the United Nations system.

In 2019, the United Nations Committee on the Peaceful Uses of Outer Space (UN COPUOS) endorsed a set of 21 voluntary guidelines, known as the UN COPUOS Guidelines for the Long-term Sustainability of Outer Space Activities [15,16]. These guidelines are categorized into four sets: policy and regulatory measures, safety in space operations, international cooperation and capacity building, and scientific and technical aspects. While the guidelines do not explicitly reference urban settlements, they broadly address the overarching objective of enhancing the quality of life of current and future generations.

Furthermore, the UN's comprehensive definition of long-term sustainability (LTS) as "the ability to sustain space activities indefinitely in a manner that ensures equitable access to the benefits of outer space exploration and utilization for peaceful purposes, meeting the needs of present generations while preserving the outer space environment for future generations" [58] encompasses the technologies and services that constitute smart cities. Activities related to the sustainability of outer space include collisions between debris and active satellites, human impacts on the space environment, and radio frequency interference originating from satellites, space weather, or terrestrial systems.

The extent to which these perceptions of space sustainability align with the principles of sustainability in smart cities remains a matter of debate.

2.4. Sustainable Smart Cities and Outer Space Sustainability: Bridging the Gap

Sustainable urban development is generally concerned with the mindful use of finite resources for the development of urban settlements for residential, commercial, or recreational purposes [59]. One important link between this and space sustainability is the concept of resource depletion. Just as cities on Earth must grapple with the challenge of resource depletion and the need to manage their resources sustainably, space-faring societies must also be mindful of the limited resources available in space. This includes everything from the resources needed to sustain human life in space to the materials needed to construct spacecraft and infrastructure.

Further, a pivotal nexus between the two realms emerges through the concept of systems thinking. Both domains embody intricate systems that demand a meticulous contemplation of the intricate interplay among diverse components. Within the context of smart cities, deliberations regarding transportation infrastructure hold the potential to reverberate throughout the urban fabric, engendering repercussions for energy consumption, air quality, and public health. Likewise, in the realm of outer space, choices concerning resource allocation and waste management bear profound ramifications for astronaut well-being, the durability of space-based infrastructure, and the overarching sustenance of extraterrestrial domains.

As our case study will later highlight, it is impossible to think about sustainable urban activities without considering outer space. An accidental collision in outer space will trigger, at the minimum, the failure of vital services and, at the maximum, the cascading collapse of the urban structures that form the current lifestyle framework.

Finally, there is the issue of innovation. Both outer space and smart cities require innovative thinking and technological advancements to address the challenges they face. In a smart city, this may involve the development of new technologies for energy generation or waste management, while in outer space, it may involve the development of new propulsion systems or life support technologies.

Overall, while the challenges of outer space sustainability and smart city sustainability may be different in nature, they share several key connections in terms of the need to manage resources sustainably, implement systems thinking, and innovate to address complex challenges. Therefore, if urban strategic decision making ignores those risks originating in outer space, smart cities will be susceptible to disruptions and failures. Thus, achieving sustainability in both cities and outer space calls for harmonized strategic approaches.

3. Materials and Methods

This study carried out a simulation of a specific catastrophic event in outer space to investigate its impact on the functioning of smart cities. The goal was to explore the interconnectedness between smart cities and disruptive space events. Understanding outer space conditions can offer valuable insights for strategic urban planning, making it possible to integrate built-in resilience into city infrastructure. In other words, an awareness of potential space disruptions can help city decision makers avoid or mitigate breakdowns in the systems that make up a sustainable urban environment. Simulation software, often used in space operation centers globally, can be valuable for both space agencies and urban planners. By predicting collision risks and taking preventative actions, operational officers in both domains can utilize the same set of data.

Using a forensic research approach, we examined a specific past event—the collision between Iridium 33 and Cosmos 2251—to assess its impact on the essential infrastructure and services in smart cities. We supplemented our understanding of this event with information from academic, operational, and news sources. Subsequently, we collected satellite orbital coordinates, known as Two-Line Element Sets (TLEs), from databases such as SOCRATES and space-track.org. Commercial software such as STK (Systems Tool Kit) could not accurately represent the outer space environment at the time of the event, so an updated simulation environment was created. Additionally, we used the open-source platform CelesTrak to search for potential satellite conjunctions over the following week by comparing a comprehensive list of satellite payloads to objects already in orbit. Determining which payloads are active versus inactive is challenging due to limited public information. We based our scenarios on data from both CelesTrak and STK (version 11.2), using STK's SGP4 propagator and STK/CAT (Conjunction Analysis Tools) to provide the minimum distances and collision probabilities when conjunctions were within 5 km. Given the computational demands and the need for additional software components to create a realistic environment, we favored CelesTrak whenever feasible.

The simulation model was developed using STK Software, with a variety of input variables considered, such as orbital altitude, velocity, mass of objects, and probability distributions for collision scenarios from platforms such as CelesTrak.org and Spacetrack.org. These variables were derived from historical collision data and publicly available datasets from agencies such as the European Space Agency and SOCRATES.

To emulate a realistic outer space environment, we incorporated factors such as gravitational pulls, drag forces, and the Earth's oblateness. The focus of the simulation was to estimate the volume, velocity, and trajectories of debris resulting from satellite collisions and their potential cascading effects on services in smart cities.

The simulation underwent a validation process using historical collision case studies, such as the Iridium 33 and Cosmos 2251 encounter, to ensure that the model's outputs were aligned with actual event outcomes.

We recognize that our focus solely on collision scenarios might have narrowed the scope of our simulation. However, collision events offer tangible and readily quantifiable examples of outer space activities that can have a significant impact on smart city infrastructure.

After compiling a list of potential collisions from Space-track.org, every payload within the ESA satellite catalogue was examined. Certain payloads identified as large debris fragments were excluded from the analysis. However, for satellites active during the analyzed events, various databases related to smart city projects were consulted. We also looked at applications that used services from potentially affected satellites to identify possible service disruptions. Due to the significance of the analyzed events and the numerous fragments still in Earth's orbit, the unavailability of comprehensive access prompted an additional analysis: we conducted a thorough assessment of objects that have remained in orbit over a period spanning several years (February 2009 to June 2022). This analysis serves as a valuable inventory of objects that have persisted in orbit for a significant duration.

Although the broader scope of this paper explored a variety of challenges and considerations concerning human-made space debris and its impact on smart cities, it is worth noting that the simulation analysis specifically focused on the collision scenario between satellites. This choice was deliberate: collisions offer a poignant and readily quantifiable example of the cascading effects outer space activities can have on urban services and infrastructure. Nevertheless, we acknowledge that space debris poses other forms of risks, such as uncontrolled satellite re-entry and fragmentation events. This article is, however, the result of a broader doctoral research project that has also tackled fragmentation and satellite re-entry events and the impact of such events on the well-functioning of urbanized areas. Future research will seek to broaden this study's simulation analysis by incorporating these and other scenarios to provide a more comprehensive assessment of the subject area.

4. Results

The collision between the military communications satellite Cosmos 2251 and the communications satellite Iridium 33, which occurred on 10 February 2009, at 11:56 a.m. Eastern Standard Time over Siberia, holds significant historical significance as the first complete payload collision. The collision resulted in the creation of tens of thousands of smaller fragments and approximately 2000 objects with a minimum diameter of 10 cm [60]. These remnants, which will persist in orbit for extended periods, pose a high risk of collision with other objects in low Earth orbit, as depicted in Figure 1. The collision transpired at a speed equivalent to 26 times the speed of sound, with Iridium 33 weighing 600 kg and Cosmos 2251 weighing 900 kg [60,61].

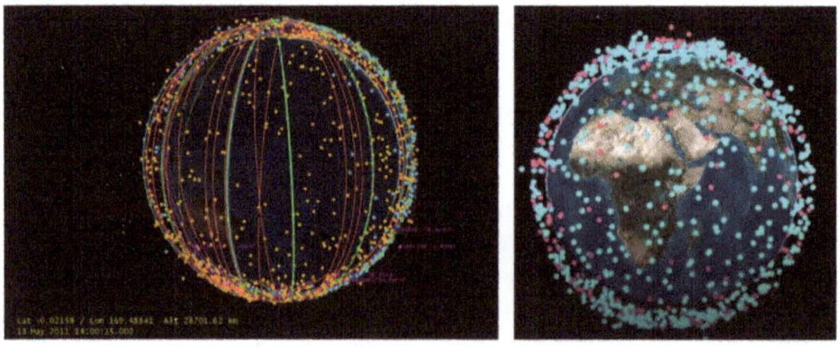

Figure 1. Fragments still in orbit on 13 May 2011, rendered with the main satellites' orbits (**left**) and fragments still in space on 24 June 2022 (**right**) (©STK).

This accidental collision between Iridium 33 and Cosmos 2251 produced the second-largest space debris event in satellite fragmentation history [62]. The immediate observations after the collision recorded 1603 fragments from Cosmos 2251 and 598 fragments from Iridium 33 [63]. Although numerous fragments from Iridium 33 were already dispersed in orbit by 10 February 2009, at 18:26:10 h, only 153 debris pieces were plotted within 35 min of the impact. The initial predictions estimated that half of the tracked debris would re-enter the Earth's atmosphere within five years; however, by 1 October 2009, fewer than 60 catalogued fragments had re-entered, leaving most newly produced debris in orbit [64]. This suggests that only 4% of the existing low Earth orbit (LEO) objects have been discovered [65].

This collision serves as a poignant example of the potential consequences when the near-Earth space environment lacks protection. Aside from the loss of one operational communications satellite, the collision generated hundreds of debris pieces larger than 1 cm, significantly increasing the long-term risk of danger to future communications and Earth resource satellites occupying similar orbits. NASA has asserted that accidental collisions between catalogued objects currently occur every one to five years, with the likelihood of such incidents predicted to increase in the future [66]. While the 66-satellite Iridium constellation experienced minimal losses, with its regular service being promptly restored [66], the damage caused by the collision in the orbital region may have been more severe. Approximately 1000 fragments from the collision have persisted in orbit for extended periods, posing a threat to other functioning satellites, including the remaining viable Iridium payloads.

As of June 2022, there were still 1485 large fragments from Cosmos 2251 and 533 large fragments from Iridium 33 in orbit that could have potentially collided with other orbiting objects, as illustrated in Figure 2. Furthermore, the remaining fragments of Cosmos 2251 in orbit could have collided with an additional 1685 debris pieces, 69 active satellites, and 61 rocket bodies. Similarly, fragments from Iridium 33 could potentially have collided with approximately 531 debris pieces, 25 active satellites, and 30 rocket bodies.

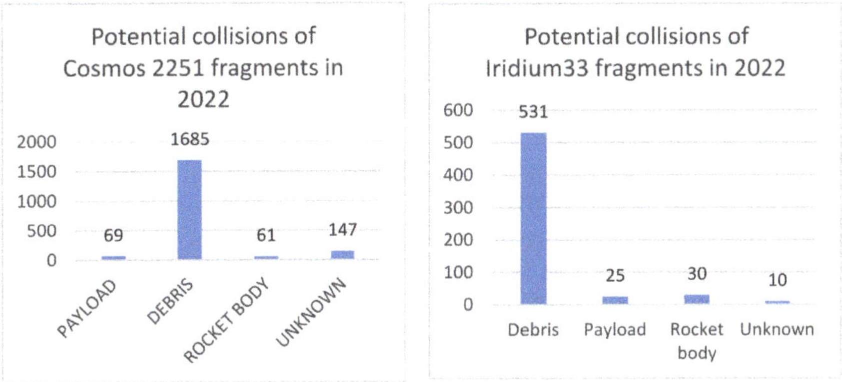

Figure 2. Potential collisions of debris caused by the collision between Iridium 33 and Cosmos 2251 with other pieces of payloads, debris, or rocket bodies in June 2022 (processed data from space-track.org).

All the catalogued fragments resulting from the 2009 collision event that are still present in orbit were accurately located and processed using the U.S. Space Track open-source database ((www.space-track.org (accessed on 24 August 2023)). The entries underwent two rounds of verification using SATCAT, a text file containing information about all the human-made objects orbiting the Earth since 1957. The analyzed data suggests that the collision between Iridium 33 and Cosmos 2251 may have impacted crucial services, as depicted in Table 1. These services include food and water management, localization services, disaster response capabilities, and research efforts.

Table 1. List of satellites that could have been potentially impacted by debris caused by the Iridium 33 and Cosmos 2251 collision in June 2022.

Collision Probability	Name 1	Name 2	Type of Satellite	Type of Service Potentially Affected
0.000274496	IRIDIUM 33 DEB	EXPLORER 7	EO and space exploration	N/A, the satellite was decommissioned at the time of research
0.001121804	IRIDIUM 33 DEB	STEX	Reconnaissance/ intelligence gathering	N/A, the satellite was decommissioned at the time of research
0.000198616	IRIDIUM 33 DEB	LANDSAT 5	EO and space exploration	▪ Food and water management ▪ Water management ▪ Climate change ▪ Ability to respond to disasters ▪ Ecosystem studies ▪ Energy sector ▪ Ability to detect forest fires ▪ Human health management ▪ Urban management ▪ Water management
0.000252395 0.000398829 0.000492121 0.001309433	IRIDIUM 33 DEB	COSMOS 1375	ASAT target—it was a target for Cosmos 1379 on 18 June 1982 Debris is still in orbit	N/A, the satellite was decommissioned at the time of research
0.002249103 0.001138301 0.000991591 0.000116312	IRIDIUM 33 DEB	DEMETER	EO (observation of geophysical parameters of the terrestrial environment)	▪ GNSS—especially localization services ▪ Earth observation ▪ Research
0.000134388	IRIDIUM 33 DEB	COSMOS 1626	ELINT (Electronic and Signals Intelligence) satellite, a military space-based system; replaced Cosmos 1408	N/A, the satellite was decommissioned at the time of research

Source: authors' calculations.

Collisions in outer space can lead to the release of debris and other materials that pose a threat to food and water storage facilities, potentially resulting in the loss of supplies. Damage to equipment used for food production or water generation due to space collisions can have significant ramifications for society's ability to sustain itself. It is imperative to establish backup systems to ensure the uninterrupted production of food and water, even in the event of a collision.

Moreover, large-scale collisions in outer space can have indirect effects on the Earth's climate due to the substantial amount of dust and debris they can introduce into the atmosphere. This debris can block the sun's rays, causing a global cooling effect. Additionally, collisions can release substantial quantities of carbon dioxide and other greenhouse gases into the atmosphere, potentially contributing to long-term climate change.

Further, collisions occurring in the vast expanse of outer space can have notable repercussions for global navigation satellite system (GNSS) services, which heavily rely on an intricate network of satellites to provide users on Earth with precise location and timing information. The presence of debris or its colliding with other satellites can cause damage to or even destroy satellites in the Earth's orbit. This inevitably leads to a reduction in coverage and a concomitant decline in accuracy in GNSS services, negatively impacting the provision of such services. Moreover, collisions can create debris fields within the Earth's orbit, thereby serving as an imminent risk to other satellites and spacecraft and critically compromising an array of indispensable services that support smart cities across the globe.

In addition, it is worth highlighting that numerous technologies harnessed in disaster response and smart city domains critically depend on satellite communications, encompassing GNSS services, satellite imaging, and satellite internet. These technologies operate as indispensable sources of vital information for emergency responders, enabling them to execute their duties promptly and effectively. At the same time, they empower city managers to make data-driven decisions aimed at improving urban infrastructure and services. If collisions in outer space inflict damage upon satellites or result in their outright destruction, the ensuing disruptions to satellite communications will consequently yield delays and curtailed functionality for disaster response and smart city technologies. For instance, the loss of GNSS services precipitates heightened challenges for emergency responders in terms of navigating and locating individuals requiring assistance. Furthermore, the proper functioning of smart city technologies, including but not limited to traffic management systems, energy grids, and public safety systems, hinges on the receipt of real-time data. Consequently, any disturbances stemming from disrupted satellite imaging or other data sources arising from collisions in outer space undeniably give rise to inefficiencies and impeded performance within these systems.

In conclusion, it is crucial to acknowledge that collisions transpiring within the realm of outer space engender indirect ramifications that impede smart cities' capacity to respond effectively to disasters and impair their optimal functioning, chiefly by disrupting satellite communications and jeopardizing essential data sources. Consequently, it becomes imperative to adopt preventive measures aimed at averting collisions altogether, coupled with the formulation of comprehensive contingency plans in the case of disruptive collisions that are designed to ensure the uninterrupted provision of vital services.

Figure 3 provides an examination of the impact of the Iridium 33 and Cosmos 2251 collision event on different categories of satellites. Our analysis of data gathered from various databases underscores the pronounced repercussions suffered by Earth Observation (EO) satellites because of this collision incident. Communication satellites also experienced a notable impact, although to a lesser extent. It is noteworthy that the least impact was observed in the case of constellation-based communication satellites, where service continuity was maintained despite the collision.

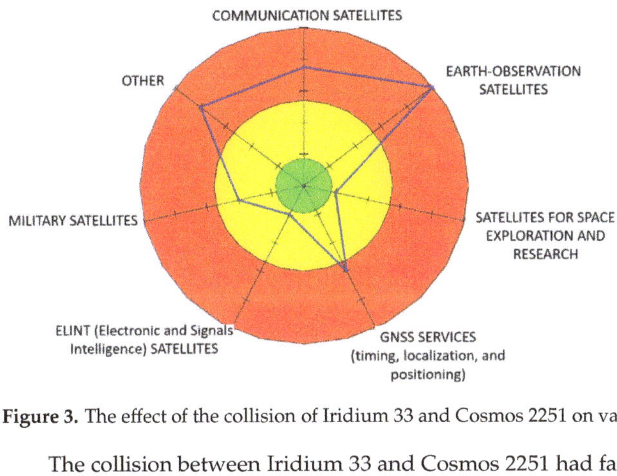

Figure 3. The effect of the collision of Iridium 33 and Cosmos 2251 on various typologies of satellites.

The collision between Iridium 33 and Cosmos 2251 had far-reaching consequences, extending beyond the destruction of a communication satellite associated with Motorola, which led to a temporary disruption of satellite telecommunication services. Thirteen years after the event, its aftermath still had the potential to compromise the continuity of critical infrastructure sectors in several smart cities, including Landsat, Demeter, and Iridium, as depicted in Figure 4.

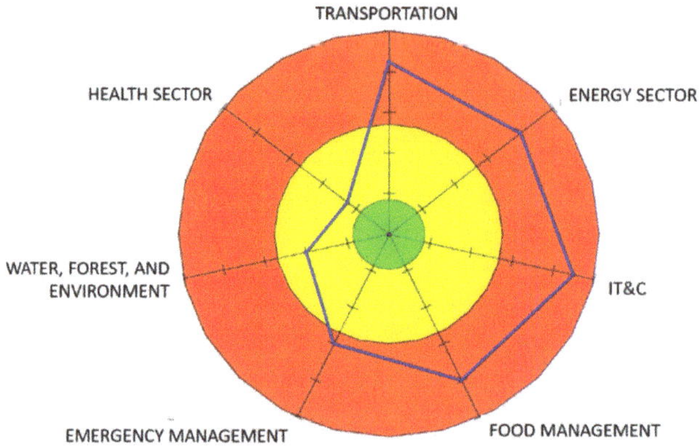

Figure 4. Critical infrastructure sectors were affected by the collision of Iridium 33 and Cosmos 2251.

In the event of additional collisions, the transportation, information, and communication technology (ICT) and energy sectors would be at risk of experiencing significant repercussions. Conversely, the health sector, particularly the emergency services, would likely face a relatively lesser impact, owing to the existing redundancies incorporated within this sector.

Our simulations indicate that space debris originating from satellite collisions has the potential to disrupt key smart city functions such as navigation, telecommunications, and even emergency services. A single collision in outer space can create a cascade of debris, with a broader impact potentially being felt across multiple city services, highlighting the need for redundant systems and contingency planning.

5. Discussion

The presence of generic space debris in outer space poses a significant threat to satellite operations, emphasizing the need for vigilant monitoring by satellite operators. This research highlights that satellite operators, including the public conjunction assessment service SOCRATES, use screening capabilities to generate close approach predictions for satellite payloads. However, as the production of debris increases, these services will face greater demands. City-level decision makers should be aware of these monitoring tools to foster a common understanding of the risks posed by outer space and their potential cascading effects on smart city technologies.

The effects of collisions in outer space extend beyond just their initial impact, as debris from such collisions poses risks to other objects in orbit. For instance, the encounter between Cosmos and Iridium produced a substantial number of visible fragments from both satellites. While efforts are being made to monitor and track such debris, observing and accounting for all potential objects remains impossible. Additionally, the mechanisms behind satellite disintegration are not fully understood, although it is acknowledged that debris fragmentation produces various fragment sizes and masses. The issue of debris in space cannot be disregarded as it does not naturally degrade into dust.

Addressing these challenges is crucial for carrying out a comprehensive analysis of the effects of orbiting debris, the identification of genuine threats, and effective communication with the public and relevant stakeholders. However, political and scientific obstacles impede progress in these areas. The scientific challenges lie in our limited knowledge regarding the scale of the issue, while the political dilemmas arise from the prolonged timeframes required for mitigating potentially harmful occurrences.

It is crucial to discuss the likelihood thresholds for satellite collision events that warrant significant attention. Currently, a collision probability of higher than 1 in 10,000 is generally considered to be of high concern and would typically trigger avoidance maneuvers if possible. Given the current technological and economic constraints, it may be justifiable to consider ignoring collision probabilities that fall below a certain threshold, such as 1 in 100,000, as the costs and efforts required to take corrective action in these cases may not be justified.

However, this is not a recommendation to dismiss low-probability events entirely but rather a suggestion to weigh them against the other pressing concerns faced by smart cities, such as immediate threats to infrastructure or public health. In terms of prioritization, decision makers should focus first on high-probability, high-impact events that can significantly disrupt urban services, while scenarios with lower collision probabilities and less immediate impacts may be accorded a lesser priority but should nevertheless not be entirely ignored.

We acknowledge that as technology improves and the costs associated with monitoring and prevention decrease, these probability thresholds should be revisited. The balancing act between economic feasibility and risk mitigation is an evolving landscape that requires ongoing assessment.

Smart cities are regarded as a solution to numerous global challenges, such as increasing energy consumption, air pollution, and urban traffic and the declining food production caused by population growth. However, cities face a multitude of issues and risks, some of which originate in space and impact vital systems and services encompassing the transportation, water, electricity, communications, healthcare, and social services sectors. To monitor service losses in real-time, a stronger link between smart cities and space infrastructure is necessary. This study proposes the establishment of an intermediate level "governing entity" that bridges small-scale urban governance structures and national security councils focused on nuclear and missile deterrence.

Specifically concerning ICT, space debris poses a latent threat to services such as navigation, telecommunications, and even weather forecasting—all of which are vital for the efficient functioning of smart cities. While outer space collisions are isolated incidents, we can still imagine the scale of impact that space debris could exert on urban services; more comprehensive measures for resilience in this regard are therefore needed.

The governing body proposed above should be responsible for building the tools required, including financial mechanisms, to ensure the deployment and delivery of operational space situational awareness (SSA) services to smart cities, as well as their ongoing monitoring. This governing entity must ensure the availability and continuity of relevant data and services, as well as their long-term sustainability. Moreover, fostering structured and mutually reinforcing relationships among stakeholders along the value chain is essential for aligning activities with the governing organization's primary objectives.

On a broader level, our research exemplifies the importance of comprehensive information for strategic smart city planning. The vulnerability of advanced technological systems, which form the backbone of smart cities, due to a lack of awareness about the risks posed by outer space underscores the direct connection between the continuity and cost of operations. Furthermore, this research establishes a direct link between the sustainability of outer space and the sustainability of smart cities, emphasizing that to avoid any disruption to ground services, outer space technologies should be leveraged responsibly for the benefit of future generations.

Within the larger context of urban resilience, the study's consideration of space debris marks a novel approach [22]. This research builds upon the primary author's earlier investigations, which initially studied the potential influence of space activities on urban amenities, by specifically exploring their potential repercussions for the functioning of smart cities.

6. Recommendations and Limitations

Collisions in outer space can have direct and indirect effects on the proper functioning of smart cities on Earth, as well as on the ability to respond to disasters and safeguard human activities in general. Urban development strategies should incorporate considerations of the potential impact of collisions in outer space and include measures to mitigate the risks they pose to the normal functioning of city services. To address these challenges, the following recommendations are proposed for urban development strategies regarding collisions in outer space.

Firstly, developing redundant communication systems is crucial. Cities should consider establishing alternative communication systems that are not solely reliant on satellite communications. Potentially involving the development of terrestrial networks or alternative satellite systems, these could ensure the continuity of communication services in cases of outer space collisions.

Secondly, cities should plan for emergency responses in the event of disruptions to satellite services. It is essential to have contingency plans in place, including alternative methods for navigation and locating individuals in need of assistance.

Thirdly, implementing resilient infrastructure is vital to withstand the potential impact of debris from collisions in outer space. Cities should reinforce buildings and critical infrastructure to reduce the risk of damage caused by falling debris.

Furthermore, collaboration with satellite operators and space agencies to monitor space debris is essential. By tracking the movement of debris and taking proactive steps to avoid collisions between debris and satellites, cities can actively contribute to reducing the risk of collisions.

Additionally, promoting sustainable practices can help mitigate the risk of collisions in outer space. Cities should encourage sustainable practices such as reducing carbon emissions and minimizing waste in order to reduce the number of debris in orbit and decrease the likelihood of collisions.

Regarding the study topic's limitations and areas for further research, several aspects require attention. Firstly, the lack of data on the impact of collisions in outer space on smart cities necessitates further research to better understand the potential impact of such collisions and to develop effective mitigation strategies regarding them.

The cost associated with the development and implementation of smart city technologies, redundant communication systems, and resilient infrastructure is a significant concern. However, research on the impact of collisions on smart technologies might prove less costly than repairing the damage caused by such disasters.

The lack of governance and regulation surrounding outer space activities, particularly regarding debris mitigation and collision avoidance, has created uncertainty for cities and satellite operators. Establishing clear governance frameworks is crucial for effectively managing the risks associated with collisions in outer space.

The integration of outer space activities with smart city technologies is still in its early stages. Further research is needed to better understand how these technologies can be integrated and how they can best support urban development goals.

Lastly, the potential inequitable distribution of benefits from smart city technologies and outer space activities needs consideration. Research should explore how these technologies can be used to address urban inequalities and promote social and environmental justice.

In conclusion, while smart city technologies and outer space activities hold significant potential to support urban development, addressing their limitations and areas for further research is essential to ensuring their responsible and equitable use. Urban development strategies should prioritize resilience and preparedness in the face of disruptions to satellite services caused by collisions in outer space. By taking proactive measures to mitigate risks, cities can minimize the potential disruptions to city services and ensure the safety and well-being of their residents.

7. Conclusions

This study explored the concept of sustainability regarding the governance of two distinct realms: outer space and smart cities. Sustainability in outer space pertains to the various means and methods by which space is dedicated to societal benefits. Conversely, sustainability in smart cities relates to the preservation of vital services and infrastructure, under any circumstances, without resource depletion. This paper identified a connection between maintaining a debris-free outer space and ensuring functional and resilient urban services.

Further, the study analyzed the accidental collision between the Iridium 33 and Cosmos 2251 satellites and its potential impact on critical services in smart cities. The findings demonstrate that this collision indirectly affected essential infrastructure such as energy grids, transportation systems, and information and communication technologies (ICT). Beyond the loss of a satellite and the temporary disruption of the associated communication network, the long-term consequences of this collision could have led to substantial disruptions across multiple critical services.

These findings establish a direct correlation between collisions in outer space and the vital services that underpin contemporary smart cities. The sustainability of outer space and the ground-level core of our cities are intricately intertwined, which necessitates the strategic engagement of urban decision-making entities with outer space sustainability, at least at an awareness level and ideally through an active engagement via intermediate governing bodies.

Consequently, urban development strategies must consider the potential hazards of space debris and ensure that urban infrastructure is resilient to its potential impacts. This may require the implementation of stricter regulations on satellite launches and the mitigation of orbital debris to safeguard the long-term sustainability of space operations.

In the context of the sustainability and smartness of cities, a temporary inverse progression has been observed [67], whereby the greater the technological interconnectedness of our urban settlements, the more technologies are placed in Earth's operational orbits. As a result, the increased presence of objects in orbit elevates the risks of collisions and compromises the time-sustainability of our lifestyles. This thought-provoking antithesis underscores the potential danger to smart cities' strategic planning posed by the lack of sustainability in outer space.

While the history of space activities is marked by a relatively low rate of incidents directly impacting ICT and other urban services, the stakes of such incidents are elevated in the context of smart cities whose infrastructural and service interdependencies are high. As the volume of space activities continues to increase, the statistical likelihood of disruptive incidents will also rise, necessitating a proactive approach to urban resilience that accounts for both terrestrial and cosmic threats.

In summary, this study emphasized the interdependence of sustainability in outer space and the effective functioning of smart cities. By recognizing this interconnectedness, urban decision making should actively include considerations of activities governed by intermediate entities. Strategic urban development planning must address the hazards posed by space debris and prioritize resilient infrastructure while advocating for stricter regulations on satellite launches and debris mitigation to ensure the long-term sustainability of space operations. Recognizing the occasional tension between the sustainability and smartness of cities underscores the need to address the risks tied to the unsustainable use of outer space when safeguarding the strategic planning of smart cities.

Author Contributions: Conceptualization, U.-E.B. and A.V.G.; methodology, U.-E.B., O.B., A.V.G. and R.D.S.; software, U.-E.B. and O.B.; validation, U.-E.B. and A.V.G.; formal analysis, U.-E.B. and R.D.S.; investigation, U.-E.B. and O.B.; resources, U.-E.B. and O.B.; data curation, U.-E.B.; writing—original draft preparation, U.-E.B.; writing—review and editing, U.-E.B., O.B., A.V.G. and R.D.S.; visualization, U.-E.B. and O.B.; supervision, U.-E.B. and A.V.G.; project administration, U.-E.B., O.B., A.V.G. and R.D.S. All authors have read and agreed to the published version of the manuscript.

Funding: This research received no external funding.

Data Availability Statement: The data presented in this study are available on request from the corresponding author.

Conflicts of Interest: The authors declare no conflict of interest.

Abbreviations

ASAT	Anti-satellite weapon
BIM	Building Information Modelling
CityGML	City Geography Markup Language—an open-source 3D city modelling format
COPUOS	United Nations Committee on the Peaceful Uses of Outer Space
ELINT	Electronic and Signals Intelligence
EO	Earth-Observation
ESA	European Space Agency
GIS	Geographic Information System
GNSS	Global Navigation Satellite System
GPS	Global Positioning System
ICT	Information and Communication Technology
LEO	Low-Earth Orbit
LTS	Long-Term Sustainability
NORAD (ID)	North American Aerospace Defense (Catalog Number)
SATCAT	Satellite Catalog Number
SOCRATES	Satellite Orbital Conjunction Reports Assessing Threatening Encounters in Space
SST	Space Surveillance and Tracking
STK	System Tool-Kit
STK/CAT	STK's Conjunction Analysis Tool (CAT)
TLE	Two-Line Element
UNESCO	United Nations Educational, Scientific, and Cultural Organization

References

1. Economist. Going off Grid: The sad decline of the oldest form of city planning. *The Economist*, 24 December 2022; Special Section; pp. 20–22.
2. Filarete. *Filarete's Treatise on Architecture: Being the Treatise by Antonio di Piero Averlino, Known as Filarete; Was Originally Composed in Milan c.1460–c.1464*; Spencer, J.R., Translator; Yale University Press: New Haven, CT, USA, 1965.
3. Leyes de las Indias. *Recopilación de Leyes de Los Reynos de las Indias: Mandadas Imprimir y Publicar por la Majestad Católica del Rey Don Carlos, Nuestro Señor*; Julián de Paredes: Madrid, Spain, 1681.
4. Aurbach, L.A. *History of Street Networks: From Grids to Sprawl and Beyond*; Pedshed Press: Hyattsville, MD, USA, 2020.
5. Howard, E. *Garden Cities of Tomorrow*; S. Sonnenschein & Co.: London, UK, 1902.
6. Le Corbusier. *Urbanisme*, 1st ed.; Les Editions G. Crès & Cie.: Paris, France, 1925.
7. Speck, J. *Walkable Cities: How Downtown Can Save America, One Step at a Time*; North Point Press: New York, NY, USA, 2012.
8. Speck, J. *Walkable City Rules: 101 Steps to Making Better Places*; Island Press: Washington, DC, USA, 2018.
9. Economist, African urbanisation. Beyond the metropolis. *The Economist*, 11 March 2023; 29–30.
10. Jacobs, J. *The Death and Life of Great American Cities*; Random House: New York, NY, USA, 1961.
11. Scarlat, C. Management ante Management: Urban planning and development strategies. 2023; *unpublished work to be submitted*.
12. Büyüközkan, G.; Ilıcak, Ö.; Feyzioğlu, O. A Review of Urban Resilience Literature. *Sustain. Cities Soc.* **2022**, *77*, 103579. [CrossRef]
13. Badea, D.; Ranf, D.E. Challenges of post-pandemic urban resilience management. *Stud. Bus. Econ.* **2023**, *18*, 37–53. [CrossRef]
14. Admiraal, H.; Cornaro, A. Future cities, resilient cities—The role of underground space in achieving urban resilience. *Undergr. Space* **2020**, *5*, 223–228. [CrossRef]
15. Cariolet, J.-M.; Vuillet, M.; Diab, Y. Mapping urban resilience to disasters—A review. *Sustain. Cities Soc.* **2019**, *51*, 101746. [CrossRef]
16. Rezvani, S.M.; Falcão, M.J.; Komljenovic, D.; de Almeida, N.M. A Systematic Literature Review on Urban Resilience Enabled with Asset and Disaster Risk Management Approaches and GIS-Based Decision Support Tools. *Appl. Sci.* **2023**, *13*, 2223. [CrossRef]
17. Huddleston, P.; Smith, T.; White, I.; Elrick-Barr, C. Adapting Critical Infrastructure to Climate Change: A Scoping Review. *Environ. Sci. Policy* **2022**, *135*, 67–76. [CrossRef]
18. Heinzlef, C.; Robert, B.; Hémond, Y.; Serre, D. Operating Urban Resilience Strategies to Face Climate Change and Associated Risks: Some Advances from Theory to Application in Canada and France. *Cities* **2020**, *104*, 102762. [CrossRef]

19. Prashar, S.; Shaw, R.; Takeuchi, Y. Community Action Planning in East Delhi: A Participatory Approach to Build Urban Disaster Resilience. *Mitig. Adapt. Strateg. Glob. Chang.* **2013**, *18*, 429–448. [CrossRef]
20. Karamouz, M.; Rasoulnia, E.; Olyaei, M.A.; Zahmatkesh, Z. Prioritizing Investments in Improving Flood Resilience and Reliability of Wastewater Treatment Infrastructure. *J. Infrastruct. Syst.* **2018**, *24*, 04018021. [CrossRef]
21. Wu, Y.; Lin, Z.; Liu, C.; Huang, T.; Chen, Y.; Ru, Y.; Chen, J. Resilience Enhancement for Urban Distribution Network via Risk-Based Emergency Response Plan Amendment for Ice Disasters. *Int. J. Electr. Power Energy Syst.* **2022**, *141*, 108183. [CrossRef]
22. Botezatu, U.E.; Bucovetchi, O.M.C. Space as integrator: From horizontal to vertical urban planning. *Smart Cities Int. Conf. (SCIC) Proc.* **2021**, *9*, 393–407.
23. Damien, S.; Charlotte, H. Assessing and mapping urban resilience to floods with respect to cascading effects through critical infrastructure networks. *Int. J. Disaster Risk Reduct.* **2018**, *30*, 235–243. [CrossRef]
24. Pickett, S.T.; Cadenasso, M.L.; Grove, J.M. Resilient cities: Meaning, models, and metaphor for integrating the ecological, socio-economic, and planning realms. *Landsc. Urban Plan.* **2004**, *69*, 369–384. [CrossRef]
25. Batty, M. The size, scale, and shape of cities. *Science* **2008**, *319*, 769–771. [CrossRef]
26. Godschalk, D.R. Urban hazard mitigation: Creating resilient cities. *Nat. Hazards Rev.* **2003**, *4*, 136–143. [CrossRef]
27. Meerow, S.; Newell, J.P.; Stults, M. Defining Urban Resilience: A Review. *Landsc. Urban Plan.* **2016**, *147*, 38–49. [CrossRef]
28. Gheorghe, A.; Katina, P. *Gamification for Resilience: Resilient Informed Decision Making*; Wiley: Hoboken, NJ, USA, 2023.
29. Lawrence, J.V. The politics of resilient cities: Whose resilience and whose city? *Build. Res. Inf.* **2014**, *42*, 191–201. [CrossRef]
30. Ruth, M.; Coelho, D. Understanding and managing the complexity of urban systems under climate change. *Clim. Policy* **2007**, *7*, 317–336. [CrossRef]
31. Tcholtchev, N.; Schieferdecker, I. Sustainable and Reliable Information and Communication Technology for Resilient Smart Cities. *Smart Cities* **2021**, *4*, 156–176. [CrossRef]
32. Gotlib, D.; Kulisiewicz, T.; Muraszkiewicz, M.; Olszewski, R. Multiagency Modeling of Transformation Strategies Towards Sustainable Smart Cities. *Appl. Sci.* **2020**, *10*, 853. [CrossRef]
33. United Nations Sustainable Development. Agenda 21. 1992. Available online: https://www.un.org/esa/dsd/agenda21/ (accessed on 1 March 2023).
34. United Nations. Transforming Our World: The 2030 Agenda for Sustainable Development. Available online: https://sdgs.un.org/2030agenda (accessed on 11 April 2023).
35. Schiavo, F.T.; Magalhães, C.F.d. Smart Sustainable Cities: The Essentials for Managers' and Leaders' Initiatives within the Complex Context of Differing Definitions and Assessments. *Smart Cities* **2022**, *5*, 994–1024. [CrossRef]
36. Kumar, A.; Rattan, J.S. A Journey from Conventional Cities to Smart Cities. In *Smart Cities and Construction Technologies*; Shirowzhan, S., Zhang, K., Eds.; InTechOpen: London, UK, 2020; pp. 3–16. [CrossRef]
37. Rotuna, C.; Cirnu, C.E.; Smada, D.; Gheorghita, A. Smart City Applications Built on Big Data Technologies and Secure IoT. *Ecoforum* **2017**, *6*, 13.
38. Townsend, A.M. *Smart Cities: Big Data, Civic Hackers, and the Quest for a New Utopia*; W.W. Norton & Company: New York, NY, USA; London, UK, 2013.
39. Carè, S.; Trotta, A.; Carè, R.; Rizzello, A. Crowdfunding for the development of smart cities. *Bus. Horiz.* **2018**, *61*, 501–509. [CrossRef]
40. Rochdane, H.; Hamdani, S.; Hanine, S.; Steils, N. Urban Crowdsourcing: Explore the role of citizens in co-creating smart cities. *Inf. Syst. Manag. Innov.* **2019**, *3*, 5–11.
41. Bastos, D.; Fernández-Caballero, A.; Pereira, A.; Rocha, N.P. Smart City Applications to Promote Citizen Participation in City Management and Governance: A Systematic Review. *Informatics* **2022**, *9*, 89. [CrossRef]
42. Ratti, C.; From "Smart City" to "Senseable City". Mobilize. Available online: https://www.mobilize.com/en/wearemobilizers/carlo-ratti-from-smart-city-to-senseable-city/ (accessed on 23 May 2023).
43. Ratti, C. Afterword: Lessons for Our Future Cities. In *Designing Smart and Resilient Cities for a Post-Pandemic World: Metropandemic Revolution*, 1st ed.; Larsson, A., Hatzigeorgiou, A., Eds.; Routledge: New York, NY, USA, 2022; pp. 365–366. [CrossRef]
44. Zanella, A.; Bui, N.; Castellani, A.; Vangelista, L.; Zorzi, M. Internet of Things for Smart Cities. *IEEE Internet Things J.* **2014**, *1*, 22–32. [CrossRef]
45. Nam, T.; Pardo, T.A. Conceptualizing Smart City with Dimensions of Technology, People, and Institutions. In Proceedings of the 12th Annual International Conference on Digital Government Research: Digital Government Innovation in Challenging Times, College Park, MD, USA, 12–15 June 2011; pp. 282–291. [CrossRef]
46. Gil-Garcia, J.R.; Pardo, T.A.; Nam, T. What Makes a City Smart? Identifying Core Components and Proposing an Integrative and Comprehensive Conceptualization. *Inf. Polity* **2015**, *20*, 61–87. [CrossRef]
47. Viale Pereira, G.; Parycek, P.; Falco, E.; Kleinhans, R. Smart governance in the context of smart cities: A literature review. *Inf. Polity* **2018**, *23*, 143–162. [CrossRef]
48. Tan, S.Y.; Taeihagh, A. Smart City Governance in Developing Countries: A Systematic Literature Review. *Sustainability* **2020**, *12*, 899. [CrossRef]
49. Tomor, Z.; Meijer, A.; Michels, A.; Geertman, S. Smart Governance For Sustainable Cities: Findings from a Systematic Literature Review. *J. Urban Technol.* **2019**, *26*, 3–27. [CrossRef]

50. Kitchin, R.; Dodge, M. The (In)Security of Smart Cities: Vulnerabilities, Risks, Mitigation, and Prevention. *J. Urban Technol.* **2019**, *26*, 47–65. [CrossRef]
51. Dodge, M.; Kitchin, R. The challenges of cybersecurity for smart cities. In *Creating Smart Cities*; Coletta, C., Evans, L., Heaphy, L., Kitchin, R., Eds.; Routledge: New York, NY, USA, 2018; pp. 205–216. Available online: https://personalpages.manchester.ac.uk/staff/m.dodge/Dodge_Kitchin_Challenges_of_Cybersecurity_for_Smart_Cities.pdf (accessed on 24 August 2023).
52. Sassen, S.; Kourtit, K. A Post-Corona Perspective for Smart Cities: 'Should I Stay or Should I Go? ' *Sustainability* **2021**, *13*, 9988. [CrossRef]
53. Greenfield, A. *Radical Technologies: The Design of Everyday Life*; Verso: London, UK, 2017.
54. Clark, J. *Uneven Innovation: The Work of Smart Cities*; Columbia University Press: New York, NY, USA, 2020.
55. Castells, M. *The Rise of the Network Society*; Wiley-Blackwell: Chichester, UK, 2010.
56. Union of Concerned Scientists, UCS. Satellite Database, 1 May 2022. Available online: https://www.ucsusa.org/resources/satellite-database (accessed on 22 November 2022).
57. United Nations Office for Outer Space Affairs. Long-Term Sustainability of Outer Space Activities. Available online: https://www.unoosa.org/oosa/en/ourwork/topics/long-term-sustainability-of-outer-space-activities.html (accessed on 11 November 2022).
58. United Nations. *Report of the Committee on the Peaceful Uses of Outer Space, Sixty-Second Session (12–21 June 2019)*; General Assembly Supplement 20; United Nations: New York, NY, USA, 2019.
59. Yigitcanlar, T. *Technology and the City: Systems, Applications and Implications*; Routledge: New York, NY, USA, 2016.
60. Weeden, B.; Billiards in Space. 23 February 2009. Available online: https://www.thespacereview.com/article/1314/1 (accessed on 11 May 2023).
61. Earl, M. Iridium 33 and Cosmos 2251: A Historic Collision. Available online: http://www.castor2.ca/08_Papers/collision.pdf (accessed on 21 April 2023).
62. Tan, A.; Zhang, T.; Dokhanian, M. Analysis of the Iridium 33 and Cosmos 2251 Collision using Velocity Perturbations of the Fragments. *Phys.—Adv. Aerosp. Sci. Appl.* **2013**, *3*, 13–25.
63. Finkleman, D. The Dilemma of Space Debris. *Am. Sci.* **2014**, *102*, 26. [CrossRef]
64. National Research Council. *Limiting Future Collision Risk to Spacecraft: An Assessment of NASA's Meteoroid and Orbital Debris Programs*; National Academies Press: Washington, DC, USA, 2011.
65. Rainbow, J.; Getting SSA off the Ground. 17 June 2022. Available online: https://spacenews.com/getting-ssa-off-the-ground/ (accessed on 20 April 2023).
66. NASA. The Collision of Iridium 33 and Cosmos 2251: The Shape of Things to Come. In Proceedings of the 60th International Astronautical Congress, Daejeon, Republic of Korea, 12–16 October 2009.
67. Botezatu, U.E.; Scarlat, C. Smart Cities Versus Sustainable development: Inverse progression? 2023; *unpublished work to be submitted*.

sensors

MDPI

Article

Measuring Heat Stress for Human Health in Cities: A Low-Cost Prototype Tested in a District of Valencia, Spain

Àlex Aduna-Sánchez [1], Antonio Correcher [2,*], David Alfonso-Solar [1] and Carlos Vargas-Salgado [1]

[1] Institute for Energy Engineering, Universitat Politècnica de València, Camí de Vera s/n, 46022 Valencia, Spain; alsanad@doctor.upv.es (À.A.-S.); daalso@iie.upv.es (D.A.-S.); carvarsa@upvnet.upv.es (C.V.-S.)
[2] Instituto de Automática e Informática Industrial, Universitat Politècnica de València, Camino de Vera s/n, 46022 Valencia, Spain
* Correspondence: ancorsal@upv.es

Abstract: Nowadays, the measurement of heat stress indices is of principal importance due to the escalating impact of global warming. As temperatures continue to rise, the well-being and health of individuals are increasingly at risk, which can lead to a detrimental effect on human performance and behavior. Hence, monitoring and assessing heat stress indices have become necessary for ensuring the safety and comfort of individuals. Thermal comfort indices, such as wet-bulb globe temperature (WBGT), Tropical Summer Index (TSI), and Predicted Heat Strain (PHS), as well as parameters like mean radiant temperature (MRT), are typically used for assessing and controlling heat stress conditions in working and urban environments. Therefore, measurement and monitoring of these parameters should be obtained for any environment in which people are constantly exposed. Modern cities collect and publish this relevant information following the Smart City concept. To monitor large cities, cost-effective solutions must be developed. This work presents the results of a Heat Stress Monitoring (HSM) system prototype network tested in the Benicalap-Ciutat Fallera district in Valencia, Spain. The scope of this work is to design, commission, and test a low-cost prototype that is able to measure heat stress indices. The Heat Stress Monitoring system comprises a central unit or receiver and several transmitters communicating via radiofrequency. The transmitter accurately measures wind speed, air temperature, relative humidity, atmospheric pressure, solar irradiation, and black globe temperature. The receiver has a 4G modem that sends the data to an SQL database in the cloud. The devices were tested over one year, showing that radio data transmission is reliable up to 700 m from the receiver. The system's power supply, composed of a Photovoltaic panel and Lithium-ion batteries, provided off-grid capabilities to the transmitter, with a tested backup autonomy of up to 36 days per charge. Then, indicators such as WBGT, TSI, and MRT were successfully estimated using the data collected by the devices. The material cost of a 12-point network is around EUR 2430 with a competitive price of EUR 190 per device.

Keywords: heat stress; WBGT; MRT; TSI; PHS; Arduino; smart city; low-cost

1. Introduction

Institutions and researchers around the world report rises in the Earth's surface temperature year by year [1–3]. Global warming concerns governments that signed the Paris Agreement in 2015 to keep the temperature rise under 2 °C. The effects of global warming are well known as extreme weather events, polar ice melting, and increases in average temperatures. Increasing the average temperature in populated areas can cause heat stress over the population at specific zones and times. That is relevant because it directly affects people's health [4,5] and has a significant economic impact [6].

Heat Stress (HS) is a known phenomenon that can be parametrized, and its monitoring can help to improve the citizens' quality of life. HS occurs when a body cannot expel excess heat. When that happens, people feel dizzy, have a high heart rate, have muscle cramps, and have an increased risk of heatstroke. HS is usually monitored to guarantee safe indoor and outdoor working conditions [7]. Still, it is also essential to study healthy living conditions in populated areas [8]. This is particularly critical in cities, as urban microclimate conditions can cause higher heat stress outdoors, as well as having a direct influence on thermal conditions inside the buildings [9,10].

Modern cities are evolving to the Smart City concept [11,12], making relevant information available to citizens and governors. In that direction, many researchers are developing sensor networks to gather details about city performance [12–15]. As stated by Bacco et al. [16], environmental monitoring is a key issue in Smart Cities as it can effectively monitor air quality [17] and thermal comfort.

However, many wireless sensors are needed to monitor a city thoroughly, so cost-effective solutions must be developed [18]. In this sense, recent research explores the cost reduction in Smart City monitoring applications. For example, Almalki et al. [19] presented a sensor architecture for Smart Cities to reduce the energy consumption of each node in the network. Another approach to lowering the cost of measuring data is using non-dedicated sensing systems (such as mobile phones) to increase the number of nodes [20] and merge this information with traditional sensing nodes. Nevertheless, some measures, such as humidity, wind speed, and wind direction (among others), are complicated to measure from non-dedicated nodes. A low-cost sensor for meteorological purposes is presented in Nan et al. [21]. The sensor can be deployed as a node in the measurement network. The variables it measures are temperature, humidity, and pressure, but the authors report low operating time and weaknesses against extreme weather conditions. Another approach oriented to low-cost sensors is presented by Hashmy et al. [22]. In this case, the authors do not show a device but rather a calibration method for low-cost sensors that increase the reliability of the measures obtained from low-cost devices.

ISO 7726 [23], ISO 7933 [24], and ISO 7243 [25] regulations standardize the methods for defining and measuring a heat stress index. The most used parameter to measure heat stress is the wet-bulb globe temperature (WBGT) [26,27]. Other commonly used indices are the Predicted Heat Strain (PHS) [28], the Universal Thermal Climate Index (UTCI) [29], the Tropical Summer Index (TSI), and the Thermal Working Limit (TWL). Other parameters such as the mean radiant temperature (MRT) [30] can also be helpful. In any case, to state the current HS, environmental variables (temperature, humidity, etc.) must be continuously measured. When studying HS in cities, a net of reliable measuring nodes will be needed.

Generally, commercial devices compute WBGT to characterize HS, although most modern HS meters give other indicators such as TWL. To calculate WBGT, they measure several magnitudes: air temperature, relative humidity, black globe temperature, radiant sunlight, or air pressure. They usually use chargers or batteries as the power source. In the standalone power case, the device autonomy can reach 16 days at the best chance. Regarding data logging, not all devices allow this feature. Those allowing data storage use custom applications to show data and do not allow real-time connection with external systems. Customers can find HS meters in the market, ranging from EUR 250 to EUR 800. Moreover, high-precision HS meters are available at prices reaching EUR 5000, turning these options too expensive if a large network with multiple nodes is needed or not suitable for outdoor urban environments. Some low-cost dedicated solutions to measure heat stress have already been developed, such as the one presented in Sulzer et al. [31]. In that case, a very cost-effective solution was obtained, but it required on-grid power supply and W-Fi connection, which are not always available.

Heat stress numerical modeling represents a different but complementary approach to measurement systems for evaluating and managing heat stress. Measurement systems directly assess physiological responses, such as core body temperature and heart rate, providing real-time and accurate data on individuals' conditions. This approach allows for immediate interventions, catering to individual variability and offering concrete validation of heat stress in specific environments. On the contrary, heat stress modeling relies on computer simulations to predict heat stress levels based on environmental factors, enabling proactive planning and cost-effective assessments across diverse scenarios. While measurement systems excel in precision and real-time action, modeling offers predictive capabilities and cost efficiency, making the choice between them dependent on the specific goals and context of the application.

Currently, several numerical models are employed to simulate the heat stress index. PALM [32] (Parallelized Large-Eddy Simulation Model) was crafted for simulating atmospheric and environmental flows. It has gained widespread usage in investigating heat stress in urban areas [33], taking into account factors such as radiation [34,35], convection, and the thermal properties of surfaces. Its capability to simulate complex urban structures adds depth to its applications. Some authors have developed works using PALM [36,37]. In contrast, ENVI-met [38] (Environmental Meteorology) is a microscale meteorological model primarily focusing on simulating urban microclimates. This model considers intricate interactions between buildings, vegetation, and the atmosphere, enabling the modeling of temperature, humidity, and other meteorological parameters at high spatial and temporal resolutions. ENVI-met proves valuable in assessing the impact of urban planning and design on heat stress, offering insights that help optimize green spaces and ventilation to create more comfortable urban environments. Works carried out using ENVI-met have been developed by [39]. Another tool is MITRAS [40] (Microscale Transport and Stream model), a comprehensive urban climate model developed to analyze the effects of different urban structures and materials on local climate conditions. MITRAS considers factors such as radiation, heat conduction, and airflow to simulate the thermal behavior of urban surfaces. It plays a pivotal role in evaluating the thermal performance of building materials and urban layouts, aiding in the design of energy-efficient and heat-resilient urban areas. Salim et al. [41] describes the developing theory and underlying processes of the microscale obstacle-resolving model MITRAS. Finally, RayMan [42], developed by the Meteorological Institute at the University of Freiburg, Germany, is designed for assessing solar radiation and thermal comfort. RayMan calculates human-biometeorological parameters, including the physiological equivalent temperature (PET), to measure thermal comfort. The RayMan model considers variables such as solar radiation, wind speed, and humidity to estimate the thermal sensation and potential heat stress experienced by individuals in outdoor environments. Human Thermal Comfort and heat stress models in urban areas using RayMan are analyzed in [43,44]. While these models significantly contribute to developing effective strategies for mitigating heat stress, it is essential to underscore the necessity for real measurements. Integrating these models with on-the-ground observations is crucial for comprehensively understanding and implementing strategies to combat heat stress in diverse environments and comprehensive means to assess such parameters.

This paper shows a cost-effective proposal for heat stress index measuring in outdoor environments. The Heat Stress Monitoring (HSM) system prototype can be deployed as a sensor network, thus covering broad areas, and it is battery-powered, featuring interesting off-grid and wireless capabilities.

2. Methodology

2.1. Heat Stress Indices

This section describes some of the heat stress indicators that can be directly calculated from environmental signals measured by the prototype proposed in this paper. It also shows other examples of the most used heat stress indices that require these parameters although they are not directly computable.

2.1.1. Wet-Bulb Globe Temperature (WBGT)

According to ISO 7243 [25], to obtain the WBGT index (in °C), there are two possible situations: WBGT index calculation for indoor (or outdoor without solar irradiance) and outdoor (with solar irradiance) environments. These two cases are calculated by Equation (1) and Equation (2), respectively.

$$WBGT_{indoor} = 0.7t_{nwb} + 0.3t_g \tag{1}$$

$$WBGT_{outdoor} = 0.7t_{nwb} + 0.2t_g + 0.1t_a \tag{2}$$

where t_{nwb} (°C) is the natural wet-bulb temperature, t_g (°C) is the globe temperature, and t_a (°C) is the dry air temperature. t_{nwb} is measured by a thermometer with its bulb covered with a wettened cotton, without shields for avoiding wind or radiation and under natural ventilation conditions. It can be calculated using Equation (3) [45].

$$t_{nwb} = t_w + 0.0021S - 0.42v_w + 1.93 \tag{3}$$

where, t_w (°C) is the psychometric temperature of wet bulb, S ($W{\cdot}m^{-2}$) is solar irradiation, and v_w ($m{\cdot}s^{-1}$) is the wind speed. t_w can be calculated with Equation (4), using air temperature (t_a) and relative humidity (RH%) measurements at standard sea level, as proposed by Stull [46].

$$t_w = t_a \tan^{-1}\left[0.151977(RH\% + 8.131659)^{1/2}\right] + \tan^{-1}(t_a + RH\%) - \tan^{-1}(RH\% - 1.676331) + 0.00391838(RH\%)^{3/2}{\cdot}\tan^{-1}(0.023101RH\%) - 4.686035 \tag{4}$$

2.1.2. Tropical Summer Index (TSI)

TSI is defined as the temperature of calm air at 50% relative humidity which imparts the same thermal sensation as the given environment [47]. TSI (in °C) is calculated using Equation (5), where t_g is the globe temperature (°C).

$$TSI = 0.745t_g + 0.308t_{nwb} - 2.06\sqrt{v_w + 0.841} \tag{5}$$

2.1.3. Other Heat Stress Indices

Other heat stress indicators require environmental parameters measured by the device proposed. For example, both the Predicted Heat Strain (PHS) model [24] and the Physiological Equivalent Temperature (PET) [48] require air temperature, mean radiant temperature (MRT), partial vapor pressure, and wind speed. Thermal Working Limit (TWL) uses dry air, wet bulb and globe temperatures, wind speed, and atmospheric pressure [49]. The Universal Thermal Climate Index (UTCI) can be derived with a model that uses air temperature, wind speed at 10 m, relative humidity, and MRT [50].

2.2. *Mean Radiant Temperature (MRT)*

MRT is a key parameter in the calculation of several of the heat stress indices described in this paper. MRT $\bar{t}_r$ (in °C) is calculated according to ISO 7726 [23] using Equations (6)–(9) depending on the globe diameter and the airflow. Equations (6) and (7) are used for forced and natural convection, respectively, when the globe diameter is less than 15 cm. Equations (8) and (9) are used when the globe diameter is equal to 15 cm. $\bar{t}_r$ depends on the globe temperature t_g (°C), the thermal emissivity of the globe ε_g, the diameter of the globe D (m), and the air temperature t_a (°C).

$$\bar{t}_{r(forced)} = \left[(t_g + 273)^4 + \frac{1.1{\cdot}10^8(v_w)^{0.6}}{\varepsilon_g{\cdot}D^{0.4}}(t_g - t_a)\right]^{1/4} - 273 \tag{6}$$

$$\bar{t}_{r(natural)} = \left[(t_g + 273)^4 + \frac{0.25{\cdot}10^8}{\varepsilon_g}\left(\frac{|t_g - t_a|}{D}\right)^{1/4}(t_g - t_a)\right]^{1/4} - 273 \tag{7}$$

$$\bar{t}_{r(forced)} = \left[(t_g + 273)^4 + 2.5{\cdot}10^8(v_w)^{0.6}{\cdot}(t_g - t_a)\right]^{1/4} - 273 \tag{8}$$

$$\bar{t}_{r(natural)} = \left[(t_g + 273)^4 + 0.4 \cdot 10^8 |t_g - t_a|^{1/4} \cdot (t_g - t_a) \right]^{1/4} - 273 \quad (9)$$

The choice between natural or forced convection equations depends on the thermal transfer coefficient value h_{cg} (W·m^{-2}·K^{-1}). h_{cg} is calculated for forced convection with Equation (10) and for natural convection with Equation (11). If the thermal transfer coefficient for forced convection is greater than that of natural convection, then the forced convection equation for MRT should be used, and vice versa.

$$h_{cg\ (forced)} = 6.3 \left(\frac{v_w{}^{0.6}}{D^{0.4}} \right) \quad (10)$$

$$h_{cg\ (natural)} = 1.4 \left(\frac{t_g - t_a}{D} \right)^{1/4} \quad (11)$$

2.3. *Prototype Requirements*

This prototype must accurately measure the environmental parameters required to calculate the described heat stress indices: air temperature, relative humidity, wind speed, atmospheric pressure, solar irradiation, and black globe temperature. It may also measure other parameters, such as precipitation and wind direction.

It must also meet the following requirements: operate in indoor and outdoor environments, be powered off-grid, and transmit the collected data wirelessly to a database in the cloud, as well as allowing for local datalogging capability. The whole system will also be designed to be replicable and low-cost.

To test the prototype, several devices will be installed in various locations within the Benicalap-Ciutat Fallera district in Valencia (Spain). The devices will be left to operate for a year, and then the results will be collected.

2.3.1. Wireless Data Transmission Test

The wireless data transmission will be carried out by a 433 MHz radio communication between the measuring device and a receiver. Several transmitters will communicate simultaneously to the same receiver every 15 min. The receiver will then upload the collected data to the cloud using a 4G connection. To test the range and efficiency of the radio system, 30 days of data will be collected. The elapsed time between transmissions will be measured to determine how many failed. Due to the slow dynamics of the studied parameters, the wireless data transmission system will be valid if at least 95% of the measures are taken at most every 35 min. The efficiency of the system will then be calculated using Equation (12).

$$\eta(\%) = \frac{N_{measures\ taken}}{N_{measures\ expected}} \cdot 100 \quad (12)$$

Finally, these parameters will be compared with the distances between every transmitter-receiver pair. Thus, the maximum range of the communication system will be obtained for urban environments. Additionally, for this test, several devices will also be installed in Universitat Politècnica de València to test the radio range in line of sight.

2.3.2. Power Supply and Autonomy Test

The energy supply system will be composed of a photovoltaic panel (PV) that will charge up to four 18650 Lithium-Ion batteries, with a capacity of 2.6 Ah each and a rated voltage of 3.7 V, so the capacity of one battery is around 962 Wh if a Depth of Discharge (DoD) of 100% is considered. To obtain the device's autonomy, it is required to estimate the battery charge level. The state of charge (SoC) is obtained using Equation (13), where V_{BAT} (V) is battery voltage.

$$\left\{\begin{array}{cc} \forall\, V_{BAT} \in [4 \quad \infty) \rightarrow & SoC = 100 \\ \forall\, V_{BAT} \in [3.1 \quad 4) \rightarrow & SoC = 100V_{BAT} - 300 \\ \forall\, V_{BAT} \in [2.7 \quad 3.1) \rightarrow & SoC = 25V_{BAT} - 67.5 \\ \forall\, V_{BAT} \in (-\infty \quad 2.7) \rightarrow & SoC = 0 \end{array}\right\} \tag{13}$$

Once the battery capacity is obtained, the device's energy consumption will be calculated. The autonomy will be estimated with this information and the time for which the device is on. This information was used to size the preliminary battery system. Once the device is installed, an analysis of the autonomy in actual operating conditions will be performed through a test where a device will be left to operate until its battery runs out.

2.3.3. Device Robustness Test

The system will be considered robust if both transmitter and receiver can operate autonomously and without interruptions in indoor and outdoor environments for 12 months. It must endure external conditions such as heavy rain, solar radiation, and extreme weather and be resilient to material degradation. This case will not consider other issues, such as energy shortage. In future developments, if the device is taken to the production phase, it will be sent to certifying entities to certify its water tightness.

3. Prototype Design

The Heat Stress Monitoring system (HSM) is a set of devices capable of periodically measuring the environmental parameters necessary to calculate the previous heat stress indices. It consists of two parts. First, a series of devices named transmitters installed in the target locations take measures periodically from their sensors. Then, they send the data through radio transmission to a receiver located in a nearby location. Finally, this receiver collects all the data and uploads it to a database in the cloud through a 4G connection. A diagram of the HSM system is shown in Figure 1.

3.1. *Transmitter*

3.1.1. Transmitter Hardware

In its most complete version, the transmitter can include up to five temperature sensors (four air temperature sensors and one 9 cm black globe temperature sensor), a relative humidity sensor, a barometric pressure sensor, two solar irradiance sensors (one facing upwards and another facing downwards), a wind speed and wind direction sensor, a raindrops sensor, and various voltage and current sensors to ensure the correct operation of the device. A list of available sensors and characteristics is shown in Table 1, and the electronics schematic is shown in Figure 2.

Although ISO 7243 [25] requires globe temperature to be measured using a standardized 15 cm diameter globe, it allows for the use of alternative globe sizes. According to Vargas-Salgado et al. [51], it is possible to use a 9 cm diameter globe in a heat stress monitor, with an acceptable margin of error. This size is cheaper and easier to handle.

An Arduino microcontroller manages this system, being compatible with both Arduino MEGA and UNO boards. Thanks to the Arduino's flexible and modular nature, the electronic circuit of the transmitter is prepared to connect additional sensors through the I2C and SPI buses. Therefore, it is compatible with a real-time clock (RTC) + SD card module that could provide offline data logging if necessary, avoiding the need for a receiver in some applications. Likewise, the transmitter also allows us to easily disconnect those sensors not needed in each case, helping to reduce manufacturing costs.

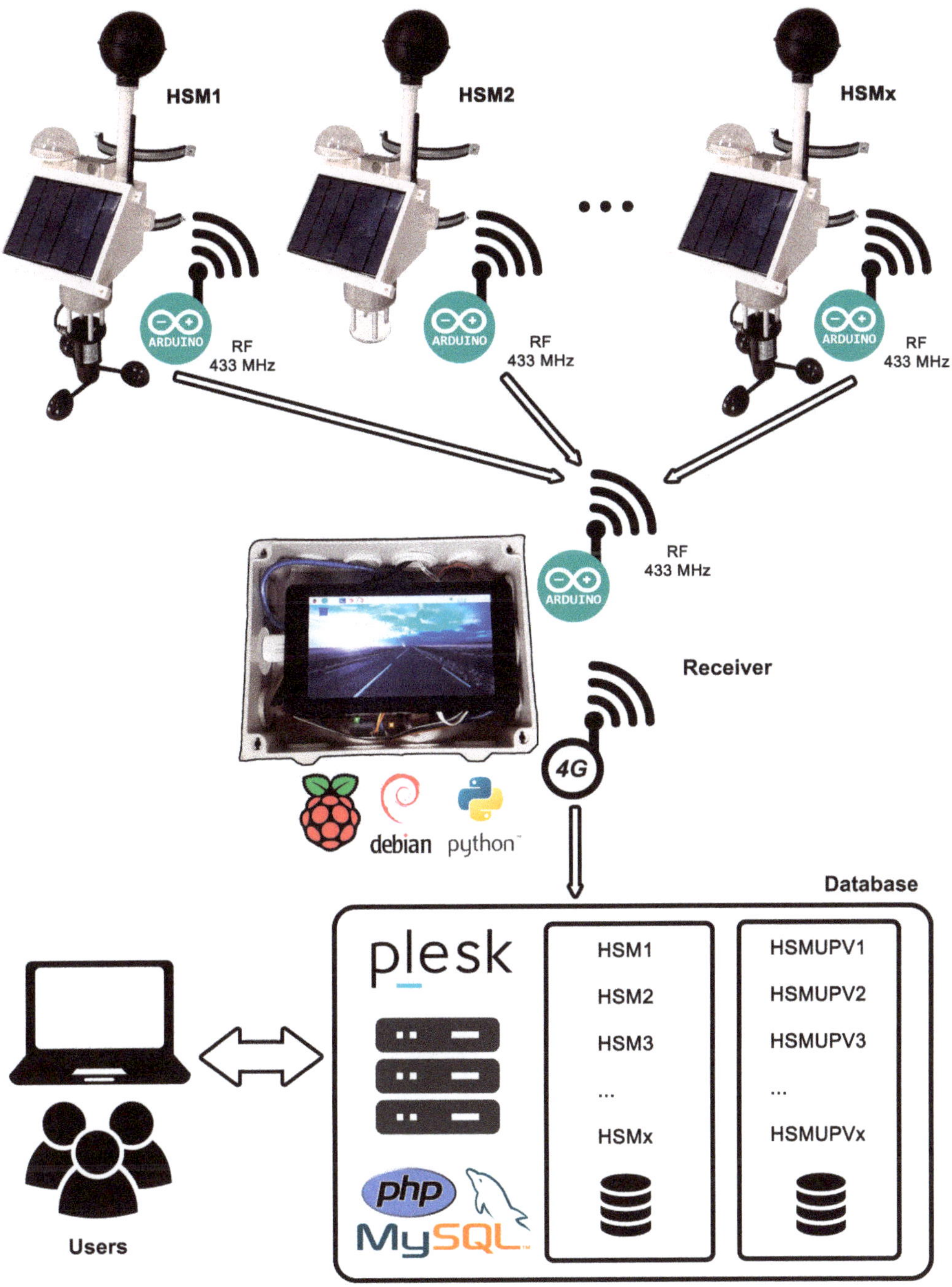

Figure 1. Complete HSM system diagram, including transmitters, receiver, and database.

Table 1. List of available sensors in the transmitter and its characteristics.

Component	Qty	Variable	Range	Output	Resolution	Accuracy	Price
DS18B20	3	Temperature	−55~125 °C	Digital	12 bits	±0.5 °C	EUR 0.90
AM2320	1	Temperature Relative Humidity	−40~80 °C 0~100%	Digital Digital	0.1 °C 0.1%	±0.5 °C ±3%	EUR 1.65
BME280	1	Air pressure Relative Humidity Temperature	300~1100 hPa 0~100% −40~85 °C	I2C Bus I2C Bus I2C Bus	0.16 Pa 0.01% 0.1 °C	±1 hPa ±3% ±1 °C	EUR 0.65
JL-FS2	1	Wind speed	0 (0.4–0.8)~30 m/s	0~5 V	0.1 m/s	±3%	EUR 32.30
HYXC-FXV	1	Wind direction	0~360° (from 0.5 m/s)	0~5 V	22.5°	±3%	EUR 32.05
INA219	3	Voltage Current	0~26 V ±3.2 A	I2C Bus I2C Bus	12 bits 0.8 mA	±1% ±1%	EUR 0.80
INA3221	1	Voltage Current	0~26 V ±3.2 A (3 Channels)	I2C Bus I2C Bus	12 bits 0.8 mA	±1% ±1%	EUR 1.95
MH-RD	1	Raindrops	0.1~2 MΩ	0~4.2 V	10 bits	N/D	EUR 0.40
Cebek C-0121	2	Solar irradiance	0~1000 W/m^2	0~300 mA	0.8 mA	±1%	EUR 32.62

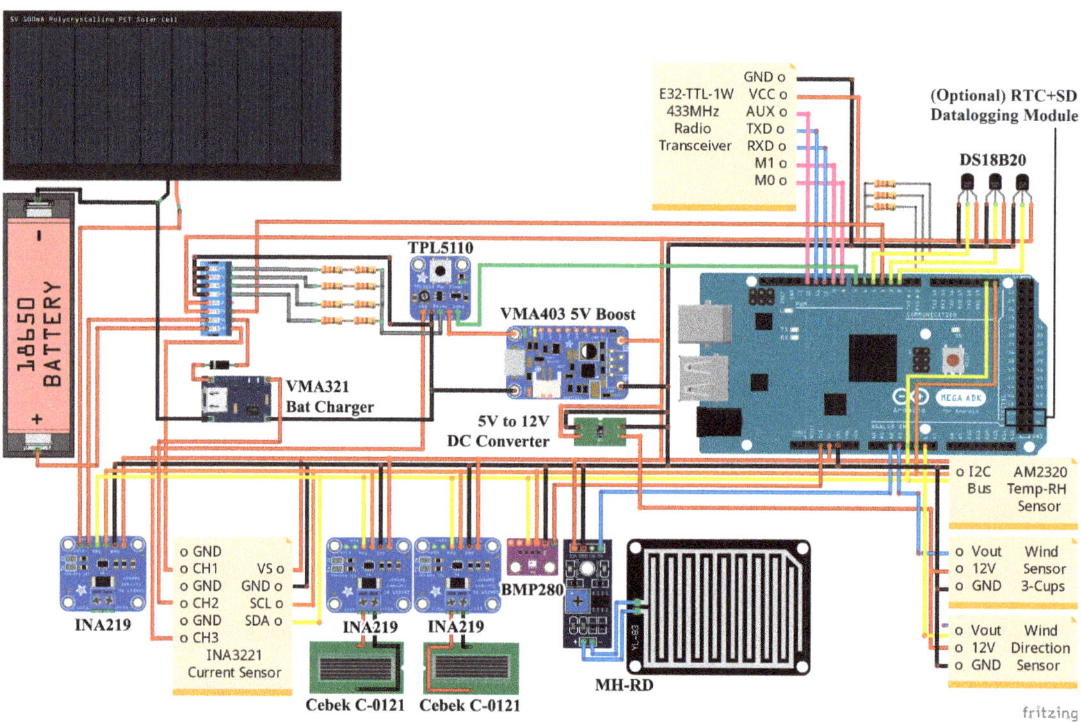

Figure 2. Transmitter electronics schematic featuring all available sensors and peripherals.

The transmitter has been designed to operate autonomously and wirelessly indoors and outdoors. It is powered off-grid by up to four 3.7 V, 2600 mAh, 18650 Lithium-ion batteries. It also includes a 3.5 W, 6 V, 165 × 135 mm photovoltaic (PV) panel that charges the batteries during the day. To save energy between measures, a TPL5110 timer switches off the device and puts it into standby mode. It can also be powered from the grid using a 5 V DC charger if needed.

3.1.2. Transmitter Firmware

The Arduino MEGA microcontroller has been programmed in C++ using the Arduino IDE 2.2.1, following the flow diagram described in Figure 3. The timer switches on the transmitter and runs the Arduino program for a certain amount of time which is 15 min by default. This interval can be selected using a switch between 1, 15, 30, and 60 min. The program takes 10 measures (one per second) from every sensor and calculates the arithmetic mean, filtering the error values. Then, it sends the final values and the device's ID through an E32-TTL-1W 433 MHz radiofrequency module with an antenna, manufactured by EBYTE and acquired in the local market in Valencia, Spain, with a maximum nominal range of 7500 m in line of sight. This system allows for unlimited transmitters to send data to one receiver within the radio range. Finally, the timer switches off the device to save energy and waits for the next cycle.

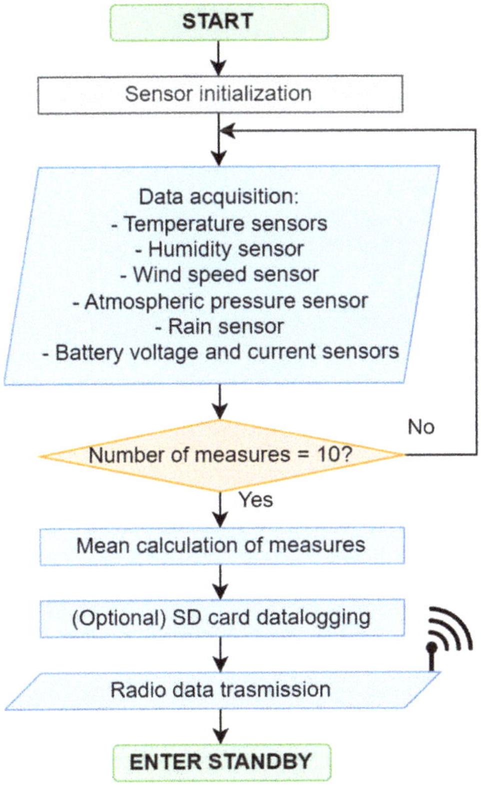

Figure 3. Transmitter firmware flow diagram.

3.1.3. Transmitter Box

The device has been assembled into a rainproof custom PLA plastic 3D-printed box. This box has been designed to accommodate all the external sensors: the anemometer, the black globe, two irradiation sensors and the raindrops sensor, as well as the radio antenna and the solar panel in an optimal tilt angle of 48°. A methacrylate semi-sphere protects the upper irradiation sensor to avoid sensor degradation and minimize dirt accumulation. It also includes a small solar radiation shield for the temperature and humidity sensors. The

DS18B20 sensors allow for two configurations to measure air temperature embedded into the device and measure air or surface temperature through an up to 100 m cable.

The box was coated with exterior acrylic paint and a clear varnish to endure extreme weather conditions and prevent PLA degradation. The joints were also sealed with silicone, and the radiation shield was covered with an insect screen. The transmitter can be installed with metal clamps and screws on walls or light poles in study locations. Figure 4 shows the different manufacturing and installation stages of a transmitter box.

Figure 4. Overview of the transmitter box in various manufacturing and installation stages.

3.2. Receiver and Database

Several parts form the receiver (Figure 5). First, an Arduino UNO microcontroller connected to another 433 MHz radio frequency module receives the data sent by every transmitter. Then, following the flow diagram in Figure 6, it sends this data via serial communication through a USB cable to a Raspberry Pi 3 board. This board works on a Debian-based operating system, which runs several Python and Linux command scripts that manage the internet connection through a 4G USB Modem and upload the data to a SQL database hosted in the cloud. Data is also saved on a microSD card for a local backup. Each radio module is configured with a unique ID that identifies the receiver. This way, every transmitter sends data to just one receiver, allowing for unlimited transmitters to send data to up to 256 receivers in the same area thus avoiding duplicate measures.

The receiver is assembled into an electrical connection box with a 7″ LCD touchscreen to navigate the OS. It needs to be installed where it can be powered from the grid with a USB power supply. The receiving radio antenna can be connected to the box through an extension cable and installed outdoors to improve range.

Figure 5. Receiver installed and running.

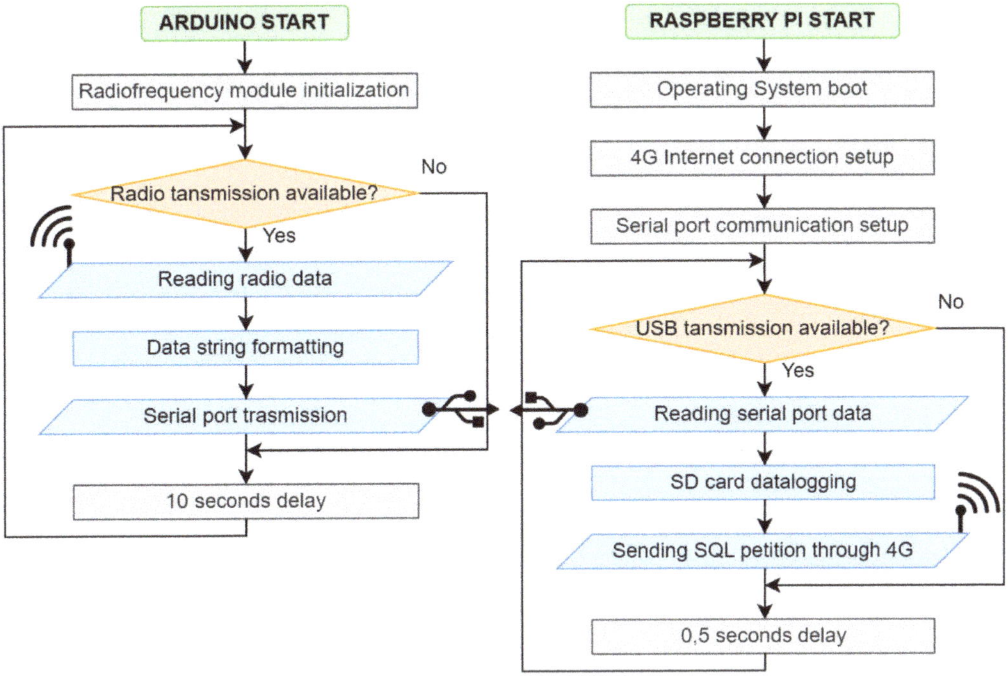

Figure 6. Receiver firmware flow diagram.

All the data collected by the transmitters' sensors is stored in an online SQL database. Users can access the database to display and download the desired data. Table 2 shows all the variables available. The transmitter and the database filter all the error values and measures. Failed radio transmissions are logged into a separate table, and incorrect or missing measures are displayed with specific error values (by default, '32,767') that can be easily filtered afterwards.

Table 2. List of variables accessible from the database.

Variable ID	Unit	Type	Description
Receiver	N/A	int	Receiver ID number (transmitters only send data to a specific receiver ID).
TX_ID	N/A	int	Transmitter ID number.
BAT_LV	%	float	Battery charge level calculated from V_BUS_V.
HR_BME	%	float	Relative humidity from BME280 sensor.
TEX_BME_C	°C	float	Air temperature from BME280 sensor.
WBGT_DS_C	°C	float	Black globe temperature from DS18B20 sensor.
TX_DS_C	°C	float	Air temperature from DS18B20 sensor.
TBOX_DS_C	°C	float	Additional DS18B20 sensor, usually measures transmitter box temperature.
HR_AM	%	float	Relative humidity from AM2320 sensor.
TEX_AM_C	°C	float	Air temperature from AM2320 sensor.
V_WIND_CUP	m/s	float	Wind speed from JL-FS2 sensor.
DIR_WIND	N/A	string	Wind direction from HYXC-FXV. Up to 16 values (N, NW, SSE, etc.).
PATM_Pa	Pa	float	Atmospheric pressure from BME280 sensor.
IRR_UP_Wm2	W/m^2	float	Solar irradiation from Cebek C-0121 sensor facing upwards.
IRR_DOWN_Wm2	W/m^2	float	Reflected solar irradiation from Cebek C-0121 sensor facing downwards.
RAIN	%	float	Estimated amount of rain (from none to full-wet) from MH-RD sensor.
V_BUS_V	V	float	Power supply voltage (batteries and solar charger) from INA3221 sensor.
V_PV_V	V	float	PV panel voltage from INA219 sensor.
I_BAT_mA	mA	float	Battery current from INA219 (negative means batteries are charging).
I_CH_mA	mA	float	Solar charging current from INA219 sensor.
I_IN_mA	mA	float	Current consumed by the Arduino and electronics from INA219 sensor.
I_PV_mA	mA	float	PV panel current from INA219 sensor.

4. Results and Discussion

4.1. Application Case: Benicalap (Valencia)

The system was tested in Valencia (Spain). Twelve transmitters and a receiver were installed in various locations within the Benicalap-Ciutat Fallera district. They were left to operate for more than one year, and the data was collected for analysis. The devices' locations are shown in Table 3 and Figure 7.

Table 3. List of HSM devices installed in the Benicalap-Ciutat Fallera district.

ID	Location	Coordinates	Distance	Installation
Receiver	Benicalap Park: Office	39°29′55.2″ N 0°23′46.7″ W	-	25 October 2018
HSM1	Benicalap Park: Entrance	39°29′54.8″ N 0°23′45.6″ W	33 m	9 January 2019

Table 3. *Cont.*

ID	Location	Coordinates	Distance	Installation
HSM2	Benicalap Park: Copse	39°29′57.0″ N 0°23′47.5″ W	70 m	9 July 2021
HSM3	Plaza Regino Mas	39°29′57.6″ N 0°23′38.3″ W	214 m	9 January 2019
HSM4	Calle Luis Braille	39°29′48.4″ N 0°23′40.2″ W	260 m	9 January 2019
HSM5	Benicalap Park: Theatre	39°29′55.5″ N 0°23′48.0″ W	34 m	11 December 2018
HSM6	Carrer del Ninot	39°30′00.0″ N 0°23′33.6″ W	345 m	9 January 2019
HSM7	Carrer del Foc	39°30′01.5″ N 0°23′38.7″ W	271 m	17 April 2019
HSM8	School: Outer wall	39°29′58.4″ N 0°23′37.8″ W	235 m	20 May 2019
HSM9	School: Inner wall	39°29′58.4″ N 0°23′37.8″ W	235 m	1 June 2019
HSM10	Senior Center: Roof	39°29′46.7″ N 0°23′43.9″ W	263 m	10 May 2019
HSM11	Senior Center: Indoor	39°29′47.1″ N 0°23′43.2″ W	258 m	10 May 2019
HSM12	Senior Center: Roof	39°29′47.1″ N 0°23′43.2″ W	258 m	1 June 2019

Figure 7. Location of HSM devices installed in Benicalap-Ciutat Fallera district. Source: Own creation based on an image obtained from Google Maps in June 2021.

As noted, in some cases, several devices are positioned in the same locations. The reason is to collect data from different environmental conditions within the same site, each serving a different purpose. For instance, in the senior center, HSM11 is located indoors. Meanwhile, HSM10 and HSM12 are installed outdoors on different roof sections. HSM12 will measure the effects of a planned green roof project, while HSM10 serves as a control. The same situation applies at the school with HSM8 outdoors and HSM9 indoors and with both devices installed on both sides of the same wall. The rest of the units are installed in various points of interest within the neighborhood, with different insolation levels or urban layout.

This application case is part of the Grow Green project, a European Union initiative. Its main objective is to tackle heat stress in urban environments through the use of nature-based solutions. Thus, several different actions have been planned in the Benicalap-Ciutat Fallera district, such as green roofs and walls, or to expand the green areas throughout the district. In this scenario, the HSM prototype plays a key role in the scientific research around the experimental validation of the solutions proposed.

4.2. Testing Results

4.2.1. Wireless Data Transmission Test

Analyzing a month of data registered in the database, given that measures should be taken every 15 min, we can determine how many radio transmissions failed. Due to the slow dynamics of the studied parameters, such as temperature, it is not critical to lose some inputs. For that, we calculated in which time intervals the measures were taken to determine when a significant amount of data was lost. As an example, a histogram of elapsed time between successful measures for HSM1 is shown in Figure 8.

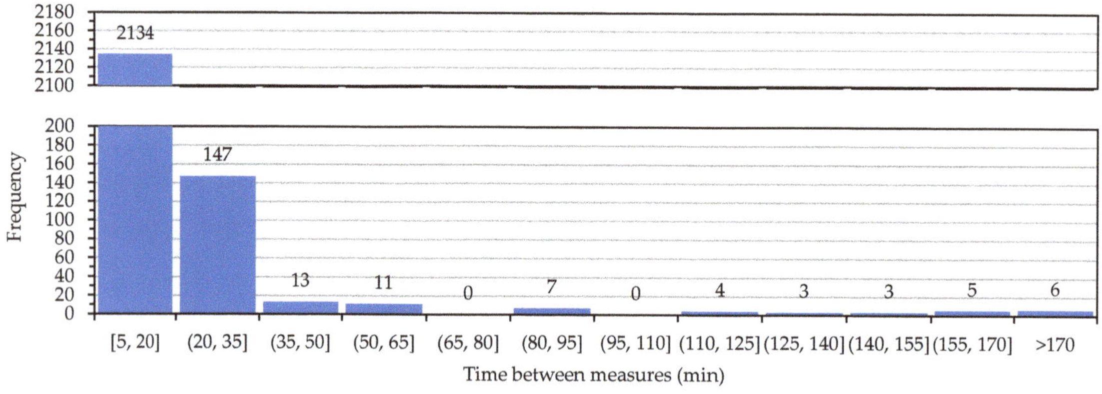

Figure 8. Histogram of elapsed time between successful measures for HSM1 in March of 2021.

Comparing these results with devices in other locations, Table 4 shows data from three devices and two additional test units installed in Universitat Politècnica de València (HSMUPV) that cover short, medium, and long-range measures, in both urban and line-of-sight settings. Data shows at least 97.5% of the radio transmissions were successful at most every 35 min (that is, every 15 or 30 min). We can therefore approve the radio system and consider it valid. Furthermore, if this data is contrasted with the distance of every device to the receiver, it can be determined that there is no significant change in the efficiency. However, there is a mean reduction of 31% in transmission errors when the transmitter is in line of sight. With that, we can establish a maximum valid tested range for the radio system of 345 m for urban environments and 700 m in line of sight.

Table 4. Relative frequency of elapsed time between successful measures and final radio transmission efficiency, depending on the distance to receiver.

Time Between Measures (min)	Relative Frequency (%)				
	HSM1	**HSM3**	**HSM6**	**HSMUPV6**	**HSMUPV7**
<20	91.47%	91.72%	91.37%	92.31%	91.32%
<35	97.77%	97.92%	97.53%	99.24%	98.19%
<50	98.33%	98.51%	98.30%	99.58%	98.74%
<65	98.80%	98.78%	98.83%	99.70%	99.34%
<80	98.80%	99.10%	99.31%	99.83%	99.73%
<110	99.10%	99.37%	99.55%	99.96%	99.95%
<140	99.40%	99.69%	99.76%	99.96%	100.00%
<170	99.74%	99.69%	99.80%	99.96%	100.00%
Distance (m)	33	214	345	570 (*)	700 (*)
Efficiency (%)	84.21%	84.31%	84.11%	89.20%	89.17%

(*) In line of sight.

4.2.2. Power Supply and Autonomy Test

The INA219 sensor measures the power consumption of the transmitter through its input current and voltage when the transmitter is switched on (about 20 s). The rest of the time, the TPL5110 timer switches off the transmitter to save energy. Two tests were carried out to obtain the transmitter power consumption and autonomy.

In the first test, a transmitter was left to operate in a laboratory, taking measures every minute until the battery was drained. As a result, it was obtained that the average power consumption of the transmitter while sending data was 0.641 W (33.33% of the time during the test) and 0.31 W in standby mode (66.66% of the time). The energy consumed to discharge the battery during the test (21.67 h) was 9 Wh. Thus, we obtain an estimated prototype battery life of around nine days per battery.

A second test was carried out to verify the device's performance and energy consumption. It used data from a device installed in its final location under actual operating conditions, taking measures every 15 min with its PV panel disconnected. The test results revealed that using all the sensors, the device could operate autonomously for up to 36 days using four batteries without recharging (Figure 9).

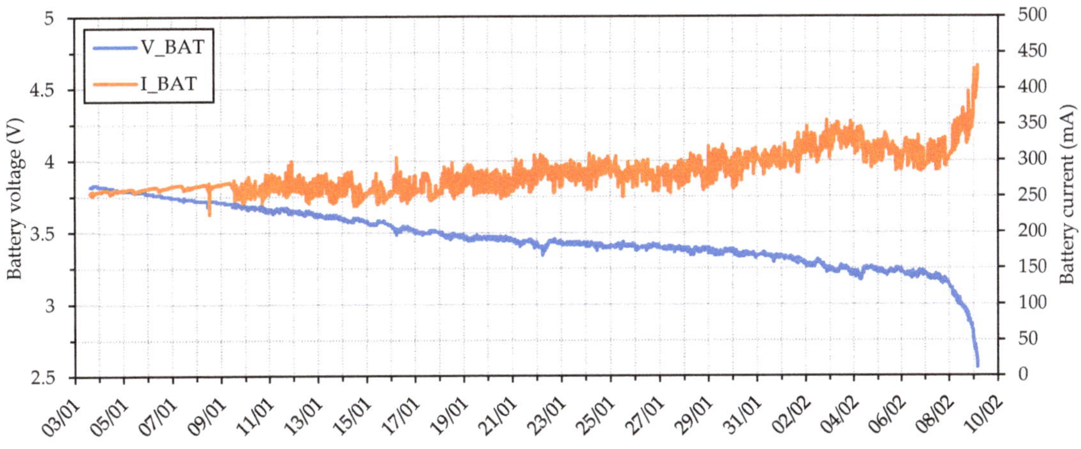

Figure 9. Battery discharge profile with four batteries measuring every 15 min without a PV panel. Data collected from HSM1 in real operating conditions from 3 January to 10 February 2019.

This supports the results in Figure 10 which shows that, when the PV panel is connected, the battery state of charge is always above 95%, as long as the daily average solar irradiation is over 10 W·m^{-2}. Even under long periods of solar irradiation under 10 W·m^{-2}, it is able to maintain the charge level above 70%. This ensures an unlimited battery life for the transmitter, working with no interruptions or energy shortages even on cloudy days, in low light regions, under dense foliage, or in the event of a faulty PV system.

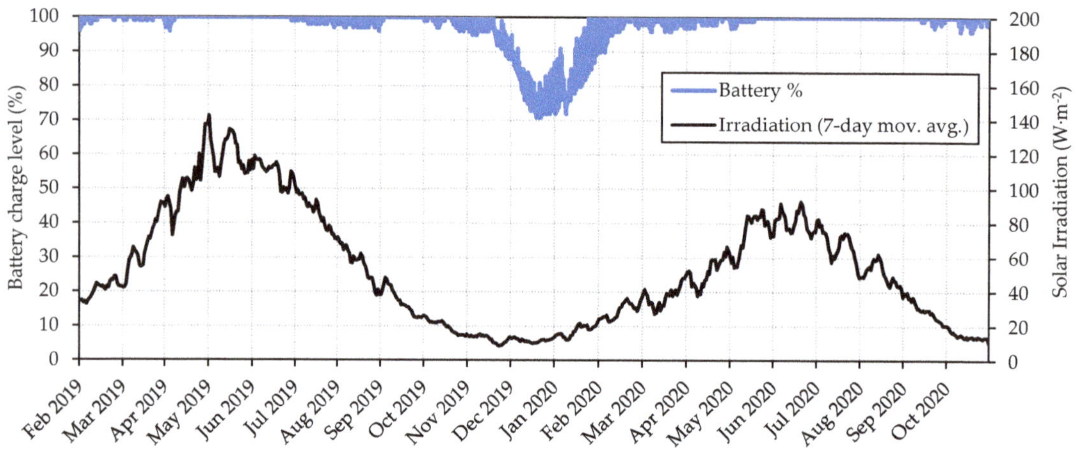

Figure 10. Battery state of charge and solar irradiation (7-day moving average). Data collected from HSM1 with PV panel in real operating conditions from February 2019 to November 2020.

4.2.3. Device Robustness

Several transmitters have been thoroughly tested in an urban environment for over a year or more. They have endured harsh weather conditions, including temperatures from −5 °C to 45 °C, strong winds, heavy rain, and long periods of intense solar irradiation.

Previous prototypes of transmitters had several flaws that required continuous updates. The most critical problems were water-related issues. Although 3D-printed PLA parts are easy to print, they are not water-tight because of the nature of fused deposition modelling. Also, PLA is not UV resistant; the material degrades after prolonged sunlight exposure [52]. Some boxes even broke when trying to open them. Thus, water filtrations during light and heavy rain were common, which led to corrosion and electrical problems. Another frequent problem was insects entering the box through the sensor radiation shield and creating dirt in the electronic circuit.

The proposed modifications included redesigning the box to minimize potential water infiltration paths. All 3D-printed components were coated with exterior acrylic paint and transparent varnish to protect them against direct sunlight exposure and enhance the seal. Critical joints were sealed using PTFE tape and silicone. An insect screen was installed within the radiation shield to deter insect entry. Subsequently, rigorous laboratory testing involving a continuous flow of water falling on the box for five minutes confirmed the absence of water infiltration issues, with no further reported problems in this regard.

It is important to note the transmitter's limitations concerning sensor robustness. Sensors were selected for accuracy, prioritizing cost-effectiveness. However, certain sensors, such as the DHT22, AM2320, and BME280 relative humidity sensors, exhibited problems after prolonged exposure to high humidity. Fortunately, these sensors are affordable and readily replaceable. Another concern is the solar irradiation sensor (Cebek C-0121), which experiences degradation over time, manifesting as yellowing and dirt accumulating on its surface, affecting measurement accuracy. To mitigate this issue, the proposed solution involves enclosing the sensor within a methacrylate dome.

No breakdowns were reported in the most recent transmitter iterations due to weather conditions (after one year of testing). However, regular sensor checks during annual maintenance and timely necessary replacements are advisable.

4.3. Sensor Data Collected

Data extracted directly from the database allows us to plot the main parameters under study. Figure 11 shows air temperature, black globe temperature, and relative humidity data, and Figure 12 shows wind speed and solar radiation data collected from HSM1 from 10 January to 12 January of 2021.

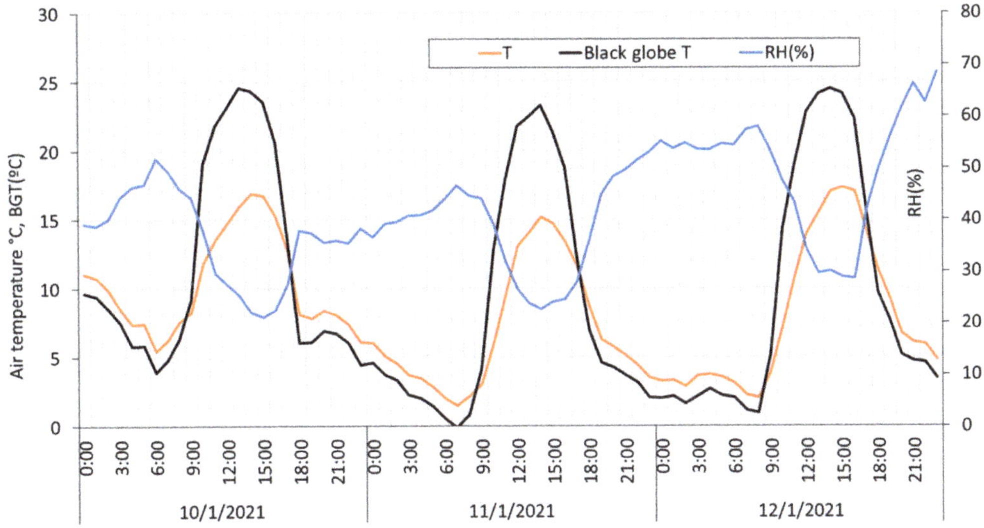

Figure 11. Air temperature, black globe temperature, and relative humidity data. Collected from HSM1 from 10 January to 12 January of 2021.

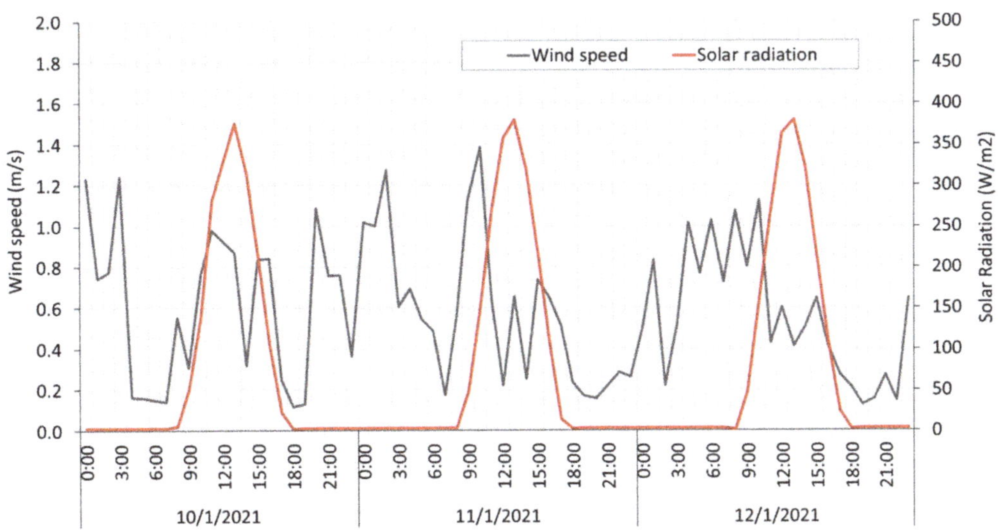

Figure 12. Wind speed and solar radiation data. Collected from HSM1 from 10 January to 12 January of 2021.

We can also plot the annual temperature variations by gathering data from various months. Figure 13 shows the monthly maximum, minimum, and average air temperature calculated from the daily averages from HSM3 from January 2020 to November 2021.

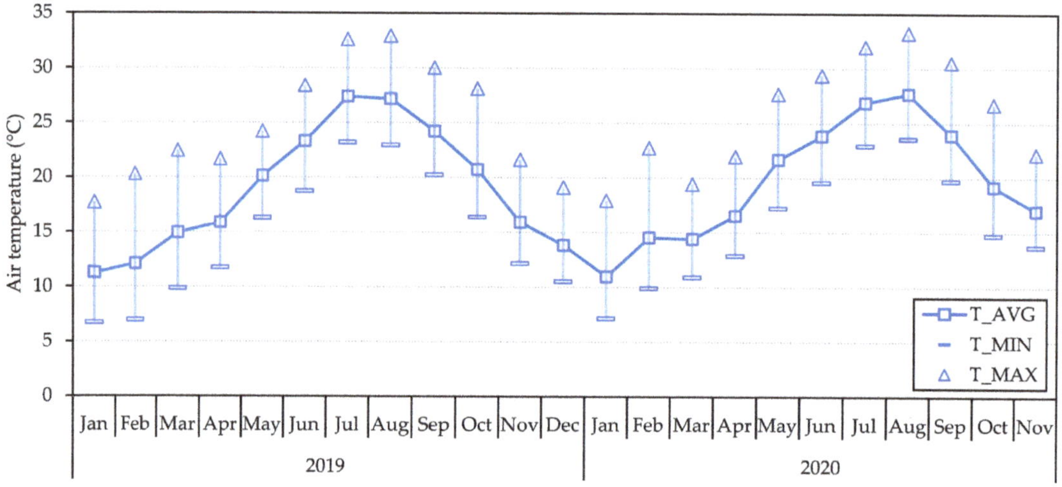

Figure 13. Monthly average of minimum, average, and maximum daily temperatures from HSM3 in 2019 and 2020.

The system successfully operated for nearly two years, diligently capturing regular daily and seasonal variations in air temperature patterns, humidity levels, black globe temperature, and solar radiation. To validate its accuracy, temperature and humidity values were meticulously compared with the data from the nearest reference weather station (Viveros, AEMet ref. 8416, N 39°28′50″ W 0°21′59″). The outcomes revealed a mean absolute error of approximately 1.15 °C for temperature and 8.01% for humidity. However, it is fundamental to acknowledge that the reference station, situated around 3 km from the HSM sensor, operates under distinct conditions. Furthermore, the reference station provides hourly data, while the HSM delivers information at a higher temporal resolution every 5 min.

Consequently, the comparison's relevance is somewhat limited due to the temporal mismatch and differing environmental contexts. Despite these considerations, a cross-verification of temperature and humidity sensors was conducted. The BME280 and AM2320 relative humidity sensors, along with the DS18B20 and AM2320 temperature sensors integrated into the same device, were scrutinized (refer to Table 1). The analysis yielded an average absolute error of 0.25 °C and 5.96% for temperature and humidity, respectively, well within the nominal accuracy range. This underscores the HSM's capabilities in ensuring measurement reliability, particularly through sensor redundancy. The redundancy mechanism significantly enhances the system's robustness, contributing to its overall effectiveness in providing accurate and dependable environmental data.

4.4. Heat Stress Indices Calculation

Using the data collected by the HSM system, we can estimate the heat stress indices described in the Methodology (see Section 2.1). Figure 14 shows the results of calculating the WBGT, MRT, and TSI index for HSM1 from 10 January to 12 January 2019.

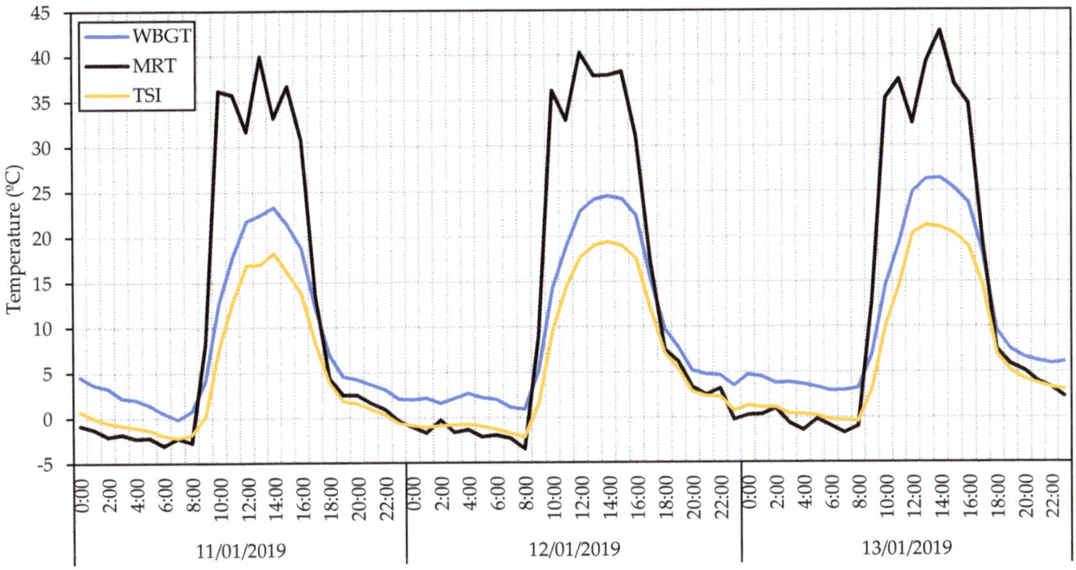

Figure 14. WBGT, MRT, and TSI heat stress index calculation from three winter days. Data collected from HSM1 from 10 January to 12 January 2019.

4.5. Total Cost of the Measurement System

Table 5 describes a detailed list of the cost of materials for an HSM system. This includes the individual cost of every transmitter and receiver and its components plus an extra 10% added for tooling use and production losses. Manufacturing costs as well as time and resources invested in design, development, and testing are not included.

Table 5. Detailed budget for the transmitter and receiver materials.

Item	Cost
Transmitter: Sensors	EUR 81.03
Transmitter: Electronic components	EUR 58.12
Transmitter: Lithium batteries (4 units)	EUR 17.55
Transmitter: Mechanical parts	EUR 17.19
Transmitter: 3D printed parts	EUR 12.87
Transmitter Total	EUR 186.76
Receiver: Electronic components and other parts	EUR 144.95
Receiver: SIM Card data plan (24 months)	EUR 26.40
Receiver: 4G USB Modem	EUR 17.07
Receiver Total	EUR 188.42

The total cost per device of the whole HSM system is reduced as more transmitters are included, as they can communicate with the same receiver. This is especially convenient for study locations like urban environments where many measuring points are needed in a reduced area. Finally, it is worth noting that the final cost of the system can be significantly reduced depending on the application. The list of sensors can be reduced to the most strictly necessary ones, the power needs may be covered with only two batteries, and the 4G modem and data plan may not be needed at all if the receiver is installed in a site with Wi-Fi connection.

As Table 6 shows, the HSM system's final cost per node is significantly lower than other commercial solutions [53,54], and it has some additional useful features that mean an advantage over its competitors.

Table 6. Comparison chart between the HSM system and other commercial solutions in terms of cost per node and other key features.

	HSM	HD32.3A-CV	Davis Vantage Pro2™ Plus
Dedicated heat stress measuring.	×	√	×
Black globe temperature sensor	√	√	√
Solar irradiation sensor	√	×	√
Customizable and modular	√	×	√
Data storage	Local + Online database	Local	Wireless display (300 m)
Off-grid power supply	Battery + PV panel	×	Battery + PV Panel
Cost per node	EUR 190	EUR 3473	EUR 2325

Finally, if we calculate the total cost of the HSM system described in the application case (see Section 4.1), with a 12-node network of transmitters and one receiver, we obtain a total of EUR 2429.54. This allows us to cover large and dense areas while maintaining a very competitive price.

5. Conclusions

This study aimed to design, build, commission, and test a low-cost Heat Stress Monitoring (HSM) system capable of measuring in an urban environment. The devices have been tested for more than one year. The HSM system consists of two parts: a series of transmitter devices that measure the environmental parameters through sensors and send the data through radio transmission, and a receiver device that receives the data from the transmitter and sends it to a cloud-based database.

The key environmental parameters the HSM system measures include air temperature, relative humidity, wind speed, atmospheric pressure, solar irradiation, and black globe temperature. The transmitter devices can be powered off-grid by a PV panel and Li-ion batteries or connected to the grid through an AC charger. They include a timer that puts the device in standby mode between measures to save energy. This gives an autonomy of up to 36 days per charge, although the PV panel can provide unlimited autonomy.

The HSM system successfully provided the parameters required to estimate heat stress indices such as WBGT, MRT, and TSI using the data collected by the devices. The HSM system can operate indoors and outdoors, and is designed to be replicable, open-source, and low-cost. The material cost for a 12-point network is around EUR 190 per measuring point, significantly lower than other commercial solutions.

The devices have demonstrated resilience in challenging weather conditions but are not without limitations. Previous prototypes were prone to water-related issues, with water infiltration during rain leading to corrosion and electrical problems. Insects entering through the sensor radiation shield caused dirt buildup in the electronic circuit. While design improvements have addressed these problems, the transmitters still have limitations related to sensor robustness. Some sensors, like relative humidity and solar irradiation, may degrade over time, particularly in high-humidity environments or due to yellowing and dirt accumulation. Replacing these sensors during periodic maintenance is recommended to maintain accuracy and functionality.

Further developments of the HSM system will focus on improving long-term durability by adopting more robust sensors, scaling up manufacturing to bring down costs even further, and exploring the possibility of giving 4G capabilities to the transmitters, eliminating the need for a receiver and making the system even more versatile.

The HSM system developed in this study is an innovative and cost-effective solution for monitoring heat stress in urban environments. The system's ability to measure key environmental parameters, estimate heat stress indices, and operate in extreme weather conditions makes it an attractive option for researchers, city planners, and public health officials concerned with mitigating the effects of heat stress.

Author Contributions: Conceptualization, À.A.-S., C.V.-S. and D.A.-S.; methodology, À.A.-S. and C.V.-S.; software, À.A.-S. and D.A.-S.; validation, À.A.-S., A.C. and D.A.-S.; investigation, À.A.-S., C.V.-S., D.A.-S., and A.C.; resources, C.V.-S. and A.C.; data curation, À.A.-S. and D.A.-S.; writing—original draft preparation, À.A.-S.; writing—review and editing, C.V.-S., D.A.-S. and A.C.; visualization, D.A.-S.; supervision, C.V.-S. and A.C.; project administration, D.A.-S.; funding acquisition, C.V.-S., D.A.-S. and A.C. All authors have read and agreed to the published version of the manuscript.

Funding: This research was funded by the European Union's Horizon 2020 research and innovation program under the project Green Cities for Climate and Water Resilience, Sustainable Economic Growth, Healthy Citizens and Environments (GROWGREEN) with reference 730283.

Institutional Review Board Statement: Not applicable.

Informed Consent Statement: Not applicable.

Data Availability Statement: Part of the data are contained within the article, the complete set of data are not publicly available.

Conflicts of Interest: The authors declare no conflict of interest.

References

1. NOAA National Centers for Environmental Information Climate at a Glance: Global Mapping. Available online: https://ncdc.noaa.gov/cag/ (accessed on 10 October 2022).
2. Ades, M.; Adler, R.; Allan, R.; Allan, R.P.; Anderson, J.; Argüez, A.; Arosio, C.; Augustine, J.A.; Azorin-Molina, C.; Barichivich, J.; et al. Global Climate. *Bull. Am. Meteorol. Soc.* **2020**, *101*, S24–S26. [CrossRef]
3. Miles, L.; Agra, R.; Sengupta, S.; Vidal, A.; Dickson, B. *Nature-Based Solutions for Climate Change Mitigation*; IUCN: Gland, Switzerland, 2021.
4. Deschenes, O. Temperature, Human Health, and Adaptation: A Review of the Empirical Literature. *Energy Econ.* **2014**, *46*, 606–619. [CrossRef]
5. Gasparrini, A.; Guo, Y.; Sera, F.; Vicedo-Cabrera, A.M.; Huber, V.; Tong, S.; de Sousa Zanotti Stagliorio Coelho, M.; Nascimento Saldiva, P.H.; Lavigne, E.; Matus Correa, P.; et al. Projections of Temperature-Related Excess Mortality under Climate Change Scenarios. *Lancet Planet. Health* **2017**, *1*, e360–e367. [CrossRef] [PubMed]
6. Fishman, R.; Carrillo, P.; Russ, J. Long-Term Impacts of Exposure to High Temperatures on Human Capital and Economic Productivity. *J. Environ. Econ. Manag.* **2019**, *93*, 221–238. [CrossRef]
7. Nassiri, P.; Monazzam, M.R.; Golbabaei, F.; Farhang Dehghan, S.; Shamsipour, A.; Ghanadzadeh, M.J.; Asghari, M. Modeling Heat Stress Changes Based on Wet-Bulb Globe Temperature in Respect to Global Warming. *J. Environ. Health Sci. Eng.* **2020**, *18*, 441–450. [CrossRef] [PubMed]
8. Castiglia, R.; Wilkinson, S.J.; Oliveira, B.; Hacon, S. Wind and Greenery Effects in Attenuating Heat Stress: A Case Study. *J. Clean. Prod.* **2021**, *291*, 125919. [CrossRef]
9. Hong, T.; Malik, J.; Krelling, A.; O'Brien, W.; Sun, K.; Lamberts, R.; Wei, M. Ten Questions Concerning Thermal Resilience of Buildings and Occupants for Climate Adaptation. *Build. Environ.* **2023**, *244*, 110806. [CrossRef]
10. Liu, S.; Kwok, Y.T.; Ren, C. Investigating the Impact of Urban Microclimate on Building Thermal Performance: A Case Study of Dense Urban Areas in Hong Kong. *Sustain. Cities Soc.* **2023**, *94*, 104509. [CrossRef]
11. Kirimtat, A.; Krejcar, O.; Kertesz, A.; Tasgetiren, M.F. Future Trends and Current State of Smart City Concepts: A Survey. *IEEE Access* **2020**, *8*, 86448–86467. [CrossRef]
12. Alavi, A.H.; Jiao, P.; Buttlar, W.G.; Lajnef, N. Internet of Things-Enabled Smart Cities: State-of-the-Art and Future Trends. *Measurement* **2018**, *129*, 589–606. [CrossRef]
13. Csaji, B.C.; Kemeny, Z.; Pedone, G.; Kuti, A.; Vancza, J. Wireless Multi-Sensor Networks for Smart Cities: A Prototype System With Statistical Data Analysis. *IEEE Sens. J.* **2017**, *17*, 7667–7676. [CrossRef]
14. Addabbo, T.; Fort, A.; Mugnaini, M.; Panzardi, E.; Pozzebon, A.; Tani, M.; Vignoli, V. A Low Cost Distributed Measurement System Based on Hall Effect Sensors for Structural Crack Monitoring in Monumental Architecture. *Measurement* **2018**, *116*, 652–657. [CrossRef]
15. Roque-Cilia, S.; Tamariz-Flores, E.I.; Torrealba-Meléndez, R.; Covarrubias-Rosales, D.H. Transport Tracking through Communication in WDSN for Smart Cities. *Measurement* **2019**, *139*, 205–212. [CrossRef]
16. Bacco, M.; Delmastro, F.; Ferro, E.; Gotta, A. Environmental Monitoring for Smart Cities. *IEEE Sens. J.* **2017**, *17*, 7767–7774. [CrossRef]

17. Awan, F.M.; Minerva, R.; Crespi, N. Using Noise Pollution Data for Traffic Prediction in Smart Cities: Experiments Based on LSTM Recurrent Neural Networks. *IEEE Sens. J.* **2021**, *21*, 20722–20729. [CrossRef]
18. Abate, F.; Carratù, M.; Liguori, C.; Paciello, V. A Low Cost Smart Power Meter for IoT. *Measurement* **2019**, *136*, 59–66. [CrossRef]
19. Almalki, K.J.; Jabbari, A.; Ayinala, K.; Sung, S.; Choi, B.-Y.; Song, S. ELSA: Energy-Efficient Linear Sensor Architecture for Smart City Applications. *IEEE Sens. J.* **2022**, *22*, 7074–7083. [CrossRef]
20. Habibzadeh, M.; Xiong, W.; Zheleva, M.; Stern, E.K.; Nussbaum, B.H.; Soyata, T. Smart City Sensing and Communication Sub-Infrastructure. In Proceedings of the 2017 IEEE 60th International Midwest Symposium on Circuits and Systems (MWSCAS), Boston, MA, USA, 6–9 August 2017; pp. 1159–1162.
21. Nan, F.; Zeng, C.; Ni, G.; Zhou, M.; Shen, H. Development and Validation of Low-Cost IoT Environmental Sensors: A Case Study in Wuhan, China. *IEEE Sens. J.* **2023**, *23*, 3069–3078. [CrossRef]
22. Hashmy, Y.; Khan, Z.U.; Ilyas, F.; Hafiz, R.; Younis, U.; Tauqeer, T. Modular Air Quality Calibration and Forecasting Method for Low-Cost Sensor Nodes. *IEEE Sens. J.* **2023**, *23*, 4193–4203. [CrossRef]
23. *AENOR UNE-EN ISO 7726*; Ergonomía de Los Ambientes Térmicos. Instrumentos de Medida de Las Magnitudes Físicas. AENOR: Madrid, Spain, 2002.
24. *AENOR UNE-EN ISO 7933*; Ergonomía Del Ambiente Térmico. Determinación Analítica e Interpretación Del Estrés Térmico Mediante El Cálculo de La Sobrecarga Térmica Estimada. AENOR: Madrid, Spain, 2005.
25. *AENOR UNE-EN ISO 7243*; Ergonomía Del Ambiente Térmico. Evaluación Del Estrés al Calor Utilizando El Índice WBGT (Temperatura de Bulbo Húmedo y de Globo). AENOR: Madrid, Spain, 2017.
26. Mirzabeigi, S.; Khalili Nasr, B.; Mainini, A.G.; Blanco Cadena, J.D.; Lobaccaro, G. Tailored WBGT as a Heat Stress Index to Assess the Direct Solar Radiation Effect on Indoor Thermal Comfort. *Energy Build.* **2021**, *242*, 110974. [CrossRef]
27. Zare, S.; Shirvan, H.E.; Hemmatjo, R.; Nadri, F.; Jahani, Y.; Jamshidzadeh, K.; Paydar, P. A Comparison of the Correlation between Heat Stress Indices (UTCI, WBGT, WBDT, TSI) and Physiological Parameters of Workers in Iran. *Weather. Clim. Extrem.* **2019**, *26*, 100213. [CrossRef]
28. Chowdhury, S.; Hamada, Y.; Ahmed, K.S. Prediction and Comparison of Monthly Indoor Heat Stress (WBGT and PHS) for RMG Production Spaces in Dhaka, Bangladesh. *Sustain. Cities Soc.* **2017**, *29*, 41–57. [CrossRef]
29. Vatani, J.; Golbabaei, F.; Dehghan, S.F.; Yousefi, A. Applicability of Universal Thermal Climate Index (UTCI) in Occupational Heat Stress Assessment: A Case Study in Brick Industries. *Ind. Health* **2016**, *54*, 14–19. [CrossRef] [PubMed]
30. Thorsson, S.; Rocklöv, J.; Konarska, J.; Lindberg, F.; Holmer, B.; Dousset, B.; Rayner, D. Mean Radiant Temperature—A Predictor of Heat Related Mortality. *Urban Clim.* **2014**, *10*, 332–345. [CrossRef]
31. Sulzer, M.; Christen, A.; Matzarakis, A. A Low-Cost Sensor Network for Real-Time Thermal Stress Monitoring and Communication in Occupational Contexts. *Sensors* **2022**, *22*, 1828. [CrossRef]
32. Maronga, B.; Banzhaf, S.; Burmeister, C.; Esch, T.; Forkel, R.; Fröhlich, D.; Fuka, V.; Gehrke, K.F.; Geletič, J.; Giersch, S.; et al. Overview of the PALM Model System 6.0. *Geosci. Model. Dev.* **2020**, *13*, 1335–1372. [CrossRef]
33. Fröhlich, D.; Matzarakis, A. Calculating Human Thermal Comfort and Thermal Stress in the PALM Model System 6.0. *Geosci. Model. Dev.* **2020**, *13*, 3055–3065. [CrossRef]
34. Krč, P.; Resler, J.; Sühring, M.; Schubert, S.; Salim, M.H.; Fuka, V. Radiative Transfer Model 3.0 Integrated into the PALM Model System 6.0. *Geosci. Model. Dev.* **2021**, *14*, 3095–3120. [CrossRef]
35. Salim, M.H.; Schubert, S.; Resler, J.; Krč, P.; Maronga, B.; Kanani-Sühring, F.; Sühring, M.; Schneider, C. Importance of Radiative Transfer Processes in Urban Climate Models: A Study Based on the PALM 6.0 Model System. *Geosci. Model. Dev.* **2022**, *15*, 145–171. [CrossRef]
36. Khan, A.; Bilal, S.; Khan, A.L.; Imran, M.; Shahzad, R.; Al-Harrasi, A.; Al-Rawahi, A.; Al-Azhri, M.; Mohanta, T.K.; Lee, I.J. Silicon and Gibberellins: Synergistic Function in Harnessing Aba Signaling and Heat Stress Tolerance in Date Palm (*Phoenix dactylifera* L.). *Plants* **2020**, *9*, 620. [CrossRef]
37. Letzel, M.O.; Krane, M.; Raasch, S. High Resolution Urban Large-Eddy Simulation Studies from Street Canyon to Neighbourhood Scale. *Atmos. Environ.* **2008**, *42*, 8770–8784. [CrossRef]
38. Liu, Z.; Cheng, W.; Jim, C.Y.; Morakinyo, T.E.; Shi, Y.; Ng, E. Heat Mitigation Benefits of Urban Green and Blue Infrastructures: A Systematic Review of Modeling Techniques, Validation and Scenario Simulation in ENVI-Met V4. *Build. Environ.* **2021**, *200*, 107939. [CrossRef]
39. Pérez-López, P.; Gschwind, B.; Blanc, P.; Frischknecht, R.; Stolz, P.; Durand, Y.; Heath, G.; Ménard, L.; Blanc, I. ENVI-PV: An Interactive Web Client for Multi-Criteria Life Cycle Assessment of Photovoltaic Systems Worldwide. *Progress. Photovolt. Res. Appl.* **2017**, *25*, 484–498. [CrossRef]
40. Schlünzen, K.H.; Hinneburg, D.; Knoth, O.; Lambrecht, M.; Leitl, B.; López, S.; Lüpkes, C.; Panskus, H.; Renner, E.; Schatzmann, M.; et al. Flow and Transport in the Obstacle Layer: First Results of the Micro-Scale Model MITRAS. *J. Atmos. Chem.* **2003**, *44*, 113–130. [CrossRef]
41. Salim, M.H.; Heinke Schlünzen, K.; Grawe, D.; Boettcher, M.; Gierisch, A.M.U.; Fock, B.H. The Microscale Obstacle-Resolving Meteorological Model MITRAS v2.0: Model Theory. *Geosci. Model. Dev.* **2018**, *11*, 3427–3445. [CrossRef]
42. Matzarakis, A.; Gangwisch, M.; Fröhlich, D. RayMan and SkyHelios Model. In *Urban Microclimate Modelling for Comfort and Energy Studies*; Springer International Publishing: Cham, Switzerland, 2021; pp. 339–361.

43. Matzarakis, A.; Rutz, F.; Mayer, H. Modelling Radiation Fluxes in Simple and Complex Environments—Application of the RayMan Model. *Int. J. Biometeorol.* **2007**, *51*, 323–334. [CrossRef] [PubMed]
44. Qureshi, A.M.; Rachid, A. Heat Stress Modeling Using Neural Networks Technique. *IFAC-Pap.* **2022**, *55*, 13–18. [CrossRef]
45. Hunter, C.H.; Minyard, O. *Estimating Wet Bulb Globe Temperature Using Standard Meteorological Measurements;* Department of Energy (DOE), Office of Scientific and Technical Information (OSTI): Oak Ridge, TN, USA, 1999.
46. Stull, R. Wet-Bulb Temperature from Relative Humidity and Air Temperature. *J. Appl. Meteorol. Climatol.* **2011**, *50*, 2267–2269. [CrossRef]
47. Bureau of Indian Standards (BIS). *National Building Code of India, Group 4;* Bureau of Indian Standards (BIS): New Delhi, India, 2005.
48. Mayer, H.; Höppe, P. Thermal Comfort of Man in Different Urban Environments. *Theor. Appl. Climatol.* **1987**, *38*, 43–49. [CrossRef]
49. Brake, D.J.; Bates, G.P. Limiting Metabolic Rate (Thermal Work Limit) as an Index of Thermal Stress. *Appl. Occup. Environ. Hyg.* **2002**, *17*, 176–186. [CrossRef] [PubMed]
50. Bröde, P.; Fiala, D.; Błażejczyk, K.; Holmér, I.; Jendritzky, G.; Kampmann, B.; Tinz, B.; Havenith, G. Deriving the Operational Procedure for the Universal Thermal Climate Index (UTCI). *Int. J. Biometeorol.* **2012**, *56*, 481–494. [CrossRef] [PubMed]
51. Vargas-Salgado, C.; Chiñas-Palacios, C.; Aguila-León, J.; Alfonso-Solar, D. Measurement of the Black Globe Temperature to Estimate the MRT and WBGT Indices Using a Smaller Diameter Globe than a Standardized One: Experimental Analysis. In Proceedings of the Proceedings 5th CARPE Conference: Horizon Europe and beyond, Universitat Politècnica de València, Valencia, Spain, 23–25 October 2019.
52. Podzorova, M.V.; Tertyshnaya, Y.V.; Pantyukhov, P.V.; Popov, A.A.; Nikolaeva, S.G. Influence of Ultraviolet on Polylactide Degradation. *AIP Conf. Proc.* **2017**, *1909*, 020173.
53. Arvind Industries Europe HD32.3A-CV Direct Reading of PMV/PPD—WBGT. Available online: https://arvindeurope.com/thermal-comfort/hd323a-cv-direct-reading-of-pmv-ppd-wbgt (accessed on 7 March 2023).
54. Davis Instruments Davis Vantage Pro2TM Plus Inalámbrica. Available online: https://www.estacionesdavis.es/es/davis-vantage-pro-2/12-davis-vantage-pro2-plus-inalambrica.html (accessed on 7 March 2023).

smart cities

MDPI

Article

Usefulness of a Civic Engagement Scale for Research on Smart Cities: Measuring Attitudes and Behavior

Jiri Remr

INESAN (Institute for Evaluations and Social Analyses), Sokolovská 351/25, 18600 Prague, Czech Republic; jiri.remr@inesan.eu

Abstract: Civic engagement plays a critical role in smart city innovation and urban development by encouraging active participation in civic activities such as volunteering, voting, community organizing, or advocacy, all of which contribute to the development of local communities. This study highlights the need to assess civic engagement in smart cities in order to improve the interactions between technology and society. The study assessed the reliability and validity of the Civic Engagement Scale (CES) in the Czech context. The results presented are based on a representative sample of 1366 respondents from the general population aged 15–74. The study included univariate statistics, tests of internal consistency, and principal component analysis. In addition, the study presents the results of confirmatory factor analysis (CFA) that was conducted to examine the fit of the proposed model to empirical data. The results indicate that the CES has excellent psychometric properties, including high internal consistency and favorable absolute and incremental indices. The Czech version of the CES can be considered a valid and reliable instrument. The findings suggest using CES to research and evaluate policy interventions aimed at developing digital platforms that enable citizens to easily participate in urban planning and smart city projects, community-driven smart city projects that ensure local needs and preferences are addressed, or implementing incentive programs for citizens.

Keywords: civic engagement; smart cities; psychometrics; cross-sectional research; participation; local communities

1. Introduction

Civic engagement involves the active participation of individuals in a range of civic activities that contribute to the overall consolidation of society. These activities include volunteering, donating, voting, and organizing community initiatives [1,2]. At the individual level, civic engagement stems from a commitment to give back to the community and embodies a set of values, knowledge, skills, and actions that generate political and social benefits for the local community [3]. Adler and Goggin [4] defined civic engagement as the way citizens participate in improving the conditions of their (local) communities. Similarly, Ehrlich [5] emphasized that civic engagement involves individual and collective actions aimed at improving the conditions of life within a given community, including support for democratic and political initiatives volunteering and donation [6,7]. In summary, civic engagement serves as a manifestation of social belonging [8], active participation [9], and the practice of active citizenship [10].

The importance of civic engagement is reflected in the growing interest of researchers and policymakers in understanding its nature, determinants, and outcomes. Deng and Fei [11] suggested that civic engagement, in addition to other constructs such as community engagement or public participation, refers to activities within social and political processes designed to enhance the learning and understanding of social issues, facilitate communication, and empower participants. This study draws attention to the need to assess civic engagement in smart cities in order to improve the interplay between technology and society. Given the objectives of this study, the focus on smart cities concerns perceptions of renewable energy, sustainable transportation, and smart public services [12].

Civic engagement is central to the successful development and functioning of modern urban environments, including smart cities. The concept of smart cities, first used in 1994 [13], depends on the interaction between citizens and urban systems, where civic engagement plays a central role in planning, decision-making, and innovation processes. Smart cities have emerged as a response to accelerating urbanization processes [12]. However, rapid urban development supported by modern information and communication technologies can lead to gaps in civic engagement and participation. Therefore, the continuous monitoring and support of civic engagement is essential to create citizen-centered smart cities, successfully implement modernization projects, and ensure the sustainability of their positive impacts [14]. In addition, interventions aimed at improving residents' quality of life through information and communication technologies can be more effective if they are developed and implemented in collaboration with citizens [1]. Smart cities aim to create environments that are responsive to the needs of residents [11]; civic engagement is the linchpin that enables this interaction and participation.

2. Literature Review

Given the importance of analyzing and measuring civic engagement, there has been a need to develop appropriate research tools. Traditionally, such efforts have involved asking respondents about their participation in various community initiatives, e.g., environmental, religious, recreational, and cultural programs [15,16]. However, given the complexity of the issue, the development of sophisticated and reliable tools in the form of scales with established reliability and assessed validity has become increasingly important [3]. The reason is that a well-designed scale can add value to both theoretical and practical knowledge about civic engagement and provide stakeholders with a conceptual framework [1,7]. A validated tool for measuring civic engagement can also help to create participatory and responsive smart cities with the capacity to transform urban life. Appropriate scales that have been developed to date fall into three main categories: the first assesses civic behavior, quantifying specific activities related to civic engagement, such as volunteering, attending public meetings, or voting [10]; the second category assesses attitudes associated with civic engagement, including altruism, interest in social issues, or a sense of social justice [1,4,8]; and the third category combines these two components by examining the relationship between attitudes and corresponding behaviors. An example of this approach is represented by the Civic Engagement Scale (CES).

Exactly a decade ago, in 2013, the CES developed by Doolittle and Faul [3] was introduced. Despite the initial lack of psychometric testing, numerous research initiatives to validate the scale and to implement it in local settings have emerged over the past decade [17,18]. However, despite the increased attention, the knowledge gap has remained regarding the psychometric characteristics of the scale, especially with respect to specific populations and framing topics (e.g., smart cities).

Therefore, the primary objective of this study was to validate the CES with respect to smart cities. Special emphasis was put on presenting the psychometric characteristics, including the reliability and convergent validity of the scale. In addition, this study investigated the applicability of CES to the adult population in Czechia. It is worth mentioning that the population studied in the original study performed by Doolittle and Faul [3] consisted of students within a reported age range of 17–63 years, with a mean age of 28.42 years. However, the data used in this study were from a representative sample of the Czech population aged 15–74, enabling the analysis of civic engagement across different segments of the population. This study aimed to deepen the understanding of civic engagement in the context of smart cities and to provide other researchers, policymakers, participation practitioners, and urban planners with a reliable tool for researching, measuring, and evaluating this critical construct. The study had three main objectives: (a) to explore the underlying dimensions (factor structures) of civic engagement with exploratory factor analysis; (b) to confirm these factor structures through confirmatory factor analysis; and (c) to assess the convergent validity of the scale.

3. Materials and Methods

3.1. Participants and Procedures

The population was defined as the general population of Czechia aged 15 to 74. In order to obtain a representative sample, an address-based sampling technique was used. First, a total of 179 sampling units were identified, and then interviewers contacted 2704 households preselected within the given sampling units. Altogether, 1391 face-to-face interviews were conducted (i.e., the response rate was 51.4%). A dataset of 1366 cases was used for analysis, as some cases had to be excluded due to incompleteness. The resulting sample described in Table 1 demonstrated a proportionate representation of the theoretical population with respect to gender, age, size of place of residence, and type of dwelling. Data collection took place in September 2022.

Table 1. Selected socio-demographic characteristics.

Variables		Theoretical Population *	Sample
Gender	Male	49.9%	49.3%
	Female	50.1%	50.7%
	Total	100.0%	100.0%
Age	15–29 years	20.1%	19.9%
	30–49 years	39.5%	39.4%
	50–74 years	40.4%	40.7%
	Total	100.0%	100.0%
Size of the place of residence	Fewer than 10,000 inhabitants	46.1%	46.3%
	10,000 to 49,999 inhabitants	22.0%	21.6%
	50,000 inhabitants and more	31.9%	32.1%
	Total	100.0%	100.0%
Type of dwelling	Family house	46.2%	46.7%
	Condominium	53.8%	53.3%
	Total	100.0%	100.0%

Note: n = 1366; * Data about the theoretical population comes from the Czech statistical office.

3.2. Ethical Considerations and Quality Assurance

Considerable attention was paid to ethical and data quality assurance issues. Several steps were taken to protect the rights of respondents, ensure their anonymity, and adhere to ethical standards. In this regard, informed consent was obtained from respondents prior to the start of each interview; this ensured that respondents were fully aware of the objectives of the study, the research process, and how the data provided would be used. When gathering data from adolescent respondents (15–18 years), the procedures were introduced and explained not only to the respondents themselves, but also to at least one parent or guardian. In addition, informed consent was obtained from both the adolescent and the guardian who was present throughout the interview process to ensure a supportive and ethical environment.

The responses obtained were rigorously anonymized, i.e., all data were processed in such a way that no specific individual could be identified. Aggregate data were presented in a way assuring that no individual could be identified. This ensured that respondents' privacy and confidentiality were fully protected. As part of data quality assurance, 35% of all interviews conducted were back-checked.

3.3. Translation of the Scale

In the first stage of the translation process, two independent translations of the original English instrument into Czech were carried out, following the procedures recommended by Sousa and Rojjanasrirat [19]. The two Czech versions were compared to identify and resolve any differences and ambiguities. Finally, a back-translation was performed to check the equivalence between the English and Czech versions [20].

Another step in the process of scale translation was a pilot test conducted with a sample of 26 respondents. The aim of this phase was to test the comprehensibility of the items for the intended population and to identify possible sources of misunderstanding. The pilot testing did not reveal any comprehension issues; therefore, the translated instrument was considered adequate and usable for the Czech population.

3.4. Measures

3.4.1. Civic Engagement Scale (CES)

The Civic Engagement Scale (CES) contains a total of 14 items rated on a seven-point Likert-type scale. Given the hypothesized gap between attitudes and behaviors [21,22], when individuals' attitudes may not directly correspond to their actions due to situational circumstances, the original study presented attitudes and behavioral patterns as two components of the scale [3]. Eight items comprised the first component focused on the attitudinal aspect of civic engagement (e.g., "I believe I should make a difference in my community."), while the remaining six items formed a second component reflecting specific behavioral patterns (e.g., "I help members of my community."). Attitudinal items were rated with the use of agreement scale ranging from definitely agree (7) to definitely disagree (1), whereas the behavioral scale was assessed by the frequency scale ranging from always (7) to never (1). This was the same format as that used by Doolittle and Faul [3] in their original study from 2013. Total scores for each component of the scale ranged from 8 to 56 for the attitudinal component and from 6 to 42 for the behavioral component, with higher scores indicating higher levels of civic engagement. This particular design of the scale has been used in other studies as well [14,17,18], and has also been carefully evaluated for reliability and validity.

3.4.2. Independent Questions

Assessing the convergent validity of a scale is an essential step in validating a research instrument because it examines whether the scale actually measures the intended construct [23]. For this reason, four specific questions were included in the research instrument. In order to decrease the response bias and minimize the burden on respondents, these indicators took the form of independent questions.

First, the belongingness to the local community was assessed by the question "I feel that I am part of the community where I live.", focusing attention on identification with the community spirit and involvement in decision-making processes. This question is relevant from the perspective of smart cities, which seeks to connect citizens within a city and encourage their active participation in social and community activities [13]. Respondents who perceive themselves as part of their community might be highly motivated to engage for the benefit of the given community, city development, and the implementation of smart city principles.

Second, the examination of the attachment to a place was another key element of civic engagement that was identified. The appropriate question "What are your feelings about the place where you currently live?" reflected the respondent's affective relationship with the place of residence and reflected the respondent's subjective perception and attachment to the place [16]. Place attachment has been linked to civic engagement by many researchers, both directly and indirectly [24–27]. Therefore, an indicator regarding the respondent's relationship with the place of residence was included to examine the convergent validity of the CES. Hereinafter, this question is reported as "place attachment".

Third, the inclusion of the question differentiating between respondents living in family houses and condominiums was important for testing the scale validity. This question was asked under the assumption that people living in family houses may have a greater need for control over their property and close environment (and therefore, a greater motivation to be active members of the community) than those living in condominiums.

Fourth, in addition to testing the convergent validity per se, a question was included to test the hypothesized relationship between CES and respondents' attitudes toward smart cities. As mentioned in the introductory session, the issue of attitudes toward smart cities was indicated by a question aimed at identifying the key areas that respondents considered desirable with respect to smart cities. These areas were renewable energy, sustainable transportation, and intelligent public services [12,13]. The corresponding multiple-choice question was "I wish my place of residence would use renewable energy, has sustainable transportation, and has intelligent public services.". The answers were transformed to an indicator of interest in smart cities as follows: high interest (all three areas checked); moderate interest (two areas checked); low interest (one area checked); no interest (no areas checked).

On top of that, standard socio-demographic indicators were also monitored. In this study, particular attention was paid to the association of CES scores with gender, age, and the size of the respondents' place of residence.

3.5. Data Analysis

Sample statistics and the characteristics of the respondents who participated in the study were analyzed in SPSS. Means (M), standard deviations (SD), skewness, and kurtosis were calculated to provide a basic description of the data distribution. The strength and direction of the relationship between variables were indicated by Pearson's correlation coefficient; the means of different subgroups were compared using analysis of variance (ANOVA). *T*-tests were used for comparison of means between two groups, whereas ANOVA was used to compare means among more than two groups. The internal consistency of the scale was assessed not only by Cronbach's alpha [28,29] but, because of correlated errors, also by McDonald's omega [30,31], and AVE along with CR [32], which were also used to inform about convergent validity.

The dimensionality of the scale was tested using exploratory factor analysis based on principal component analysis [33]. In accordance with classical test theory, the psychometric properties of the scale were examined using confirmatory factor analysis (CFA), performed in AMOS 24. Given the need to follow explorative and confirmative strategies simultaneously, the total dataset of 1366 cases was randomly divided into two equivalent subsets. The first subset was used for exploratory factor analysis, while the latter was used for confirmatory factor analysis. Similar procedures have also been used successfully in previous studies [34,35]. The sample size was sufficient even after this split [36].

Missing data were assumed to be missing at random [37]. For no variable did the proportion of missing values exceed 3%. In the case of EFA and CFA, the listwise method was used; thus, the analyses in question were performed only with complete datasets; on the other hand, the independent questions and their analyses may have different numbers of valid cases.

4. Results

4.1. Descriptive Statistics

All items had similar mean scores, with attitudinal components ranging from 4.04 to 5.19 (with standard deviations in a relatively narrow range from 1.437 to 1.710) and behavioral components ranging from 2.91 to 3.87 (with standard deviations in a range from 1.720 to 1.923). Table 2 also shows the skewness of the individual items, which range from −0.119 to −0.555 for the attitude items and from 0.004 to 0.611 for the behavior items. Moreover, the kurtosis of these items ranged from −0.165 to −0.732 or −0.659 to −1.017, respectively. Skewness and kurtosis for both scale components ranged from −1.5 to +1.5, indicating a normal distribution [38]. Additionally, the floor and ceiling effects reached acceptable values (1.2% for the floor, 3.2% for the ceiling in the case of attitude items; 9.6% for the floor and 1.2% for the ceiling for behavior items), which is consistent with the recommendation of Cain et al. [39] that these values should be less than 50%.

Table 2. Descriptive statistics of the Civic Engagement Scale (CES).

	Attitude Items	n	Mean	SD	Skewness	Kurtosis	ITC
1	I feel responsible for my community.	683	4.54	1.679	−0.367	−0.669	0.799
2	I believe I should make a difference in my community.	683	4.64	1.590	−0.428	−0.427	0.845
3	I believe that I have a responsibility to help the poor and the hungry.	683	4.83	1.437	−0.420	−0.165	0.757
4	I am committed to serve in my community.	683	4.50	1.592	−0.280	−0.557	0.861
5	I believe that all citizens have a responsibility to their community.	683	4.89	1.574	−0.484	−0.325	0.812
6	I believe that it is important to be informed of community issues.	683	5.19	1.479	−0.555	−0.238	0.759
7	I believe that it is important to volunteer.	683	4.04	1.694	−0.172	−0.712	0.739
8	I believe that it is important to financially support charitable organizations.	683	4.15	1.710	−0.119	−0.732	0.726
	The whole attitude component	683	36.78	10.715	−0.349	−0.187	
	Behavior Items						
1	I am involved in structured volunteer position(s) in the community.	683	2.91	1.811	0.611	−0.774	0.789
2	When working with others, I make positive changes in the community.	683	3.23	1.766	0.300	−0.963	0.856
3	I help members of my community.	683	3.40	1.720	0.251	−0.831	0.794
4	I stay informed of events in my community.	683	3.87	1.792	0.004	−0.971	0.805
5	I participate in discussions that raise issues of social responsibility.	683	3.28	1.923	0.410	−1.017	0.817
6	I contribute to charitable organizations within the community.	683	2.97	1.763	0.523	−0.659	0.777
	The whole behavior component	683	19.66	9.356	0.347	−0.648	

Table 2 also shows the summary data for the two components of the CES; the mean score for the attitudinal component of the scale was 36.87, with a standard deviation of 10.715, while the mean score for the behavioral component was 19.66, with a standard deviation of 9.356.

4.2. Internal Consistency

Cronbach's alpha coefficient was calculated to assess the internal consistency of the scale. Its value reached significant values of 0.940 (attitude component) and 0.934 (behavioral component), indicating that the items consistently measured the same construct. These values not only confirmed the internal consistency of the present form of the scale; they even exceeded the original scores reported by Doolittle and Faul [3], of 0.91 and 0.85. In this context, the McDonald's omega coefficient was also calculated, which reached a value of 0.940 for the attitude component and 0.935 for the behavioral component.

Internal consistency was also tested by the item-total correlations (ITCs) [40]. These coefficients ranged from 0.726 to 0.861 for the attitudinal component and from 0.777 to 0.856 for the behavioral component. All values were significantly higher than the critical value of 0.4 recommended by Furr [41]. Such results supported the hypothesis of a unidimensional structure of civic engagement attitudes and civic engagement behaviors, as measured by this scale [42].

4.3. Dimensionality of the CES

Exploratory factor analysis was conducted to assess the structure of the scale [33]. The results showed that both components were unidimensional when the principal component analysis for each part of the scale showed only one factor with an eigenvalue greater than 1. This result was further confirmed by the scree plots, which showed that the optimal number of factors for each component is one. According to Henson and Roberts [43], higher factor loadings indicate a greater contribution of the items to a given factor, with factor loadings greater than 0.4 indicating a significant contribution. Table 3 shows the factor loadings of the items to the extracted factor; it is obvious that all items achieved factor loadings greater than 0.4. Table 3 also shows the obtained communalities (h^2), which ranged from 0.62 to 0.82 across both components being above the critical value recommended by Velicer and Fava [44].

Table 3. Exploratory factor analysis.

		n	FL	h^2
CES—attitude items				
1	I am committed to serve in my community.	683	0.901	0.81
2	I believe I should make a difference in my community.	683	0.888	0.79
3	I believe that all citizens have a responsibility to their community.	683	0.863	0.74
4	I feel responsible for my community.	683	0.852	0.73
5	I believe that it is important to be informed of community issues.	683	0.819	0.67
6	I believe that I have a responsibility to help the poor and the hungry.	683	0.818	0.67
7	I believe that it is important to volunteer.	683	0.797	0.64
8	I believe that it is important to financially support charitable organizations.	683	0.785	0.62
CES—behavior items				
1	When working with others, I make positive changes in the community.	683	0.905	0.82
2	I participate in discussions that raise issues of social responsibility.	683	0.876	0.77
3	I stay informed of events in my community.	683	0.868	0.75
4	I help members of my community.	683	0.860	0.74
5	I am involved in structured volunteer position(s) in the community.	683	0.855	0.73
6	I contribute to charitable organizations within the community.	683	0.846	0.72

Exploratory factor analysis also brought some other important results: the coefficient of determination, i.e., the proportion of variance in the identified factor explained by the items, was 70.8% for the attitudinal component and 75.4% for the behavioral component. For the attitudinal component, the KMO reached 0.932 and Bartlett's test of sphericity showed a result of $\chi^2 = 4417.764$ (df = 28, $p < 0.001$). Similarly, for the behavioral component, the KMO was 0.920 and Bartlett's test was $\chi^2 = 3232.478$ (df = 15, $p < 0.001$). These results supported the hypothesis of a unidimensional structure of each of the CES components [36].

4.4. Confirmatory Factor Analysis (CFA)

The psychometric properties of the CES were assessed using confirmatory factor analysis (CFA) with maximum likelihood estimation [45]. Given $\chi^2 = 27.397$ with df = 13, $p < 0.05$, for the attitude component and $\chi^2 = 11.411$ with df = 6, $p < 0.05$, for the behavior component, other indices such as the root mean square error of approximation (RMSEA), standardized root mean square residuals (SRMR), comparative fit index (CFI), Tucker–Lewis index (TLI), and goodness of fit index (GFI) were used to evaluate the model fit. Results for both components of the CES are shown in Figure 1, which depicts the tested models together with standardized coefficients (factor leadings) and communalities.

The values of absolute and incremental indices along with their critical values are presented in Table 4, which also presents the results of the initial and improved models.

Table 4. Absolute and incremental indices.

	RMSEA	SRMR	GFI	CFI	TLI	NFI
CES—attitude component						
Original model	0.139	0.0393	0.911	0.940	0.916	0.936
Improved model	0.040	0.0149	0.990	0.997	0.993	0.994
CES—behavior component						
Original model	0.120	0.0261	0.956	0.971	0.952	0.968
Improved model	0.036	0.0098	0.994	0.998	0.996	0.996
Critical values	<0.07 [45]	<0.08 [46]	>0.95 [45]	>0.90 [45]	>0.95 [45]	>0.95 [46]

The process of assessing the fit of the proposed model to the data involved several iterations, during which the model was refined to reflect errors that include unobserved variables and capturing variance not included in the latent construct.

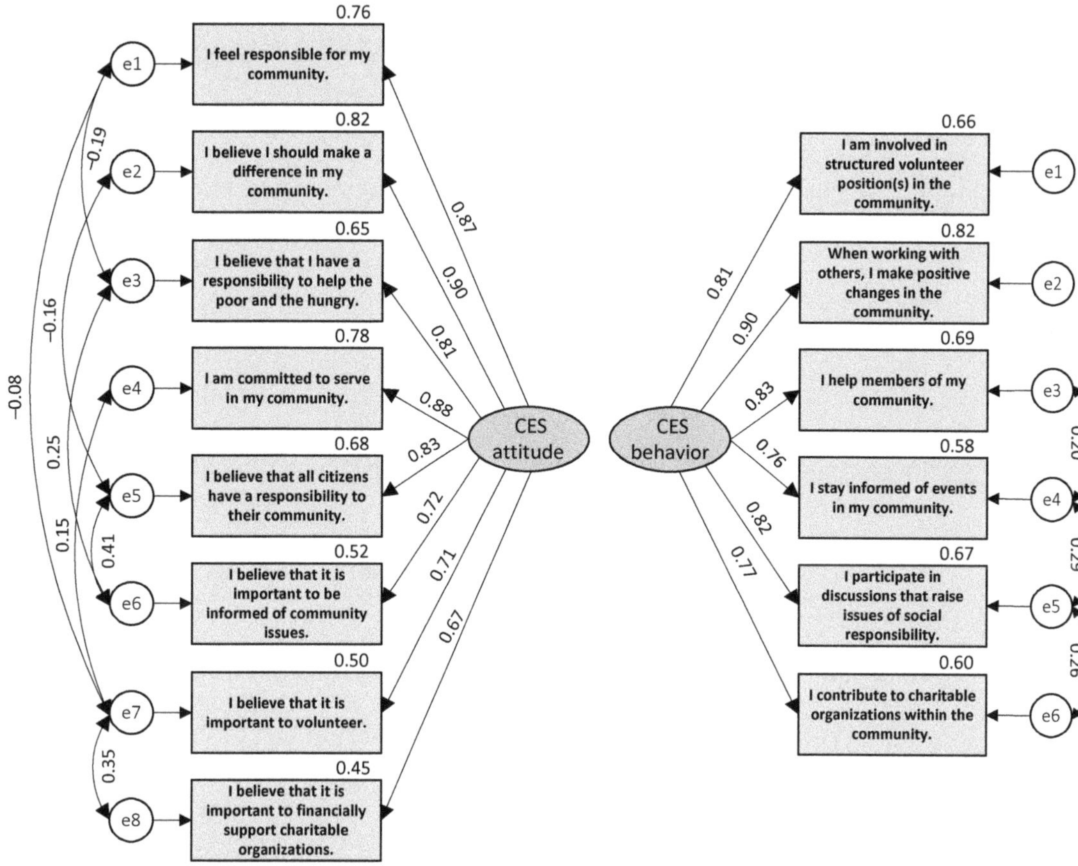

Figure 1. Confirmatory factor analysis of the improved CES models.

4.5. Construct Convergent Validity

Convergent validity of the Czech version of the CES was assessed through average variance extracted (AVE) and composite reliability (CR) [32,47]. The scale achieved an AVE of 0.71 for the attitudinal component and 0.75 for the behavioral component, which means that, on average, 71% and 75% of the indicators' variance was explained by the latent variable. Therefore, the results suggested that the latent variable was a fairly accurate representation of the construct being measured. Composite reliability (CR) for the Czech data was 0.95 for both components, supporting the assumption that the CES adequately measured the civic engagement.

More evidence with respect to the construct validity is the extent to which the scale items are correlated with each other [48]. This is because high correlation can be interpreted as a consistent way of measuring the same construct. The correlation matrix in Table 5 provides strong evidence of such construct validity, as it shows significant associations between all items.

Differences in mean scores on the attitudinal component of the scale between different subgroups were examined. Table 6 shows that older respondents scored more highly on the CES than younger respondents. Additionally, significantly higher CES scores were observed among respondents living in single-family homes compared with those living in condominiums. In addition, respondents with higher place attachment achieved significantly higher CES scores.

Table 5. Correlational matrix.

	Attitude Items	**1**	**2**	**3**	**4**	**5**	**6**	**7**	**8**
1	I feel responsible for my community.	1.000							
2	I believe I should make a difference in my community.	0.693 **	1.000						
3	I believe that I have a responsibility to help the poor and the hungry.	0.529 **	0.581 **	1.000					
4	I am committed to serve in my community.	0.683 **	0.705 **	0.616 **	1.000				
5	I believe that all citizens have a responsibility to their community.	0.624 **	0.617 **	0.549 **	0.678 **	1.000			
6	I believe that it is important to be informed of community issues.	0.511 **	0.577 **	0.570 **	0.555 **	0.631 **	1.000		
7	I believe that it is important to volunteer.	0.509 **	0.531 **	0.471 **	0.574 **	0.503 **	0.471 **	1.000	
8	I believe that it is important to financially support charitable organizations.	0.504 **	0.528 **	0.454 **	0.527 **	0.491 **	0.460 **	0.607 **	1.000
	Behavior Items	**1**	**2**	**3**	**4**	**5**	**6**		
1	I am involved in structured volunteer position(s) in the community.	1.000							
2	When working with others, I make positive changes in the community.	0.656 **	1.000						
3	I help members of my community.	0.567 **	0.645 **	1.000					
4	I stay informed of events in my community.	0.529 **	0.633 **	0.603 **	1.000				
5	I participate in discussions that raise issues of social responsibility.	0.596 **	0.640 **	0.557 **	0.637 **	1.000			
6	I contribute to charitable organizations within the community.	0.604 **	0.611 **	0.550 **	0.524 **	0.614 **	1.000		

** Correlation is significant at the 0.01 level (2-tailed); $n = 683$; Kendall's tau_b.

As hypothesized, civic engagement, as measured by the CES, was found to be significantly related to belongingness. In this regard, Table 6 shows that among respondents with a low feeling of belongingness (i.e., those who did not identify with their local community, and therefore strongly disagreed with the statement "I feel that I am part of the community where I live."), the CES score reached a low value of 22.73. In contrast, individuals who strongly identified with their local community had a significantly higher CES score (43.04). A similar pattern was observed in the case of place attachment. In this regard, respondents who indicated that they could hardly imagine living anywhere else achieved a CES score of 39.88, while those who had planned to move away because they did not like their place of residence had a CES score of 23.42. Thus, the CES appears to be a sensitive indicator in this regard. In addition, respondents who attributed more importance to smart city innovations showed higher CES scores (41.36), followed by 37.59 among those with moderate perceived importance, and 29.73 among respondents for whom the smart cities innovations had low importance.

Table 6. Associations of CES—attitude with independent questions.

	CES Mean	**SD**	**F/t**	**df**	***p*-Value ***	**Eta**
Gender						
Male	36.18	10.936	−1.434	684	0.152 **	0.055
Female	37.35	10.485				
Age						
15–29 years	34.13	11.845	8.500	2	<0.001	0.156
30–49 years	36.98	10.032				
50–74 years	38.52	10.293				
Size of the place of residence						
Fewer than 10,000 inhabitants	35.85	11.890	1.489	2	0.226	0.066
10,000 to 49,999 inhabitants	36.79	8.946				
50,000 inhabitants and more	37.57	11.079				

Table 6. *Cont.*

	CES Mean	SD	F/t	df	*p*-Value *	Eta
I feel that I am part of the community where I live (belongingness).						
Definitely agree	43.04	8.818				
Agree	38.24	8.461	87.902	3	<0.001	0.532
Disagree	28.69	10.099				
Definitely disagree	22.73	10.003				
Place attachment						
I can hardly imagine living anywhere else.	39.88	9.197				
There are some other places where I could live.	35.01	10.681	35.137	3	<0.001	0.366
It is nothing special here, I can live elsewhere as well.	28.31	11.207				
I plan to move away because I do not like this place.	23.42	11.843				
Type of dwelling						
Family house	38.31	10.22	3.081	684	0.002 **	0.117
Condominium	35.76	10.93				
Importance of smart cities innovations						
High	41.36	9.236				
Moderate	37.59	9.435	43.990	3	<0.001	0.409
Low	29.73	10.638				
None	32.73	10.725				

* ANOVA; ** *t*-test.

Table 7 shows the results for the behavioral component of the CES. There were statistically significant associations of CES scores with belongingness: the CES score for respondents who strongly disagreed with the statement "I feel that I am part of the community where I live" was 14.27, while the CES score for those who strongly agreed with this statement was 22.76. There were also statistically significant associations with age, with CES scores being highest for respondents aged 50 or older. However, unlike the attitudinal component, the differences based on type of dwelling were not statistically significant. In addition, the results summarized in Table 7 show that there were no statistically significant differences based on the size of the place of residence, and gender.

Differences reflecting the sensitivity of the CES to place attachment are evident. In this regard, Table 7 shows that people who strongly identified with their place of living (i.e., those who said they could not imagine living anywhere else) had a CES score of 21.10, compared with 11.17 for respondents who planned to move away because they disliked their place of living. As expected, there were statistically significant differences between respondents from family houses and those living in condominiums on the attitudinal component (38.31 versus 35.76); however, no such association was demonstrated in the case of the behavioral component, where the CES scores were 20.36 and 19.19, respectively. Finally, the results showed a statistically significant positive relationship between behavioral components of the CES and the importance of smart city innovations.

Table 7. Associations of CES—behavior with other indicators.

	CES Mean	SD	F/t	df	*p*-Value *	Eta
Gender						
Male	19.85	9.779	0.504	660	0.614 **	0.019
Female	19.49	8.946				
Age						
15–29 years	16.52	8.982				
30–49 years	20.22	9.231	14.338	2	<0.001	0.201
50–74 years	21.32	9.260				
Size of the place of residence						
Fewer than 10,000 inhabitants	19.77	9.858				
10,000 to 49,999 inhabitants	18.95	8.786	1.084	2	0.339	0.056
50,000 inhabitants and more	20.21	9.405				

Table 7. *Cont.*

	CES Mean	SD	F/t	df	*p*-Value *	Eta
I feel that I am part of the community where I live (belongingness).						
Definitely agree	22.76	10.226				
Agree	20.51	8.459	21.344	3	<0.001	0.296
Disagree	15.56	8.085				
Definitely disagree	14.27	8.972				
Place attachment						
I can hardly imagine living anywhere else.	21.10	9.303				
There are some other places where I could live.	18.86	9.128	9.783	3	<0.001	0.203
It is nothing special here, I can live elsewhere as well.	16.19	9.301				
I plan to move away because I do not like this place.	11.17	5.167				
Type of dwelling						
Family house	20.36	9.568	1.601	684	0.110 **	0.061
Condominium	19.19	9.193				
Importance of smart cities innovations						
High	21.73	9.911				
Moderate	20.43	8.708	12.243	3	<0.001	0.230
Low	16.01	8.456				
None	18.05	8.400				

* ANOVA; ** *t*-test.

5. Discussion

The purpose of this study was to evaluate the psychometric properties of the CES. A combination of several analytical approaches was used, including internal consistency testing, principal component analysis, and confirmatory factor analysis. Based on data from a sample representative of the Czech population, the results showed that the CES performed well. The scale had a mean score of 36.78 for the attitudinal component and 19.66 for the behavioral component; it had acceptable values of skewness and kurtosis [38]. Each of the scale components showed a high degree of internal consistency when the correlation between the items and the total score ranged from 0.633 to 0.847 (attitudinal component) and from 0.586 to 0.826 (behavioral component), meeting the criteria for acceptability [48].

Scale validation results further demonstrated a good fit of the proposed model to the Czech data. The RMSEA reached 0.040 for the attitudinal component and 0.036 for the behavioral component; the SRMR values were 0.0149 and 0.0098, respectively; and the GFI reached excellent values of 0.990 and 0.994 [45–47]. Similarly satisfactory results were obtained for the CFI (0.997/0.998), TLI (0.993/0.996), and NFI (0.994/0.996).

Convergent validity, as expressed by the average variance extracted (AVE) and composite reliability (CR), exceeded the recommended thresholds of 0.5 and 0.6, respectively [32,47,49], and showed significant associations with all independent questions in this study. CR usually provides a more accurate estimate of internal consistency than Cronbach's alpha, but in this case, the results were similar due to the fact that most of the statistical assumptions of Cronbach's alpha were met.

The results obtained are consistent with other studies indicating that civic engagement does not show statistically significant differences based on gender [50,51]. Some studies have also concluded that civic engagement does not differ by age [2,3,50,52]; however, this study showed the opposite result for both the attitudinal and behavioral components. This opposite finding can largely be explained by the fact that the other studies focused primarily on student populations, while this study considers a much broader range of ages.

It is important to note that due to the cross-sectional nature of this study, the direction of the relationship between civic engagement and self-reported attitudes cannot be clearly established [53]. It is unclear whether higher levels of civic engagement influenced respondents' place attachment or whether, instead, civic engagement in local communities was based on these feelings. Both perspectives have merit, but a different research design would be needed to determine the direction of causality. Similarly, it remains unclear whether attitudes within the CES can be considered predictors of behavior (engagement practices) or whether the two components are of equal importance. The reliance on self-reported data is also a limitation of this study [53]. Although the study confirmed statistically significant associations between variables, it is important to note the limitations associated with using self-reported data rather than independently validated objective indicators.

The CES can help identify potential barriers to active participation in smart city initiatives [50,54] and can provide early warning of potential problems, allowing decision makers to improve their mitigation strategies [55]. However, some interventions may bring additional risks, such as widening the digital divide [11,14,56]. Therefore, it would be useful to consider ICT literacy, perceptions of smart services, and the specificities of urban versus rural populations [14,55–57]. From a practical point of view, it seems that the CES can contribute positively to the development of smart cities by providing useful evidence to measure the impact of appropriate policies and interventions to stimulate civic engagement. Using CES, civic engagement can be tracked consistently over time or compared across cities, regions [15,27,51], and target groups [58].

6. Conclusions

The implications of this study for smart city research and practice are at least twofold. First, it is a novel contribution to validation of the CES, which aims to provide a standardized tool for thorough examinations of civic engagement. In this regard, the study highlights the importance of civic engagement attitudes and behaviors as two key components related to smart cities and shaping the urban environment. The results confirmed the good psychometric properties of the CES and suggested that this scale could be a valuable tool for measuring civic engagement in the context of smart cities. Future research could aim to use the CES for detailed case studies or comparative analyses of different smart city projects to understand how civic engagement varies and what factors influence it. In addition, researchers could use the CES to examine the impact of civic engagement on the effectiveness of particular smart city initiatives, including exploring the relationship between civic participation and outcomes in terms of improved city services, sustainability, or even quality of life. Second, policymakers and smart city practitioners can use the results of this study to measure the level of civic engagement and design targeted interventions that can encourage citizens to participate in smart city initiatives. Based on the CES findings, they could develop targeted interventions to increase civic engagement, including community workshops, digital platforms for citizen feedback, and educational programs on smart city technologies. In addition, policymakers and smart city practitioners could foster partnerships between city governments, academic institutions, and technology companies to develop innovative solutions that promote civic engagement and address urban challenges.

Future research could examine the invariance of the CES and assess its performance within specific smart city initiatives and interventions around the world. This would provide a comprehensive perspective on how civic engagement translates into different urban settings and how it supports the innovation and development of smart cities.

Funding: This submission was funded by Operation Program Research, Development and Education, European Structural and Investment Funds, and by the Ministry of Education, Youth and Sports of the Czech Republic, grant number/registration number CZ.02.2.69/0.0/0.0/18_054/0014660 as part of the project "Setting the conditions and the environment for international and cross-sector cooperation".

Institutional Review Board Statement: The study was conducted in accordance with the Declaration of Helsinki of 1975 (https://www.wma.net/what-we-do/medical-ethics/declaration-of-helsinki/, accessed on 20 September 2023) and followed the AAPOR ethical code (https://www.aapor.org/Standards-Ethics/AAPOR-Code-of-Ethics.aspx, accessed on 20 September 2023). The research design and the research instrument (the questionnaire) were approved in INESAN by the Research Ethics Board (IREBA/2022/923). The institute holds an HRS4R HR Excellence in Research award (https://inesan.eu/en/hrs4r-2/, accessed on 20 September 2023), which acknowledges the highest standard of ethics adhered to by researchers at this institute (https://www.euraxess.cz/jobs/hrs4r, accessed on 21 March 2023).

Informed Consent Statement: Informed consent was obtained from all subjects involved in the study. This study involved data collected from anonymous respondents. All subjects gave their informed consent for inclusion before their participation in the survey.

Data Availability Statement: The data used to support the findings of this study will be available from the corresponding author upon reasonable request.

Acknowledgments: The author would thanks to all respondents and interviewers engaged in this study, and to all members of the supportive research team.

Conflicts of Interest: The author declares no conflict of interest. The funders had no role in the design of the study; in the collection, analyses, or interpretation of data; in the writing of the manuscript; or in the decision to publish the results.

References

1. Talò, C.; Mannarini, T.; Rochira, A. Sense of community and community participation: A meta-analytic review. *Soc. Indic. Res.* **2014**, *117*, 1–28. [CrossRef]
2. Jacoby, B. Civic engagement in today's higher education: An overview. In *Civic Engagement in Higher Education: Concepts and Practices*; Jacoby, B., Ed.; Jossey-Bass: San Francisco, CA, USA, 2009; pp. 5–30.
3. Doolittle, A.; Faul, A.C. Civic Engagement Scale: A Validation Study. *SAGE Open* **2013**, *3*, 1–7. [CrossRef]
4. Adler, R.P.; Goggin, J. What Do We Mean By "Civic Engagement"? *J. Transform. Educ.* **2005**, *3*, 236–253. [CrossRef]
5. Ehrlich, T. Civic learning: Democracy and education revisited. *Educ. Rec.* **1997**, *78*, 57–65.
6. Sherrod, L.R.; Lauckhardt, J. The development of citizenship. In *Handbook of Adolescent Psychology: Contextual Influences on Adolescent Development*, 3rd ed.; Lerner, R.M., Steinberg, L., Eds.; Wiley: Hoboken, NJ, USA, 2009; Volume 2, pp. 372–408.
7. Barrett, M.; Zani, B. *Political and Civic Engagement: Multidisciplinary Perspectives*; Routledge: London, UK, 2014.
8. Lasker, R.D.; Weiss, E.S. Broadening participation in community problem solving: A multidisciplinary model to support collaborative practice and research. *J. Urban Health* **2003**, *80*, 14–47. [CrossRef] [PubMed]
9. Grillo, M.C.; Teixeira, M.A.; Wilson, D.C. Residential Satisfaction and Civic Engagement: Understanding the Causes of Community Participation. *Soc. Indic. Res.* **2010**, *97*, 451–466. [CrossRef]
10. Andolina, M.; Jenkins, K.; Keeter, S.; Delli Carpini, M.X.; Zukin, C. *A New Engagement? Political Participation, Civic Life, and the Changing American Citizen*; Oxford University Press: Oxford, UK, 2006.
11. Deng, G.; Fei, S. Exploring the factors influencing online civic engagement in a smart city: The mediating roles of ICT self-efficacy and commitment to community. *Comput. Hum. Behav.* **2023**, *143*, 107682. [CrossRef]
12. Albino, V.; Berardi, U.; Dangelico, R.M. Smart cities: Definitions, dimensions, performance, and initiatives. *J. Urban Technol.* **2015**, *22*, 3–21. [CrossRef]
13. Senior, C.; Temeljotov Salaj, A.; Johansen, A.; Lohne, J. Evaluating the Impact of Public Participation Processes on Participants in Smart City Development: A Scoping Review. *Buildings* **2023**, *13*, 1484. [CrossRef]
14. Zait, A.; Andrei, A.G. Civic Engagement at the Crossroads of Online and Offline Spaces: A PLS-SEM Assessment. *Sci. Ann. Econ. Bus.* **2019**, *66*, 559–572. [CrossRef]
15. Hoskins, B.L.; Mascherini, M. Measuring Active Citizenship through the Development of a Composite Indicator. *Soc. Indic. Res.* **2009**, *90*, 459–488. [CrossRef]
16. Irani, M.; Aghdam, S.R.; Ghasemzadeh, B. Investigating the link between place attachment, civic engagement, and pro-environmental behaviors. *Environ. Dev.* **2023**, *47*, 100897. [CrossRef]
17. Belay, A.; Tefera, B. Validation of Adolescent Civic Engagement Scale in the Ethiopian Context. *J. Posit. Sch. Psychol.* **2022**, *6*, 8306–8320.
18. Akin, A.; Fatih, U.; Umran, A. The Validity And Reliability of Turkish Version of the Civic Engagement Scale. *J. Educ. Instr. Stud. World* **2014**, *4*, 55–59.
19. Sousa, V.D.; Rojjanasrirat, W. Translation, adaptation and validation of instruments or scales for use in cross-cultural health care research: A clear and user-friendly guideline. *J. Eval. Clin. Pract.* **2011**, *2011*, 268–274. [CrossRef] [PubMed]
20. Yu, D.S.; Lee, D.T.; Woo, J. Issues and challenges of instrument translation. *West. J. Nurs. Res.* **2004**, *26*, 307–320. [CrossRef] [PubMed]
21. Padel, S.; Foster, C. Exploring the gap between attitudes and behaviour: Understanding why consumers buy or do not buy organic food. *Br. Food J.* **2005**, *107*, 606–625. [CrossRef]
22. Terlau, W.; Hirsch, D. Sustainable consumption and the attitude-behaviour-gap phenomenon-causes and measurements towards a sustainable development. *Proc. Food Syst. Dyn.* **2015**, *6*, 199–214.
23. Flake, J.K.; Fried, E.I. Measurement schmeasurement: Questionable measurement practices and how to avoid them. *Adv. Methods Pract. Psychol. Sci.* **2020**, *3*, 456–465. [CrossRef]
24. Forsyth, D.R.; van Vugt, M.; Schlein, G.; Story, P.A. Identity and sustainability: Localized sense of community increases environmental engagement. *Anal. Soc. Issues Public Policy* **2015**, *15*, 233–252. [CrossRef]

25. Scannell, L.; Gifford, R. Defining place attachment: A tripartite organizing framework. *J. Environ. Psychol.* **2010**, *30*, 1–10. [CrossRef]
26. Anton, C.E.; Lawrence, C. Home is where the heart is: The effect of place of residence on place attachment and community participation. *J. Environ. Psychol.* **2014**, *40*, 451–461. [CrossRef]
27. Hesari, E.; Moosavy, S.M.; Rohani, A.; Besharati Kivi, S.; Ghafourian, M.; Saleh, S.; Pour, B. Investigation the relationship between place attachment and community participation in residential areas: A structural equation modelling approach. *Soc. Indic. Res.* **2020**, *151*, 921–941. [CrossRef]
28. Tavakol, M.; Dennick, R. Making sense of Cronbach's alpha. *Int. J. Med. Educ.* **2011**, *2*, 53–55. [CrossRef] [PubMed]
29. Raykov, T. Scale reliability, Cronbach's coefficient alpha, and violations of essential tau-equivalence with fixed congeneric components. *Multivar. Behav. Res.* **1997**, *1997*, 329–353. [CrossRef] [PubMed]
30. Revelle, W.; Zinbarg, R. Coefficients alpha, beta, omega, and the GLB: Comments on Sijtma. *Psychometrika* **2009**, *74*, 145–154. [CrossRef]
31. Hayes, A.F.; Coutts, J.J. Use omega rather than Cronbach's alpha for estimating reliability. But.... *Commun. Methods Meas.* **2020**, *14*, 1–24. [CrossRef]
32. Kalkbrenner, M.T. Alpha, omega, and H internal consistency reliability estimates: Reviewing these options and when to use them. *Couns. Outcome Res. Eval.* **2023**, *14*, 77–88. [CrossRef]
33. Comrey, A.L.; Lee, H.B. *A First Course in Factor Analysis*; Psychology Press: London, UK, 2013.
34. Remr, J. Validation of the Health Consciousness Scale among the Czech Population. *Healthcare* **2023**, *11*, 1628. [CrossRef]
35. Remr, J. Assessing System Justification in Czech Population Using the System Justification Scale. *Eur. J. Investig. Health Psychol. Educ.* **2023**, *13*, 1801–1814. [CrossRef]
36. Hogarty, K.Y.; Hines, C.V.; Kromrey, J.D.; Ferron, J.M.; Mumford, K.R. The Quality of Factor Solutions in Exploratory Factor Analysis: The Influence of Sample Size, Communality, and Overdetermination. *Educ. Psychol. Meas.* **2005**, *65*, 202–226. [CrossRef]
37. Graham, J.W. Missing data analysis: Making it work in the real world. *Annu. Rev. Psychol.* **2009**, *60*, 549–576. [CrossRef] [PubMed]
38. Ho, A.D.; Yu, C.C. Descriptive statistics for modern test score distributions: Skewness, kurtosis, discreteness, and ceiling effects. *Educ. Psychol. Meas.* **2015**, *75*, 365–388. [CrossRef] [PubMed]
39. Cain, M.K.; Zhang, Z.; Yuan, K.H. Univariate and multivariate skewness and kurtosis for measuring nonnormality: Prevalence, influence and estimation. *Behav. Res. Methods* **2017**, *49*, 1716–1735. [CrossRef] [PubMed]
40. Wieland, A.; Durach, C.F.; Kembro, J.; Treiblmaier, H. Statistical and judgmental criteria for scale purification. *Supply Chain. Manag. Int. J.* **2017**, *22*, 321–328. [CrossRef]
41. Furr, R.M. *Scale Construction and Psychometrics for Social and Personality Psychology*; Sage Publications: London, UK, 2011.
42. Kline, R.B. *Principles and Practice of Structural Equation Modeling*; Guilford Publications: New York, NY, USA, 2023.
43. Henson, R.K.; Roberts, J.K. Use of exploratory factor analysis in published research: Common errors and some comment on improved practice. *Educ. Psychol. Meas.* **2006**, *66*, 393–416. [CrossRef]
44. Velicer, W.F.; Fava, J.L. Affects of variable and subject sampling on factor pattern recovery. *Psychol. Methods* **1998**, *3*, 231. [CrossRef]
45. Schreiber, J.B.; Nora, A.; Stage, F.K.; Barlow, E.A.; King, J. Reporting structural equation modeling and confirmatory factor analysis results: A review. *J. Educ. Res.* **2006**, *99*, 323–338. [CrossRef]
46. Hu, L.; Bentler, P.M. Fit indices in covariance structure modeling: Sensitivity to underparameterized model misspecification. *Psychol. Methods* **1998**, *3*, 424–453. [CrossRef]
47. Byrne, B.M. Structural Equation Modeling with AMOS. In *Basic Concepts, Applications, and Programming*, 2nd ed.; Routledge: New York, NY, USA, 2001.
48. Nunnally, J.C.; Bernstein, I.H. Validity. *Psychom. Theory* **1994**, *3*, 99–132.
49. Field, A. *Discovering Statistics Using SPSS*, 5th ed.; Sage Publications: London, UK, 2017.
50. Barnason, S.; Li, C.J.; Hall, D.M.; Wilhelm Stanis, S.A.; Schulz, J.H. Environmental Action Programs Using Positive Youth Development May Increase Civic Engagement. *Sustainability* **2022**, *14*, 6781. [CrossRef]
51. Choudhary, R.; Gupta, D.S. Civic Engagement a Precursor to Well-being. *Indian J. Health Wellbeing* **2017**, *8*, 152–156.
52. Zaff, J.; Boyd, M.; Li, Y.; Lerner, J.V.; Lerner, R.M. Active and engaged citizenship: Multi-group and longitudinal factorial analysis of an integrated construct of civic engagement. *J. Youth Adolesc.* **2010**, *39*, 736–750. [CrossRef] [PubMed]
53. DeVellis, R.F. *Scale Development: Theory and Applications*, 4th ed.; SAGE: London, UK, 2016.
54. Baccarne, B.; Evens, T.; De Marez, L. Understanding civic crowdfunding as a mechanism for leveraging civic engagement and urban innovation. *Technol. Innov. Manag. Rev.* **2020**, *10*, 51–66. [CrossRef]
55. Park, J.; Fujii, S. Civic Engagement in a Citizen-Led Living Lab for Smart Cities: Evidence From South Korea. *Urban Plan.* **2023**, *8*, 93–107. [CrossRef]
56. Albalá Genol, M.Á.; Etchezahar, E.; Maldonado Rico, A.; Gómez Yepes, T. Representations of Social Justice and Digital Civic Engagement: The Influence of Psychosocial Variables in Teacher Training. *Sustainability* **2022**, *14*, 7096. [CrossRef]

57. Fung, A. Varieties of participation in complex governance. *Public Adm. Rev.* **2006**, *66*, 66–75. [CrossRef]
58. Veeh, C.A.; Plassmeyer, M.; Nicotera, N.; Brewer, S.E. A combined measure of civic engagement for use among emerging adults. *J. Soc. Soc. Work Res.* **2019**, *10*, 13–34. [CrossRef]

urban science

MDPI

Article

The Role of Performance in Smart Meter's Acceptance: A Survey in Joinville, Brazil

Diego Fettermann [1,*], Pedro Christoffel [1], Jaime Castillo [2] and Angelo Sant'Anna [3]

1 Laboratory of Entrepreneurship and Innovation, Department of Production Engineering, Federal University of Santa Catarina, Florianopolis 88040900, Brazil; christoffelpedro@gmail.com
2 Department of Industrial Processes, Universidad Catolica de Temuco, Temuco 8331150, Chile; jcastill@uct.cl
3 Polytechnic School, Federal University of Bahia, Salvador 40210630, Brazil; angelo.santanna@ufba.br
* Correspondence: d.fettermann@ufsc.br

Abstract: The incorporation of renewable energy sources necessitates the upgrade of the electrical grid to a smart grid, which involves the implementation of smart meters. Although smart meters provide benefits to users, many smart meter implementation projects have failed to be accepted by users. This article assesses the factors influencing the acceptance of household smart meters in Joinville, a city in the south of Brazil. Based on the Unified and Extended Theory of Acceptance and Use of Technology (UTAUT2), a structural equation model was estimated using data from a sample of 136 respondents in the city of Joinville. The results indicate that Performance Expectancy, Hedonic Motivation, and Social Influence constructs have a more substantial effect on the Intention to Use smart meters. The results provide evidence for planning the upgrade of the electrical grid by implementing smart meters in southern Brazil.

Keywords: smart meter; acceptance; energy; technology acceptance model

1. Introduction

Global warming is a pressing concern that garners attention from nations worldwide, with this issue intensifying in recent years primarily due to human activities. If substantial efforts to reduce greenhouse gas (GHG) emissions are not mobilized in the coming decades, the impacts of climate change are projected to escalate gradually and become increasingly severe [1]. In light of this predicament, one of the pivotal measures to confront this challenge is the adoption of smart grids. Smart grids are designed to facilitate the efficient utilization of renewable energy sources [2]. The grid update aims to harmonize the demands and resources of all stakeholders, including providers, operators, end users, and others, in the electrical energy market [3]. The smart grid is built upon digital resources and information technology, enabling a two-way exchange of information between the energy provider and consumers. Moreover, smart energy grids allow for comprehensive monitoring, analysis, control, and communication across the energy system [4].

The literature underscores that innovations within the energy sector, including those integrated into smart grids, require social acceptance to succeed [5,6]. In simpler terms, stakeholders, such as the general public and consumers, must be open to using these technologies or acknowledging their utilization by others [4]. Despite the significance of taking residential customer acceptance into account, various initiatives aimed at implementing smart meters have encountered issues concerning their acceptance by residential consumers [7–9]. Therefore, proposing an acceptance model for residential smart meters is paramount to improving implementation processes to enhance acceptance among householders and, consequently, move towards the development of the present energy system to smart grids. Over the past decade, numerous studies have delved into the factors influencing households' acceptance of smart meters, e.g., [10–13]. Despite this, several smart meter implementation initiatives continue to assume the universal acceptance of smart meters by consumers, even though such assumptions face widespread rejection in the literature, e.g., [12–16].

Like numerous other countries, Brazil is steadily gearing up for an extensive smart meter implementation program, which involves replacing 64 million meters with an investment of 91 billion reais by 2030 [17]. Despite a growing body of literature on smart meter acceptance, a notable gap exists in understanding the factors influencing this acceptance within South America, including Brazil. A comprehensive review of existing research findings, encompassing literature reviews [18–24] and smart meter acceptance studies conducted in over 40 locations [25], confirms this scarcity of research within the South American context.

Brazil's population has already surpassed 213.3 million [26]. Against the backdrop of this population growth, the Brazilian government is keen on modernizing its energy grid for several compelling reasons, such as the increasing share of renewable energy sources in its energy matrix [27], the implementation of variable energy tariffs, comprehended locally as the white tariff [28], the increase in the number of consumers that are producing their own energy, also named of "prosumers" [29], and the decrease of non-technical energy losses [29,30]. With these foundational principles in mind, the Brazilian government actively promotes public policies and investments in smart meter adoption, thereby replacing conventional electricity meters with smart meters [31]. Despite the national plan for updating the electrical system, the replacement of traditional meters by smart meters is decentralized and carried out by the various energy utilities. This complexity occurs due to the inherent diversity and complexity of the Brazilian electrical system, which encompasses a variety of entities, ranging from private companies to public–private partnerships and fully public concessionaires. Though implementing smart meters offers numerous benefits to energy utilities, including improved data accuracy, the elimination of manual reading, and the potential to introduce various customer-centric services, the cost of replacing traditional meters with smart ones is paid for by residential consumers, as per the National Electric Energy Agency (ANEEL) [31]. Nevertheless, energy utilities have undertaken several pilot projects for smart meter implementation. Nevertheless, transitioning from pilot projects to widespread smart meter adoption necessitates a deeper examination of user needs and perspectives [8]. Consequently, understanding the factors that influence the acceptance of smart meters by the population is instrumental in shaping more effective public policy implementation. Furthermore, this knowledge helps to mitigate the delays and setbacks frequently encountered during the implementation process [32], which often arise from the oversight of consumer perspectives in the transition to smart grids, as noted in the literature, e.g., [12–14].

In light of this context, two fundamental inquiries guided this research. Firstly, what are the factors of smart meter acceptance? Secondly, how do these factors impact the acceptance of smart meters? Within this framework, the primary objective of this research is to assess the factors influencing the acceptance of residential smart meters in a specific city in the south of Brazil, specifically in the city of Joinville. Joinville is the largest urban center and industrial epicenter in Santa Catarina state [33]. It was selected as the focal point of this study due to its pivotal role in the Brazilian economic landscape, being the third wealthiest city in the southern region of the country [33]. Moreover, focusing the research on a single region ensures a more dependable and representative sample within the designated population [34]. The article's results unveil critical insights for the formulation of public policies concerning the deployment of smart residential meters in the study area. Additionally, the findings aim to furnish valuable evidence regarding consumer behavior in the installation of smart meters in the southern region of Brazil.

2. Materials and Methods

The methodological approach is organized into three distinct steps. Initially, a model for evaluating the acceptance of new smart energy meter equipment is proposed, drawing upon the technology acceptance theory derived from the literature [35]. The second step elucidates the development of the research instrument. Finally, the third step expounds upon the estimation method and the metrics employed to validate the model, encompassing the measurement and structural models.

2.1. Theoretical Foundation to Model Proposal

In the realm of estimation theories for assessing the acceptance of new technologies, particularly those employed in the context of smart meters, there is a notable predilection for utilizing the Unified Theory of Acceptance and Use of Technology (UTAUT) [36] and the Unified and Extended Theory of Acceptance and Use of Technology (UTAUT2) [35] in the existing literature, e.g., [37–40]. UTAUT2, as proposed by Venkatesh et al. [35] stands out as one of the most comprehensive and versatile models for technology acceptance. This model represents an extension of the original UTAUT [36], which was developed through the integration of eight distinct acceptance models previously scattered throughout the literature. The Unified and Extended Theory of Acceptance and Use of Technology (UTAUT2) [35] includes the constructs Performance Expectancy, Effort Expectancy, Social Influence, Facilitating Conditions, Hedonic Motivation, Price Value, and Habit as predictors of the Behavioral Intention to Use construct, and the constructs Habit, Facilitating Conditions, and Behavioral Intention to Use as predictors of the Use Behavior construct.

The UTAUT2 model was used as the base estimation model for this analysis because it is a recent theory and developed from previous theories, in addition to being considerably utilized to estimate the acceptance of smart meters in the literature, e.g., [37–39]. Even so, it was necessary to adapt the UTAUT2 model since the original model includes the relationship between the constructs Behavioral Intention > Use Behavior. The Use Behavior construct refers to consuming or using a specific technological product or service after purchasing it [35]. Users need to utilize the smart meter to measure the Use Behavior construct. Unfortunately, there are still no users in the population target who already use the smart meter, undermining User Behavior construct estimations. Moreover, the Habit, Price Value, and Facilitating Conditions constructs within the UTAUT2 model have been adapted and substituted with the Environmental Awareness, Associated Costs, and Violation of Privacy constructs. Environmental Awareness measures the extent to which individuals are concerned and informed about environmental shifts and the global warming issue [41]. In parallel, Violation of Privacy assesses the level of apprehension regarding the safeguarding of privacy and the security of consumer data [42]. Although not initially featured in the conventional UTAUT2 model, the inclusion of these constructs is imperative in acceptance models for new technologies that intersect sustainable energy and information technology, e.g., [4,42]. Associated Costs, another construct that diverges from the original UTAUT2 framework yet holds relevance in the context of smart meter acceptance, encompasses individual and social costs (subsidies) essential for the success of the implementation [41], particularly in Brazil, where the householders are responsible for paying for the meter [31]. The items used to build the AC and other constructs included in the proposed model were sourced from the literature (Appendix B). Some items were adapted for smart meter purposes and translated into Portuguese. Figure 1 displays the model proposed in this study.

The extant literature on smart meter acceptance encompasses various studies that utilize survey methods to construct and estimate acceptance models. These models are often grounded in established theories, such as the Theory of Reasoned Action (TRA) [43], the Theory of Planned Behavior (TPB) [44], the Technology Acceptance Model (TAM) [45], the Unified Theory of Acceptance and Use of Technology (UTAUT) [36] and its revised version, UTAUT2 [35]. Employing a compatibility analysis of the constructs proposed by Venkatesh et al. [36] makes it possible to gather the magnitude of the relationships in the proposed model (Figure 1) previously estimated in the literature. This analysis revealed 44 relationships that were also estimated using data from 15 different studies. A comprehensive compilation of these estimations of smart meter acceptance is presented in

Appendix A. The findings of this literature review contribute to the analysis and discussion of the results in the present study.

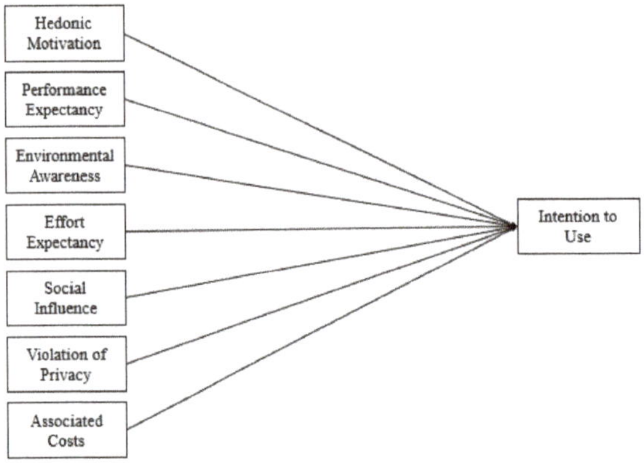

Figure 1. The model proposed to estimate the acceptance of smart meters.

2.2. *Questionaire Building*

The questionnaire is divided into two main sections. In the first part, the socio-demographic characteristics of the respondents are collected, including information, such as age, gender, education level, type of housing, number of bedrooms, city, and neighborhood. The second part of the questionnaire measures respondents' agreement or disagreement with 43 statements (Appendix B), each corresponding to measuring the constructs within the estimation model. For each item, respondents use a five-point Likert scale, where 1 represents "Strongly Disagree," and 5 signifies "Strongly Agree." All items used in the questionnaire are listed in Appendix B. A preview of the original mobile version of the questionnaire interface (for mobile devices) can be seen in Figure 2. The complete questionnaire is available at https://www.questionpro.com/a/TakeSurvey?tt=1vpL5VvhoAk= (accessed on 19 December 2023).

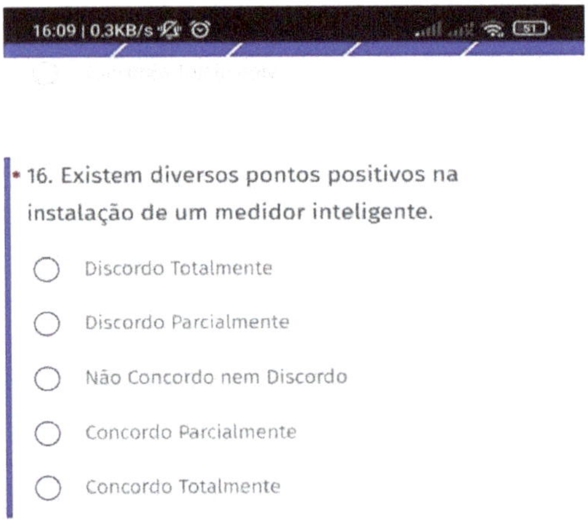

Figure 2. Interface of the mobile version of the questionnaire (in Portuguese).

2.3. *Estimation Model*

Structural equation modeling (SEM) is a pivotal statistical tool in behavioral and social sciences, enabling the simultaneous examination of numerous interdependencies encapsulated by abstract concepts with statistical efficiency [46]. Two primary approaches exist for implementing SEM: the Partial Least Squares Structural Equation Modeling—PLS-SEM) and the Covariance Structural Equation Modeling—CB-SEM) [47]. PLS-SEM is the preferred choice for studies geared towards causal-predictive analysis, often applied in exploratory research, while CB-SEM is more suited for testing well-established theories, as it demands a more intricate theoretical foundation for its application [48]. Additionally, PLS-SEM excels in analyzing and resolving complex models, encompassing a multitude of variables, constructs, and data that do not conform to a multivariate normal distribution [49,50]. The utilization of PLS-SEM has made significant strides due to the introduction of new metrics for model evaluation, offering more options for assessing the consistency of estimated models [47–49]. Recent metrics aimed at evaluating discriminant validity and the overall model fit have enhanced the reliability of estimates [51]. In light of these considerations, the present study will use the PLS-SEM method and Smart PLS® v.3 software to estimate the proposed model.

In the assessment of convergent validity, it is essential to ensure that factor loadings exceed 0.707, revealing that at least 50% of the original variable (item) variance can be attributed to the corresponding latent variable [48,51]. For the reliability of constructs, the Composite Reliability (CR) should be greater than 0.7 for each construct [48,50]. Cronbach's alpha is often recommended with values greater than 0.6 [46,48] to ensure the internal consistency of the constructs. Additionally, the Average Variance Extraction (AVE) should be greater than 0.5 [48,50,51]. These criteria collectively provide validity of the measurement model.

Discriminant validity is crucial as it measures the degree to which a construct is empirically distinct from other constructs within the structural model [51]. The conventional approach to evaluating discriminant validity is the Fornell and Larcker [52] approach, which suggests that the Average Variance Extraction (AVE) of each construct should be greater than the squared correlation between that construct and other constructs [53]. In PLS-SEM models, assessing discriminant validity is also recommended using the Heterotrait–Monotrait (HTMT) approach [48–51]. HTMT values below 0.85 indicate that the constructs are empirically distinct [49]. This multifaceted approach to assessing discriminant validity helps ensure that the model's constructs are distinct and not unduly correlated.

The literature suggests using the Standardized Root Mean Square (SRMR) to evaluate the structural model, as it is a reference to measure the approximate fit and gather empirical evidence for the proposed model [48]. The SRMR values should be below 0.08, according to the literature [50]. In addition, this model will provide R^2 values, Bayesian Information Criteria (BIC), and Akaike's Information Criterion (AIC) as part of the assessment process. These metrics collectively help in evaluating the fit and performance of the structural model.

3. Results

This section is divided into five subsections. The first four subsections present results from the sample, common method bias, the measurement model, and the structural model, respectively. The fifth subsection presents the estimation results. The data analysis was conducted using Smart PLS® v.3 software. Further testing of the model's predictive ability using alternative tools was not undertaken.

3.1. Sample

The questionnaire was disseminated through various social media platforms, text message groups, and professional and personal networks within the target population. We received a total of 203 responses to our questionnaire, with a response rate of 54.13%. Of the 203 responses, 65 were incomplete and two were from outside the geographic region of Joinville, resulting in 136 valid questionnaires (acceptance rate of 67%). Respondents took an average of 9 min to complete the valid questionnaires. Table 1 presents the sociodemographic characteristics of the sample, compared with data from the last available Census of Joinville's population.

Table 1. Descriptive analysis of the sample.

Sample Characteristics	Sample		Census (IBGE, 2010)
	N	%	%
Age			
15 to 24 years old	34	25.00	18.00
25 to 39 years old	37	27.21	26.30
40 to 59 years old	40	29.41	25.20
60 years old or more	25	18.38	8.80
Education			
Incomplete elementary	0	0	35.02%
Complete elementary	3	2.21%	20.48%
Complete high school	43	31.62%	31.64%
Complete higher education	49	36.03%	12.30%
Postgraduate	41	30.15%	*
Gender			
Female	71	52.21%	50.40%
Male	64	47.06%	49.60%
Not identified	1	0.74%	*
Type of housing			
House	65	47.79%	83.53%
Apartment	71	52.21%	15.76%
Number of bedrooms			
1 bedroom	6	4.41%	22.36%
2 bedrooms	29	21.32%	40.63%
3 bedrooms	69	50.74%	31.60%
4 or more bedrooms	32	23.53%	5.41%

* Data not available in the Census [26].

Table 1 highlights slight discrepancies between the sample and the general population regarding age and gender. The sample, on the whole, exhibits a higher quality of life, resulting in differences in variables, such as Education and Number of Bedrooms. Similarly, the difference in the Type of House variable can be attributed to a significant portion of the population residing in homes with substandard infrastructure, commonly referred to as favelas, which impacts the number of people living in houses. New technologies are typically adopted by this demographic segment first, so the higher income within the sample does not compromise the estimations, especially compared to similar studies [53]. Despite variations in Education, Type of Housing, and Number of Bedrooms between the sample and the population, the sample can be deemed suitable, especially compared to analogous studies, e.g., [53,54]. The similarity between the sample and the census data suggests that our findings are generalizable.

3.2. Common Method Bias

A typical challenge in survey data is Common Method Bias (CMB), which can bias the results [55]. The presence of CMB is especially concerning for self-report surveys [56]. To ensure that the survey instrument does not bias the estimations, we carefully examined all relationships between the constructs, and the VIF values within the models remained below 1.569, well below the limit of 3.3 recommended by the literature, e.g., [57]. The VIF values suggest that the survey instrument and/or method did not bias the model estimations.

3.3. Measurement Model

Table 2 summarizes the convergent validity of the proposed model. Factor loadings should be significant and greater than 0.707 [48,50,51]. Nineteen items were removed from the model due to low convergent validity. The proposed Associated Costs construct was also removed due to its low convergent validity. Validating the Associated Costs construct is challenging when assessing smart meter acceptance, as also found in the literature, e.g., [58]. All constructs had Composite Reliability (CR) values exceeding 0.70, as suggested in the literature [48]. Similarly, Cronbach's alpha values were also satisfactory, above 0.7 [48,50]. All constructs had Average Variance Extraction (AVE) values exceeding 0.5, which is considered satisfactory [48,50], except for the Associated Costs construct, which was removed.

Table 2. Convergent validity.

Construct	Item	Factor Loading	Cronbach's Alpha	CR	AVE
Perform Expectancy (PE)	PE1 PE2 PE3 PE4 PE5	0.789 0.879 0.856 removed removed	0.828	0.885	0.660
Hedonic Motivation (HM)	MH1 MH2 MH3 MH4 MH5	0.855 0.795 0.834 0.712 removed	0.828	0.877	0.641
Environmental Awareness (EA)	EA1 EA2 EA3 EA4 EA5	0.955 removed removed 0.765 removed	0.700	0.854	0.748
Effort Expectancy (EE)	EE1 EE2 EE3 EE4 EE5	0.840 0.779 removed 0.851 removed	0.764	0.864	0.679
Social Influence (SI)	IS1 IS2 IS3 IS4 IS5	removed 0.707 0.788 0.911 removed	0.736	0.846	0.650
Associated Costs (ACs)	AC1 AC2 AC3 AC4 AC5	removed removed removed removed removed			
Violation of Privacy (VP)	VP1 VP2 VP3 VP4 VP5	0.827 removed removed 0.827 removed	0.727	0.875	0.778
Intention to Use (IU)	IU1 IU2 IU3 IU4 IU5	removed 0.798 0.816 0.837 0.740	0.863	0.901	0.645

Discriminant validity is presented in Table 3 using the traditional approach [52] and in Table 4 the Heterotrait–Monotrait (HTMT) [48–50] approach. In Tables 3 and 4, the bold type indicates the latent variables. In Table 3, the bold type also shows the square root of AVE in the diagonal, which must be greater than the correlation values to assure discriminant validity. The results indicate that all latent variables are sufficiently distinct, supporting the discriminant validity of the proposed model.

Table 3. Fornell and Larcker approach (Discriminant Validity).

	EA	EE	HM	IU	PE	SI	VP
EA	**0.865**						
EE	0.090	**0.824**					
HM	0.015	0.389	**0.801**				
IU	0.268	0.403	0.426	**0.803**			
PE	0.314	0.449	0.247	0.668	**0.813**		
SI	0.311	0.392	0.354	0.532	0.481	**0.806**	
VP	−0.002	0.140	0.023	0.087	0.158	−0.047	**0.882**

Table 4. Heterotrait–Monotrait (HTMT) approach (Discriminant Validity).

	EA	EE	HM	IU	PE	SI
EE	0.140					
HM	0.098	0.487				
IU	0.316	0.484	0.422			
PE	0.395	0.560	0.277	0.769		
SI	0.425	0.498	0.408	0.616	0.571	
VP	0.101	0.197	0.149	0.113	0.212	0.148

3.4. Structural Model

Table 5 presents the structural model evaluation results. The SRMR value is slightly higher than the 0.08 recommended in the literature [48,50], suggesting that the current model could be improved by adding more variables or moderating demographic factors. However, the study is valid, and further studies are needed to better understand the acceptance of smart meters in the analyzed population. The R^2 values for the Intention of Use (IU) construct of UTAUT2 have moderate explanatory power, similar to those reported in other studies [59–61].

Table 5. Structural Model.

	SRMR	R^2	R^2 Adjusted	BIC	AIC
Model Estimated	0.096	0.550	0.529	−75.286	−95.675

3.5. Estimations and Discussion

The relationships among the constructs, coefficients, and other indicators of the estimated model are presented in Table 6, derived from a bootstrap simulation of 5000 samples based on the dataset of 136 respondents from Joinville, Brazil. The coefficient value (β) indicates the strength of the relationship between constructs, and *p*-values determine the probability of confirming this relationship in the population.

The results of the relationships between the constructs of the estimated model are also represented graphically in Figure 3.

Table 6. Model estimated.

Relationships	Coefficient (β)	T Values	*p*-Value	VIF
EA → IU	0.047	0.704	0.482	1.178
EE → IU	0.008	0.113	0.910	1.455
HM → IU	0.230	3.981	0.000 ***	1.263
PE → IU	0.497	6.010	0.000 ***	1.569
SI → IU	0.195	2.543	0.011 **	1.556
VP → IU	0.011	0.165	0.869	1.061

** significant at 5%, and *** significant at 1%.

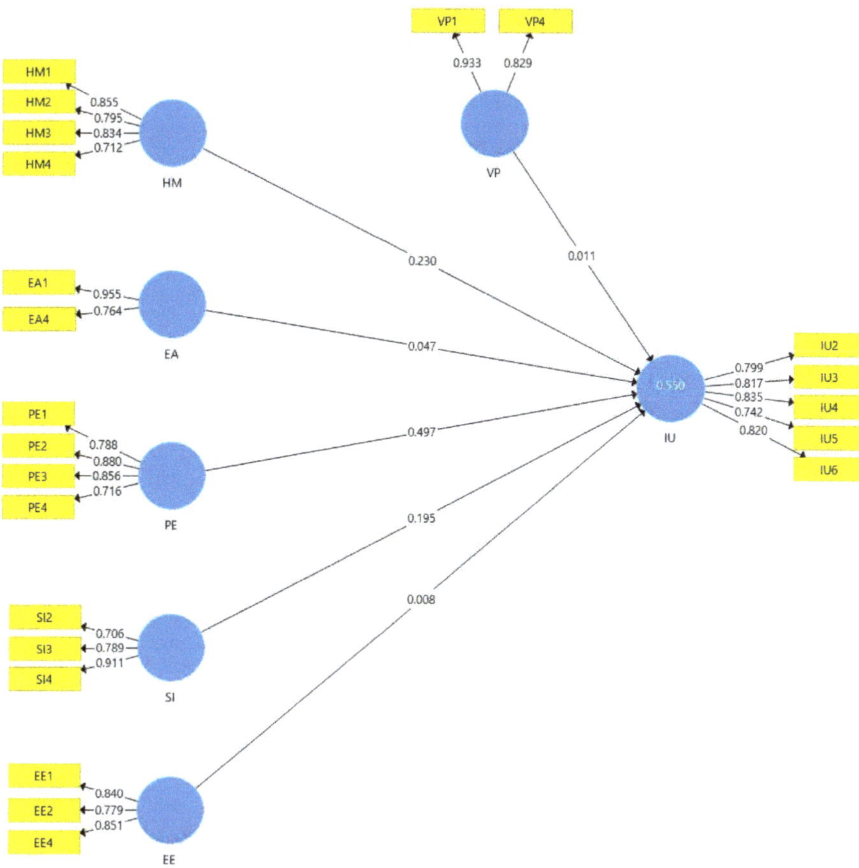

Figure 3. Estimations of the proposed model.

4. Discussion

Performance Expectancy (PE) refers to how an individual believes using a particular technology will help them perform better at a given task or job [36]. In the case of smart meters, feedback provided to consumers is considered a critical factor in smart meter adoption [42,62] because it helps users achieve goals, such as reducing their daily energy consumption [63]. The relationship between PE and IU (β = 0.497; *p*-value < 0.001) was significant and the strongest of the seven constructs in the model. This result is consistent rk with the literature and previous studies on smart meter adoption in Republic of Korea and the USA [42,64,65].

Hedonic Motivation (HM) refers to the enjoyment and satisfaction that using a new technology provides rather than its functional benefits [35]. HM could influence the Intention to Use (IU) smart meters, reflecting an individual's desire to use the device. As with Performance Expectancy (PE), the relationship between HM and IU is also significant in Joinville (β = 0.230; p-value < 0.001). However, the relationship is notably weaker than what has been reported in examinations conducted in Republic of Korea (β = 0.630) [11], Vietnam (β = 0.435) [11], and Indonesia (β = 0.505) [62]. Despite this, consumers in Joinville who find smart meters to be enjoyable to use will be more likely to accept them. Although there are few studies on this relationship, those suggest a positive effect of HM on IU, mainly when gamification interfaces are applied, e.g., [66,67]. Fensel et al. [66] also suggest that designing a user-friendly and intuitive platform to monitor and control energy consumption is paramount for improving the acceptance of smart meters. Given the limited number of studies on this relationship, further research is needed to investigate the effect of HM on the acceptance of smart meters.

Social influence (SI) refers to the degree to which an individual is susceptible to using a new technology because of the influence of other people or organizations that they consider essential [36]. Although this factor has already been frequently tested in the literature, it has become more critical due to the increasing use of social media [68]. SI was significantly related to IU (β = 0.195; p-value < 0.05) for the Joinville, SC sample. The estimated value is similar to those obtained by Warkentin et al. [38] (β = 0.208; p-value < 0.001) in the USA and Guerreiro et al. [9] (β = 0.108; p-value < 0.05) in Évora, Portugal. SI may be more important in Brazil because, according to a 2021 report by Hootsuite and WeAreSocial, Brazil is the third most active country on social media in the world, with users aged 13 or older spending an average of 3 h and 42 min per day on social media. This is more than in any other country except Colombia and the Philippines. Furthermore, 82.2 percent of Brazilians aged 13 or older are active social media users, compared to a global average of 53.3 percent [69]. The COVID-19 pandemic has accelerated the digital inclusion of older adults, who are now using smartphones and social media more to stay in touch with friends and family and to search for information and products that they are interested in [70]. This indicates that SI is becoming increasingly crucial for the acceptance of smart meters by the general population, not just younger individuals. Therefore, one strategy to boost the acceptance of smart meters is to use digital media as an ally, both through social networks and e-commerce, investing in the creation of content and dissemination on the topic.

Environmental Awareness (EA) is the degree to which people are concerned and aware of environmental changes and the problem of global warming [41]. Although not part of the traditional UTAUT2 model, EA is recommended for inclusion in acceptance models for new sustainable energy technologies, such as smart meters [4,42]. In this study, EA did not have a significant effect on UI (β = 0.047; p-value = 0.482). This result contradicts previous research on smart meter acceptance in Malaysia [4,40,42] and the USA [4,42]. Chen et al. [42] and Bugden and Stedman [4] note that a limitation of their studies is that they assume a positive relationship between EA and smart meter acceptance. For example, Chen et al. [43] report that 39.9% of their sample participants were liberal-leaning Democrats, who are more likely to have pro-environmental attitudes. Similarly, Bugden and Stedman [4] found that a large portion of their sample was from wealthier segments of the North American population, who are generally more favorable towards the environment and the benefits of smart meters. The Joinville sample also has higher purchasing power than the average population. However, the results suggest that the environmental factor is not yet decisive for smart meter acceptance. The non-significant EA may indicate that the estimated model lacks a better fit, perhaps requiring more variables or the moderation of socio-demographic factors to explain this factor.

Effort Expectancy (EE) encompasses concepts such as familiarity and perceived ease of use. It is defined as the individual's perceived difficulty or ease in learning to use a specific technology [36]. Interestingly, in the present study, EE had no significant effect on Intention to Use (IU) (β = 0.008; p-value = 0.910), unlike all previous studies on smart meter acceptance, e.g., [4,9,40,65]. This result is contradictory, as EE comprises factors, such as ease of use and familiarity, which are recurrent in the literature as influencers of smart meter acceptance [19,59,71]. People who know little about this technology tend to judge it as complex [19]. Conversely, the easier a new technology seems to be to use, the greater its acceptance [58,65]. Again, the fact that the sample in this study had a higher income than the population average suggests they likely have access to many other smart devices. This greater familiarity with smart devices may increase familiarity with smart meters, making the relationship between EE and UI non-significant. Despite this, the result reinforces the need to test moderation with demographic variables for this relationship.

Violation of Privacy (VP) represents the degree of concern about ensuring the privacy and security of consumer data [58]. Like Environmental Awareness (EA), despite not being part of the traditional UTAUT2 model, these constructs should be incorporated into smart meter acceptance models [4,42]. Similar to Effort Expectancy (EE), the effect of VP on Intention to Use (IU) was not significant ($\beta = 0.011$; p-value = 0.869), which is opposite to the results of studies conducted in Malaysia [40] and the USA [4,42]. The issue of privacy violation is considered relevant in the digital world despite a lack of consensus in the literature on its effect on the intention to use residential smart meters. Recent studies provide evidence that concerns about hacker invasion, and the leakage and distribution of personal data are significant factors in the Intention to Use smart meters, e.g., [38,40,42,62]. However, other studies have not found significant results in this relationship, e.g., [72]. For example, Wunderlich et al. [72] reported that smart meter technology is still in its early stages in Germany, which suggests that people may evaluate privacy violations as less important than the potential benefits that this new technology can offer. The specific situation in Joinville and other Brazilian cities, where physical violence and property crime are high, may also contribute to survey respondents being less concerned about privacy and data protection. The non-significance of the relationship between VP and IU associated with the target sample/population reinforces the need to test moderation with demographic variables for this relationship.

Finally, of the other constructs analyzed, only Associated Costs (ACs) were not validated, as they did not achieve satisfactory results in the Measurement Model, and it was removed from the estimated model. Although AC is not originally incorporated into UTAUT2, it is a relevant factor for smart meter acceptance, as it is defined as the personal financial and social costs (subsidies) required to make the initial investment effective [41]. However, as Gumz et al. [58] have noted, validation problems related to the AC construct are common when estimating smart meter acceptance. This suggests that more consistent items should be included to enable the estimation of this relationship in future research.

5. Conclusions

This article assessed the factors influencing the acceptance of residential smart meters in Joinville, a city in the south of Brazil. The PLS-SEM model was estimated using 136 responses from the population. Focusing on a particular city provides more reliable estimations by eliminating specific factors that could affect the results.

This article provides several managerial implications for the more efficient implementation of smart meters in Brazil, particularly in the city of Joinville. Although the sample is limited to one city, the cultural and climate similarity between other cities in southern Brazil suggests that these results can be generalized to a much wider region. Performance Expectancy, Hedonic Motivation, and Social Influence were found to have a significant impact on the acceptance of smart meters by the study population. Developing public policies and communication strategies focused on these factors is essential to reduce potential consumer resistance, as seen in several other smart meter implementation projects. Moreover, the results contribute to the smart grid implementation, which depends entirely on the smart meter.

Among the theoretical contributions of this study is the limited amount of research on smart meter acceptance in South American populations. Furthermore, some of the estimated relationships suggest new directions for future research, such as the non-significance of the relationships between Violation of Privacy (VP) and Associated Costs (ACs) constructs and the Intention to Use (IU) construct. The higher level of violence in Brazil may reduce the perceived risk of personal data exposure, unlike what is seen in populations with higher levels of security, such as in the United States, e.g., [4,42]. Understanding the factors contributing to lower perceptions of privacy-related problems in Brazil can enable the development of solutions to facilitate the use of consumption data from smart meters.

The study's findings are derived from a specific sample, and the generalizability of these results to the broader population may be limited. The study's limitations acknowledge the potential for biases arising from the sampling procedure and the need for further research to assess the predictive accuracy of the proposed model.

Author Contributions: Conceptualization, D.F. and P.C.; methodology, D.F., P.C. and A.S.; formal analysis, D.F. and P.C.; investigation, D.F. and P.C.; resources, D.F. and P.C.; writing—original draft preparation, D.F. and P.C.; writing—review and editing, D.F., P.C., J.C. and A.S.; visualization, D.F., P.C., J.C. and A.S.; funding acquisition, D.F. All authors have read and agreed to the published version of the manuscript.

Funding: This research was funded by National Council for Scientific and Technological Development: 401432/2022-0 and Fundação de Amparo à Pesquisa e Inovação de Santa Catarina: 2022TR001415 and the APC was waived by MDPI.

Data Availability Statement: The data presented in this study is unavailable due to the privacy procedures and legal terms.

Conflicts of Interest: The authors declare no conflict of interest.

Appendix A. Previous Estimations of Smart Meter Acceptance in the Literature

Reference	Sample Size	Country	Coefficient (β)
Performance Expectancy → Behavioral Intention to Use			
[9]	515	Portugal	0.128
[37]	453	Germany	0.140
[37]	453	Germany	0.150
[40]	318	Malaysia	0.254
[40]	318	Malaysia	0.275
[42]	711	USA	0.140
[42]	711	USA	0.670
[65]	255	Republic of Korea	0.614
[64]	287	Republic of Korea	0.551
[71]	227	Saudi Arabia	−0.00151
[11]	270	Taiwan	0.263
[11]	211	Republic of Korea	0.351
	292	Taiwan	0.160
Effort Expectancy → Behavioral Intention to Use			
[13]	609	EUA	0.350
[9]	515	Portugal	0.100
[40]	318	Malaysia	0.219
[40]	318	Malaysia	0.228
[40]	318	Malaysia	0.135
[65]	255	Republic of Korea	0.209
[64]	287	Republic of Korea	0.195
[11]	270	Taiwan	0.364

Reference	Sample Size	Country	Coefficient (β)
Social Influence → Behavioral Intention to Use			
[9]	515	Portugal	0.108
[38]	229	EUA	0.208
[40]	318	Malaysia	−0.016
[73]	1035	USA	0.08
[74]	292	Taiwan	0.087
Hedonic Motivation → Behavioral Intention to Use			
[62]	301	Indonesia	0.505
[73]	211	Republic of Korea	0.630
[73]	211	Vietnam	0.435
Associated Costs → Behavioral Intention to Use			
[13]	609	USA	0.210
[9]	515	Portugal	−0.166
[38]	229	USA	−0.145
[42]	711	USA	−0.150
[65]	255	Republic of Korea	−0.198
[11]	270	Taiwan	0.472
[73]	1035	USA	−0.343
[74]	292	Taiwan	−0.179

Appendix B

Code	Ref.	Item Utilized in the Questionnaire (Portuguese)	Original Item in English
HM1	[75]	Gosto de experimentar novas tecnologias.	I like to experiment with new (information) technologies.
HM2	[75]	Eu gosto de testar coisas novas.	I like to try new things.
HM3	[75]	Quero ser sempre o primeiro a testar novas tecnologias.	Among my fellows, I am usually the first to try out new information technology.
HM4	[67]	Eu gosto de usar novos aplicativos e novos dispositivos.	I like using new gadgets and apps.
HM5	[76]	Eu gosto de acompanhar desenvolvimentos tecnológicos na tv e na internet.	I like to follow the technical developments in newspapers and TV
PE1	[65]	O medidor inteligente fornece informações importantes para mim.	The smart grid provides useful information for me.
PE2	[77]	O medidor inteligente ajuda no meu controle do consumo de energia.	Because of the programme, I will have a better overview ofmy electricity consumption.
PE3	[77]	O medidor inteligente vai contribuir para reduzir minha conta de luz.	The programme will reduce my electricity bill.
PE4	[11]	Os medidores inteligentes vão ajudar a melhorar a qualidade do fornecimento de energia.	I expect, a smart meter would improve the reliability and quality of energy supplied and service from the utility.
PE5	[77]	Os medidores inteligentes ajudam a diminuir as quedas de energia.	Because of the programme, there will be fewer blackouts in the future.
EA1	[78]	O problema das mudanças climáticas é muito importante para mim.	I put emphasis on the issue of climate change.
EA2	[78]	A substituição dos combustíveis fósseis por energias renováveis é muito importante para mim.	I put emphasis on the issue of renewable energy.
EA3	[71]	Gostaria de consumir energia elétrica de fontes renováveis.	You would like to buy "green" electricity

Code	Ref.	Item Utilized in the Questionnaire (Portuguese)	Original Item in English
EA4	[78]	Me importo com as emissões de gases do efeito estufa na atmosfera.	I put emphasis on the issue of global CO_2 emission.
EA5	[42]	É importante considerar o impacto ambiental da geração de energia elétrica.	I am concerned about energy shortage.
EE1	[35]	O medidor inteligente parece ser fácil de se usar.	Learning to operate the system would be easy for me.
EE2	[11]	Parece ser fácil aprender sobre o que são os medidores inteligentes.	I expect, learning about and understanding smart meters would be easy for me.
EE3	[67]	Uso equipamentos inteligentes (smart) no meu dia-a-dia.	I use office electronic devices (computer, printer, etc.) for my work or at home on a daily basis.
EE4	[62]	É fácil acompanhar o consumo de energia no display do medidor inteligente.	I would have no difficulty reading the information on smart meter in-house display.
EE5	[67]	É fácil utilizar equipamentos inteligentes (smart).	I like using new gadgets and apps.
SI1	[36]	Pessoas com as quais eu me importo recomendariam o uso do medidor inteligente.	People who are important to me think that I should use the system.
SI2	[11]	Eu instalaria um medidor inteligente se isso fosse recomendado pelo governo.	I will install a smart meter in my house if it is a government policy
SI3	[11]	Eu instalarei um medidor inteligente se as pessoas das minhas redes sociais também instalarem.	I will install a smart meter in my house if people in my social network do.
SI4	[73]	Meus amigos vão gostar se eu usar o medidor inteligente.	My friends want me to use the most advanced technologies available.
SI5	[36]	Eu instalarei um medidor inteligente se meus colegas de trabalho acharem que é uma boa ideia.	I use the system because of the proportion of coworkers who use the system.
VP1	[79]	Quero que minhas informações privadas estejam seguras.	I am concerned that a person can find private information about me on the Internet.
VP2	[65]	Quero estar seguro contra ataques de hackers.	The smart grid can be attacked by cyber hackers.
VP3	[11]	Quero que minha casa esteja segura contra invasões e arrombamentos.	I expect that my privacy would not be compromised by a smart meter in my house.
VP4	[38]	A privacidade dos meus dados na internet é muito importante para mim.	It is very important to me that I am aware and knowledgeable about how my personal electrical usage information will be used.
VP5	[38]	Quero que meus equipamentos inteligentes protejam meus dados.	I trust that my electric company would keep my best interests in mind when dealing with my electrical usage data.
AC1	[80]	Estou ciente do custo para se implementar novas tecnologias.	I am aware of the cost of deploying the technologies.
AC2	[77]	A economia de energia não compensa o custo do medidor inteligente.	The programme will increase my electricity bill.
AC3	[74]	Ter que pagar para ter o medidor inteligente me incomodaria.	Using the SM device will have much additional cost.
AC4	[11]	Há custos para a implementação de medidor inteligente em minha casa.	I expect there are no additional cost are associated with installing a smart meter in my house
AC5	[9]	A mudança para a rede elétrica inteligente gera custos financeiros e problemas de saúde.	The EB may bring more risks to my health and my family
IU1	[36]	Existem diversos pontos positivos na instalação de um medidor inteligente.	Using the system is a bad/good idea.
IU2	[42]	Pretendo usar o medidor inteligente quando ele estiver disponível.	I intend to use a smart meter when the opportunity arises.

Code	Ref.	Item Utilized in the Questionnaire (Portuguese)	Original Item in English
IU3	[74]	Eu gostaria de utilizar um medidor inteligente em minha residência.	I intend to use the SM device.
IU4	[38]	Gostaria que a companhia elétrica instalasse um medidor inteligente em minha residência.	Given these circumstances, installing smart meters at my home would be a good idea.
IU5	[13]	As pessoas deveriam usar medidores inteligentes.	Smart meters would benefit my community
IU6	[13]	Estou empolgado para ter um medidor inteligente em minha residência.	I would be excited to have a smart meter in my home

References

1. IPCC. Intergovernmental Panel on Climate Change; Paris. 2021. Available online: https://www.ipcc.ch/2021/08/09/ar6-wg1-20210809-pr/ (accessed on 30 November 2023).
2. Hossain, M.S.; Madlool, N.A.; Rahim, N.A.; Selvaraj, J.; Pandey, A.K.; Khan, A.F. Role of Smart Grid in Renewable Energy: An Overview. *Renew. Sustain. Energy Rev.* **2016**, *60*, 1168–1184. [CrossRef]
3. Ellabban, O.; Abu-Rub, H. Smart Grid Customers' Acceptance and Engagement: An Overview. *Renew. Sustain. Energy Rev.* **2016**, *65*, 1285–1298. [CrossRef]
4. Bugden, D.; Stedman, R. Unfulfilled Promise: Social Acceptance of the Smart Grid. *Environ. Res. Lett.* **2021**, *16*, 034019. [CrossRef]
5. Goulden, M.; Bedwell, B.; Rennick-Egglestone, S.; Rodden, T.; Spence, A. Smart Grids, Smart Users? The Role of the User in Demand Side Management. *Energy Res. Soc. Sci.* **2014**, *2*, 21–29. [CrossRef]
6. Gumz, J.; Fettermann, D.C. Better Deployments Come with Acceptance: An Investigation of Factors Driving Consumers' Acceptance of Smart Meters. *Curr. Sustain./Renew. Energy Rep.* **2023**, *10*, 23–35. [CrossRef]
7. Hess, D.J.; Coley, J.S. Wireless Smart Meters and Public Acceptance: The Environment, Limited Choices, and Precautionary Politics. *Public Underst. Sci.* **2014**, *23*, 688–702. [CrossRef]
8. Jegen, M.; Philion, X.D. Power and Smart Meters: A Political Perspective on the Social Acceptance of Energy Projects. *Can. Public Adm.* **2017**, *60*, 68–88. [CrossRef]
9. Guerreiro, S.; Batel, S.; Lima, M.L.; Moreira, S. Making Energy Visible: Sociopsychological Aspects Associated with the Use of Smart Meters. *Energy Effic.* **2015**, *8*, 1149–1167. [CrossRef]
10. Krishnamurti, T.; Schwartz, D.; Davis, A.; Fischhoff, B.; de Bruin, W.B.; Lave, L.; Wang, J. Preparing for Smart Grid Technologies: A Behavioral Decision Research Approach to Understanding Consumer Expectations about Smart Meters. *Energy Policy* **2012**, *41*, 790–797. [CrossRef]
11. Chou, J.-S.; Kim, C.; Ung, T.-K.; Yutami, I.G.A.N.; Lin, G.-T.; Son, H. Cross-Country Review of Smart Grid Adoption in Residential Buildings. *Renew. Sustain. Energy Rev.* **2015**, *48*, 192–213. [CrossRef]
12. Buchanan, K.; Banks, N.; Preston, I.; Russo, R. The British Public's Perception of the UK Smart Metering Initiative: Threats and Opportunities. *Energy Policy* **2016**, *91*, 87–97. [CrossRef]
13. Bugden, D.; Stedman, R. A Synthetic View of Acceptance and Engagement with Smart Meters in the United States. *Energy Res. Soc. Sci.* **2019**, *47*, 137–145. [CrossRef]
14. Bertoldo, R.; Poumadère, M.; Rodrigues, L.C., Jr. When Meters Start to Talk: The Public's Encounter with Smart Meters in France. *Energy Res. Soc. Sci.* **2015**, *9*, 146–156. [CrossRef]
15. Throndsen, W. What Do Experts Talk about When They Talk about Users? Expectations and Imagined Users in the Smart Grid. *Energy Effic.* **2017**, *10*, 283–297. [CrossRef]
16. Kumar, A. Beyond Technical Smartness: Rethinking the Development and Implementation of Sociotechnical Smart Grids in India. *Energy Res. Soc. Sci.* **2019**, *49*, 158–168. [CrossRef]
17. Dranka, G.G.; Ferreira, P. Towards a Smart Grid Power System in Brazil: Challenges and Opportunities. *Energy Policy* **2020**, *136*, 111033. [CrossRef]
18. Alkawsi, G.A.; Bte. Ali, N. A Systematic Review of Individuals' Acceptance of IOT-Based Technologies. *Int. J. Eng. Technol.* **2018**, *7*, 136. [CrossRef]
19. Boudet, H.S. Public Perceptions of and Responses to New Energy Technologies. *Nat. Energy* **2019**, *4*, 446–455. [CrossRef]
20. Buchanan, K.; Russo, R.; Anderson, B. The Question of Energy Reduction: The Problem(s) with Feedback. *Energy Policy* **2015**, *77*, 89–96. [CrossRef]
21. Darby, S. Smart Metering: What Potential for Householder Engagement? *Build. Res. Inf.* **2010**, *38*, 442–457. [CrossRef]
22. Jaramillo, N.C.; Cardona, C.J.F.; Henao, J.D.V. Smart Meters Adoption: Recent Advances and Future Trends. *Dyna* **2014**, *81*, 221–230. Available online: https://revistas.unal.edu.co/index.php/dyna/issue/download/3691/402#page=225 (accessed on 30 November 2023). [CrossRef]
23. van de Kaa, G.; Fens, T.; Rezaei, J.; Kaynak, D.; Hatun, Z.; Tsilimeni-Archangelidi, A. Realizing Smart Meter Connectivity: Analyzing the Competing Technologies Power Line Communication, Mobile Telephony, and Radio Frequency Using the Best Worst Method. *Renew. Sustain. Energy Rev.* **2019**, *103*, 320–327. [CrossRef]

24. Yang, B.; Liu, S.; Gaterell, M.; Wang, Y. Smart Metering and Systems for Low-Energy Households: Challenges, Issues and Benefits. *Adv. Build. Energy Res.* **2019**, *13*, 80–100. [CrossRef]
25. Gumz, J.; Fettermann, D.C. What Improves Smart Meters' Implementation? A Statistical Meta-Analysis on Smart Meters' Acceptance. *Smart Sustain. Built Environ.* **2022**, *11*, 1116–1136. [CrossRef]
26. IBGE (Instituto Brasileiro de Geografia e Estatistica). *Estimativas Da População*; IBGE: Brasilia, Brazil, 2021. Available online: https://ftp.ibge.gov.br/Estimativas_de_Populacao/Estimativas_2021/estimativa_dou_2021.pdf (accessed on 5 December 2022).
27. REN21 (Renewable Energy Policy Network for the 21st Century). *Renewables 2018: Global Status Report*; REN21: Paris, France, 2018. Available online: https://opus.lib.uts.edu.au/bitstream/10453/142688/2/GSR2018_Full-Report_English.pdf (accessed on 19 December 2023).
28. Aneel—Agência Nacional de Energia Elétrica. *Tarifa Branca*; ANEEL: Brasilia, Brazil, 2015. Available online: https://www.gov.br/aneel/pt-br/assuntos/tarifas/tarifa-branca (accessed on 19 December 2023).
29. Rivera, R.; Esposito, A.S.; Teixeira, I. *Redes Elétricas Inteligentes (Smart Grid): Oportunidade Para Adensamento Produtivo e Tecnológico Local*; BNDES: Rio de Janeiro, Brazil, 2013. Available online: https://web.bndes.gov.br/bib/jspui/bitstream/1408/2927/1/RB%2040%20Redes%20el%C3%A9tricas%20inteligentes_P.pdf (accessed on 19 December 2023).
30. Gumz, J.; Fettermann, D.C.; Frazzon, E.M.; Kück, M. Using Industry 4.0's Big Data and IoT to Perform Feature-Based and Past Data-Based Energy Consumption Predictions. *Sustainability* **2022**, *14*, 13642. [CrossRef]
31. de, A. Dantas, G.; de Castro, N.J.; Dias, L.; Antunes, C.H.; Vardiero, P.; Brandão, R.; Rosental, R.; Zamboni, L. Public Policies for Smart Grids in Brazil. *Renew. Sustain. Energy Rev.* **2018**, *92*, 501–512. [CrossRef]
32. Moro, S.R.; Imhof, A.C.; Fettermann, D.C.; Cauchick-Miguel, P.A. Barriers to Bicycle Sharing Systems Implementation: Analysis of Two Unsuccessful PSS. *Procedia CIRP* **2018**, *73*, 191–196. [CrossRef]
33. Foltran, M. Santa Catarina Tem Cinco Municípios Entre Os 100 Maiores PIBs Do Brasil. 2022. Available online: https://estado.sc.gov.br/noticias/santa-catarina-tem-cinco-municipios-entre-os-100-maiores-pibs-do-brasil/ (accessed on 18 December 2023).
34. Forza, C. Survey Research in Operations Management: A Process-based Perspective. *Int. J. Oper. Prod. Manag.* **2002**, *22*, 152–194. [CrossRef]
35. Venkatesh, V.; Thong, J.Y.L.; Xu, X. Consumer Acceptance and Use of Information Technology: Extending the Unified Theory of Acceptance and Use of Technology. *MIS Q.* **2012**, *36*, 157. [CrossRef]
36. Venkatesh, V.; Morris, M.G.; Davis, G.B.; Davis, F.D. User Acceptance of Information Technology: Toward a Unified View. *MIS Q.* **2003**, *27*, 425. [CrossRef]
37. Gerpott, T.J.; Paukert, M. Determinants of Willingness to Pay for Smart Meters: An Empirical Analysis of Household Customers in Germany. *Energy Policy* **2013**, *61*, 483–495. [CrossRef]
38. Warkentin, M.; Goel, S.; Menard, P. Shared Benefits and Information Privacy: What Determines Smart Meter Technology Adoption? *J. Assoc. Inf. Syst.* **2017**, *18*, 758–786. [CrossRef]
39. Alkawsi, G.A.; Ali, N.; Baashar, Y. An Empirical Study of the Acceptance of IoT-Based Smart Meter in Malaysia: The Effect of Electricity-Saving Knowledge and Environmental Awareness. *IEEE Access* **2020**, *8*, 42794–42804. [CrossRef]
40. Alkawsi, G.A.; Ali, N.; Mustafa, A.S.; Baashar, Y.; Alhussian, H.; Alkahtani, A.; Tiong, S.K.; Ekanayake, J. A Hybrid SEM-Neural Network Method for Identifying Acceptance Factors of the Smart Meters in Malaysia: Challenges Perspective. *Alex. Eng. J.* **2021**, *60*, 227–240. [CrossRef]
41. Huijts, N.M.A.; Molin, E.J.E.; Steg, L. Psychological Factors Influencing Sustainable Energy Technology Acceptance: A Review-Based Comprehensive Framework. *Renew. Sustain. Energy Rev.* **2012**, *16*, 525–531. [CrossRef]
42. Chen, C.; Xu, X.; Arpan, L. Between the Technology Acceptance Model and Sustainable Energy Technology Acceptance Model: Investigating Smart Meter Acceptance in the United States. *Energy Res. Soc. Sci.* **2017**, *25*, 93–104. [CrossRef]
43. Ajzen, I.; Fishbein, M. *Understanding Attitudes and Predicting Social Behavior*, 1st ed.; Prentice-Hall: Englewood Cliffs, NJ, USA, 1980.
44. Ajzen, I. The Theory of Planned Behavior. *Organ. Behav. Hum. Decis. Process.* **1991**, *50*, 179–211. [CrossRef]
45. Davis, F.D. Perceived Usefulness, Perceived Ease of Use, and User Acceptance of Information Technology. *MIS Q.* **1989**, *13*, 319. [CrossRef]
46. Hair, F.J., Jr.; Sarstedt, M.; Hopkins, L.; Kuppelwieser, V.G. Partial Least Squares Structural Equation Modeling (PLS-SEM). *Eur. Bus. Rev.* **2014**, *26*, 106–121. [CrossRef]
47. Ringle, C.M.; Sarstedt, M.; Straub, D.W. Editor's Comments: A Critical Look at the Use of PLS-SEM in "MIS Quarterly". *MIS Q.* **2012**, *36*, iii. [CrossRef]
48. Hair, J.F.; Risher, J.J.; Sarstedt, M.; Ringle, C.M. When to Use and How to Report the Results of PLS-SEM. *Eur. Bus. Rev.* **2019**, *31*, 2–24. [CrossRef]
49. Henseler, J.; Ringle, C.M.; Sarstedt, M. A New Criterion for Assessing Discriminant Validity in Variance-Based Structural Equation Modeling. *J. Acad. Mark. Sci.* **2015**, *43*, 115–135. [CrossRef]
50. Hair, J.F.; Astrachan, C.B.; Moisescu, O.I.; Radomir, L.; Sarstedt, M.; Vaithilingam, S.; Ringle, C.M. Executing and Interpreting Applications of PLS-SEM: Updates for Family Business Researchers. *J. Fam. Bus. Strategy* **2021**, *12*, 100392. [CrossRef]
51. Benitez, J.; Henseler, J.; Castillo, A.; Schuberth, F. How to Perform and Report an Impactful Analysis Using Partial Least Squares: Guidelines for Confirmatory and Explanatory IS Research. *Inf. Manag.* **2020**, *57*, 103168. [CrossRef]

52. Fornell, C.; Larcker, D.F. Evaluating Structural Equation Models with Unobservable Variables and Measurement Error. *J. Mark. Res.* **1981**, *18*, 39–50. [CrossRef]
53. Fettermann, D.C.; Borriello, A.; Pellegrini, A.; Cavalcante, C.G.; Rose, J.M.; Burke, P.F. Getting Smarter about Household Energy: The Who and What of Demand for Smart Meters. *Build. Res. Inf.* **2021**, *49*, 100–112. [CrossRef]
54. Fettermann, D.C.; Cavalcante, C.G.S.; Ayala, N.F.; Avalone, M.C. Configuration of a Smart Meter for Brazilian Customers. *Energy Policy* **2020**, *139*, 111309. [CrossRef]
55. Podsakoff, P.M.; MacKenzie, S.B.; Lee, J.-Y.; Podsakoff, N.P. Common Method Biases in Behavioral Research: A Critical Review of the Literature and Recommended Remedies. *J. Appl. Psychol.* **2003**, *88*, 879–903. [CrossRef]
56. Podsakoff, P.M.; Organ, D.W. Self-Reports in Organizational Research: Problems and Prospects. *J. Manag.* **1986**, *12*, 531–544. [CrossRef]
57. Kock, N. Common Method Bias: A Full Collinearity Assessment Method for PLS-SEM. In *Partial Least Squares Path Modeling*; Springer International Publishing: Cham, Switzerland, 2017; pp. 245–257. [CrossRef]
58. Gumz, J.; Fettermann, D.C.; Sant'Anna, Â.M.O.; Tortorella, G.L. Social Influence as a Major Factor in Smart Meters' Acceptance: Findings from Brazil. *Results Eng.* **2022**, *15*, 100510. [CrossRef]
59. Sheikh, Z.; Islam, T.; Rana, S.; Hameed, Z.; Saeed, U. Acceptance of Social Commerce Framework in Saudi Arabia. *Telemat. Inform.* **2017**, *34*, 1693–1708. [CrossRef]
60. Owusu Kwateng, K.; Osei Atiemo, K.A.; Appiah, C. Acceptance and Use of Mobile Banking: An Application of UTAUT2. *J. Enterp. Inf. Manag.* **2019**, *32*, 118–151. [CrossRef]
61. Senyo, P.; Osabutey, E.L.C. Unearthing Antecedents to Financial Inclusion through FinTech Innovations. *Technovation* **2020**, *98*, 102155. [CrossRef]
62. Chou, J.-S.; Gusti Ayu Novi Yutami, I. Smart Meter Adoption and Deployment Strategy for Residential Buildings in Indonesia. *Appl. Energy* **2014**, *128*, 336–349. [CrossRef]
63. Avancini, D.B.; Rodrigues, J.J.P.C.; Martins, S.G.B.; Rabêlo, R.A.L.; Al-Muhtadi, J.; Solic, P. Energy Meters Evolution in Smart Grids: A Review. *J. Clean. Prod.* **2019**, *217*, 702–715. [CrossRef]
64. Park, E.-S.; Hwang, B.; Ko, K.; Kim, D. Consumer Acceptance Analysis of the Home Energy Management System. *Sustainability* **2017**, *9*, 2351. [CrossRef]
65. Park, C.-K.; Kim, H.-J.; Kim, Y.-S. A Study of Factors Enhancing Smart Grid Consumer Engagement. *Energy Policy* **2014**, *72*, 211–218. [CrossRef]
66. Fensel, A.; Tomic, S.; Kumar, V.; Stefanovic, M.; Aleshin, S.V.; Novikov, D.O. SESAME-S: Semantic Smart Home System for Energy Efficiency. *Inform.-Spektrum* **2013**, *36*, 46–57. [CrossRef]
67. Wemyss, D.; Castri, R.; Cellina, F.; De Luca, V.; Lobsiger-Kägi, E.; Carabias, V. Examining Community-Level Collaborative vs. Competitive Approaches to Enhance Household Electricity-Saving Behavior. *Energy Effic.* **2018**, *11*, 2057–2075. [CrossRef]
68. Chawla, Y.; Kowalska-Pyzalska, A.; Oralhan, B. Attitudes and Opinions of Social Media Users Towards Smart Meters' Rollout in Turkey. *Energies* **2020**, *13*, 732. [CrossRef]
69. Kemp, S. Digital 2021: Brazil. 2021. Available online: https://datareportal.com/reports/digital-2021-brazil (accessed on 19 December 2023).
70. Barbosa, M. 97% Dos Idosos Acessam a Internet, Aponta Pesquisa Da CNDL/SPC Brasil. 2021. Available online: https://cndl.org.br/varejosa/numero-de-idosos-que-acessam-a-internet-cresce-de-68-para-97-aponta-pesquisa-cndl-spc-brasil/ (accessed on 18 December 2023).
71. Düştegör, D.; Sultana, N.; Felemban, N.; Al Qahtani, D. A Smarter Electricity Grid for the Eastern Province of Saudi Arabia: Perceptions and Policy Implications. *Util. Policy* **2018**, *50*, 26–39. [CrossRef]
72. Wunderlich, P.; Veit, D.J.; Sarker, S. Adoption of Sustainable Technologies: A Mixed-Methods Study of German Households. *MIS Q.* **2019**, *43*, 673–691. [CrossRef]
73. Hmielowski, J.D.; Boyd, A.D.; Harvey, G.; Joo, J. The Social Dimensions of Smart Meters in the United States: Demographics, Privacy, and Technology Readiness. *Energy Res. Soc. Sci.* **2019**, *55*, 189–197. [CrossRef]
74. Chen, K.-Y.; Yeh, C.-F. Factors Affecting Adoption of Smart Meters in the Post-Fukushima Era in Taiwan: An Extended Protection Motivation Theory Perspective. *Behav. Inf. Technol.* **2017**, *36*, 955–969. [CrossRef]
75. Cheng, D.; Liu, G.; Qian, C.; Song, Y.-F. Customer Acceptance of Internet Banking: Integrating Trust and Quality with UTAUT Model. In Proceedings of the 2008 IEEE International Conference on Service Operations and Logistics, and Informatics, Beijing, China, 12–15 October 2008; IEEE: Beijing, China, 2008; pp. 383–388. [CrossRef]
76. Matschoss, K.; Kahma, N.; Heiskanen, E. Pioneering Customers as Change Agents for New Energy Efficiency Services—an Empirical Study in the Finnish Electricity Markets. *Energy Effic.* **2015**, *8*, 827–843. [CrossRef]
77. Moser, C. The Role of Perceived Control over Appliances in the Acceptance of Electricity Load-Shifting Programmes. *Energy Effic.* **2017**, *10*, 1115–1127. [CrossRef]
78. Lai, H.C.; Yu, Y.C.; Tuan, Y.M. Electricity-Saving Behavior Antecedents: Electricity-Saving Motivations, Constraints, Knowledge and Beliefs. In Proceedings of the 2016 IEEE International Conference on Industrial Engineering and Engineering Management (IEEM), Bali, Indonesia, 4–7 December 2016; IEEE: Bali, Indonesia, 2016; pp. 567–571. [CrossRef]

79. Dinev, T.; Hart, P. An Extended Privacy Calculus Model for E-Commerce Transactions. *Inf. Syst. Res.* **2006**, *17*, 61–80. [CrossRef]
80. Nambisan, S.; Agarwal, R.; Tanniru, M. Organizational Mechanisms for Enhancing User Innovation in Information Technology. *MIS Q.* **1999**, *23*, 365. [CrossRef]

smart cities

MDPI

Article

Benchmarking the Functional, Technical, and Business Characteristics of Automated Passenger Counting Products

Cristina Pronello *, Luca Baratti and Deepan Anbarasan

Interuniversity Department of Regional and Urban Studies and Planning, Politecnico di Torino, 10125 Turin, Italy; deepan.anbarasan@polito.it (D.A.)
* Correspondence: cristina.pronello@polito.it

Abstract: Urban transport planning and the integration of various mobility options have become increasingly complex, necessitating a thorough understanding of user mobility patterns and their diverse needs. This paper focuses on benchmarking different Automatic Passenger Counting (APC) technologies, which play a key role in Mobility as a Service (MaaS) systems. APC systems provide valuable data for analysing mobility patterns and informing decisions about resource allocation. Our study presents a comprehensive data collection and benchmark analysis of APC solutions. The literature review emphasises the significance of passenger counting for transport companies and discusses various existing APC technologies, such as pressure sensors, wireless sensors, optical infrared sensors (IR), and video image technology. Real-world applications of APC systems are examined, highlighting experimental results and their potential for improving accuracy. The methodology outlines the data collection process, which involved identifying APC companies, conducting interviews with companies and customers, and administering an ad hoc survey to gather specific information about APC systems. The collected data were used to establish criteria and key performance indicators (KPIs) for the benchmarking analysis. The benchmarking analysis compares APC devices and companies based on ten criteria: technology, accuracy, environment, coverage, interface, interference, robustness (for devices), price, pricing model, and system integration (for companies). KPIs were developed to measure performance and make comparison easier. The results of the benchmarking analysis offer insights into the costs and accuracy of different APC systems, enabling informed decision making regarding system selection and implementation. The findings fill a research gap and provide valuable information for transport companies and policy makers, and we offer a comprehensive analysis of APC systems, highlighting their strengths, weaknesses, and business strategies. The paper concludes by discussing limitations and suggesting future research directions for APC technologies.

Keywords: automatic passenger counting; transport; MaaS; benchmarking; accuracy

1. Introduction

Planning urban transport and managing transport supply and demand are complex tasks that require in-depth knowledge of user mobility patterns and needs, which can be influenced by several factors including but not limited to residential density and land-use mix levels [1–3]. Integrating traditional public transport services with on-demand and shared mobility options, known as Mobility as a Service (MaaS), plays a crucial role in this context [4,5]. Intelligent Transport System (ITS) tools, using technologies like camera recognition, Wi-Fi networks, and GPS, facilitate data collection and analysis of mobility needs [6,7].

APC systems provide valuable data for identifying mobility patterns, obtained through continuous measurement that can yield timely and accurate information [8]. Various APC technologies exist and have given rise to a wide range of systems on the market [9].

Counts of passengers on board can be obtained through manual or automatic methods [10] and are crucial data for transport companies. Automatic Fare Collection (AFC)

technologies such as smartcards are able to calculate space–time passenger volumes as passengers board and alight [11]. In contrast, APC systems use electronic devices placed near vehicle doors to count passengers without any explicit action on their part. The collected data are sent to a backend service for analysis and storage [10]. APC and AFC systems are closely related to the Automatic Vehicle Positioning (AVL) system, which allows passenger data to be matched to information about routes and actual vehicle stops (taking into account deviations in routes and stop skips) [12]. The implementation of APC systems is beneficial for both transport systems and federal or state funding agencies. For transport systems, APC data help to identify factors affecting ridership and passenger flows. Federal agencies, on the other hand, require ridership reports in compliance with government legislation [13].

The common strategy involves connecting the APC to the on-board computer or router, which can then transmit or store the collected data. These data are then analysed and processed by the central control system.

In general, two categories of typically collected data can be summarised:

- Service planning data: they include data on daily variations in the distribution of boarding and alighting, total number of passengers, overcrowding during peak hours, and actual versus scheduled journey times. These data are useful for inferring possible seasonal variations in ridership and journey times. A significant amount of data are required to ensure statistical significance.
- Scheduling data: they include data on individual stops, trips and dwell times, boarding and alighting, and punctuality performance. The objective is to monitor and, if necessary, modify travel schedules to better align supply with demand [14].

In addition to the established advantages of Automatic Passenger Counting (APC) systems, a further significant benefit is the prediction of bus arrival times. This is achieved by analysing APC data with neural network processing, as shown by Chen et al. [15], Cheng et al. [16], and Zhou et al. [17]. This additional functionality enhances public transport by improving scheduling accuracy, optimising operational plans, and consequently increasing user satisfaction.

Various studies have explored ways of improving existing APC systems using new technologies, often including assessments of accuracy and precision and presenting experimental results [18–21]. This section provides a review of technologies used in APC system development.

A number of studies have separated APC technologies into different categories according to the methodologies used. Olivo et al. [11] make a distinction between indirect and direct traveller number estimation, while Grgurević et al. [8] classify APC systems as being either integrated or independent counting systems. Integrated systems, such as those using RFID or gate-based technologies, face accuracy challenges due to potential fraud. APC systems typically involve cameras or sensors placed over each bus door, connected to a counting unit that runs some kind of software. Processed data, along with time, date, and GIS information, are transmitted to the on-board computer for analysis. Different data transmission mechanisms, cable-based or wireless, may be used for this purpose. The main APC technologies employed are pressure sensors, wireless sensors, optical sensors (infrared technology), and video image technology.

Weight-based systems using pressure sensors seek to count passengers either by sensing their footsteps on mats or estimating passenger volumes based on the total weight of the vehicle. The former method requires careful installation and may have edge insensitivity issues, while the latter lacks information on boarding and alighting and relies on assumptions about passengers' average weight [14].

Wireless sensors are APC systems that detect passengers' mobile devices using wireless sensors including Wi-Fi and Bluetooth. Wi-Fi-based systems detect devices within a certain radius using probe request signals, which contain useful data like MAC addresses and timestamps. The collected data are transmitted to a backend service for computing passenger numbers, but reliability and precision may vary due to factors such as passengers carrying more than one device, MAC address randomisation, and irregular probe request intervals [22].

Optical sensors (infrared technology) employ infrared (IR) sensors for passenger counting. Active IR technology uses transmitter and receiver sensors on each side of the door to monitor the flow of people boarding or alighting. When a person passes through the door, the interruption of the IR signal indicates their presence. Passive IR technology relies on pyroelectric sensors to count individuals based on detected temperature. However, distinguishing multiple individuals on the basis of their temperature requires the use of pyroelectric sensor arrays [23]. Although the use of IR sensors during periods of congestion has its limitations, these sensors are often combined with other types of sensors in order to improve counting accuracy [24].

Video image technology uses cameras with optical sensors to measure passenger volumes. Using recognition software to process recorded images, passengers can be categorised based on their shapes and distinguished from other objects [25]. This technology can identify the direction of movement of passengers and distinguish between boarding and alighting [26]. However, camera-based systems face challenges relating to lighting conditions, privacy concerns, and variations in visibility.

APC systems are typically evaluated through laboratory tests, but real-world scenarios present numerous variables that cannot be simulated. Kotz et al. [27] combine IR sensors, onboard cameras, and pressure sensors to improve accuracy, achieving 97.62% accuracy compared to 82.5% with IR alone. Pu et al. [6] propose a low-cost method based on body kinematics and pressure sensors, achieving 90% accuracy. Nitti et al. developed a Wi-Fi-based APC system with 100% accuracy in static scenarios and 94% in dynamic cases. Li et al. [28] use RGB video technologies with accuracy up to 96.5%, but video systems are costly. Overall, video systems are promising despite the cost.

This study benchmarks different APC solutions, analysing their strengths, weaknesses, and business strategies. The overall aim is to help transport providers make informed decisions about resource allocation. The paper, after having discussed the used systems, starts by presenting the methodology for collecting the data needed to benchmark the APC systems described, and our results are presented. Finally, the conclusions highlight some limitations of the present study and outline future work.

2. Materials and Methods

To benchmark the APC devices, we analysed their characteristics and the marketing strategies of the companies behind them. The benchmarking was performed using a three-step methodology (Figure 1): (a) collecting data relating to the companies producing and selling APC systems and to the characteristics of the APC products; (b) defining criteria and key performance indicators (KPIs) for the benchmarking; and (c) evaluating an APC system according to the KPIs thus defined.

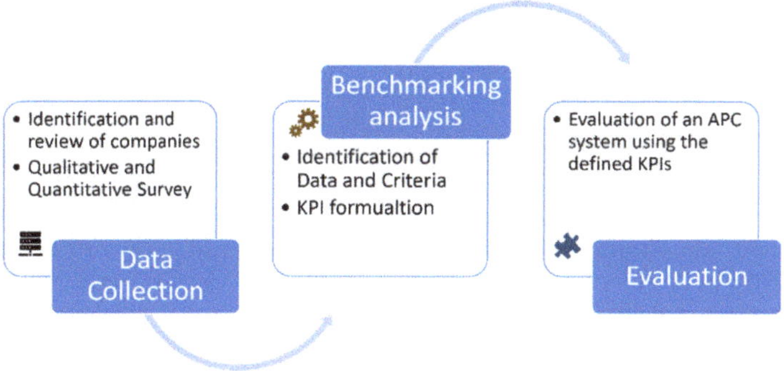

Figure 1. Methodology of research.

Our objective in this benchmarking was to help transport operators to identify the best-performing products.

2.1. Data Collection

Data collection involved the following: (1) identification and review of companies producing and selling APC systems; (2) a survey in two parts; *qualitative*, based on interviews with companies producing APC systems and with system integrators; and *quantitative*, based on a questionnaire to characterise the products and the needs of buyers, administered to transport operators.

The first step distinguished companies as either manufacturers or system integrators, with further analysis on their business models and integration levels. The term "integration levels" refers to the classification of components provided to clients for installation and use, including APC hardware units, AVL/AVM on-board gateways, and IoT platforms for data processing.

The process for identifying target companies involved in producing or selling APC systems utilised a custom-built smart search tool. This tool employed a circular methodology, where its output—a collection of various online resources such as links, articles, and social media posts, sorted by relevance—could be re-inputted for further research. The tool's structure is depicted in Figure 2. Keywords were selected based on characteristics of both the device (e.g., technology, accuracy, and environmental impact) and the company (e.g., pricing strategies). Initially, three reference companies (selected based on simple key-word search and prior knowledge about such companies) provided 'seed' keywords and sentences, which were processed using the Rapid Keyword Extraction (RAKE) Algorithm [29]. This approach mitigated the limitations of human-only keyword searches.

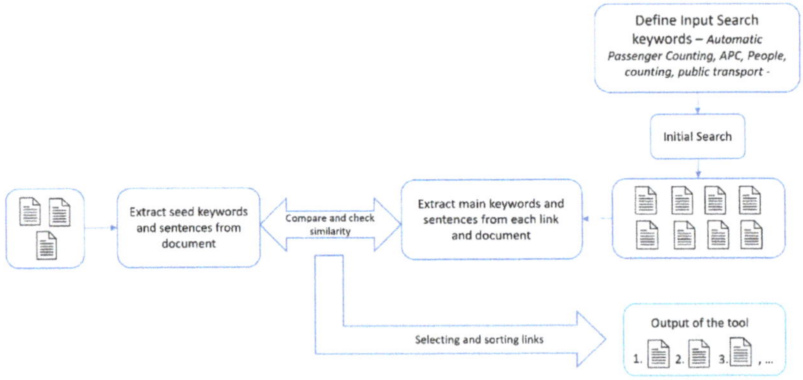

Figure 2. Smart search tool searching and sorting approach.

The identified keywords initiated the search for relevant companies, using a comprehensive set of documents from Google Search, Google Scholar, and Scopus. The smart search tool transformed unstructured text into structured, normalised data. Each company's data were then classified based on these keywords and sentences. A string-matching algorithm compared this data with the reference material from the initial phase. The resulting collection of resources, ordered by relevance, was manually verified. The process was iteratively repeated, incorporating additional data from previous iterations and interviews, to identify more companies.

As the information retrieved online was not complete and not all the characteristics that we needed were available on companies' websites, a survey was created including the following two parts performed in chronological order:

- first, a qualitative survey based on interviews, targeting the companies found on the web where there were information gaps. For producers, the interview was structured around the following elements:
  - characteristics of the company, business strategy, and revenue model;
  - device-specific details, such as interface and integration levels;
  - real-world accuracy, environmental factors, robustness, and potential issues;
  - economic and commercial information like pricing and maintenance contracts.

  For system integrators, the interview focused on contract details and the distinction between manufacturers and integrators.

  The sample of respondents was selected using LinkedIn to identify the employees with skills in APC technology and pricing (mainly commercial profiles);
- second, a quantitative survey to address the lack of data and diversify the sample, enlarging the target respondents. As the interviewees in the previous step did not necessarily provide all the information that we wished to use for our analysis, a second survey was necessary to collect the missing information. To this end, a different sample was targeted, namely transport operators. The reason is that operators purchase APC systems through public tenders where they define the specifications as well as the reference price for the tender. This means that they are quite well informed regarding pricing systems and various other aspects that producers do not usually disclose. To select the operators, the "snowball" sampling plan was used, selecting an initial random group of companies and associations including transport operators (e.g., ASSTRA and ANAV), who were asked to randomly identify other operators falling within the target population [30]. The quantitative survey was designed to collect the following information:
  - real-world accuracy, as measured in real world by the company;
  - components and level of integration of the system;
  - perception of price;
  - business model;
  - maintenance contract.

To evaluate the above information, a number of statements were drawn up, which respondents were invited to endorse or reject, with their responses scoring from 1 to 6 on a Likert scale. The survey was administered using the Lime Survey platform, and the link to the questionnaire was sent to selected technicians working for Italian public transport companies, as well as to ANAV (Associazione Nazionale Autotrasporto Viaggiatori) and ASSTRA (Azienda di trasporti, trasporti internazionali merci), two trade bodies linking transport companies that in turn disseminated the link to their members.

The data analysis was designed according to the two survey typologies. For the qualitative survey, the interviews were recorded and transcribed. A textual analysis of the interviews was performed, extracting the elements investigated, which were then classified in a table.

The data collected through the questionnaire were directly provided by Lime survey in table format. The data were cleaned, and a database was constructed to allow a statistical analysis of the sample using IBM SPSS software.

2.2. *Benchmarking Analysis*

Based on the collected data, the benchmarking consisted of a comparison of the devices and characteristics of the companies. The analysis was structured in three steps: (a) identification of data to be used for the comparison and the criteria relevant for the benchmarking; (b) determination of the KPIs; and (c) analysis of the results in relation to a reference standard [31]. A benchmarking analysis seeks to improve organisational, process, or product performance.

The criteria that we selected in the first step are listed below, separated into those relating to device characteristics and those relating to company characteristics.

Device-related criteria:

- technology: different approaches (e.g., Wi-Fi and infrared) with opportunities and challenges;
- accuracy: quantification of performance of available technologies;
- environment: requirements and standards for system implementation;
- coverage: number of devices needed and cost variance;
- interface: communication interfaces for product selection;
- interference: potential interference during device operation;
- robustness: hardware components' ability to withstand adverse conditions.

Company-related criteria:

- price: price paid by buyers to manufacturers or system integrators;
- pricing model: business strategies selected by manufacturers and integrators in selling their products;
- system integration: level of integration provided to customers.

Key Performance Indicators

KPIs are essential for benchmarking analysis, measuring progress made towards the realisation of strategic goals. They are quantifiable indicators that may be used to compare the performance of different companies [25,32]. Two KPIs were built based on the survey responses and focussed on two key elements for the transport operators: cost and accuracy.

The first *KPI* (KPI_1) combines the internal and external costs of APC systems to compute the total cost useful to rank and classify them in terms of cost incurred by operators to purchase or develop an APC system. Equation (1) shows how this indicator was computed:

$$KPI_1 = \sum_{i=1}^{res} a_i \times H \times \frac{€}{\text{hour}} + \sum_{n=1}^{comp} C_n \times w_n \tag{1}$$

where:

- i is the counter for the different tasks involved (integration, maintenance, etc.), considering only in-house processes. It ranges from zero to the number of different tasks (*res*);
- a is an integer corresponding to the number of resources allocated for the ith task (number of employees, etc.);
- H is an integer corresponding to the number of hours required for the completion of the ith task;
- EUR/hour is the cost per hour of the ith task;
- n is the counter for the system components (hardware unit, on-board gateway, IoT platform, etc.). It ranges from zero to the number of system components (*comp*);
- C is the price of the nth component;
- w is a normalised value weighting the critical nature (complexity) of the nth component.

The cost of APC systems was determined by a linear combination of internal (allocated resources) and external (components) costs, weighted by factors a and w. The weights (w_n) were obtained from the survey. The averages of the responses on the Likert scale (1 to 6) were calculated and then normalised using Equation (2).

$$w_i = \frac{x_i - x_{min}}{x_{max} - x_{min}} \tag{2}$$

where:

- x_i is the mean/median value of the data collected by the survey;
- x_{min} is the minimum score of the Likert scale;
- x_{max} is the maximum value used for the normalisation.

These weights were then applied to Equation (1).

The second indicator (KPI_2) brings the costs obtained from Equation (1) into relation with the accuracy of the system as a whole, and it is useful to rank and classify different solutions available on the market. It was computed using Equation (3):

$$KPI_2 = \frac{w_1 \times KPI_1}{w_2 \times ACCURACY} \quad (3)$$

where:

- w_1 is a normalised value weighting the costs of the system;
- KPI_1 is the result of Equation (1) for the system;
- w_2 is a normalised value weighting the accuracy of the system;
- ACCURACY is the accuracy of the system, expressed as a simple number (not percentage) ranging between 0 and 100. Note that this value may refer to data available in the company's product data sheet. If real-field accuracy is available, this information can be used in Equation (2).

Equation (3) can be used to compute KPI_2 in cases where both cost and accuracy are available, expanding the scope of KPI_1 to include system performance in addition to the purely financial aspect. The weights (w_1 and w_2) were derived from the survey responses; the average values corresponding to the levels of importance ascribed to accuracy and cost were normalised using Equation (2), similarly to the weights in KPI_1. These values were then used to obtain KPI_2. Therefore, there is likely to be a high correlation between the two KPIs if the accuracy of the compared APC systems is similar. Although many companies report similar high accuracy values, ideal scenarios often diverge from the actual accuracy in the field [33]. This discrepancy, influenced by product-specific factors and different city contexts, can lead to significant differences between KPI_2 and KPI_1.

3. Results

Here we present the results of the data collection, starting with the information about companies and devices, followed by the qualitative and quantitative data collected through the two surveys. Finally, the benchmarking results (see Section 3.4 below) are presented separately, with Section 3.4.1 examining the data collected according to the predefined criteria and Section 3.4.2 reporting the KPIs obtained, supported by relevant figures and tables.

3.1. Review of Companies Producing and Selling APC

The web search identified 27 APC companies from Europe, Asia, and the United States. Data from the companies' websites and device datasheets are presented, although we do not include system integrators that use APC products from other manufacturers or that did not provide specifications.

Table 1 summarises the results according to the criteria used for the benchmarking analysis:

- **Technology used**: Some systems use stereoscopic 3D cameras with depth, colour, and texture pattern algorithms, while others use binocular cameras and the Time of Flight (ToF) technique. IR methods based on ToF estimation and wireless solutions that sniff MAC addresses are also represented in the market [28];
- **Accuracy** is a device characteristic that customers pay particular attention to when selecting an APC solution. Regional legislation in Italy requires systems to attain a certain accuracy threshold before they may be purchased using public funding. Table 1 shows that most companies offer an accuracy of 95% or higher, with the exception of Softailor and TripPeopleCounter, which report an accuracy of 90%;
- **Environment** and device coverage: The different technologies used in APC devices must meet specific operating conditions and environmental standards. For example, ISO 16750 provides guidance on environmental conditions encountered by electrical and electronic systems in vehicles [34]. Infrared-based solutions are reliable in some

environments, but may have difficulty counting passengers accurately in crowded conditions. Video imaging technology, on the other hand, performs well in crowded vehicles, but performance may be impaired in situations of excessive brightness. Wi-Fi approaches are not affected by lighting conditions but are more susceptible to external interference. Wired solutions offer robustness against interference, but may require additional resources for installation and maintenance;

- **Interface**: The interface of an APC system is crucial for data transmission and device integration. The most common interfaces are RS-485, Ethernet, PoE, WLAN, and Wi-Fi. Companies need to take into consideration the advantages and limitations of the interfaces on which their devices rely. Some devices, such as Eurotech's DynaPCN, offer more than one interface option. Some devices lack interference information, which highlights the need for comprehensive communication capabilities;
- **Robustness**: The robustness of APC devices is evaluated according to various international standards, the most common being the Ingress Protection Code (IP) [35]. The IP Code, defined by the IEC, assesses how well a device can withstand dust, intrusion, accidental contact, and water. Certifications like IP30, IP45, IP65, and IP67 indicate a device's protection levels against solid particles and liquids. Not all companies advertise their certifications on their websites, but companies will provide them to customers on request;
- **Pricing model**: This is different for manufacturers and system integrators. Information about companies' pricing models is not generally available online and has to be obtained through interviews and surveys;
- **System integration**: System integration refers to the extent to which a company offers customers integrated components, including APC hardware units, on-board gateways, and IoT platforms. While both manufacturers and system integrators provide integration, manufacturers sell components to system integrators, who customise solutions according to customer requirements, resources, and the desired level of integration;
- **Price**: information about price, which is difficult to collect, was mainly obtained through interviews and surveys due to companies' reluctance to reveal such details other than to prospective customers.

Table 1. Products of the companies selected and criteria used for the benchmarking.

Company	Solution	Technology	Accuracy	Environment	Coverage	Interface	Robustness
Dilax Intelcom	DILAX IRS-400	Stereoscopic 3D	99%	-	1 per door	GSM/GPRS/Wi-Fi	IP65
Foorir		Binocular camera (BC)	98%	-	1 per door	RJ45/RS485/WI-FI	T: −20 °C to 40 °C, Hum: 10–80%
Ermetis	Zyklop People	BC	98%	EN50155	1 per door	RS485, Wi-Fi	IP65
Eurotech	DynaPCN 10-20-01	Stereoscopic vision and IR	98–99%	EN50155	1 per door	RS-485, Ethernet PoE Ethernet	IP65
Footfallcam	FootfallCam 3D Pro2	Stereoscopic 3D vision	99.5%	min 1 lux	1 per door	Ethernet or PoE Ethernet/WI-FI	T: −10 °C to 45 °C, Hum: 10–90%
Hella Aglaia	APS-B	Stereoscopic vision and IR	99%	EN 50125-1	1 per door	Ethernet or PoE Ethernet	IP65
IRIS Intelligent Sensing	IRMA Matrix	3D IR matrix sensor (ToF)	99%	EN50155	1 per door	CAN/PoE Ethernet Wi-Fi IBIS-IP	IP65/IP67
Microlog	-	Stereoscopic 3D	95%	-	1 per door	-	n.a.
MRB make recording better	HCP-168-HCP088	BC-3D vision	95–98%	0 lux	1 per door	RJ45/RS45 or RS485/WI-FI	IP43
Softailor		BC	90%	-	1 per door	-	n.a.
Stremax	APC-P3	BC-3D vision	99%	-	1 per door		n.a.

Table 1. *Cont.*

Company	Solution	Technology	Accuracy	Environment	Coverage	Interface	Robustness
Xovis	PTC-1	3D light sensor technology	99%	EN50155	1 per door	Ethernet or PoE Ethernet/WI-FI	IP65
Exeros	-	3D light sensor	99%	-	1 per door	-	n.a.
Infodev	DA-400	3D light directional sensor	97.5–99.5%	-	1 per door	RS-485/PoE Ethernet/Wi-Fi, GSM	n.a.
Interautomation	VDV457 v2.1	2D/3D sensor	99%	-	1 per door	-	n.a.
Masstrans	-	BC-3D vision ToF	97%	-	1 per door		n.a.
R2P	-	Stereoscopic 3D vision	98%	-	1 per door	Wi-Fi, Ethernet, RS485, RS232, LoRaWAN	n.a.
RETAIL Sensing	-	BC	98%	-	1 per door	Wi-Fi, Ethernet, RS485, RS232,	n.a.
Trip people counter	-	Wi-Fi solution	% error: 10%	EN50155	vehicle-size based	Teltonika RUT955	IP30
V-Count	Ultima GO	Stereoscopic 3D vision	98%	min 30 lux	1 per door	Ethernet or PoE Ethernet	T: −15 °C to 40 °C, Hum: 10–90%

Note: n.a.—not available.

3.2. Outcomes of the Interviews

The interviews involved employees with different roles in organisations including GTT, ASP, Eurotech, IVECObus, Leonardo S.p.A., Dilax Intelcom, IRIS Intelligent Sensing, TRISolutions, and Aesys. LinkedIn was particularly helpful as regards setting up interviews. The results of the interviews relevant for the calculation of the KPIs were classified and analysed according to the benchmarking criteria:

- **Accuracy**: Regional legislation in the Piedmont region (in Northwest Italy) has imposed a minimum accuracy of 95% for APC systems used in public transport [18]. However, there is often a significant gap between the nominal accuracy declared in device datasheets and the actual accuracy observed in real use, with a discrepancy that has been estimated to be 20–30% [18]. Concerns have been raised about the effectiveness of in-house testing and about inconsistencies in the numbers of passengers recorded as boarding and alighting;
- **Pricing**: Hardware-only companies have low profit margins and often outsource to low-cost labour companies. SaaS (Software as a Service) companies, on the other hand, can enjoy margins of up to 70–80%. The main costs in the SaaS model are product development, sales activities, customer service, and software hosting. By implementing third-party processing, companies can reduce production costs, ensure expert maintenance, and benefit from just-in-time supplies. It is also important to distinguish between CAPEX (capital expenditure) and OPEX (operational expenditure), where CAPEX includes the supply of hardware/software and integration services, while OPEX covers operational management contracts;
- **Warranty**: Repairs of hardware components are handled by the integrator, who coordinates with the manufacturer for replacements. Warranty and maintenance contracts for software components are concluded online. Warranty contracts are included in CAPEX and typically last 24–36 months. Maintenance variables include repair time, bus line size, and deployment;
- **System integration**: The process of integrating APC components is examined. System integrators typically purchase APC hardware, firmware, and cables from manufacturers, while on-board gateways and IoT platforms are often developed by the integrators themselves, although they can also be purchased from hardware manufacturers or their authorized suppliers. Whether integrators develop these components themselves

or source them externally will depend on factors such as their degree of knowledge, their resources, and the costs involved. In most cases, integrators provide comprehensive data management, processing, and storage tailored to the customer's needs, either through a central control system at the customer's headquarters or through outsourcing;

- **Price**: Regarding price, it is important to distinguish between APC solutions according to the different technologies used and the different settings in which they are implemented. The transport systems that we consider are:
  - closed systems with access after validation (e.g., metro);
  - closed systems with access after ticket reservation (e.g., high-speed train);
  - open systems: systems where there are no turnstiles or reservations (e.g., bus).

Information was obtained on the various commercial offers made by manufacturers to integrators, with a focus on the costs to the integrator. The data collected refer to 2015. The Consumer Price Index (CPI) in Italy increased significantly from a base of 100 in 2015 to 113.2 points by 2023. At the same time, the Industrial Producer Prices Index, also based on the 2015 baseline, rose to 128.7 points [36]. This suggests an inflation-adjusted cost increase of between 13% and 28%. However, data collected from respondents indicate that the actual price variation was within a range of ±10%. It is crucial for an accurate and relevant comparison to compare different Automatic Passenger Counting (APC) systems from the same time period. This approach ensures that the comparison is relative and contextually appropriate, facilitating the selection of the most suitable APC system for a given time and circumstance.

The first offer concerns the Eurotech DynaPCN 10-01-00 APC hardware unit. The prices quoted are per unit. The offer includes:

- DynaPCN 10-01-00: EUR 510;
- Cable connector (optional): EUR 15;
- 12-month warranty contract from commissioning.

The following commercial offers concern fleet monitoring systems, which provide the necessary on-board gateways for the proper functioning of APC systems. This offer refers to the supply of AVM units to support a bus fleet monitoring system based on smart ticketing. The prices quoted are per unit. The offer includes:

- AVM unit: EUR 1670;
- GPS/GPRS antenna: EUR 73;
- System installation and testing: EUR 380;
- Data communication contract: charged to the customer;
- Central control system (IoT platform) at customer's headquarters:
  - Web software: EUR 80,000;
  - Annual maintenance fee: EUR 12,000;
  - Hardware components: charged to the customer;
  - Software management component: charged to the customer;
- Central control system (IoT platform) in outsourcing mode:
  - Web software: EUR 6000;
  - Web service settings (access, fleet, and configuration): EUR 5000;
  - Annual maintenance fee: included;
- One-shot additional services (design, assistance, customisation, etc.): EUR 60,000;
- 12-month warranty contract from commissioning.

It is important to note the price difference between the two strategies for the central control system. The customer in-house installation would cost EUR 92,000, while the outsourcing mode is EUR 11,000.

Moreover, the installation of the central control system, including hardware and software components, is typically borne by the customer because vendors do not consider them to be profitable processes. Customers are free to choose the best solution according to their specific needs.

Manufacturers typically generate profits from the sale of AVM units, with a selling price of EUR 1670. The cost price of these units is around EUR 600, indicating that the manufacturer's markup is around EUR 1000 for the proprietary license.

The one-shot additional services are stand-alone services concerning the configuration, activation of the central control system, etc. These costs should only be incurred once.

In the following, we present tenders referring to data from 2021. The data refer to four supply contracts for Eurotech. Below is a Eurotech tender for supplying a DynaPCN 10-20-01 with a PoE interface. The prices quoted are per unit. The offer made by the manufacturer to the system integrator includes:

- DynaPCN 10-20-01: EUR 581;
- Cable connector (optional): EUR 37;
- Multi-band communication antenna: EUR 149;
- Training course (1 day, mandatory for programmers, installers, and maintainers) that is free;
- 24-month warranty contract from commissioning.

In contrast, the offer made by the authorized supplier to the system integrator includes:

- DynaPCN 10-20-01: EUR 758;
- Cable connector (optional) and multi-band communication antenna not provided.

The supplier's mark-up is EUR 177, and shipping costs and fleet size affect the price of the system. Large supplies often include the sale of hardware components and allow for the amortisation of software and service costs, but this can create a lack of understanding in the purchasing process arising from the data management costs that are included in the AVL system costs.

The interviews allowed us to establish possible values for the KPI_1 variables relating to internal costs. These values do not necessarily correspond to the actual values, but reflect the best knowledge of the interviewees employed by system integrators. However, they can be used to provide an example of the sort of calculations made by contracting authorities, who are in possession of the actual values. The values that we established are:

- tasks involved (cardinality i):
  - data management and maintenance;
- resources allocated to tasks (a_i): two persons for each task;
- hours necessary for work completion (H):
  - 40 h for weekly data management, not considering overtime, on-call services, etc.;
  - 2 h for average maintenance work:
- cost per hour of work (EUR/hour):
  - data management: EUR 45/hour;
  - maintenance: EUR 35/hour.

3.3. Results of the Questionnaire

The data collected concern 12 public transport companies, one organisation comprising 14 transport operators, a second organisation comprising three operators, and two further companies that answered the questionnaire anonymously. The survey drew responses from all over Italy, with a number of companies being based in Northern Italy (Figure 3). Some of the companies are owned by public administrations (municipalities, etc.), and others are privately owned. Among the respondents, catchment managers, IT managers, innovation and technology development managers, technical directors, general managers; IT programmers, maintenance managers, and engineers are all represented.

Figure 3. Geographic locations of the respondent companies.

The respondents were asked to provide information about attitudes within their companies to APC systems and about how far these systems are accepted by the personnel. The data concern perceptions of APC systems as fundamental tools for planning and operating services (Figure 4), of the reliability of the data collected by APC systems (Figure 5), and of the effort required to correctly understand and manage the data (Figure 6). Figure 4 indicates that APC systems are perceived as useful tools, with results polarised towards acceptance (median Likert value of 5, and mean equal to 4.3).

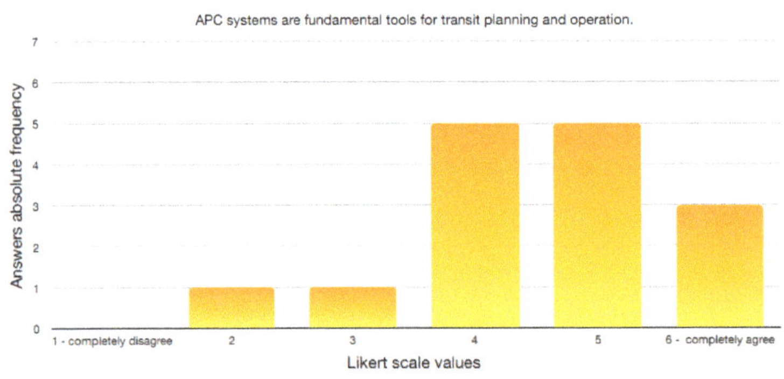

Figure 4. Utility of APC systems for planning and operation.

The median value for the perception of data reliability (Figure 5) is 3 (mean equal to 2.8). It emerged clearly from the interviews with transport companies that data reliability depends on the type of installed solution and on the technical skills of the company.

Respondents perceived APC data as easy to use and understand (Figure 6), with a median Likert value of 2 (mean equal to 2).

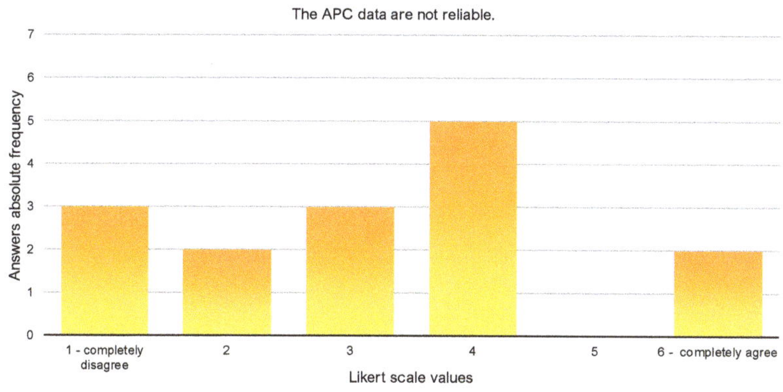

Figure 5. APC data reliability.

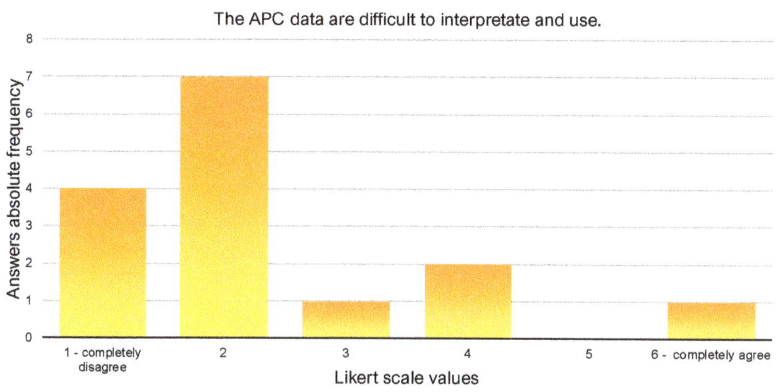

Figure 6. Level of effort required to understand and use APC data.

Accuracy: Accuracy emerges as a pivotal determinant among the possible features of an APC system. Participants unanimously assigned the highest rating to the importance of accuracy (median of 6 and mean of 4). They were also asked about whether the minimum acceptable accuracy level is determined by third parties. The responses are shown in Figure 7, revealing a lack of polarization in respondents' perceptions and a broad distribution across the values of the Likert scale, which may be attributable to the diverse geographical locations of the respondents (in various Italian regions). Notably, transport companies operating in the Piedmont region, subject to regional legislation, uniformly awarded the maximum score, suggesting that there are indeed regional factors at play here.

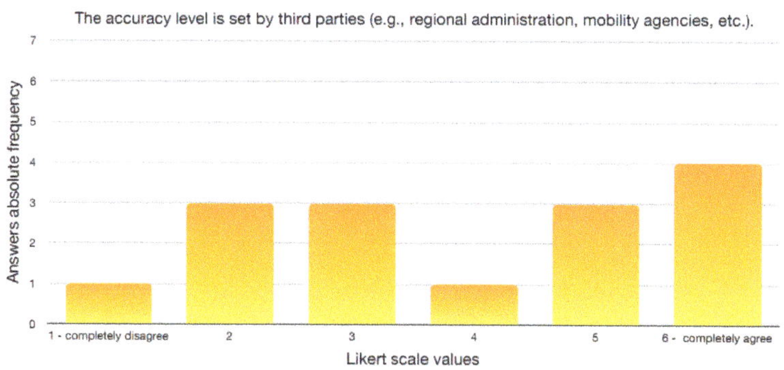

Figure 7. Accuracy defined by third parties.

Interface: The type of interface used by a device is seen as being of medium/high importance. Almost two-thirds of responses were around the middle of the Likert scale, while for the remaining third of respondents, the type of interface was of prime importance.

Pricing: The survey sheds light on attitudes concerning different forms of APC system procurement contracts. These contracts can include both the product and associated assistance. A majority of the survey participants expressed a preference for the annual full-service mode (Figure 8). The median score for the annual service arrangement is 5 (mean equal to 4.4), while the median score for the 'one-shot' full-service model is 4 (mean of 3.6).

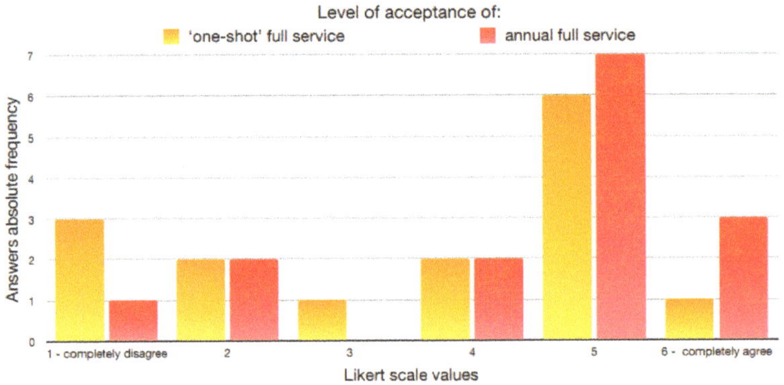

Figure 8. Level of acceptance of different contract modes.

The results also show that the 'one-shot' full-service approach is considered favourable, even though a notable portion of respondents consider it less convenient. Furthermore, respondents were asked to evaluate two possibilities of service inclusion within the contract structure. The first option includes the product along with both data analytics and reporting services. The second option consists of two separate contracts: one for the product and assistance and another for data management (analytics + reporting). As shown in Figure 9, the first option, which includes comprehensive services, registers the highest level of acceptance (median = 5 and mean = 4.4). In contrast, the concept of separate contracts for these services is deemed inconvenient by almost all respondents (median = 2 and mean = 2.4). However, it is worth noting that certain transport companies might still lean towards separate contracts, possibly due to pre-existing integrated service arrangements.

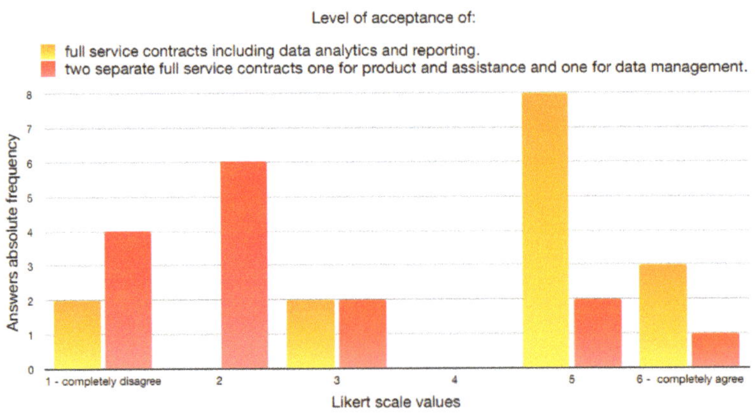

Figure 9. Level of acceptance of single or separate service contracts.

The survey also addressed the question of who handles the end data. Respondents were asked to say which is the final entity managing the data, a crucial point in the OPEX. Most of the companies are the end data managers. Indeed, it is unusual to leave APC collected data to third parties, external to the company.

Finally, supply methodologies were analysed and the results are shown in Figure 10. The possibilities analysed are the following:

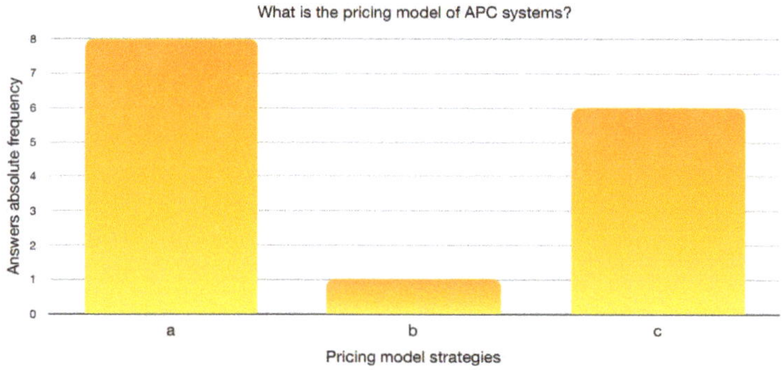

Figure 10. Pricing model of APC systems.

a: one-shot supply of hardware components without additional services;

b: supply of hardware components with annual contract for data management;

c: supply of hardware components with annual contract for data management and processing.

A near-even distribution can be seen between the one-shot supply of hardware components and the supply of hardware components accompanied by an additional annual data management and processing contract. This suggests a prevailing trend where data management and processing are intertwined and provided as a unified service. The chosen supply model appears to be heavily contingent on the requirements of the contracting authority.

System Integration: With regard to system integration, it was found that manufacturers generally do not supply vehicles with pre-integrated APC systems, but leave the specification and integration to the contracting authorities. Integration efforts for non-prepared vehicles were also examined, and this revealed that transport companies perceive this integration as expensive, on par with the integration of newly acquired vehicles.

As for the challenges of integrating APC systems in vehicles that are not initially equipped to support them, the survey indicates that transport companies consider this integration process to be very expensive, or at least as expensive as the integration efforts required for newly manufactured vehicles.

Among the various types of contracts, transport companies prefer integration and the supply of hardware components only. Finally, critical elements among APC system components were examined. The results are shown in Figure 11.

Respondents were asked to identify the most "critical" system components, that is to say those whose continued successful operation consumes the most resources. The element seen as having the greatest "criticality", in this broad sense of the term, is the IoT platform (median = 5 and mean = 4.3), necessary for the correct planning and operation of the transport services. In contrast, the element deemed to be the least "critical", requiring little maintenance effort, is the AVM/Gateway unit (median = 2 and mean = 2.6). These results are unsurprising, considering that AVM has been mandatory for a number of years, and during that time, it has been thoroughly tested, improved, and refined. Finally, the hardware component (APC device) is the only one perceived as having a low-to-medium criticality (median = 3 and mean = 3.3). This may be taken as confirmation that perceived efficiency and reliability are highly dependent on the particular equipment used.

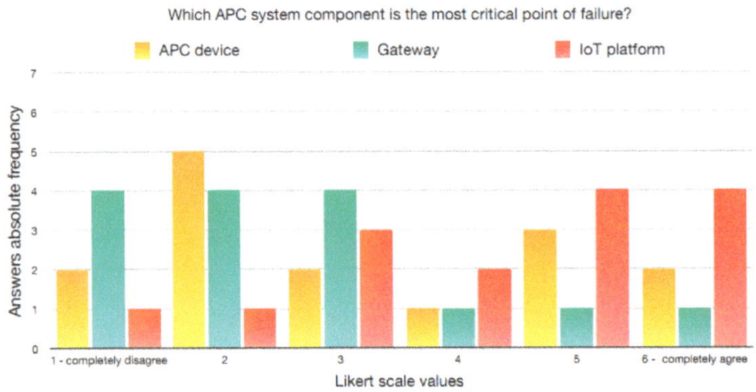

Figure 11. Components of APC systems seen as most likely to fail.

Price: As we might expect, price is seen as important by all respondents (median = 5). To give us a clearer idea of the respondents' attitudes towards price, they were also asked to what extent they could envisage accepting a trade-off between accuracy and price, in a hypothetical scenario, where price was halved in exchange for a reduction in accuracy from 100% to 80% (Figure 12).

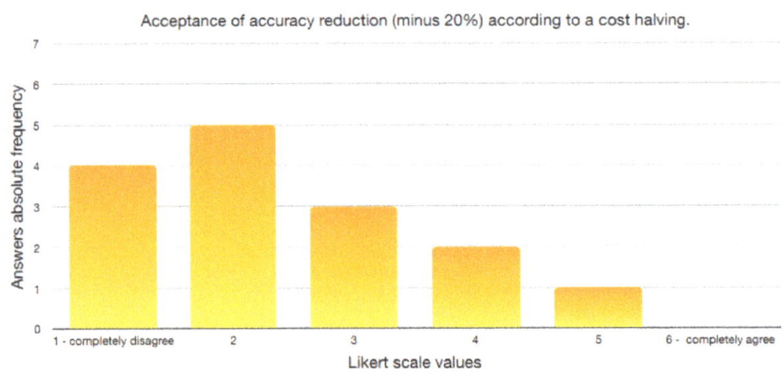

Figure 12. Acceptance of accuracy reduction versus cost halving.

Most of the companies consider accuracy more important than the price of a system. This probably reflects the availability of state subsidies for the purchase of the APC systems, which can be a motivating factor when manufacturers and system integrators reply to tenders.

To help us better understand the dynamics of state subsidies and how they are perceived by public transport companies, respondents were asked about their own company's source of finance for the purchase of APC systems. The proposed options are:

- company capital;
- public subsidies;
- company capital and public subsidies.

The results are shown in Figure 13.

Most of the companies purchased their systems by drawing both on their own capital funds and on public subsidies. This result is consistent with what has been previously stated concerning the importance of public finance for public transport companies. Regarding perceptions of how prices break down, the respondents were asked what percentages of the total price they would expect to be due to the hardware unit, to the on-board gateway, and to the IoT platform. The results are the following:

- hardware unit: 35%;

- on-board gateway: 45%;
- IoT platform: 20%.

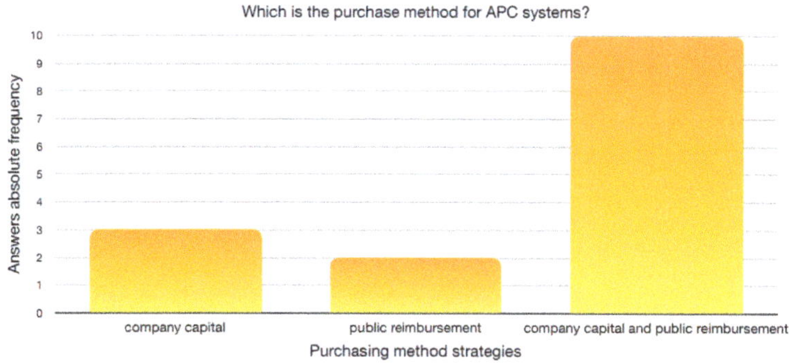

Figure 13. Purchase method of APC systems.

The on-board gateway is therefore perceived as being the most expensive component of the system, followed by the hardware unit and the IoT platform. Interestingly, the oldest component of the system, which has been mandatory for years and around which a robust market has developed, is deemed to be the most expensive. However, it is important to remember that these are the opinions of the respondents and that these data need to be analysed carefully, because they may be biased depending on the specific company supply/case. We also note that the component that inspires the least confidence is judged to be the least expensive.

Weights: With regard to the KPI weights defined in the methodology, for KPI_1, the component criticality mean values are as follows: hardware unit weight = 3.3, on-board gateway weight = 2.6, and IoT platform weight = 4.3.

For KPI_2, the mean values for the levels of importance of price versus accuracy are 4.8 and 6, respectively. The mean values for component criticality are the same as those used for KPI_1, with the following median values: accuracy weight = 6, hardware unit weight = 3, on-board gateway weight = 2, and IoT platform weight = 5.

3.4. Benchmarking Analysis

This section presents the results of the benchmarking analysis, organised into two sections. The first section examines the data collected according to the predefined criteria, while the second presents the KPIs obtained, supported by relevant figures and tables.

3.4.1. Criteria for the Benchmarking

The collected data were characterised and categorised according to predefined criteria for the purposes of the benchmarking analysis. The analysis focused on device and company characteristics, summarising the most relevant information for each criterion.

Implemented technology: The market offers several solutions for APC systems, classified according to the technologies implemented. The most common technology is stereoscopic optical technology, often combined with infrared sensors for greater accuracy. Other methods are also available, such as Time-of-Flight (ToF) technology and wireless solutions.

Accuracy: When choosing an APC solution, customers prioritise accuracy as a crucial feature of the device. Although online accuracy tends to be consistently high (98% or higher) across companies, regional legislation imposes minimum accuracy levels in order for a device to be eligible for public funding. However, our interview results indicate that real-world accuracy is often 20–30% lower than advertised values, highlighting a significant gap between datasheet claims and real performance.

Environment and device coverage: Environmental standards such as EN50155, EN 50125-1, and ISO 16,750 are important for the development of APC devices. Regional legislation in Piedmont requires compliance with EN50155. Visual solutions generally require one device per door, while Wi-Fi solutions vary according to vehicle size, external interference, and software calibration. The corresponding standards for each device are listed in Table 1, in the "Environment" column.

Interfaces and interference: Various communication interfaces are available on the market for APC systems, including RS-485, Fast Ethernet, PoE Ethernet, WLAN, 3G/4G/LTE, and Wi-Fi, each with its own pros and cons. Transport companies can use this information to choose the most suitable solution. Regional legislation suggests the purchase of APC devices with PoE interfaces because of the advantages that these interfaces offer. Interference problems depend on the technology implemented: wired solutions are more robust but require more resources for installation and maintenance, and wireless solutions are easier to install but susceptible to interference, while Wi-Fi solutions require software calibration and the use of routers.

Robustness: Robustness and certification standards, such as IP Code (IP30, IP65, and IP67), play a key role in ensuring the durability of APC devices and protection from adverse conditions. Regional legislation often requires compliance with the IP65 standard, while it emerges from interviews that box breakages and related problems occur only rarely.

Criteria relating to company characteristics are pricing, system integration, and price, as discussed below.

Pricing: The decision to operate as a manufacturer or system integrator is influenced by various factors, including experience, technical skills, and the challenges associated with the hardware industry. The interviews highlight the advantages of an SaaS model for system integrators, with higher profit margins and the use of third-party processing to optimise production costs and ensure timely supplies. The installation process involves training the integrator as a "trainer" through courses provided by the manufacturer, while the involvement of dealers as intermediaries between manufacturers and customers can facilitate negotiations. Warranty contracts and maintenance efforts are typically included in the capital expenditure (CAPEX) of the contracting authority, with considerations such as repair times and the size and distribution of bus lines.

System integration: The system integration criterion focuses on the level of integration offered to customers and the criticality of the APC system components. Transport companies prefer the provision of hardware components, with the IoT platform as the most critical element (median = 5), followed by the hardware component (median = 3), and the least critical AVM unit (median = 2).

Price: When pricing APC systems, it is essential to consider the different technologies and deployment scenarios. The interviews show that Eurotech products are perceived as efficient and reliable, despite being more expensive. Integration with existing vehicle systems is emphasised by both system integrators and transport companies to reduce implementation costs. In addition, increasing the quantity of components supplied can help reduce shipping costs.

3.4.2. Key Performance Indicators

The values of variable C (component price) were set according to the prices from tenders mentioned above. For the APC hardware unit price, we used Eurotech's 2021 offer concerning the supply of hardware to system integrators by authorised dealers. For the gateway and IoT platform price, we used tender data from 2015 (see Section 3.2) relating to a smart-ticket bus fleet monitoring system. For the IoT platform, outsourcing was chosen. The component prices are:

- hardware unit: EUR 758;
- on-board gateway: EUR 1670;
- IoT platform: EUR 11,000.

The weights for the typical component criticalities were established according to the average results of the survey (see "Weights" in Section 3.3) and then normalised. Hence, the values of weights obtained using Equation (2) are:

- hardware unit weight: 0.46;
- on-board gateway weight: 0.32;
- IoT platform weight: 0.66.

Using these values and Equation (1), KPI_1 was computed using Equation (4).

$$KPI_1 = 2 \times 40 \times 45 + 2 \times 2 \times 35 + 0.46 \times 758 + 0.32 \times 1670 + 0.66 \times 11.000 \quad (4)$$

The resulting value for KPI_1 is EUR 11,883.08.

Note that this value is expressed in euros (EUR), as it refers to a cost. It is a cost that weights the internal and external costs of APC systems, giving them lesser or greater prominence according to the issues or allocated resources that each component or work requires. In this example, only two types of internal costs were considered: data management and maintenance. In real projects, the number of types of internal work is much higher, resulting in a higher total cost. Therefore, this cost value cannot be considered a realistic expense for the possible purchase of the system. This indicator is intended to compare the different solutions available on the market and not to obtain the price of the system.

From this preliminary result, the second *KPI* (KPI_2) could then be estimated with reference to Equation (3). KPI_2 refers to the linear combination of the cost and accuracy of the system, thus assessing the overall value proposition of different APC systems in terms of their accuracy. Instead, KPI_1 quantifies the relative cost efficiency between various APC products. In order to be as realistic as possible, the accuracy value was set according to the data published in the Eurotech DynaPCN 10-20-01 datasheet, that is to say 98%, and not according to real values (much lower). In this way, the knowledge and the data used as settings were the same as for a hypothetical customer who uses these indicators to estimate the performance of a potential solution without knowing the real values. Hence, in accordance with the steps and instructions presented in the methodology, the accuracy value is 98%.

In addition, to estimate KPI_2, the two weights needed to balance the linear combination of the variables cost and accuracy are:

- cost weight: 0.76 $\frac{1}{€}$;
- accuracy weight: 1.

Using these weight values and Equation (3), KPI_2, computed using Equation (5), is equal to 92.15 and is a dimensionless value; it is a score that ranks systems based on the weighting of cost and accuracy.

$$KPI_2 = \frac{0.76 \times 11,883.08}{1 \times 98} = 92.15 \quad (5)$$

This value makes it possible to compare different solutions available on the market, offering a clear and immediate ranking of available systems. Unfortunately, as in the case of KPI_1, it is not possible to define an upper limit for KPI_2 due to the cost that can vary between different tenders, because their upper limit is unknown. To achieve the best possible solution, KPI_2 must be minimised, such that the best solution is the one with the lowest value, in other words, the tender with the lowest cost (numerator) and the highest accuracy (denominator).

4. Discussion

The results obtained are intended to be useful to stakeholders, enabling them to compare APC systems and to make informed decisions on the allocation of resources.

First, the analysis revealed that most APC systems advertise accuracies of 95% or higher according to company datasheets. Regional legislation in some Italian regions, such as Piedmont, Lombardy, Friuli-Venezia Giulia, and Apulia, imposes minimum accuracy levels established by third parties in order for systems to be eligible for state subsidies.

However, the interviews reveal a significant disparity between the accuracy of APC systems as advertised in product datasheets and their actual performance under real conditions, with a drop of 20–30% in real accuracy. A manager from a regional transport company explained that the process of verifying the accuracy of APC systems is mainly conducted in the laboratory, with simulated passenger counts that may require passengers to make specific gestures, or other adjustments to be made, for results to be accurate.

The disparity between stated accuracy and actual accuracy represents a critical concern within the context of APC systems. Remarkably, although purchasers of systems are aware of this disparity, they mostly choose not to bring it to the forefront for contractual and funding reasons. At the same time, the survey reveals their insistence on retaining data acquired from APC systems for their exclusive use. Considering that companies are aware of the unreliability of these systems (the 20–30% discrepancy mentioned above), it is hard to see how the use of data that they know to be unreliable can help them to design and plan effective transport systems.

About one-third of respondents consider both 'one-shot' and annual full-service contracts to be convenient, with a preference for annual contracts as fleet size increases. The common trend is to prefer a single full-service contract, although in the future, the market may shift towards offering full IoT services. Transport companies with high fleet penetration prefer separate full-service contracts for items, service, and data management, considering the IoT platform and APC hardware as critical system components.

That being said, a contradiction between company preferences and reality is also evident in this case. The survey has brought to light how companies secure a portion of their funding for the procurement of these systems through public subsidies. These subsidies are primarily allocated on an annual basis and can fluctuate from year to year. In most instances, these public subsidies can be utilized for system acquisition in the current fiscal year. Consequently, there exists a disparity between what is pragmatically favoured by companies in terms of entrepreneurial advancement (fleet expansion and software technological innovations) and what is economically and commercially advantageous. From an entrepreneurial perspective, opting for an annual full-service contract is certainly preferable due to the aforementioned reasons. However, in the realm of economics, adopting a 'one-shot' contract approach allows for the substantial utilization of public subsidies designated for such investments, thereby exerting a lesser financial burden on the corporate balance sheet. Frequently, this approach is favoured by company management.

Concerning the hardware component, nearly half of the companies consider APC hardware units to be critical. Interviews with local transport companies revealed a significant issue of device unbalancing on buses due to jolting, requiring almost constant maintenance, a fact often under-represented in online data. Typically, transport companies outsource the supply and configuration of APC hardware to system integrators, who provide comprehensive services including component supply, installation, testing, system design, and documentation, as well as warranty and maintenance contracts. Expenditure for these authorities is classified into Capital Expenditure (CAPEX) and Operational Expenditure (OPEX), with CAPEX covering the supply of hardware and the services of integrator and OPEX covering operational management contracts. Despite the importance of maintenance contracts in OPEX, their influence on the choice of APC manufacturer or model is limited. Post-warranty, hardware maintenance costs are often transferred to the customer as part of separate maintenance contracts. These contracts generally transfer the hardware costs directly to the customer and charge the work performed on an hourly basis; we have tried to reflect this in the cost calculation.

The costs linked to these units are seen as being mostly due to the maintenance effort required by installed devices, rather than the cost of installing them in the first place. In addition, price considerations emerged, where the analysis of tender data revealed the importance of price in distinguishing and ranking different solutions. Particularly noteworthy is the revelation that the perceived costs concerning the three main components of APC systems are not in line with reality. The survey shows that the costs perceived by companies for the hardware unit, on-board gateway, and IoT platform components are 35%, 45%, and 20%, respectively. However, insights gleaned from worker interviews, specifically regarding pricing, reveal that the perceived cost for the hardware component is significantly higher than the actual cost (inclusive of maintenance expenses). The synthesis of these two aspects—namely, the critical perception of the hardware component and the much lower actual percentage cost compared to the perceived cost—requires thoughtful consideration.

It would be prudent for companies to consider investing in APC systems over a time frame of no less than five to ten years. This approach underscores the fact that hardware cost is a one-time expense and, thus, marginal in relation to the cost of the other two components. The implication of this logic is that companies might find it advantageous to consider acquiring more robust hardware systems (potentially at a higher initial cost) for two reasons. Firstly, the costs of hardware components would nonetheless remain marginal in the context of the entire system expenditure. Secondly, hardware components play a critical role and often require maintenance, which can lead to temporary unavailability of the vehicle and the loss of significant data when the system is non-operational while the vehicle remains in service. Opting for more robust hardware components would alleviate both problems without unduly impacting the long-term economic outlook.

The method for calculating KPI_1 provides a benchmark for obtaining the prices proposed by integrators, but considers only the costs of data management and maintenance, whereas real projects involve various other types of internal work that increase the total cost. In addition, fleet size also influences the price of the system, with a different cost distribution for software and services than for hardware. The lack of price specifications further complicates the collection of price data and hampers the understanding of the purchasing process.

Defining a price for an APC system is challenging because of the many factors to be considered. For transport companies, it is recommended to iterate the calculation of KPI_1 using the data and solutions from the quotations, selecting the solution with the lowest value, to minimise the cost in order to choose the optimal solution. However, setting an upper limit for KPI_1 is impossible due to variable costs between tenders.

5. Conclusions

This research aims to shed light on APC technologies and support transport companies purchasing these systems in choosing the best solutions on the market through specific metrics. To this end, we have attempted to provide a sound framework through the use of key performance indicators (KPIs). The data underpinning these KPIs may vary depending on the specific circumstances of each transport authority, tender, and business requirements. KPI_2 implies a choice between using the manufacturer's advertised accuracy or conducting independent tests to check the accuracy in real field. The drawback is the cost of the test but allowing a more customised and potentially more accurate assessment of APC technologies.

This study identifies major APC manufacturing companies and system integrators, analyses APC products from a technical and functional perspective, and examines the business systems of these companies. Data were collected through website searches, interviews, and a quantitative survey, from which we obtained a wealth of information on APC products, but only limited information on activities, prices, and pricing models. The work provides interested stakeholders with an overview of APC products and companies, an interpretation of accuracy, insights into contract possibilities and preferences, and benchmarks for APC system costs.

In continuing the work begun, it is our intention to broaden the survey audience, to collect more data on pricing policies and components, and to monitor the evolution of technology and the appearance of new systems on the market. This is particularly important because in the literature, to the best of our knowledge, there are no indicators allowing for the comparison of APC systems from a financial point of view. Thus, we need to collect more data to try to identify thresholds to facilitate the comparison.

Author Contributions: Conceptualisation, C.P.; methodology, C.P.; validation, C.P., L.B. and D.A.; formal analysis, L.B. and D.A.; investigation, C.P., L.B. and D.A.; resources, C.P.; data curation, C.P., L.B. and D.A.; writing—original draft preparation, C.P. and D.A.; writing—review and editing, C.P.; supervision, C.P.; project administration, C.P.; funding acquisition, C.P. All authors have read and agreed to the published version of the manuscript.

Funding: This research received no external funding.

Informed Consent Statement: Informed consent was obtained from all subjects involved in the study.

Data Availability Statement: The data are not publicly available due to privacy of the companies and individuals involved.

Acknowledgments: This research study was made possible through the collaboration and support of several transport companies that responded to the questionnaire and to the interviewed people form the selected companies cited in the paper. We also thank Mobyforall, the company where this work was developed.

Conflicts of Interest: The authors declare no conflict of interest.

References

1. Ben-Akiva, M.E.; Lerman, S.R. *Discrete Choice Analysis: Theory and Application to Travel Demand*; MIT Press Series in Transportation Studies; MIT Press: Cambridge, MA, USA, 1985; ISBN 978-0-262-02217-0.
2. Cervero, R. Mixed Land-Uses and Commuting: Evidence from the American Housing Survey. *Transp. Res. Part Policy Pract.* **1996**, *30*, 361–377. [CrossRef]
3. Banister, D. The Sustainable Mobility Paradigm. *Transp. Policy* **2008**, *15*, 73–80. [CrossRef]
4. Butler, L.; Yigitcanlar, T.; Paz, A. How Can Smart Mobility Innovations Alleviate Transportation Disadvantage? Assembling a Conceptual Framework through a Systematic Review. *Appl. Sci.* **2020**, *10*, 6306. [CrossRef]
5. Cruz, C.O.; Sarmento, J.M. "Mobility as a Service" Platforms: A Critical Path towards Increasing the Sustainability of Transportation Systems. *Sustainability* **2020**, *12*, 6368. [CrossRef]
6. Pu, Z.; Zhu, M.; Li, W.; Cui, Z.; Guo, X.; Wang, Y. Monitoring Public Transit Ridership Flow by Passively Sensing Wi-Fi and Bluetooth Mobile Devices. *IEEE Internet Things J.* **2020**, *8*, 474–486. [CrossRef]
7. Halyal, S.; Mulangi, R.H.; Harsha, M.M. Short-Term Passenger Demand Modelling Using Automatic Fare Collection Data: A Case Study of Hubli-Dharwad BRTS. *Adv. Transp. Stud.* **2023**, *59*, 103–122.
8. Grgurević, I.; Juršić, K.; Rajič, V. Overview of Wi-Fi-Based Automatic Passenger Counting Solutions in Public Urban Transport. In *EAI/Springer Innovations in Communication and Computing*; Springer Science and Business Media Deutschland GmbH: Berlin/Heidelberg, Germany, 2022; pp. 181–196.
9. Qureshi, K.; Abdullah, A. A Survey on Intelligent Transportation Systems. *Middle-East J. Sci. Res.* **2013**, *15*, 629–642.
10. Nitti, M.; Pinna, F.; Pintor, L.; Pilloni, V.; Barabino, B. Iabacus: A Wi-Fi-Based Automatic Bus Passenger Counting System. *Energies* **2020**, *13*, 1446. [CrossRef]
11. Olivo, A.; Maternini, G.; Barabino, B. Empirical Study on the Accuracy and Precision of Automatic Passenger Counting in European Bus Services. *Open Transp. J.* **2019**, *13*, 250–260. [CrossRef]
12. Osama, A.; Mahdy, H.; Kandil, K.; Elhabiby, M. Monitoring the Impact of Unscheduled Stops and Unscheduled Route Deviations on Bus Service Using Automatic Vehicle Location System: A Case Study. *Adv. Transp. Stud.* **2016**, *39*, 37–48.
13. United States Department of Transportation. ITS Transit Fact Sheet—Automatic Passenger Counter—ITS Professional Capacity Building Program. Available online: https://www.pcb.its.dot.gov/factSheets/apc/apc_overview.aspx#page=tech (accessed on 10 October 2022).
14. Baltes, M.; Rey, J. The "Ins and Outs" of APCs: An Overview of Automatic Passenger Counters. *J. Public Transp.* **1999**, *2*, 47–64. [CrossRef]
15. Chen, M.; Liu, X.; Xia, J.; Chien, S.I. A Dynamic Bus-Arrival Time Prediction Model Based on APC Data. *Comput. Aided Civ. Infrastruct. Eng.* **2004**, *19*, 364–376. [CrossRef]
16. Cheng, S.; Liu, B.; Zhai, B. Bus Arrival Time Prediction Model Based on APC Data. In Proceedings of the 6th Advanced Forum on Transportation of China (AFTC 2010), Beijing, China, 16 October 2010; pp. 165–169.
17. Zhou, P.; Zheng, Y.; Li, M. How Long to Wait? Predicting Bus Arrival Time with Mobile Phone Based Participatory Sensing. In Proceedings of the 10th International Conference on Mobile Systems, Applications, and Services, Association for Computing Machinery. New York, NY, USA, 25 June 2012; pp. 379–392.
18. Regione Piemonte, D.G.R. n. 40-5004. Available online: https://trasparenza.regione.piemonte.it/media/7241/download (accessed on 9 January 2022).
19. Diyasa, I.G.S.M.; Purbasari, I.Y.; Setiawan, A.; Winardi, S. Smart Passenger Information System Based On IoT. In Proceedings of the 2019 TRON Symposium (TRONSHOW), Tokyo, Japan, 11–13 December 2019; pp. 1–5.

20. IVECO BUS Home Page. Available online: https://www.iveco.com/ivecobus/es-es/Pages/Home-Page.aspx (accessed on 11 December 2022).
21. Nasir, A.S.A.; Gharib, N.K.A.; Jaafar, H. Automatic Passenger Counting System Using Image Processing Based on Skin Colour Detection Approach. In Proceedings of the 2018 International Conference on Computational Approach in Smart Systems Design and Applications (ICASSDA), Kuching, Malaysia, 15–17 August 2018; pp. 1–8.
22. Kimpel, T.J.; Strathman, J.G.; Griffin, D.; Callas, S.; Gerhart, R.L. Automatic Passenger Counter Evaluation: Implications for National Transit Database Reporting. *Transp. Res. Rec.* **2003**, *1835*, 93–100. [CrossRef]
23. Ragland, D. *Estimating Pedestrian Accident Exposure: Automated Pedestrian Counting Devices Report*; University of California: Berkeley, CA, USA, 2007.
24. Jalali, S. Estimating Bus Passengers' Origin-Destination of Travel Route Using Data Analytics on Wi-Fi and Bluetooth Signals. Ph.D. Thesis, University of Ottawa, Ottawa, ON, Canada, 2019.
25. Håkegård, J.E.; Myrvoll, T.A.; Skoglund, T.R. Statistical Modelling for Estimation of OD Matrices for Public Transport Using Wi-Fi and APC Data. In Proceedings of the 21st International Conference on Intelligent Transportation Systems (ITSC), Maui, HI, USA, 4–7 November 2018; ISBN 978-1-72810-323-5.
26. Li, L. Time-of-Flight Camera—An Introduction. Available online: https://www.ti.com/lit/pdf/sloa190 (accessed on 13 January 2023).
27. Kotz, A.J.; Kittelson, D.B.; Northrop, W.F. Novel Vehicle Mass-Based Automated Passenger Counter for Transit Applications. *Transp. Res. Rec.* **2016**, *2563*, 37–43. [CrossRef]
28. Li, G.; Chen, A. Frequency-Based Path Flow Estimator for Transit Origin-Destination Trip Matrices Incorporating Automatic Passenger Count and Automatic Fare Collection Data. *Transp. Res. Part E Logist. Transp. Rev.* **2022**, *163*, 102754. [CrossRef]
29. Rose, S.; Engel, D.; Cramer, N.; Cowley, W. Automatic Keyword Extraction from Individual Documents. In *Text Mining: Applications and Theory*; John Wiley & Sons, Inc.: Hoboken, NJ, USA, 2010; pp. 1–20. [CrossRef]
30. Naderifar, M.; Goli, H.; Ghaljaie, F. Snowball Sampling: A Purposeful Method of Sampling in Qualitative Research. *Strides Dev. Med. Educ.* **2017**, *14*. [CrossRef]
31. Sobral, T.; Galvão, T.; Borges, J. Visualization of Urban Mobility Data from Intelligent Transportation Systems. *Sensors* **2019**, *19*, 332. [CrossRef]
32. Kistowski, J.V.; Arnold, J.A.; Huppler, K.; Lange, K.-D.; Henning, J.L.; Cao, P. How to Build a Benchmark. In Proceedings of the 6th ACM/SPEC International Conference on Performance Engineering, Association for Computing Machinery. New York, NY, USA, 31 January 2015; pp. 333–336.
33. Pronello, C.; Garzón Ruiz, X.R. Evaluating the Performance of Video-Based Automated Passenger Counting Systems in Real-World Conditions: A Comparative Study. *Sensors* **2023**, *23*, 7719. [CrossRef]
34. ISO 16750. Available online: https://en.wikipedia.org/w/index.php?title=ISO_16750&oldid=1088542724 (accessed on 30 August 2023).
35. IP Code. Available online: https://en.wikipedia.org/w/index.php?title=IP_code&oldid=1188610955 (accessed on 12 August 2023).
36. ISTAT. Nic—Medie Annue Dal 2016 (Base 2015). Available online: http://dati.istat.it/Index.aspx?QueryId=23063 (accessed on 7 January 2022).

smart cities

MDPI

Article

Bibliometric Study on the Conceptualisation of Smart City and Education

Debora Scala [1,*], Ángel Ignacio Aguilar Cuesta [2], Maria Ángeles Rodríguez-Domenech [1] and María del Carmen Cañizares Ruiz [1]

1 Facultad de Letras, Universidad de Castilla-La Mancha, Avenida Camilo José Cela s/n, 13071 Ciudad Real, Spain; mangeles.rodriguez@uclm.es (M.Á.R.-D.); mcarmen.canizares@uclm.es (M.d.C.C.R.)

2 Facultad de Filosofía y Letras, Universidad Autónoma de Madrid, Calle Francisco Tomás y Valiente 1, 28049 Madrid, Spain; angel.aguilar@uam.es

* Correspondence: debora.scala@alu.uclm.es

Abstract: In recent years, research in the smart city sector has experienced exponential growth, establishing itself as a fundamental and multifaceted field of study. Education is one of the sectors of interest in smart cities. Concurrently, the extensive academic literature on smart cities makes identifying the main areas of interest related to education, leading institutions and authors, potential interconnections between different disciplines, and existing gaps more complicated. This article maps the knowledge domain of education in smart cities through a bibliometric analysis to identify current trends, research networks, and topics of greatest interest. A total of 88 articles, published between 2000 and 2023, were examined using an interdisciplinary approach. The leading countries are mainly located in Europe and North America and include China. Bibliometrics provides an intellectual configuration of knowledge on education in smart cities; a co-word analysis identifies conceptual sub-domains in specific themes. In general, education within smart cities represents a universal challenge that requires a structured and interdisciplinary approach at all levels. Finally, this paper offers some suggestions for future research, adopting a more comprehensive view of the areas of investigation through a holistic analysis of stakeholders.

Keywords: smart city; education; smart education; educational trends; bibliometrics

1. Introduction

The concept of smart cities in relation to education is an important area of research in many fields and disciplines and has established itself as a contemporary field of study that still needs to be developed and deepened.

The complexity of smart city education is analysed from different perspectives. Smart city solutions are increasingly being used to address the challenges that cities face, such as energy consumption and traffic management. Education is an integral part of smart cities, and research in this area has recently expanded [1].

In this article, we will explore the emerging field of education in the context of smart cities, using a bibliometric approach to identify and classify the key concepts that emerge at this intersection.

Smart cities represent an environment in which technology, urban planning, architecture, engineering, economics, governance, social interactions, and education converge to create an efficient and sustainable urban ecosystem. In this scenario, education plays a critical role in preparing citizens to live and interact effectively in these complex environments [2].

As more and more people move to urban centres—according to a 2017 United Nations report, it is estimated that by 2050, 70% of the world's population will live in urban areas—smart cities are emerging as innovative solutions to address challenges that arise [3]

and enrich the lives of citizens [4]. As highlighted by Albino [5], the smart city concept has become increasingly prominent in academic debate and global policy initiatives. To fully grasp the concept of a smart city, a brief exploration is necessary. In the literature, various definitions of a smart city can be examined, emphasizing different aspects ranging from socio-economic to technological. Caragliu et al. [6] mention a series of characteristics drawn from Hollands [7], who used an economic approach to identify them. He underscores that smart cities utilise a network of infrastructures to enhance economic and political efficacy and are oriented towards cultural, social, and urban development. Caragliu et al. [6] themselves provide a widely accepted definition of a smart city: "A city is smart when investments in human and social capital and traditional (transport) and modern (ICT) communication infrastructure fuel sustainable economic growth and a high quality of life, with wise management of natural resources, through participatory governance" [6].

In the current context, a smart city continues to be defined as an urban environment that applies technology to enhance the benefits and reduce the deficiencies of urbanization for its citizens [8].

Giffinger [9] have developed six characteristics for the smart city: the smart economy, smart people, smart governance, smart mobility, smart environment, and smart living. These characteristics indicate the areas of urban life in which the concept of a smart city can be utilised to improve existing infrastructures through the use of data and information and communication technologies (ICTs) to provide efficient services to citizens.

In this dynamic context, education emerges as a key sector. Within smart cities, education takes on a new form, often linked to sustainable development, active citizenship, and learning to use technology [10], and others have revitalised their urban economies to foster, attract, and retain innovation capabilities [11].

The importance of education goes far beyond its traditional role, becoming fundamental to understanding, interpreting, and addressing the challenges of modern society [12]. In addition, education is increasingly recognised as a critical element in empowering citizens and enabling them to play more active roles in the initiatives that characterise smart cities [13]. Elements such as the presence of a creative class and high levels of education are proving to be significant factors in the growth of urban intelligence [6]. Consequently, without creativity, education, knowledge, and learning, a city cannot aspire to a true state of intelligence [14].

Some authors talk about creativity and apply this concept as a key driver of smart cities; therefore, people, education, learning, and knowledge are central in smart cities [15]. Education is a fundamental element that contributes to the attractiveness of a city, attracting companies, organizations, and people from different backgrounds thanks to a dynamic learning environment.

The importance of the human element in smart cities is critical. To move towards more advanced and intelligent cities, it is fundamental to start with people and human capital. In this context, education is essential rather than relying solely on technology, which cannot automatically transform and improve cities alone; it is strongly argued that people and their interactions are critical factors for the success of any city [7].

Research on smart cities continues to attract the interest of a wide range of professionals, academics, public officials, managers, politicians, and educators. This field of study is very broad, providing fertile ground for research and an opportunity for complementary investigations [16]. At the same time, the expansion of the academic literature on the subject makes identifying main areas of interest, leading institutions, key authors, possible links between different disciplines, and existing gaps more complex.

In this context, a bibliometric analysis of the academic literature is particularly useful for mapping the domain of knowledge, identifying current trends, research networks [17], and relevant topics, especially in a rapidly evolving field such as smart cities and education. Studies conducted even a few years ago may not be able to capture new streams of thought or emerging collaborations. Although bibliometric analyses are based on quantitative and rigorous approaches, they can enrich the results of a systematic literature review by analysing in more detail the characteristics of the literature on a given topic and are valuable tools for evaluating the performance of institutions, journals, or individual scholars [18]. Supported by effective textual analyses and data mining, they can also provide new perspectives and insights into the field of smart cities and education.

As interest in smart city research has grown, several literature reviews have emerged which cover different aspects of the sector [19]. However, only a limited number of these publications have focused specifically on the educational aspect within smart cities. For example, Rico-Bautista [20] studied the use of the Internet of Things (IoT) on smart university campuses. Others have compared two different databases, providing an overview of authors contributing to the field and places where their research has been published [21], while others [1] have provided an overview of existing educational initiatives and the challenges faced in smart city research.

This paper examines education in smart cities, taking an interdisciplinary approach and using Scopus as a database. While some researchers have conducted bibliometric analyses on smart cities, focusing on the network of publications that comprise the intellectual structure of this field of research, there is notable fragmentation and a lack of cohesion in the topic of education in smart cities [22]. Others, through bibliometric analyses, highlight the importance of implementing appropriate technologies to address the challenges of sustainable urban education, thus reinforcing the role of smart cities as catalysts for unconventional innovation [23].

This study aims to fill existing gaps by mapping the knowledge domain and creating new research opportunities, thus providing a clear and updated framework on education in smart cities, considering the growing academic literature on the topic. It highlights the temporal and geographical distribution of research, identifies the most influential authors and journals, and reveals the conceptual structure of education in smart cities through the mapping of the knowledge domain. Furthermore, the analysis illustrates the interdisciplinarity of education in smart cities and highlights some challenges and techniques for effective textual data mining. Finally, while this manuscript makes a significant contribution to the analysis of studies on education in smart cities, it also opens new perspectives for future research in this still partially unexplored field.

The next section of this document defines the conceptual framework; the following sections explain how the data were collected and analysed. The findings are then discussed, and future challenges and gaps are identified. The final section draws some conclusions.

2. Research Questions

This paper provides important insights into education in the context of smart cities and asks three fundamental research questions:

- How is the analysis of education in smart cities evolving?
- What is the intellectual configuration of knowledge on education in smart cities?
- What are the main areas of interest and possible gaps in the literature on education in smart cities?

The first research question aims to determine the volume and geographical distribution of studies on education in smart cities over the years to identify growth trends, reference countries, or journals [24]. The second question focuses on recognizing scholars and articles that have had a significant impact in the field, exploring their nature, relevance and evolution; in practical terms, this means defining their impact on the academic literature through conceptual mapping. The third question focuses on the topics that have received the most attention from scholars and those that have been less explored, with the aim of identifying areas for future investigation.

3. Materials and Methods

A typical bibliometric analysis was applied to the selected data in order to answer the research questions related to education in smart cities, focusing on the evolution, intellectual configuration, trends, and gaps in this field through an interdisciplinary approach.

The choice of a suitable database is a crucial point and is still a matter of debate [25]. In general, large databases such as Scopus, Emerald, Web of Science, ACM, and IEEE [26] are used, but there is no consensus on which is the most appropriate. However, other authors point out that "broad coverage is essential to ensure accurate analysis of the field of study" [27].

In the initial context of this research project, it was planned to use a bibliometric approach that integrated analyses of two major scientific databases, Scopus and Web of Science (WoS). This strategy aimed to ensure comprehensive coverage of the relevant scientific literature. However, a preliminary analysis revealed significant overlap in the content of the two databases, with a high incidence of documents present in WoS that were already present in Scopus. This overlap led to a proliferation of duplicate documents, reducing the effectiveness of the bilateral analytical approach.

A concrete example of this problem was observed during an initial exploratory search using the keywords "smart city" and "education". The results showed a total of 557 documents in WoS, while in Scopus, the number rose to 1648. The comparative analysis suggested that most of the documents identified in WoS were already present in the Scopus data set.

In light of this finding, it was decided to revise the research methodology and to use Scopus exclusively as the main database for the bibliometric analysis. This decision was based on the finding that Scopus offered a broader and more representative coverage of the corpus of literature relevant to the selected keywords. The decision to focus the analysis on a single database allowed for the optimization of resources and data collection efficiency while ensuring an accurate and comprehensive representation of the field of study under investigation.

Consequently, all disciplines that interact in some way with education in smart cities were examined, so in addition to geography, sociology, and architecture, the study also analysed contributions from fields such as engineering, economics, marketing, and urban planning.

In terms of procedure, this study was based on an initial phase of collection, followed by an analysis aimed at mapping the knowledge domain (MKD), also known as a science mapping analysis [28], using bibliometric techniques.

The collection phase used a systematic review methodology [29] to minimise bias and ensure completeness of information; it also used the PRISMA (Preferred Reporting Items for Systematic Reviews and Meta-Analyses) methodology to ensure a rigorous and reproducible approach to the collection of articles (SM) [30,31].

From the resulting 1648 documents, different screening criteria were applied according to the study phenomenon (Figure 1). These criteria were in accordance with the statement "Preferred Reporting Items for Systematic Reviews and Meta-Analyses", better known in the scientific community by its acronym (PRISMA). The PRISMA framework favours a structure and organization that helps authors to better publicise their implemented review process [32].

Firstly, after reviewing the abstracts of the resulting documents, it was realised that the inclusion of the commands in the fields covered (the title, abstract, and keywords) included publications that were not in line with the focus of the study. This led to a partial modification of the search engine so that the command for descriptors related to smart cities and education ("Smart City" AND "Education" OR "Teaching" OR "Learning") fulfilled the condition of appearing in the title of the publication. This restriction reduced the sample to 1642 documents.

We then excluded from the search results documents that were not articles, proceedings, books, or book chapters. This reduced the sample to 378 documents. On the other hand, we included all documents that were published between 2000 and 2023 according to the previous filter. In order to meet the objective of the study, we excluded those published before the beginning of the millennium and those published after. Thus, after applying the various filters, the final sample consisted of 88 documents (32 articles, 49 communications to congresses, 7 book chapters, and 0 books). The results were exported as comma-separated values (.csv) for a subsequent analysis of the sample using bibliometric techniques.

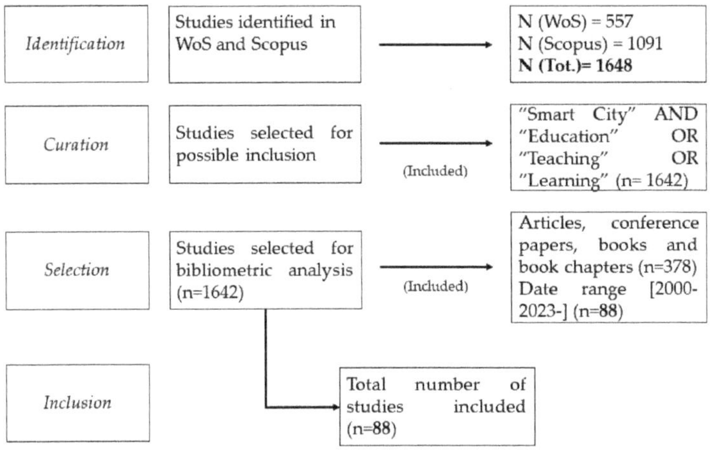

Figure 1. Flowchart of the study selection process based on the PRISMA statement.

Several bibliometric analysis techniques were used for the analysis:

(a) Scientific production in order to study the evolution of this production as a function of the variables identified.
(b) Bibliographic coupling to determine the degree of influence of an article in the scientific field based on its similarity (common references) with other related publications.
(c) Co-authorship, using countries as the unit of analysis in this study. The aim was to examine existing co-operations in relation to the topic under study.
(d) Co-citation to know the frequency with which different articles are cited together.
(e) Co-occurrence for the identification of the most frequently used keywords in the study articles.

An accurate data collection phase is essential for the quality of data and subsequent analyses. As highlighted by several authors [32], the correct selection of keywords for the search query is a crucial step for the accurate analysis of the chosen topic. This aspect is often overlooked in various systematic reviews and bibliometric analyses despite its importance for more rigorous research [33].

In this research, data were selected and then exported in RIS format for use with Zotero (v 1.0), a bibliographic management software. They were then integrated into SciMat (v.1.1) [32], a software specialised for bibliometric analyses. This process allowed for the extraction of specific data required for a bibliometric analysis.

A study of the relational nodes formed between the articles in the sample was carried out using VOSviewer (v.1.6.20), a free but not open-source software that allows for the construction and visualisation of bibliometric networks. It displays relational maps generated at the levels of bibliographic coupling, co-authorship, co-citation, and co-occurrence [28].

For the analysis of scientific production, the following variables were defined: year, in order to study the evolution of the articles over time; field of knowledge, in order to know in which thematic areas the articles are indexed; journals, in order to identify the most productive publications with respect to the object of interest of this study; country, in order to know in which countries the most research on the phenomenon under study is carried out; affiliation, in order to study the institutions with the greatest production; and publications with greatest impact, in order to know which were the most relevant articles in the field of study at the time this research was carried out. Not all results were considered.

However, exclusion criteria were established according to the variables considered, which are listed below (Table 1).

Table 1. Study variables and exclusion criteria.

Variables	Exclusion Criteria
Year	All publications outside the data range 2000–2023
Area of knowledge	All areas with fewer than 10 publications
Journals	All journals with fewer than 3 articles
Country	All countries with less than 4 publications
Affiliation	All institutions with less than 4 publications
Publications with more impact	All publications with less than 20 citations

Source: authors' own elaboration.

4. Results

In order to carry out an analysis on the subject of this study, the results were organised according to the different bibliometric techniques applied. In this way, the research questions were answered. The analyses of bibliographic coupling, co-authorship, co-citation, and co-occurrence began with an analysis of scientific production.

4.1. Scientific Production Analysis

The analysis was conducted on all 88 publications that constituted the study's corpus.

4.1.1. Year

Considering that the publications were included in the data range 2000–2023 (Figure 2), both included, a discontinuous evolution is observed, with continuous increases and decreases. This generates a range of 17 publications between the most productive year (17 in 2019) and the year with the lowest number of documents (0 in 2000). In any case, there is an increase in production in the last period (2013–2023) compared to the first period studied (2000–2012).

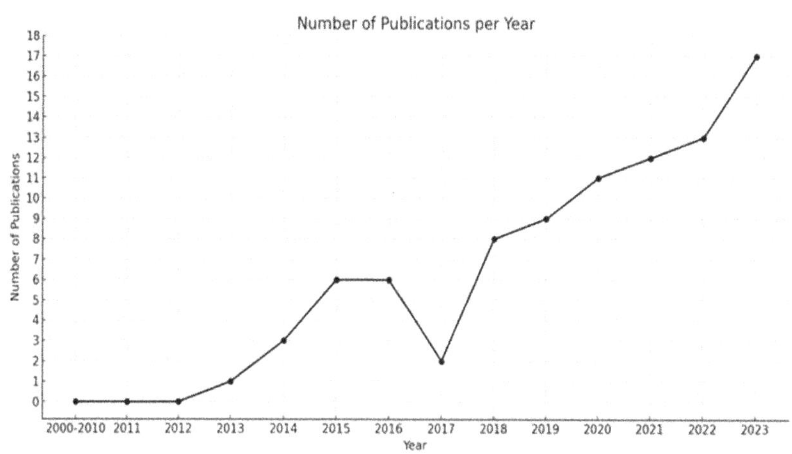

Figure 2. Number of papers per year.

Analysing the data on the annual production of articles in the period between 2000 and 2023, a significant picture emerges in relation to the topic of education in smart cities. Looking at the graph, we can see that before 2013, there is no scientific publication specifically dedicated to this topic.

However, from that year onwards, there is a gradual increase in the number of studies, with a slight upward trend consolidating around 2015. This phenomenon can be attributed to the increased attention of the scientific community to the concepts of smart cities and the Sustainable Development Goals outlined in the United Nations 2030 Agenda [34].

Interestingly, despite this positive trend, a significant decline is observed in 2017, a year in which only two publications are counted. This decline could be due to several factors, such as changes in research priorities.

This was followed by a resurgence of interest and a consequent increase in scientific production in recent years. This growth can be interpreted as a response to the increasing relevance of issues related to sustainability and technological innovation in the field of smart city education.

4.1.2. Area of knowledge

For this variable, the exclusion criterion was not to reach 10 publications. It should also be noted that the publications were subject to a multi-classification criterion with regard to their inclusion in the fields of knowledge, with the result that the same publication can be included in one or more fields of knowledge, depending on the subject it deals with. This means that the sum of all the fields in which the publications are included (Table 2) is greater than the 88 documents that make up the final sample.

Table 2. Documents by subject area of knowledge.

Area	Number of Publications
Computer Science	56
Social Sciences	43
Engineering	40
Energy	19
Business, Management and Accounting	13
Environmental Science	12

Source: authors' own elaboration.

The field of computer science registered the highest production (56), somewhat more than social sciences, which is in second place with 43 publications. The plurality in the fields is striking, distinguishing two areas: fields of knowledge, more rooted in the content of smart cities, and the educational field, with the variable ICT and educational technology as tools/instruments conveyed to the educational process.

In the context of research on education in smart cities, an analysis of the extract-ed data reveals a multidisciplinary panorama, highlighting how different fields of knowledge contribute to this emerging theme.

As can be seen, the field of computer science stands out as the most productive, reflecting the crucial importance of information and communication technologies (ICTs) in contemporary research. The dominance of computer science shows that ICT is not only a field of study in its own right but also a catalyst for progress in other fields and an indispensable tool for innovation and understanding urban dynamics in smart cities.

The area of social sciences shows a strong interest in the study of social behaviour, structures, and human interactions and plays a central role in understanding social phenomena in smart cities.

The focus on engineering shows an interest in technological innovation and the practical application of science in the urban context.

4.1.3. Journals

The journal *Sustainability* ranks first based on the exclusion criterion of less than three publications on the topic of analysis (Table 3). An analysis of the data in the table shows that the journal *Sustainability* stands out as the main source of publications on the topic of education in smart cities, with a total of five articles. This fact highlights the relevance of *Sustainability* as a reference point in the scientific debate on sustainability issues applied to the educational context of smart cities.

Table 3. Top eight journals indexed with the greatest numbers of publications.

Journal Name	Number of Publications
Sustainability (Switzerland)	6
Advances in Intelligent Systems and Computing	2
Security and Communication Network	2
Internet design and Architecture(s)	2
Journal of Higher Education Theory and Practice	2

Source: authors' own elaboration.

Next is a group of journals, each with two publications, which also contribute to the literature in this field, although to a lesser extent compared to *Sustainability*. This secondary group shows a more varied distribution of research, suggesting that despite the predominant concentration in a single journal, there is widespread interest across different editorial platforms.

Finally, it can be observed that many other articles were published in journals that only dealt with the topic once. This aspect reflects a dispersion of the scientific discourse on this topic across a wide range of publications. Such a dispersion can be interpreted both as a sign of the nascent and still consolidating nature of this field of study and as an indication of a growing interest that cuts across different disciplines and academic contexts.

4.1.4. Country

Having less than four publications on the topic under study (Table 4) was the exclusion criterion for the country variable.

Table 4. Top ten countries with the most indexed publications.

Country	Documents by Country
United State	10
China	9
Spain	9
Saudi Arabia	6
United Arab Emirates	6
United Kingdom	6
Italy	5
Austria	4
Indonesia	4
Mexico	4

Source: authors' own elaboration.

A geographical analysis was carried out in order to identify the leading countries in the field of science and the evolution of these nations over the years. The data on the geographical distribution of scientific production, presented in Table 4, show a clear predominance of publications from European and North American countries and the People's Republic of China (PRC). Specifically, the top three countries in terms of the volume of publications are the United States, the PRC, and Spain.

Given that contributions on the topic of smart cities can span various research areas, we delved deeper into trends in scientific contributions from the United States, China, and Europe. The United States shows a marked inclination towards social sciences, accounting for 30.4% of contributions, followed by computer sciences at 21.7% and business management at 17%. This profile suggests a strong interest in understanding and managing the social and entrepreneurial dynamics associated with smart cities, as well as in technological solutions.

In contrast, China demonstrates a predominance of technological and mathematical research, with a significant 36% of contributions in computer science, followed by engineering at 23% and mathematics at 23.5%. This reflects the country's significant focus on technological and innovative development in the context of smart cities.

An analysis of European publications, focusing on the top three countries, namely Spain, Italy, and Austria, reveals a more balanced distribution of research areas: computer science accounts for 35%, social sciences for 25%, engineering for 20%, and environmental sciences for 12%. This suggests a European trend towards a holistic approach in smart city research, integrating technology, social aspects, engineering, and environmental issues.

These regional differences in research focus reflect both the specific priorities and competencies of individual countries as well as the varied cultural, economic, and political approaches towards the development and integration of smart cities.

In order to obtain a more detailed and comprehensive analysis, it was later considered necessary to complement the study of the number of publications per country with other bibliometric metrics, such as a citation analysis and co-authorship and network visualisations.

A VOSviewer graph was generated from a total of 44 countries with at least two publications, of which 21 countries met this criterion (Figure 3). The results of this visual analysis show well-defined clusters, many of which are not connected to each other.

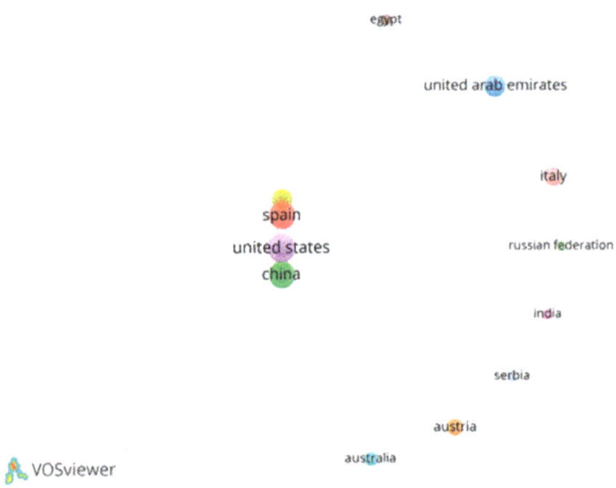

Figure 3. Co-authorship for countries. Purple: United States, Mexico. Green: China, Indonesia, South Korea, United Kingdom. Red: Spain, Brazil, Greece, Lithuania. Yellow: Saudi Arabia, Pakistan. Blue: United Arab Emirates, Canada. Pink: Italy. Brown: Egypt. Orange: Austria. Light green: Russia. Sky blue: Australia. Lilac: Serbia.

4.1.5. Affiliation

As in the case of the country variable, in order to take into account institutions of affiliation, it was established that a minimum of two publications should be registered (Table 5).

Table 5. Top ten institutions of affiliation with the most indexed publications.

Institution	Documents by Institution
Tecnológico de Monterrey	2
Pharos University in Alexandria	2
Fachhochschule Technikum Wien	2
Universidad de Jaén	2
Universidad de Granada	2
Universidade Estadual de Campinas	2
Universidade Estadual de Londrina	2
Matematički Institut SANU	2
Universitas Islam Negeri Maulana Malik Ibrahim Malang	2
Pharos University	2

Source: authors' own elaboration.

When analysing the table of publications on the topic of interest, a significant data point emerges concerning the institutions that have contributed in terms of the volume of documents. It can be seen that the top 10 institutions listed have each produced two publications relevant to the topic studied.

In this context, it is particularly relevant to note the geographical composition of these institutions. Two Spanish universities stand out, the University of Granada and the University of Jaén, confirming the country's active role in research on this topic. Moreover, the presence of an Austrian university underlines the important contribution of European institutions to the scientific literature in this field.

However, most of the institutions listed in the top 10 are located outside the European context. This geographical distribution indicates a global and diversified participation in research in the specific field, with a notable contribution from institutions located in different parts of the world.

4.1.6. Publications with the Highest Impact

In terms of the most relevant publications, based on the number of citations, published papers with a total of less than 20 citations were excluded (Table 6).

The table below looks at publications with at least 20 citations. At the top of the list is a French study with 93 citations. This paper explores how the smart city concept, which encompasses aspects such as the environment, mobility, and everyday life, influences "digital native" students. It highlights the potential of smart home technologies to enrich the educational and everyday experiences of these students [35]. Others focus on the role of ICT in higher education, highlighting how technology-enhanced learning (TEL) can be a key factor for sustainable socio-economic development, and raising questions about the conditions necessary for its effectiveness [36].

We see how others analyse the development of "smart classrooms" in universities, providing an overview of their potential and the ethical dilemmas they raise. This study highlights how advanced technologies can transform the educational environment but require careful regulation [37].

Finally, another line of research focuses on the importance of education as a universal human right and proposes a model based on computational intelligence to personalise education. This approach is in line with the concept of smart education, which uses data mining techniques to improve the learning experience of students [38].

Table 6. The top five publications with the highest scientific impact.

Authors	Year	Title	Journal	Cited by	Average Number of Appointments per Year
Baudier, P., Ammi, C., and Deboeuf-Rouchon, M. [35]	2020	Smart home: Highly-educated students' acceptance	*Technological Forecasting and Social Change*	93	23.3
Daniela, L., Visvizi, A., Gutiérrez-Braojos, C., and Lytras, M.D. [37]	2018	Sustainable higher education and Technology-Enhanced Learning (TEL)	*Sustainability* (Switzerland)	86	14.3
Kwet, M. and Prinsloo, P. [36]	2020	The 'smart' classroom: a new frontier in the age of the smart university	*Teaching in Higher Education*	78	19.5
Gomede, E., Gaffo, F.H., Briganó, G.U., de Barros, R.M., and Mendes, L.S. [38]	2018	Application of computational intelligence to improve education in smart cities	*Sensors (Switzerland)*	45	7.5
Musa, S. [39]	2018	Smart cities-a road map for development	*IEEE Potentials*	38	6.3
Mircea, M., Stoica, M., Ghilic-Micu, B. [40]	2021	Investigating the Impact of the Internet of Things in Higher Education Environment	Mircea, M., Stoica, M., and Ghilic-Micu, B.	35	11.7
Ahmed, V., Alnaaj, K.A., and Saboor, S. [41]	2020	An investigation into stakeholders' perception of smart campus criteria: The American University of Sharjah as a case study	Ahmed, V., Alnaaj, K.A., and Saboor, S.	35	8.8
Williamson, B. [42]	2015	Educating the smart city: Schooling smart citizens through computational urbanism	Williamson, B.	35	3.9

Table 6. *Cont.*

Authors	Year	Title	Journal	Cited by	Average Number of Appointments per Year
Bracco, S., Delfino, F., Laiolo, P., and Morini, A. [43]	2018	Planning & open-air demonstrating smart city sustainable districts	*Sustainability* (Switzerland)	32	5.3
Molnar, A. [1]	2021	Smart cities education: An insight into existing drawbacks	*Telematics and Informatics*	22	7.3
Nuseir, M.T., Basheer, M.F., and Aljumah, A. [44]	2020	Antecedents of entrepreneurial intentions in smart city of Neom Saudi Arabia: Does the entrepreneurial education on artificial intelligence matter?	Nuseir, M.T., Basheer, M.F., and Aljumah, A.	20	5.0

Source: authors' own elaboration.

4.2. Bibliographic Linking

From the application of this bibliometric methodology, we can derive some key information about the structure of research and collaboration between authors in the field of education in smart cities. The image of the graph generated by VOSviewer shows a bibliometric coupling based on authors who have produced at least one document and who have received at least 20 citations each. In total, 251 authors were considered, of which only 30 met the established criteria for inclusion in this study and were further grouped into eight different clusters.

The clusters, represented by different colours, indicate groups of authors who are more closely linked by mutual citations, suggesting common areas of research. The fragmented nature of these clusters, with only a few authors per cluster, suggests that education in the context of smart cities is not only a large field but also highly specialised and fragmented. This fragmentation may be the result of different sub-fields of research developing independently.

The red cluster, represented by Bracco's publication, has a total link strength of 157. The green cluster has a total link strength of 192, the dark blue cluster has a total link strength of 210, and the brown cluster represented has a link strength of 80. The violet cluster has a link strength of 127. Finally, the orange and blue clusters appear to form an interconnected core, indicating that the authors in these groups are often cited together and suggesting a strong thematic correlation.

4.3. Co-Authorship Analysis

The use of VOSviewer to visualise citation networks by country provides an immediately interpretable representation of the main national clusters. As shown in Figure 4, each nation is represented by a circle whose size and label are proportional to its weight in the field of study. The lines connecting the different circles indicate the citation relationships between different nations, while the spatial distance between two circles provides a visual measure of their thematic and citation correlation.

In order to understand the existing collaborative networks between countries in the scientific production under study, a co-authorship analysis was applied (see Figure 5). Countries with at least two co-authored publications with at least 20 citations were included. In total, 11 countries met this criterion. A further restriction was applied to the graph as only seven countries had a co-authorship relationship, as reflected in the graph generated using VOSviewer.

Therefore, three clusters were formed: a red cluster consisting of Saudi Arabia, Lithuania, and Greece, a green colour marking the collaboration cluster between Spain and Brazil, and finally, a blue cluster representing the close collaboration between the United States and the United Kingdom. It is clear that Spain plays an important role as it is the only country that has links with all the countries studied.

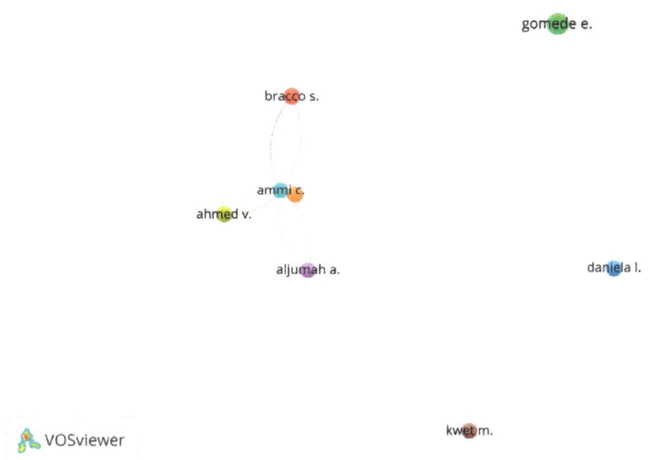

Figure 4. Co-authorship with "countries" as the unit of analysis. Purple: Ahmed V.; Yellow: Aljumah A.; Sky Blue: Ammi C.; Orange: Ghilic-Micu B.; Red: Bracco S.; Brown: Kwet M. Pink: Musa S.; Green: Gomede E.; Dark Blue: Daniela.

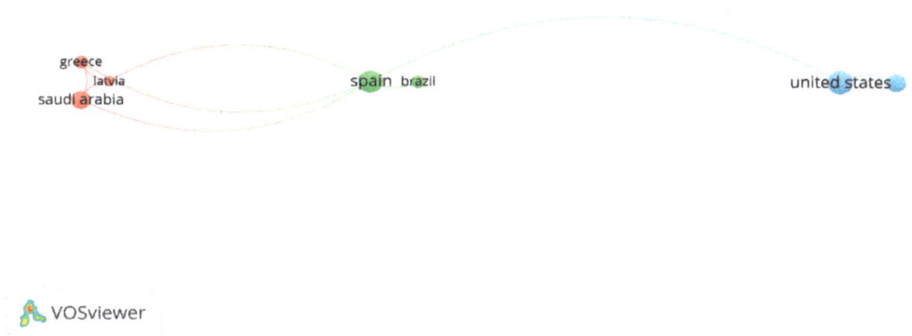

Figure 5. Network visualisation of citations by country. Red: Saudi Arabia, Greece, Lithuania. Green: Spain, Brazil. Blue: United Kingdom, United States.

4.4. Analyzing Co-Citation and Co-Occurrence

In order to investigate the main trends and developments in the field of smart city education, two bibliometric techniques were used: co-citation analysis and keyword co-occurrence analysis. These techniques allowed us to trace connections and emerging themes in this area of study.

In the co-citation analysis, we used a criterion that required a minimum of five citations for each paper, a requirement that was met by 91 publications (see Figure 6). This approach identified seven main groups of co-citations, based on publications that were frequently cited together. This highlighted the importance of certain key authors in the context of education in smart cities; in particular, the works of Nam. This bibliometric review aims to document and synthesise research trends in the field of education in smart cities over the last 25 years.

Through a bibliographic analysis of 2877 Scopus-indexed documents, this review found that this is a very recent topic in the literature, with over 80% of relevant documents published since 2010. It is a global subject in the literature, with significant contributions from both economically developed and developing societies. A citation analysis identified the key authors and documents that have shaped the development of this literature. Author co-citation analysis, used to identify the intellectual structure of the SCON knowledge base, revealed four "schools of thought" or dominant lines of inquiry. In order of size and importance, these were alternative materials for sustainable construction, sustainable construction management, recycling and waste reduction, and social sustainability in construction management. The results of the keyword co-occurrence analysis confirmed these findings regarding the conceptual structure of the SCON knowledge base, including the conclusions that "social sustainability" represents the "weakest" dimension and "alternative materials" the leading edge of the sustainability literature. Container title Journal of Cleaner Production [29] (total link strength 359), Pardo (total link strength 234), Gil-Garcia (total link strength 245), Chourabi (total link strength 226), Molina (total link strength 279), and Yigitcanlar (total link strength 261) showed a high frequency of co-citation. This suggests that their works have become fundamental references in the field of smart cities and, consequently, education in smart cities, contributing significantly to the understanding of the different applications and facets of education in these advanced urban environments.

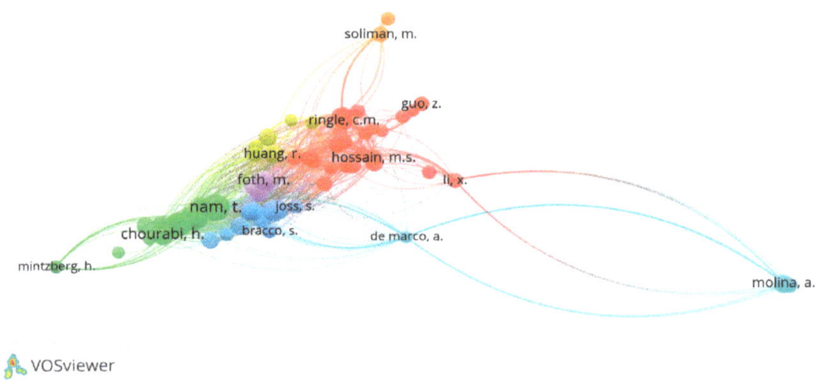

Figure 6. Co-citation of cited authors. Red: Guo Z., Hossain M. S., Li X.; Green: Nam T., Chourabi H., Scott, L.; Blue: Albino V., Giffinger R., Bibri S. E., Caragliu A.; Yellow: Huang R., Lee K., Williamson B.; Violet: Visvizi A., Molnar A., Daniela L.; Sky blue: de Marco, Molina. Orange: Soliman M., Guetl C.

When analysing the co-occurrence of keywords in the 88 publications selected for our sample (see Figure 7), we identified 313 keywords proposed by the authors. In addition, the documents were indexed with 578 different keywords, making a total of 783. Among these, 20 keywords appeared at least five times, as highlighted in Figure 6. We then grouped these keywords into four clusters based on the descriptors that most frequently co-occurred. The dominant keyword is "smart city", which emerges as the central node, the most representative and significant concept within the analysed dataset. Other recurring keywords representing different clusters include terms such as "education", "students", and "sustainable development". Within the green cluster, the term "educational computing" stands out due to its higher frequency.

The keyword analysis allows us to identify the most relevant research areas. For example, if we look at the green cluster, we can see that it is an area related to computer science, highlighted by specific terms from this field. The blue cluster, on the other hand, focuses more on the school context, including words such as curricula, students, teaching, and e-learning, suggesting a strong correlation with educational methodologies and teaching technologies. Finally, the red cluster focuses on issues of sustainability and sustainable development, including terms related to environmental education, reflecting an interest in research on environmental awareness and sustainable practices.

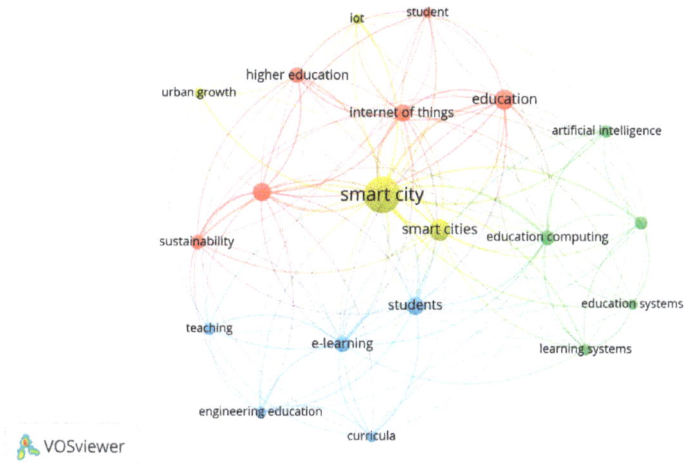

Figure 7. Concurrence of keywords in scientific production. Yellow: smart city, smart cities; Blue: Students, e-learning, teaching; Red: Education, sustainability, internet of things; Green: AI, Education computing, education system.

5. Discussion

Using bibliometric and text mining techniques, this article examines education in the context of smart cities. In this section, we interpret the overall findings, analyse the implications, provide suggestions for future research, and highlight the limitations of this study. The bibliometric analysis is based on a literature review with a final corpus of 88 publications.

The bibliometric work has revealed that the scientific production on education in smart cities is not yet fully developed, leaving ample room for future growth, dissemination, and the creation of collaborative networks. As can be seen, the field of study shows a discontinuous evolution in terms of the proliferation of publications, with the first works analysed dating back to 2013. There is an increase in interest after 2015, indicating greater attention to environmental issues, sustainable development, and the application of information and communication technologies in urban centres.

The geographical analysis of the literature corpus shows that the topic of education in smart cities is of global interest. In addition to a broad sample of studies from various European and North American countries, there is a significant expansion of the literature in other regions of the world. China in particular stands out as one of the leading countries in smart city research, with a significant impact on the education sector.

Overall, these examples show that the correct interpretation of results requires a combined analysis of articles, authors, and keywords to identify the main strands of a research area. In this sense, scientific mapping offers a unique opportunity to visualise the conceptual structure of a topic [29].

The analysis of the bibliographic sources of the articles shows a marked prevalence of publications in the journal *Sustainability*, which deals with issues related to environmental, cultural, economic, and social sustainability. There are also numerous journals in the field of engineering and technology, such as *Sensors* or *IEEE Access*, a multidisciplinary journal, and the *Journal of Higher Education Theory and Practice*. It is also worth noting that about 36% of the articles examined were communications presented at conferences. This indicates that the topic of education in smart cities is not only highly topical but also in a dynamic phase of development.

The collaborative networks between different countries in the field of education in smart cities show some interesting trends. It can be observed that most of the collaborations take place between researchers affiliated with institutions in the United States. However, collaborations between researchers from the United Kingdom and the United States are particularly productive. Spain stands out in this context, with significant collaborations with Brazil. Co-operation between Saudi Arabia, Greece, and Lithuania also stands out.

With regard to the main lines of research, the multidisciplinary character of smart cities and education is clearly evident. On one hand, some publications include environmental sciences as well as computer sciences, engineering, and social sciences [45]. On the other hand, there are various articles on engineering and computer science, while Bracco focuses on environmental sciences and engineering [46]. This diversity reflects the complex and multifaceted nature of the smart cities field.

This bibliometric analysis has some limitations. First, the selected literature is exclusively in English, which means that the final corpus may not fully represent all existing research networks and current trends. For example, some geographical areas have significant production in languages such as Spanish or Arabic, so a correct geographical or authorial analysis should also include these publications.

Furthermore, the decision to use only the Scopus database does not guarantee full coverage of all academic production available in English. Future research could therefore benefit from including articles in other languages from other databases to avoid missing potentially important documents.

Although the search query was carefully formulated and includes a wide range of keywords, there is always a risk that some relevant terms will be missed, especially given the extensive academic production on smart cities. Furthermore, this study does not evaluate the quality of the selected articles, as such an evaluation would require specific criteria that may not be applicable to all the disciplines involved. Given that the aim of this manuscript is a holistic analysis of the topic, the inclusion and exclusion criteria adopted are considered adequate for the selection of the final literary corpus.

As highlighted by some authors, "bibliometrics as a method has the advantage of reducing elements of judgement and producing quantitative results that tend to be the sum of many small judgments and assessments made by many people" [45].

6. Considerations

The findings of this study may have significant implications for the practical development of education in smart cities. The identified trends and research networks provide a foundation to inform educational policies and urban development strategies.

The convergence of educational technologies, smart city management, urban infrastructure, e-learning, and sustainable urban development, as highlighted in this study, suggests an integrated and multidisciplinary approach to addressing the challenges of contemporary education. In particular, educational institutions can leverage this information to design curricula that integrate digital literacy and environmental awareness, preparing students to become active and informed citizens in smart cities. The integration with cutting-edge technologies such as artificial intelligence (AI), big data, cloud computing, blockchain, and nanotechnology can catalyse an educational revolution characterised by the development of dynamic teaching management systems, interactive learning environments, and innovative pedagogical approaches.

Concurrently, urban administrators can collaborate with academic institutions to develop urban infrastructures that support a connected and interactive learning environment, promoting universal access to education and encouraging the use of environmental IoT technologies for more efficient resource monitoring and management. This interdisciplinary collaboration is crucial to holistically addressing the complexities of smart cities and ensuring that the evolution of smart cities is driven by an inclusive and sustainable educational vision.

For instance, the city of Turin in Italy undertakes innovative initiatives in the field of smart cities, establishing a consolidated collaboration with the Polytechnic University of Turin, leveraging academic expertise to promote intelligent and sustainable urbanization. The Polytechnic University of Turin, through various projects and partnerships, significantly contributes to transforming Turin into a smart city, focusing efforts on key aspects such as sustainable mobility, energy, building design, and the quality of life of its citizens, with a focus on reducing carbon dioxide emissions. An example of this collaboration is the "Torino City Lab", an initiative aimed at creating economic development policies based on innovation and new technologies, offering new opportunities to the city through citizen participation.

This not only indicates a growing interest in the field but also creates new opportunities for future studies that can translate theoretical knowledge into innovative practices, tangibly contributing to the development of smart cities that are equally attentive to the educational dimension and the training of citizens. In the European context, the implementation of school programs aimed at education for active citizenship demonstrates the potential of this educational approach and lays the groundwork for future educational investigative initiatives.

7. Conclusions

This article analyses academic production on education and smart cities through a bibliometric approach; it provides a clear and updated framework of the knowledge structure in this specific sector, demonstrating its relevance and the contemporaneity of the topic treated, highlighting that it is an expanding and evolving subject. Indeed, this document makes a substantial contribution to the literature on education and smart cities, identifying trends and gaps and suggesting directions for future research. In particular, it identifies thematic areas such as educational technologies, smart city management, urban infrastructures, e-learning, and sustainable urban development. It also highlights a little explored but timely line of inquiry, providing a valuable and up-to-date overview of the literature on education and smart cities. As such, it is useful for scholars approaching the fields of education and smart cities for the first time. Indeed, while highlighting some of the limitations of existing research, it also proposes several suggestions for future studies, such as a more comprehensive analysis of education that integrates the perspectives of students and teachers or the potential of environmental IoT technologies and smart city management. Further analysis is needed to increase citizens' awareness of their role in sustainable urban development.

This leads to the need for further research at an interdisciplinary level, taking into account the need to improve education and urban management at all levels and to make citizens more involved in sustainable development and aware of their environmental impact. In this respect, the creation of interdisciplinary research teams would be very beneficial for future research on these topics and would favour the integration of different theoretical approaches from different disciplines.

Finally, given that this article highlights the growth trajectory of the literature on education and smart cities, a significant increase in academic production is predicted for the coming years.

Author Contributions: Conceptualization and methodology, D.S. and M.Á.R.-D.; software, D.S., Á.I.A.C. and M.Á.R.-D.; validation, formal analysis and investigation, D.S., Á.I.A.C. and M.d.C.C.R.; writing—original draft preparation, D.S.; writing—review, editing and visualization D.S., Á.I.A.C., M.Á.R.-D. and M.d.C.C.R.; supervision, M.Á.R.-D. and M.d.C.C.R. All authors have read and agreed to the published version of the manuscript.

Funding: University of Castilla-La Mancha (Spain). Next generation participatory urban regeneration in Spanish medium-sized cities: service learning and citizen participation." Ref. 2022-GRIN-34264.

Data Availability Statement: These studies are available within Web of Science and Scopus.

Conflicts of Interest: The authors declare no conflicts of interest.

References

1. Molnar, A. Smart cities education: An insight into existing drawbacks. *Telemat. Inform.* **2021**, *57*, 101509. [CrossRef]
2. Angelidou, M.; Psaltoglou, A.; Komninos, N.; Kakderi, C.; Tsarchopoulos, P.; Panori, A. Enhancing sustainable urban development through smart city applications. *J. Sci. Technol. Policy Manag.* **2018**, *9*, 146–169. [CrossRef]
3. Chourabi, H.; Nam, T.; Walker, S.; Gil-Garcia, J.R.; Mellouli, S.; Nahon, K.; Pardo, T.A.; Scholl, H.J. Understanding smart cities: An integrative framework. *Proc. Annu. Hawaii Int. Conf. Syst. Sci.* **2012**, *45*, 2289–2297. [CrossRef]
4. Yin, C.; Xiong, Z.; Chen, H.; Wang, J.; Cooper, D.; David, B. A literature survey on smart cities. *Sci. China Inf. Sci.* **2015**, *58*, 1–18. [CrossRef]
5. Albino, V.; Berardi, U.; Dangelico, R.M. Smart Cities: Definitions, Dimensions, Performance, and Initiatives. *J. Urban Technol.* **2015**, *22*, 3–21. [CrossRef]
6. Caragliu, A.; Del Bo, C.; Nijkamp, P. Smart Cities in Europe. *J. Urban Technol.* **2011**, *18*, 65–82. [CrossRef]
7. Hollands, R. Will the Real Smart City Please Stand Up? *City* **2008**, *12*, 303–320. [CrossRef]
8. IMD. Smart City Index 2021. October 2021. Available online: https://imd.cld.bz/Smart-City-Index-2021/2/ (accessed on 30 December 2023).
9. Giffinger, R.; Gudrun, H. Smart cities ranking: An effective instrument for the positioning of the cities. *ACE Archit. City Environ.* **2010**, *4*, 7–26. [CrossRef]
10. Zhu, H.; Shen, L.; Ren, Y. How can smart city shape a happier life? The mechanism for developing a Happiness Driven Smart City. *Sustain. Cities Soc.* **2022**, *80*, 103791. [CrossRef]
11. Hollands, R.G. Critical interventions into the corporate smart city. *Camb. J. Reg. Econ. Soc.* **2015**, *8*, 61–77. [CrossRef]
12. Visvizi, A.; Lytras, M.D.; Daniela, L. (Re) defining smart education: Towards dynamic education and information systems for innovation networks. In *Enhancing Knowledge Discovery and Innovation in the Digital Era*; IGI Global: Hershey, PA, USA, 2018; pp. 1–12.
13. Hudson, L.; Wolff, A.; Gooch, D.; van der Linden, J.; Kortuem, G.; Petre, M.; ten Veen, R.; O'Connor-Gotra, S. Supporting urban change: Using a MOOC to facilitate attitudinal learning and participation in smart cities. *Comput. Educ.* **2019**, *129*, 37–47. [CrossRef]
14. Arroub, A.; Zahi, B.; Sabir, E.; Sadik, M. A literature review on Smart Cities: Paradigms, opportunities and open problems. In Proceedings of the 2016 International Conference on Wireless Networks and Mobile Communications (WINCOM), Fez, Morocco, 26–29 October 2016; pp. 180–186.
15. Nam, T.; Pardo, T. Conceptualizing smart city with dimensions of technology, people, and institutions. In Proceedings of the 12th Annual International Conference on Digital Government Research, College Park, MD, USA, 12–15 June 2011; p. 291. [CrossRef]
16. Ismagilova, E.; Hughes, L.; Dwivedi, Y.K.; Raman, K.R. Smart cities: Advances in research—An information systems perspective. *Int. J. Inf. Manag.* **2019**, *47*, 88–100. [CrossRef]
17. Glänzel, W.; Schubert, A. A new classification scheme of science fields and subfields designed for scientometric evaluation purposes. *Scientometrics* **2003**, *56*, 357–367. [CrossRef]
18. Guo, Y.-M.; Huang, Z.-L.; Guo, J.; Li, H.; Guo, X.-R.; Nkeli, M.J. Bibliometric Analysis on Smart Cities Research. *Sustainability* **2019**, *11*, 3606. [CrossRef]
19. Sharifi, A.; Allam, Z.; Feizizadeh, B.; Ghamari, H. Three Decades of Research on Smart Cities: Mapping Knowledge Structure and Trends. *Sustainability* **2021**, *13*, 7140. [CrossRef]
20. Rico-Bautista, D.; Medina-Cárdenas, Y.; Guerrero, C.D. Smart university: A review from the educational and technological view of internet of things. In *International Conference on Information Technology & Systems: ICITS 2019: Information Technology and Systems*; Springer: Berlin/Heidelberg, Germany, 2019; pp. 427–440.
21. Durán-Sánchez, A.; Álvarez-García, J.; de la Cruz Del-Río-Rama, M.; Sarango-Lalangui, P.O. Analysis of the scientific literature published on smart learning. *Rev. Espac.* **2018**, *39*. Available online: https://www.scopus.com/inward/record.uri?eid=2-s2.0-85044159030&partnerID=40&md5=850d9588cca23773e1a1159c0ea7a0b0 (accessed on 18 December 2023).
22. Mora, L.; Deakin, M.; Reid, A. Combining co-citation clustering and text-based analysis to reveal the main development paths of smart cities. *Technol. Forecast. Soc. Change* **2019**, *142*, 56–69. [CrossRef]
23. Secinaro, S.; Brescia, V.; Lanzalonga, F.; Santoro, G. Smart city reporting: A bibliometric and structured literature review analysis to identify technological opportunities and challenges for sustainable development. *J. Bus. Res.* **2022**, *149*, 296–313. [CrossRef]
24. Udomsap, A.D.; Hallinger, P. A bibliometric review of research on sustainable construction, 1994–2018. *J. Clean. Prod.* **2020**, *254*, 120073. [CrossRef]
25. Mongeon, P.; Paul-Hus, A. The journal coverage of Web of Science and Scopus: A comparative analysis. *Scientometrics* **2016**, *106*, 213–228. [CrossRef]
26. da Silva, A.P.F.; do Nascimento, A.N.; de Pinho, M.A.B.; Falk, J.A. Estudo bibliometrico sobre custo em organizações da construção civil: Contribuições do congresso brasileiro de custo de 1996 a 2010. Anais Do Congresso Brasileiro De Custos - ABC. Available online: https://anaiscbc.emnuvens.com.br/anais/article/view/363 (accessed on 8 February 2023).
27. Boyack, K.W.; Klavans, R. Co-citation analysis, bibliographic coupling, and direct citation: Which citation approach represents the research front most accurately? *J. Am. Soc. Inf. Sci. Technol.* **2010**, *61*, 2389–2404. [CrossRef]
28. Cobo, M.J.; López-Herrera, A.G.; Herrera-Viedma, E.; Herrera, F. Science mapping software tools: Review, analysis, and cooperative study among tools. *J. Am. Soc. Inf. Sci. Technol.* **2011**, *62*, 1382–1402. [CrossRef]

29. Petticrew, M.; Roberts, H. *Systematic Reviews in the Social Sciences: A Practical Guide*; John Wiley & Sons: Hoboken, NJ, USA, 2008.
30. Liberati, A.; Altman, D.G.; Tetzlaff, J.; Mulrow, C.; Gøtzsche, P.C.; Ioannidis, J.P.A.; Clarke, M.; Devereaux, P.J.; Kleijnen, J.; Moher, D. The PRISMA Statement for Reporting Systematic Reviews and Meta-Analyses of Studies That Evaluate Health Care Interventions: Explanation and Elaboration. *Ann. Intern. Med.* **2009**, *151*, W-65. [CrossRef] [PubMed]
31. Moher, D.; Liberati, A.; Tetzlaff, J.; Altman, D.G.; The PRISMA Group. Preferred Reporting Items for Systematic Reviews and Meta-Analyses: The PRISMA Statement. *PLoS Med.* **2009**, *6*, e1000097. [CrossRef] [PubMed]
32. Cobo, M.J.; López-Herrera, A.G.; Herrera-Viedma, E.; Herrera, F. SciMAT: A new science mapping analysis software tool. *J. Am. Soc. Inf. Sci. Technol.* **2012**, *63*, 1609–1630. [CrossRef]
33. de Carvalho, G.D.G.; Sokulski, C.C.; da Silva, W.V.; de Carvalho, H.G.; de Moura, R.V.; de Francisco, A.C.; da Veiga, C.P. Bibliometrics and systematic reviews: A comparison between the Proknow-C and the Methodi Ordinatio. *J. Informetr.* **2020**, *14*, 101043. [CrossRef]
34. United Nations General Assembly. *Transforming Our World: The 2030 Agenda for Sustainable Development, Resolution Adopted by the General Assembly on 25 September 2015, A/RES/70/1*. 2015. Available online: https://undocs.org/A/RES/70/1 (accessed on 8 February 2023).
35. Baudier, P.; Ammi, C.; Deboeuf-Rouchon, M. Smart home: Highly-educated students' acceptance. *Technol. Forecast. Soc. Change* **2020**, *153*, 119355. [CrossRef]
36. Kwet, M.; Prinsloo, P. The "smart" classroom: A new frontier in the age of the smart university. *Teach. High. Educ.* **2020**, *25*, 510–526. [CrossRef]
37. Daniela, L.; Visvizi, A.; Gutiérrez-Braojos, C.; Lytras, M.D. Sustainable higher education and Technology-Enhanced Learning (TEL). *Sustainability* **2018**, *10*, 3883. [CrossRef]
38. Gomede, E.; Gaffo, F.H.; Briganó, G.U.; de Barros, R.M.; Mendes, L.d.S. Application of computational intelligence to improve education in smart cities. *Sensors* **2018**, *18*, 267. [CrossRef]
39. Musa, S. *Smart Cities—A Road Map for Development*; IEEE Potentials, vol. 37, no. 2; Institute of Electrical and Electronics Engineers Inc.: Piscataway, NJ, USA, 2018; pp. 19–23. [CrossRef]
40. Mircea, M.; Stoica, M.; Ghilic-Micu, B. *Investigating the Impact of the Internet of Things in Higher Education Environment*; IEEE Access, vol. 9; Institute of Electrical and Electronics Engineers Inc.: Piscataway, NJ, USA, 2021; pp. 33396–33409. [CrossRef]
41. Ahmed, V.; Alnaaj, K.A.; Saboor, S. An investigation into stakeholders' perception of smart campus criteria: The American University of Sharjah as a case study. *Sustainability* **2020**, *12*, 5187. [CrossRef]
42. Williamson, B. Educating the smart city: Schooling smart citizens through computational urbanism. *Big Data Soc.* **2015**, *2*. [CrossRef]
43. Bracco, S.; Delfino, F.; Laiolo, P.; Morini, A. Planning & Open-Air Demonstrating Smart City Sustainable Districts. *Sustainability* **2018**, *10*, 4636. [CrossRef]
44. Nuseir, M.T.; Basheer, M.; Aljumah, A. Antecedents of entrepreneurial intentions in smart city of Neom Saudi Arabia: Does the entrepreneurial education on artificial intelligence matter? *Cogent. Bus. Manag.* **2020**, *7*, 1825041. [CrossRef]
45. Dos Santos, A.F.; Rausch, R.B. Perícia Contábil na Revista Brasileira de Contabilidade: Uma análise bibliométrica do período de 1992 a 2008. Available online: https://anaiscbc.emnuvens.com.br/anais/article/view/1038 (accessed on 8 February 2023).
46. Castellon, G.D.; Palacios, C.B.S.; Fernandez, F.L.R. Accessibility training from the university. Reflections for community intervention in the face of population aging. *Noved. Poblac.* **2020**, *16*, 116–128.

smart cities MDPI

Review

Contextualizing the Smart City in Africa: Balancing Human-Centered and Techno-Centric Perspectives for Smart Urban Performance

Nessrine Moumen [1,*], Hassan Radoine [1], Kh Md Nahiduzzaman [2] and Hassane Jarar Oulidi [3]

1 School of Architecture, Planning and Design (SAP+D), University Mohammed VI Polytechnic (UM6P), Ben Guerir 43150, Morocco
2 School of Engineering, University of British Columbia (UBC), Okanagan, Kelowna, BC V1V1V, Canada
3 Hassania School for Public Works Engineering (EHTP), Casablanca 20230, Morocco
* Correspondence: nessrine.moumen@um6p.ma

Abstract: The continuous growth of urban populations and the complexities of their current management in Africa have driven local governments to explore new technologies to optimize their urban and territorial performance. These governments and related stakeholders' resort to the term "smart city" to orient the current urban planning policies and practices to be more efficient and adequate. Nevertheless, the issue that remains is how to contextualize this global term that has not yet been fully adopted by African cities that have claimed to be "Smart". This contextualization becomes more complex in this critical context, where the city has not yet reached an ideal performance. Therefore, to reach this prospective African smart city, a critical review of how it would be both human-centered and techno-centered is imperative. This paper would review accordingly the above argument and set key performance indicator-based methodology on how to evaluate the smartness of a city in the African context.

Keywords: Smart City; sustainable; urban; approach; ICT; Africa; assessment; key performance indicators

1. Introduction

Over the last few decades, urban populations have grown at a breakneck pace around the world. More than 55 percent of the world's population now lives in cities, and this rate is expected to rise to 70 percent in the next 30 years, with an additional 25 billion people expected to relocate to cities by 2050 [1]. The exponential expansion of the global population has reached an unprecedented level. This, coupled with the rapid urbanization process, leads to a variety of socio-technical, organizational, and economic factors that create numerous challenges. These challenges pose a threat to both the environmental integrity and economic resilience of cities [2]. As a result, most governments are purposefully concerned about implementing "smart" concepts to enhance the utilization and control of both material and immaterial resources and properties.

In this context, the concept of a "Smart City" is introduced, encompassing the intelligent and strategic use of new technological advances in communication and information (ICT to establish a sustainable urban environment and improve the Quality of Life (QoL). Smart cities have become increasingly prominent in modern societies due to their diverse range of applications [3,4]. In this context, it is essential to provide relevant references to substantiate the significance of smart cities. To set the stage for our research direction, we can draw upon published opinions, statements, and reflections in the literature.

In the realm of infrastructure, smart building management systems have been developed to effectively regulate temperature and lighting, leveraging digital technologies to optimize energy consumption [5]. This energy-efficient approach aligns with the broader goal of creating sustainable urban environments as noted in studies by experts [5–7].

Moreover, the healthcare sector has witnessed significant advancements within smart cities. Innovations in healthcare diagnostics have paved the way for improved medical services [8]. Published opinions from healthcare professionals emphasize the transformative impact of smart city technologies on patient care and health outcomes [9]. Along with these advancements, smart cities leverage cutting-edge technologies to enable intelligent edge processing and network connectivity. The integration of these technologies has been explored in recent research by [10], who highlight the potential for enhancing communication networks and data processing within urban environments [3,11].

Education has not been left behind in the smart city revolution, with modern technologies being used to facilitate education systems and provide interactive and personalized learning experiences for students. This is evidenced by a large number of publications that highlight the benefits of digital education [12,13].

Efficient governance is an essential component of urban innovation, with the inclusion of digital technologies to facilitate administrative procedures [4,14]. Insights from governmental experts, such as [15–18], underscore the role of digital governance in promoting transparency and citizen engagement.

Lastly, the security issue is tackled by the deployment of smart security measures [19]. These measures aim to minimize security risks and protect valuable assets, sensitive data, and individuals [20,21]. This ensures a safe and secure environment for smart city inhabitants.

In addition to these positive aspects, the implementation of smart security measures is crucial for minimizing security risks and safeguarding valuable assets, sensitive data, and individuals. A comprehensive approach to security is vital, as outlined in the report of COO and co-founder, VivaCity, from Open Access Government news [22] and cybersecurity experts [23–27]. By connecting these threads of research and expert opinions, we can gain a more coherent understanding of the multifaceted applications and significance of Smart Cities in contemporary society. This approach ensures that our research direction is well grounded and informed by the broader discourse on smart city applications and their impact [28–30].

One of the reasons for the popularity of this paradigm, according to [31], is that the Smart City (SC) idea is accompanied by the promise of enhanced competitiveness and economic prosperity. This vision has captivated municipal authorities worldwide, transcending geographical boundaries and encompassing regions that may experience significant urbanization. As a consequence of its increasing popularity, the concept of the Smart City has arisen as a prevalent topic of scientific scrutiny and analysis. It has become a model for smart urban development and sustainable socio-economic progress around the world [32,33]. Similar to other frequently employed notions, a smart city lacks a single definition because it is an interdisciplinary concept, leading to ongoing debates [34].

The authors classify Smart City definitions into two different concepts [35]. The first relies on technology areas that make cities more efficient and functional, such as Big Data and Internet of Things (IoT).

The second strategy focuses on citizens and emphasizes soft aspects like commitment and participation, education, culture, public policy, social innovation, and governance. In this context, several examples from around the world demonstrate that certain cities are described as smart, but the definition ascribed to this term varies. Since the Smart City idea has developed out of observational practice, a systemic theoretical review of this phenomenon is lacking. Before delving into the specifics of a Smart City as an invention, we must first grasp its fundamental conceptual features. In fact, understanding the significance of a Smart City is the first step. A brief examination of the extant literature reveals that the term of Smart City is quite debatable [36–38]. Indeed, the emergence of comparable concepts like digital cities, learning cities, virtual cities, and augmented cities has introduced a certain level of ambiguity to the conceptual term. Hence, academic, industrial, and governmental sectors have all contributed to the gradual development of the Smart City paradigm (Table 1).

Table 1. Smart City definitions from the literature: recapitulation and consensus findings.

Definition Number	Smart City Definition	Key References for Definitions	Consensus Finding
1	"A smart city is an urban area that utilizes digital technology to improve performance, well-being, and efficiency while reducing costs and resource consumption."	[39,40]	Smart cities exploit digital technology to improve performance and quality of life and to minimize the consumption of resources.
2	"A smart city is distinguished by the use of information and communication technologies (ICT) to improve quality, well-being and reduce its resource utilization."	[41–43]	ICT play a central role in smart city initiatives, with the aim of enhancing quality of life and resource efficiency.
3	"In a smart city, digital technologies are utilized to advance infrastructure, services, and the quality of life for its residents."	[44,45]	Digital technologies in smart cities enhance infrastructure and services for residents, leading to improved quality of life.
4	"A smart city is one that promotes sustainability, efficiency, and innovation by the use of ICT."	[15,46]	Using ICT, smart cities prioritize efficiency, environmental responsibility, and innovation.
5	"Data and technology serve a purpose in smart cities to optimize urban operations and resources, thereby improving the living conditions of its inhabitants."	[47,48]	Data and technology optimization in smart cities leads to improved urban operations and living conditions for residents.
6	The term "smart city" denotes an advanced urban environment where residents integrate informational and urban features with emerging technologies to foster sustainable, environmentally conscious, competitive businesses, and superior living standards.	[49]	Integration of information and technology for sustainability and improved living standards.
7	Smart Cities are the result of cutting-edge, knowledge-based techniques targeted at improving urban centers' competitiveness, support, and environmental, cultural, and socio-economic functioning. These smart cities are built with an evolving combination of human, technological, and social capital, with entrepreneurship assets.	[50]	Knowledge-driven strategies for enhanced urban competitiveness and socio-economic functioning.
8	The notion of the "smart city" is perceived as a unique intellectual capacity that encompasses multiple aspects of technical, social, and innovative economic advancement. This concept is shaped by the interplay of these factors. The term "interconnected" signifies the expansion of the broadband economy, while "green" denotes urban infrastructure that fosters environmental conservation and carbon gas emissions. On the other side, the collaboration of "creative and knowledge-based cities" seeks for enhancement of the city's capacity for innovation through the development of creative and knowledge-based human capital. The term "intelligent or smart" highlights the capacity to generate valuable information from real-time urban data processed through sensor technologies.	[51]	Intellectual capacity integrating social, technical, and economic factors for innovation and environmental conservation.

Table 1. *Cont.*

Definition Number	Smart City Definition	Key References for Definitions	Consensus Finding
9	The advancement of the smart city paradigm has been shaped by the collaborative efforts of three pivotal sectors: academia, industry, and government. Within the context of the smart city idea, the city is perceived as a system comprising multiple interconnected subsystems. The system's overall capacity to exhibit intelligent and cohesive behavior relies on the effective coordination of these subsystems. Broadly speaking, a smart city encompasses a multidimensional framework with diverse objectives.	[52]	Collaborative efforts of academia, industry, and government for an interconnected and multidimensional smart city framework.
10	The expression "smart city" corresponds to a city that makes use of ICT, technology, and innovation developments to address urban concerns, such as enhancing livability, fostering economic development, creating a sustainable, safe environment, and facilitating the implementation of efficient urban management strategies.	[53]	Utilization of ICT, technology, and innovation for addressing urban concerns and efficient urban management.
11	A smart city is an organized, networked, and intelligent area. It frequently possesses several intelligent characteristics, including economics, governance, environment, people, and mobility. This concept consists of a governance dimension and a technology dimension, with the latter serving as the foundation. Everything about modern technology, gadgets, etc., falls under the technological category. The technological component serves as a basis for the management component, consisting of novel strategies and unconventional approaches that are put into practice using digital technologies and other technological advances.	[54]	Organized, networked, and intelligent area with characteristics in economics, governance, environment, people, and mobility.
12	To establish a smart city with high production and value creation, we need people with advanced degrees and innovative planning techniques. By promoting ongoing information generation, sharing, evaluation, renewal, and updating, this city strives for knowledge-based development. This can be accomplished by fostering a constant connection between the city's residents and, concurrently, between them and residents of other cities. These contacts are supported by both the knowledge-sharing culture of the population and the well-designed technological networks and infrastructures of the city.	[55–57]	Knowledge-based development through ongoing information generation, sharing, and innovative planning.

This table now includes a "Consensus Finding" column that summarizes the shared themes or key concepts found in these definitions.

2. Smart City-Related Concept and Dimensions

Faced with environmental, economic, and social challenges, our cities must evolve into smart, resilient, inclusive, and sustainable future cities.

2.1. Smart as a Concept

The conceptual framework underpinning the notion of "smart" within the context of sustainable cities is intricately woven into the fabric of three pivotal developments. The first pertains to the pervasive adoption of sustainability principles, which has been increasingly acknowledged as fundamental to urban development [58]. The second catalyst is the rise in urbanization rates, a phenomenon widely observed and documented in the works of urban scholars such as [59,60]. The third catalyst, pivotal in this paradigm shift, is the rapid advancements in ICTs [40]. The intricate interplay of these three forces has ushered in a transformative era for urban landscapes, giving rise to the emergence of sustainable cities.

The concept of smart, sustainable cities hinges on the deliberate integration of innovative and extensive ICT applications in the pursuit of sustainable urban planning. This notion finds resonance in the work of [40], which posits that the effective use of ICT is an integral part of achieving sustainability goals in urban settings. The strategic integration of ICT is argued to enhance efficiency, resource management, and overall urban functionality [47]. However, it is crucial to critically examine and reflect upon these assertions.

Among the myriad ICT applications in urban settings, the Internet of Things (IoT) and its concomitant utilization of big data emerge as particularly prominent examples. The IoT facilitates real-time data collection and communication among interconnected devices, offering unprecedented insights into urban dynamics [61]. Additionally, the utilization of Big Data in tandem with IoT technologies enable data-driven decision-making processes, fostering a more informed and adaptive approach to sustainable urban planning [62].

However, to ensure the robustness of these claims, it is imperative to engage in a rigorous scholarly inquiry. This involves verifying the empirical evidence supporting the positive impacts of sustainability principles [7,63], questioning assumptions about the uniform benefits of rapid urbanization [64], and critically reflecting on the potential challenges and unintended consequences associated with extensive ICT integration in urban planning [65]. By triangulating these concepts with a diverse array of references, this discourse endeavors to contribute to a nuanced and well-substantiated comprehension of the intricate linkage between urbanization and sustainability, ICT, and the realization of smart, sustainable cities.

2.2. Smart City Dimensions

Within this part, we present one of this paper's key sections: a conceptualization of the Smart City dimensions.

To establish a collective set of multidimensional aspects, we need to look at Smart City concepts in depth [49,52]. The three aspects of technology, economy and society, and governance are illustrated in the following (Table 2).

Table 2. Dimensions and consensus concepts in Smart Cities.

Dimensional Aspect	Sub-Concepts	Core Findings	Common Consensus among Scholars	Sources
Technology and Data	Data Generation, Sources, Analytics, Infrastructure	– Smart cities rely on connected environments for data generation. – Data are sourced from IoT devices, social media, and public records. – Advanced analytics processes extract insights and inform decision-making. – A robust digital infrastructure, including high-speed internet and wireless networks, is vital for seamless communication and data exchange.	Scholars commonly agree on the significance of connected environments, diverse data sources, and advanced analytics for informed decision-making in smart cities.	[66]
Economy/ Society	Economic Goals, Techno-centric Approach, Social Involvement, Private Sector Collaboration	– Economic goals include enhancing quality of life and optimizing resource utilization. – A techno-centric approach fosters competitiveness, sustainable development, and economic growth. – The social dimension involves citizens in a passive consumer role; the private sector collaborates for innovative solutions.	Scholars widely concur on economic goals, the techno-centric approach, and the changing role of citizens and the private sector in the social dimension of smart cities.	[66]

Table 2. *Cont.*

Dimensional Aspect	Sub-Concepts	Core Findings	Common Consensus among Scholars	Sources
Governance	Technological Aids, Data-Driven Decision-Making, Active Government Support	– Governance involves a preference for technological aids in city management. – Decisions are increasingly data-driven, relying on smart devices, IoT infrastructures, and big data. – Active government support fosters technological innovations.	Scholars commonly acknowledge the reliance on technological aids, data-driven decision-making, and government support for technological innovations in the governance of smart cities.	[66]

This table (Table 2) provides a highlight of the sub-key concepts associated with each dimensional aspect of smart cities, supported by core findings across the dimensions of smart cities emphasized by common consensus among scholars. This tabular presentation format is intended to provide an in-depth comprehension of the smart city literature's consensus, and to present the information in a concise and relevant manner.

3. Key Features of Smart City Development

In the dynamic landscape of Smart City development, the conceptualization of "smart people" (Table 3) takes center stage, an idea substantiated by [40]. Their research highlights the indispensable role of citizens equipped with proficient ICT skills and a substantial educational background, positing that these demographics form the basis of significant social and human capital. This assertion prompts critical reflection on the societal implications of smart city initiatives, raising questions about inclusivity, accessibility, and potential inequalities among citizens.

Table 3. Key characteristics associated with the development of a smart city [67].

Feature	Smart City Development's Content
Smart people	In a "smart city", citizens possess proficient ICT skills alongside a substantial level of education. Such a city is recognized for its significant social and human capital.
Smart infrastructure	A "smart city" is characterized by its infrastructure, which is built upon intelligent systems, IoT, and other cutting-edge Industry 4.0 technologies.
Smart living	In a "smart city", every citizen assumes a more participatory role within the community, actively involving themselves in the utilization of public and private services and efficiently evaluating their appropriateness. By considering aspects such as cultural amenities, health conditions, personal security, housing standards, and other factors, a "smart city" strives to offer a high quality of life.
Smart economy	In a "smart city", the economy is built upon the foundation of ICTs and various Industry 4.0 technologies. This encompasses not only ICT-related sectors but also encompasses "smart" enterprises that integrate ICT and new technologies into their manufacturing processes and business models.
Smart mobility	A "smart city" is a city that leverages contemporary transportation technology, logistics, and innovative transport systems to enhance urban mobility and improve the overall quality of life.
Smart management	The concept of a "smart city" denotes a city that incorporates intelligent management and administration practices, employing novel approaches to citizen engagement and communication, such as "e-management" and "e-democracy".
Smart environment	A "smart city" is one that has a safe, "green", and long-term environment, with modern technology incorporated into natural settings that are free of environmental and health risks.

The complex nature of "smart infrastructure" is illuminated through the lens of [68]. Their work explores the incorporation of smart systems and Industry 4.0 technologies, raising crucial considerations about the integration challenges, cybersecurity concerns, and societal implications of heavy reliance on sophisticated infrastructure.

The concept of "smart living" intersects with the work of [69], who explore the participatory role of citizens in assessing their quality of life as part of smart cities. This prompts a critical examination of the balance between technological integration and human-centered urban planning, raising questions about privacy, autonomy, and the potential for technological solutions to enhance rather than replace human experiences.

The authors of Ref. [15] contribute to the discourse by highlighting the fundamental role of ICT in shaping the "smart economy". Their research calls for a critical examination of the economic implications, particularly regarding potential disruption, job displacement, and the equitable distribution of economic benefits in the context of smart city transformations.

The expansion of "smart mobility", as explored by [70], invites critical thinking about the societal implications of modern transportation technology. Considerations about accessibility, the potential for increased surveillance, and the societal tradeoffs associated with prioritizing efficiency in urban mobility emerge.

The authors of Ref. [71] delve into the complexities of "smart management" and introduce novel approaches to citizen engagement and communication. This sparks critical discussions about democratic implications, potential biases in decision-making algorithms, and the overarching question of who holds power and influence in the Smart City governance model. The vision of a "smart environment" intricately interwoven with modern technology, as discussed by [47], requires critical reflection on the environmental sustainability of Smart City initiatives. Considerations arise regarding the environmental footprint of technology integration, potential unintended consequences, and the long-term environmental viability of Smart City developments.

These critical reflections, grounded in scientific research, underscore the nuanced and complex nature of Smart City development. They encourage thoughtful consideration of the broader societal, economic, and environmental implications and urge stakeholders to navigate these transformations with a keen awareness of the potential challenges and opportunities embedded in the smart city paradigm.

4. Smart City Approaches in Literature Studies

4.1. Conceptualizing the Smart City's Approach

This paper aims to categorize present approaches to understanding the Smart City concept and to organize relevant theoretical insights. A comprehensive and descriptive research methodology was used to accomplish this purpose.

To achieve our objective, we undertook a thorough literature analysis, focusing on the exploration of concerns associated with Smart Cities (SC). During this process, we extensively reviewed the works of researchers from academic and expert backgrounds to identify multiple definitions of the term. To broaden our understanding, we conducted thorough searches for the terms "smart cities" and "smart city" in collections such as Web of Science (WoS), Scopus, and Google Scholar. The authors of the definitions and the works that were selected for inclusion were chosen based on a personal appraisal of their particular significance. This evaluation considered the works' citation numbers, the authors' recognition for their work in the field of Smart Cities, as well as how innovative and original the chosen works are. To eliminate redundancy in Table 1, which collects the definitions and terminologies of Smart City, there are also rejected references where a high degree of similitude was found with the definitions of Smart City previously established by other authors.

Finally, the literature of more than 95 articles was reviewed to find the most relevant definitions of Smart City provided by the contributors. Hence, based on a compilation of 30 distinct definitions, the rationale behind Smart City rhetoric was explored. The additional definitions identified (typically representing current research on Smart City) enlarged the descriptions of approaches for understanding the Smart City and thereby broadened the representation of perspectives and viewpoints on it. This definition-based discussion refers to the conceptual framework of Smart City and not to real-life cases.

Therefore, additional contextual research is required as will be explained in the next section of this document. A qualitative content study of the gathered definitions allowed for the separation of keywords and the recognition of the key attributes of the Smart City notion. In the accompanying Table 4, the selected keywords and studied definitions are listed. The identified keywords in the two approaches could be divided into three divisions in accordance with their technological, economic/social, and governance dimensions, allowing for the differentiation of four dimensions of the Smart City paradigm. The introduced method helped to discern the two approaches to deduce existing research and analysis on assimilating the Smart City paradigm (Figure 1). Based on the aforementioned qualitative definition analysis, two approaches to conceptualizing the smart city were noted (Table 4).

Table 4. Smart city definition classified by approach. Source: created by the author.

	Definition	Author/Year	Keywords
Techno-centric Approach	In its core, a Smart City denotes an urban setting that leverages ICTs and relevant technological progress to optimize the efficient execution of standard municipal functions and improve the quality of services delivered to urban residents.	[72]	urban space; ICT; technological innovations; improving performance of city functions; enhancing the QoS provided to citizens
	Smart Cities are all about integrating ICT into every element of human life, smart devices, sensor networks, and real-time data collection.	[73]	sensor networks, intelligent devices, gathering real-time data, integration of ICT
	Being a Smart City entails employing resources and technology intelligently and with coordination, with the aim to create an inclusive, livable, and sustainable urban environment.	[74]	intelligent use of resources and technologies; inclusive; livable sustainable urban areas
	A Smart City (SC) is demarcated as the application of computing technologies aimed at enhancing the intelligence, connectivity, and efficiency of critical infrastructure and services inherent to urban environments. This encompasses a spectrum of domains, including but not limited to city administration, education, healthcare, real estate, public safety, and utilities.	[75]	smart technologies ; intelligent connectivity; infrastructure; services
	Smart and sustainable cities are usually based on the realization of various ICT implementations of ubiquitous computing, particularly the IoT, wherein connected objects work together via various distributed computing systems to offer data and services to urban organizations and residents, enabling seamless collaboration.	[76]	ICT implementations; IoT, distributed computing systems; urban organizations; urban residents
	The concept of a Smart City embraces advanced principles that seek to manage cities and urban areas in a contemporary manner, leveraging the technical tools offered by the latest innovations, such as information technology. It aims to align with ecological standards, while simultaneously prioritizing resource preservation and achieving intended outcomes.	[77]	cutting-edge concept; modern management; use of technical tools; the standards of ecology
	Smart sustainable cities integrate ICT to manage resources more intelligently. This reduces costs and energy use, improves the provision of services and the standard of living, and causes less of an ecological impact, all which support innovation and a green economy.	[78–80]	ICT; smart sustainable cities, reducing costs and energy use, provision of services; ecological impact; green economy

Table 4. *Cont.*

	Definition	Author/Year	Keywords
Techno-centric Approach	A smart city is a city where new digital technologies are used to coordinate and combine ICT with conventional infrastructure.	[15]	conventional infrastructure; coordinating and combining ICT; digital technologies
	The author indicates that by an "instrumented, interconnected, and intelligent city", we mean a city that is managed by ICT. Smart technologies have the potential to improve the intelligence, interconnection, and efficiency of a city's essential infrastructure and services, including city management, health services, education, public safety, residential and commercial properties, transportation, and amenities [75].	[32]	instrumented and interconnected city; city management; transportation and amenities; efficiency of the city's infrastructure
	With the growing impact of urbanization, governments, corporations, and communities are increasingly turning to technology to tackle associated challenges. This has led to the emergence of "smarter" cities that prioritize connectivity among seven vital components: municipal governance, educational systems, healthcare services, public safety initiatives, real estate management, transportation networks, and utility provisions. To achieve this, a smart city utilizes software systems, server, and network infrastructures, as well as consumer devices collectively known as Smart Computing and Information Technology, ensuring the seamless integration and efficient management of these crucial aspects.	[75]	Smart Computing technologies; services; Rapid Urbanization; city infrastructure; Smart City Planning; City Management
Human- and social-centric approach	A smart city is a community that actively and sustainably supports the general well-being of all its members while maintaining a high stability level to become a consistently a better place for living, working, and playing.	[81]	community, general well-being; high stability level; living, working, and playing place
	Socially conscious smart cities must engage with citizens and social capital, rather than simply trusting that software can alter and enhance cities entirely on its own. The impactful feature of information technology is not its ability to inherently build smart societies, but rather its accessibility to be used socially in ways that embolden and promote public awareness, as well as engage the public in a participatory democracy relating to the urban environment in which they live.	[41]	individuals; human resources; socially harnessed information technology; people's empowerment; education; people's engagement in political discourse
	A city achieves smart status when social and human capital investments, along with traditional and new (ICT) communication infrastructures, contribute to sustained economic growth and a high standard of living.	[82,83]	human capital; social investments; development of modern infrastructure; sustainable economic development; high standard of living
	The smart city is an integrated approach wherein citizens and social capital interact extensively by means of technology-based solutions. Based on a multi-stakeholder, municipality-based collaboration, the smart city concept aspires to effectively accomplish sustainable and resilient development along with an improved standard of living.	[5]	integrated approach; technology-driven solutions; resilient and sustainable development; high standard of living; collaborative efforts of multiple stakeholders

Table 4. *Cont.*

	Definition	Author/Year	Keywords
Human- and social-centric approach	Smart city is a complex combination of social, infrastructural, human, and business assets that is fused, managed, and integrated into city fabrics through advanced technologies, in order to tackle social, economic, and environmental challenges, integrating multi-actor, multi-sector, and multi-layer perceptions.	[84]	complex combination of assets; multi-layer perceptions; city fabrics; advanced technologies; social, economic, and ecological environmental challenges
	Fundamentally, the smart city vision is an individualistic or human-centric approach, with the goal of delivering a high-tech, secure, and sustainable future. It relies on strong governance among its government entities and other formal and informal organizations at the organizational level. It also emphasizes the development and dissemination of knowledge management procedures, the development of national missions, goals, and targets, and the establishment of financial models for national progress.	[85]	individualistic or human-oriented approach; sustainable future; knowledge management procedures; financial models
	A smart city is a well-organized and functional city that attends to the needs of its residents, manages public spaces, and coordinates processes, establishing optimal conditions for enhancing the quality of life for its inhabitants, and fulfilling the objectives of relevant stakeholders.	[86]	effective process organization; efficient management of public space; maximizing growth potential; optimizing conditions to enhance citizens' quality of life; aligning with stakeholders' objectives; fostering a well-structured city
	A smart city is a place that uses ICT-based solutions in collaboration with several governmental stakeholders to tackle social challenges.	[87]	ICT-based solutions; public challenges; multi-stakeholder collaboration; civic foundation
	A smart city is defined by its excellence across six key domains: smart government, smart economics, smart transport, smart buildings, smart people, and smart environment. It is a community that thrives on a well-balanced blend of resources and the proactive initiatives of self-sufficient, informed, and independent residents.	[88]	future-oriented; abilities and behaviors of self-determining, independent, and conscious citizens
	Suitable and dependable governing structures, along with open-minded and innovative individuals, are anticipated to provide the necessary support for smart cities. By collaborating effectively, they can enhance local productivity, which serves as a fundamental requirement for driving accelerated economic growth.	[33]	physical infrastructure; human capital; social capital; urban performance; urban architecture, planning, and governance

This section focuses on the different approaches to understanding the concept of smart cities based on a literature review. Its objective is not to assess the various perspectives of smart cities, but to describe and illustrate them using specific definitions.

The table lists two approaches to the concept of Smart Cities (SCs): techno-centric and human social centric. The techno-centric approach views Smart Cities as urban spaces where advanced technologies, such as sensors and real-time data networks, have become an integral part of every dimension of human life, permeating various aspects and activities to optimize resource use and service delivery. In contrast, the human social approach sees SCs as more than just the use of modern technologies for improved efficiency and cost savings. Instead, it highlights the importance of cross-sector collaboration, investing in social capital, and implementing participatory governance to foster sustainable economic growth and enhance the well-being of city residents (Figure 1). The Smart City concept spans across diverse dimensions, incorporating human, infrastructural, social, and entrepreneurial capital. It strategically utilizes emerging technologies to effectively address prevailing challenges in the social, economic, and environmental spheres.

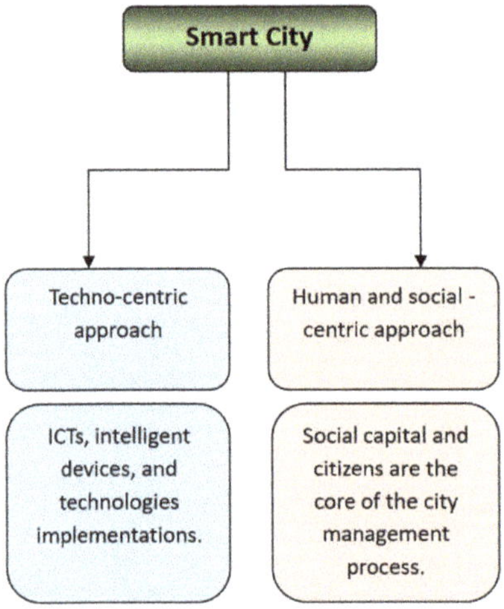

Figure 1. Approaches for understanding the smart city concept. Source: created by the author.

In the literature first discussed, the notion of smart cities and their technocratic approach underscore the utilization of ICTs [78] to improve urban management and efficiency. The goal of SCs, according to authors adhering to this approach, is to integrate smart devices, real-time data networks, and sensors [73] into every facet of human existence to maximize the efficient utilization of resources [78], optimize the provision of services, and enhance well-being and living standards [76,89]; in addition, the authors of Ref. [90] propose a multidimensional framework for understanding smart cities that integrates development drivers (such as technology, governance, social inclusion, and sustainability) with desired outcomes (such as livability, workability, sustainability, resilience, innovation, and equity). The authors argue that this framework can help policymakers and practitioners make informed decisions about the development and execution of smart city initiatives by identifying the relationships between development drivers and desired outcomes.

In Ref. [73], an extreme illustration is given of the technology-centric Smart City paradigm, Smart City transitions, in which sensors, digital devices, and shared data networks are incorporated with various elements of human existence. In this sense, incorporating the most cutting-edge technological solutions into the city space may appear to be a goal in and of itself; however, as Ref. [78] points out, Smart Cities utilize information and communication technologies (ICTs) to enhance their intelligence and resource utilization efficiency. Therefore, this approach leads to cost and energy reductions, enhanced service provision, and an overall improved quality of life. According to [76], as well as [90–93], the effective implementation of information and communication technology (ICT) has become the key to optimal governance and an essential requirement for achieving urban intelligence [92]. In Ref. [75], the authors have the perception that a city achieves smart status when ICT technologies enhance the knowledge, efficiency, and connectivity of critical infrastructure and service components, including governance, healthcare, educational institutions, safety and security, property, logistics, and community services.

On this point, modern cities, which are brimming with intelligent sensors, have the capability to autonomously optimize resource utilization, control traffic and street lights, along with maintaining surveillance on safety conditions, promptly notify emergency services of accidents, and even provide information regarding necessary road and building maintenance [76,94]. According to [74], a city is smart when it intelligently leverages all available technologies and resources in a coordinated manner, in order to establish interconnected and environmentally sustainable urban hubs that are beneficial for its residents. A Smart City should be maintained based on the obtained input data and processed by these "smart" systems and devices, according to [95,96].

In Ref. [77], the authors emphasize the significance of technology in the creation of Smart Cities. The authors use the term "SC" to refer to a creative idea that takes advantage of the technological resources available today. Its purpose is to help modern city management adhere to ecological and resource conservation principles. According to [15], a Smart City is not only planned and coordinated utilizing newer technologies, but ICTs and classical infrastructure are also converging in a Smart City.

Thus, given the fact of the techno-centric approach, the definition of smart, sustainable cities is based on incorporating the innovation and widespread use of Information and Communication Technology (ICT) in the transition to sustainable urban planning. The Internet of Things (IoT) and associated big data applications are one of the most prominent ICT applications in urban environments. This literature includes other various scholars such as, [15,74–77,92–94], who highlight the importance of technology in creating SCs that are coordinated, integrated, and sustainable. The use of ICTs is not only about making infrastructure and service components more intelligent and efficient, but also merging them with traditional infrastructure. The Smart City concept is not solely dependent on the use of modern technologies [68], but it revolves around fostering collaboration across sectors and investing in the development of social capital. According to [9], the paradigm of a Smart City involves the establishment and utilization of interrelationships and linkages between human and social capital, alongside information and communication technologies (ICTs), with the aim of generating sustainable economic growth and improving the quality of life. The management of natural resources is based on participatory governance. The Smart City strategy aims to focus on a multifaceted strategy that considers all city actors rather than just the use of ICT. In Ref. [97], the Smart City is depicted as a hub that can harmonize competitiveness and sustainable development through the integration of diverse developmental dimensions and infrastructure investments. It represents a multidimensional amalgamation of human, infrastructural, social, and entrepreneurial capital, which is seamlessly blended, coordinated, and integrated into the urban fabric using innovative technologies. This approach aims to address social, economic, and environmental challenges, from a multi-actor, multi-sector, and multi-level perspective [84]. Technological advancements in the digital realm support community-led urban development and participation-based governance [98].

4.2. The Smart City in Africa: A New and Fast-Growing Research Area

To respond to the challenges posed by these exponential population growths, African cities have already created technopoles and are currently implementing initiatives borrowing from the smart cities concept. In terms of growth, the top ten fastest growing cities in the world—as well as their projected growth by 2025—are all African (Figure 2), a sign of how swiftly the continent's population is expanding. As urbanization is a recent and very rapid phenomenon in Africa, cities are not yet equipped with complex and high-tech systems, as some cities in Europe, America, and Asia can be. Despite their difficulties linked to a very strong demographic explosion, significant inequalities, healthcare, and education systems that are still underperforming, and sometimes unstable political contexts, African cities are gradually beginning to take innovative initiatives.

African metropolises, being comparatively younger and more adaptable than cities in developed nations, face fewer limitations imposed by existing infrastructure. This affords them numerous opportunities to undertake projects with substantial technological potential, including the direct adoption of mobile and fiber optics and the implementation of Smart Grids for efficient energy management. These localized initiatives signify progress towards the establishment of smart cities in Africa, leading to the gradual emergence of the smart city phenomenon on the African continent.

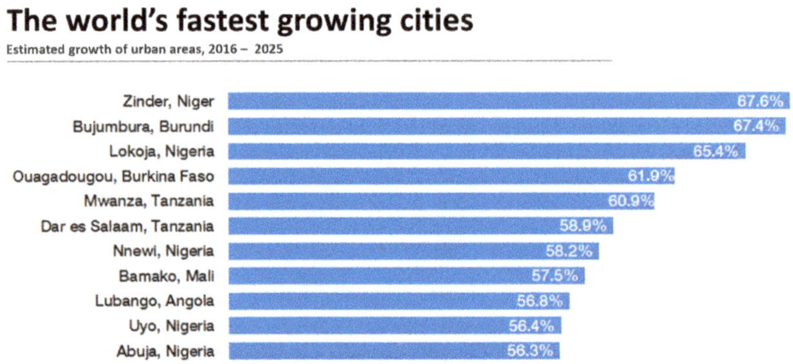

Figure 2. World's fastest growing cities: estimated growth of urban areas 2016–2025. Source: UN World Urbanization Prospects 2014 Revision by the UN's Department of Economic and Social Affairs.

The concrete applications of smart cities in Africa are currently concentrated in about ten of the fifty-four countries of the continent. Even if North African cities (Cairo, Tunis, Algiers, and Casablanca) are heading the list, many regional megacities such as Accra, Lagos, Abidjan, or Nairobi are also very successful in terms of economic attractiveness for investors and growth for their middle class. New cities such as Johannesburg and Kigali are also showing amazing speed and maturity thanks to the emergence of a dynamic and connected middle class.

Due to the distinctive geography, history, and culture of each city in Africa, a uniform smart city model is not applicable. Instead, diverse projects are being established in various metropolises, and each year, multinational corporations, recognizing the continent's potential for growth, actively participate in ambitious initiatives alongside African cities.

Prior to the global economic crisis in 2009, several African countries had initiated efforts in building Smart Cities, with consideration for the challenges involved, such as Konza Technopolis in Kenya, Cité du Fleuve in the Democratic Republic of Congo, Eko Atlantic City in Nigeria, and the Lanseria Smart City project in South Africa [99].

Rwanda leads the way in developing smart city infrastructure in Africa. Modernization in Kigali is part of the government's strategy to improve access to public services [100]. The Irembo platform allows citizens to do tasks online. Rwanda works with the private sector to develop smart cities. They partnered with Nokia to enhance social sustainability. The Vision City project in Kigali aims to create a technologically advanced neighborhood equipped with streetlamps powered by solar energy and offering complimentary Wi-Fi in the central town square [101].

The Nigerian Smart City Initiative launched in 2017 aimed to boost ICT innovation in Nigeria and demonstrate how it could be integrated with physical infrastructure to improve service delivery. Similarly, Addis Ababa, Ethiopia, addressed its parking shortage by implementing smart parking, which uses Chinese technology to park cars in a steel-structure building using an automated elevator, and IBM sent a Smarter Cities Challenge team to Accra, Ghana, to explore ways to leverage technology for economic and social development [102].

The challenges faced by Morocco in establishing new cities and dealing with rural exodus are rooted in the adoption of a "Smart City" approach. This is highlighted by Morocco being the first country in North Africa to embrace 3G technology. The Casablanca Smart City initiative has played a major role in transforming Casablanca into Africa's premier financial hub, with a focus on software development, e-commerce, and ICT companies [103]. According to the Smart City Index Report 2023, published by the International Institute for Management Development, Rabat ranks among the top five smart cities in Africa [104]. This ranking is based on factors such as integrating artificial intelligence in telemedicine, electric mobility, the digital economy, and the ecological transition. These developments demonstrate the country's rapid progress.

While officials and city stakeholders propose building new smart cities as a solution to abandoning old towns, it is important to acknowledge that Africa has limited Smart City developments due to the high costs and complexities associated with infrastructure and socio-environmental factors. Nevertheless, with careful planning and investment, it is possible to overcome these challenges and create sustainable smart cities that meet the needs of the urban population. Therefore, implementing Smart City initiatives in existing towns and cities is a more feasible and confident strategy to accommodate the increasing urban mass.

4.3. The Smart City Approach in an African Context

An African "Smart City" approach involves using technology and data to improve the quality of life for residents, increase efficiency and sustainability, and promote economic development. There are several key elements to an African smart city approach, including (Figure 3):

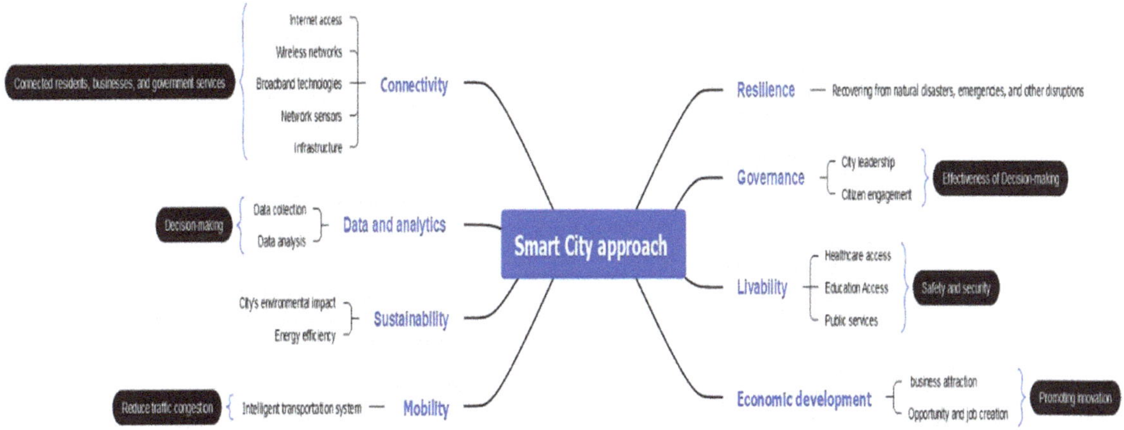

Figure 3. Smart City approach flowchart. Source: created by the author.

Connectivity: building and maintaining high-speed internet and wireless networks to connect residents, businesses, and government services.

Data and analytics: collecting and analyzing data to inform decision-making and improve city services.

Sustainability: using technology and data to reduce the city's environmental impact and promote energy efficiency.

Mobility: improving transportation options and reducing congestion with smart transportation systems.

Livability: using technology and data to improve access to healthcare, education, and other public services.

Economic development: using technology and data to attract and retain businesses, create jobs, and promote innovation.

Governance: using technology and data to improve the effectiveness of city leadership and decision-making processes, as well as citizen engagement and participation.

Safety and security: using technology and data to improve safety and security for residents and visitors.

Resilience: using technology and data to prepare for, withstand, and recover from natural disasters, emergencies, and other disruptions.

A smart city approach typically involves collaboration between the city government, private sector companies, and citizens. It also involves a holistic approach where all the aspects of the city are considered like environment, social, economic, technological, and governance. It also requires a clear vision and long-term strategic plan for the development of the city.

There is no generic solution to the development of smart cities in Africa, as each city presents a specific array of challenges and opportunities. However, there are some general principles that can be followed to ensure that smart city initiatives are successful in Africa:

Engage with stakeholders: it is important to involve a wide range of stakeholders in the development of smart city initiatives, including government officials, business leaders, and community organizations. This will help to ensure that the initiatives are aligned with the needs and priorities of the community.

Focus on improving the delivery of public services: smart city initiatives can help to improve the delivery of public services, such as healthcare, education, and transportation, which can have a significant impact on the lives of residents.

Promote economic development: smart city initiatives can help to drive economic development and job creation in the city, by attracting new businesses and promoting innovation.

Foster innovation and the use of technology: smart city initiatives should involve the use of innovative technologies and approaches to address local challenges and opportunities.

Ensure sustainability: smart city initiatives should be designed to be financially sustainable over the long term and should aim to reduce the environmental impact of the city.

Promote social equity: smart city initiatives should aim to promote equal access to services and opportunities for all members of the community, regardless of their socio-economic status or background.

5. Developing and Assessing Afro-Smart City Methodology

Afro-Smart Cities refer to urban centers in Africa that are designed to incorporate technology and innovation to enhance the quality of life for citizens. Before approaching Afro-Smart Cities, it is essential to have a clear understanding of the local context (Figure 4) and learn about the cultural, social, economic, and political factors that impact the city. This will help to identify the unique challenges and opportunities that need to be addressed.

This schema (Figure 4) illustrates the interconnected steps involved in approaching Afro-Smart Cities. Understanding the local context leads to the identification of key stakeholders and priority areas [105]. Leveraging technology and innovation, along with involving citizens, contributes to the development and implementation of sustainable and inclusive Afro-Smart Cities. The final step emphasizes the importance of fostering partnerships and collaborative networks to achieve the desired outcomes.

To effectively implement a smart city approach in Africa, it is important to prioritize financing the infrastructure that supports the informal economy and manages population growth in highly dynamic urban areas. Furthermore, it is necessary to consider the geopolitical situation and stability of the region to determine the feasibility and scalability of implementing the Afro-Smart City approach. Rational investment in human capital resources can also boost the readiness of a smart African city in terms of skill development. Because without the right skills and capacities, it would be almost impossible to fully realize the outputs of implementing a successful African smart city strategy.

There are several ways to assess Afro-Smart Cities strategies and initiatives performance; this includes first defining what we mean by "Afro-Smart City", as this can vary widely. As a result, it is essential to have a comprehensive understanding of the local context and what we are looking for, which will help to ensure that we are assessing the right things and that we are comparing apples to apples. Secondly, it is necessary to conduct a comprehensive review of existing smart cities initiatives in Africa by looking for examples of successful projects and identifying common themes or best practices. Subsequent to this step, the focus shifts to creating a set of criteria or metrics to assess the performance of smart city initiatives. These should be in line with the smart city objectives and might include factors such as sustainable economic development, environmental impact, and social inclusion, as well as gathered insights and perspectives on the effectiveness of the initiatives from key stakeholders including government officials, business leaders, and community organizations through interviews, focus groups, or surveys. The information gathered can be used then to assess the advantages and disadvantages of the current initiatives and make recommendations for improvement [106]. The recommendations can be followed by a comparison against similar initiatives in other cities, both inside and outside the country. This can provide valuable insights and best practices that can be adopted to improve the initiatives being assessed.

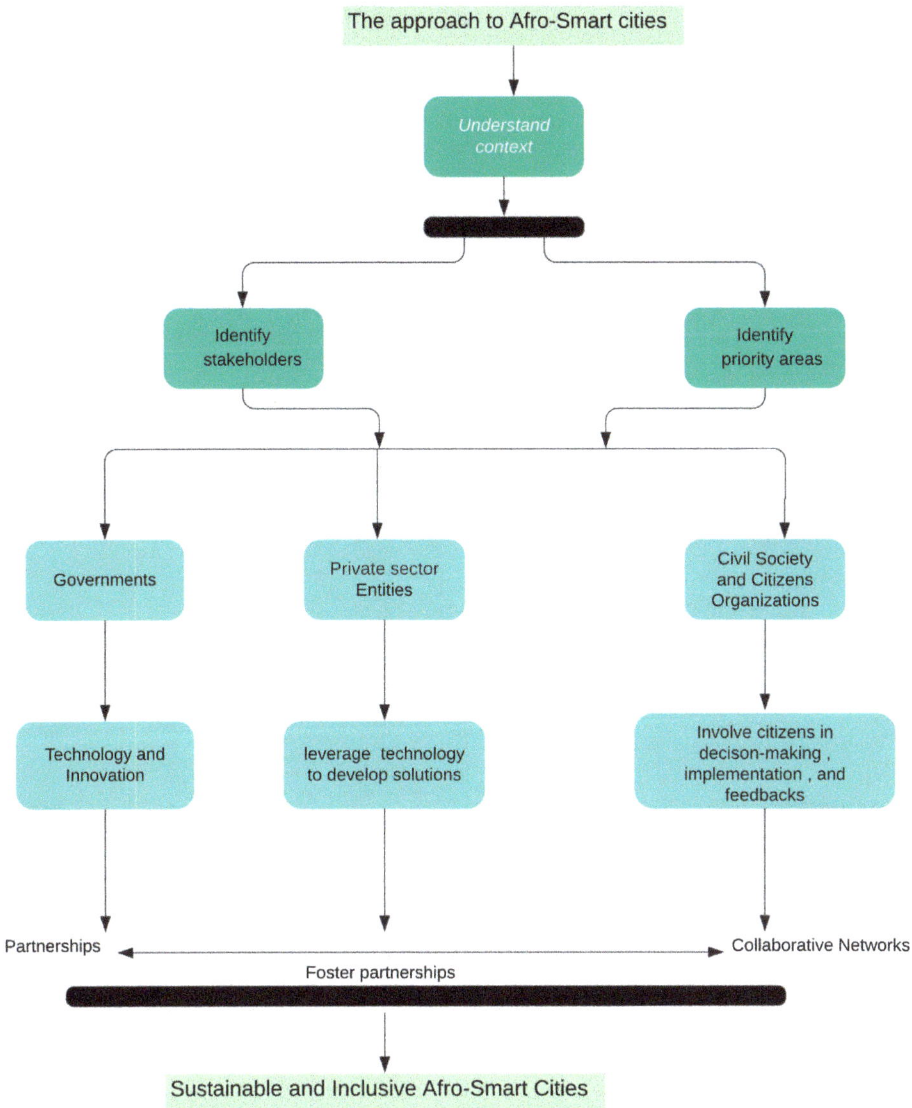

Figure 4. Schema representing the approach to Afro-Smart Cities. Source: created by the author.

It should be noted that the continuous monitoring and evaluating of the progress of the initiatives over time [11] will allow us to identify any problems early on and adjust as needed. As a final point, it is necessary to generate a report on the results, by summarizing the findings and results of the assessment, and to make recommendations for improvement [107]. This report should be easily accessible and understandable to all stakeholders, including the public. One of the most common methods for assessing Smart Sustainable City (SSC) Performance is the Key Performance Indicators (KPIs) [108]; these are a set of metrics that are used to measure the performance of an Afro-Smart City. These metrics can be utilized to evaluate various aspects of a city's performance, such as energy efficiency and consumption, transportation and mobility, air quality, digital infrastructure, quality of life and citizen engagement and participation [109,110]. These KPIs can help evaluate the effectiveness of ICT interventions in driving smart and sustainable development in cities [108,111]. It is important to adapt and customize these indicators based on specific African city contexts and goals.

The figure below (Figure 5) represents a matrix of KPIs for assessing the role of ICT in rendering African cities more sustainable and smarter [112].

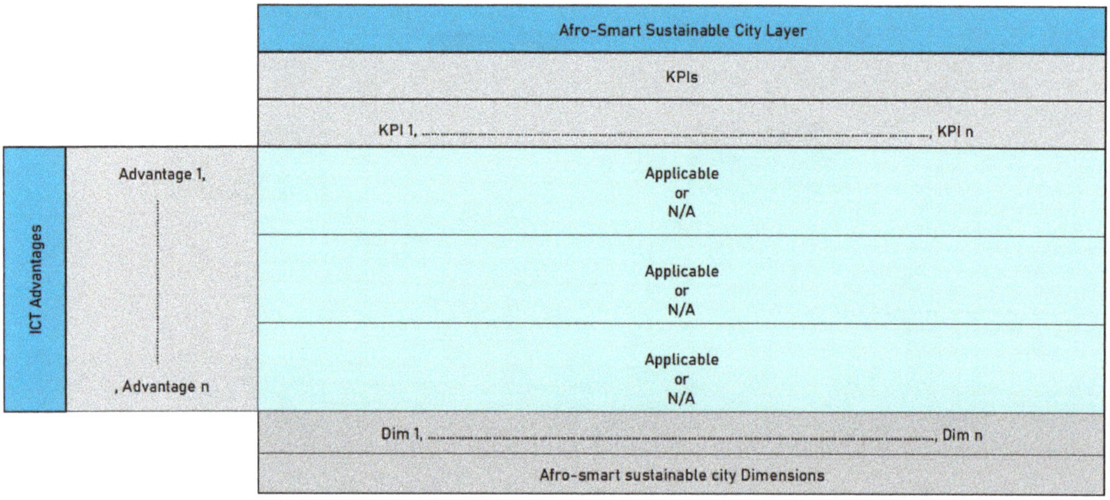

Figure 5. Matrix with the KPIs for Afro-Smart Sustainable Cities (ASSC). Source: created by the author.

This matrix provides an overview of the main KPIs for assessing Afro-Smart and Sustainable Cities in relation to the contributions of ICT-based solutions.

The methodology used for the formulation of the Key Performance Indicators matrix was derived from an in-depth examination of various collections of smart city index systems, the Smart Sustainable Cities standard, and the ICT development index (IDI), particularly those offered by the Global City Indicators Facility, UN-Habitat, the International Telecommunication Union (ITU), The United Nations Economic Commission for Europe (UNECE), and the International Organization for Standardization (ISO) [112]. Additionally, the indicators of European ranking of medium-sized cities (ERMC) were also taken into consideration during this process. This matrix, which serves as the foundation for the evaluation of Smart City initiatives in Africa, is primarily based on the technological advancements that these initiatives bring forth, subsequently generating additional value. The evaluation of this additional value is then conducted using key performance indicators. Moreover, the advantages of ICT are classified according to the different dimensions of an Afro-Smart Sustainable City. For each dimension, there are one or more key performance indicators that have been defined to assess these advantages qualitatively or quantitatively.

This inclusive matrix representation plays a crucial role in facilitating the integration of all dimensions of sustainability, as visually depicted in Figure 5.

Depending on the specific requirements and priorities, the matrix can be adjusted and enlarged as illustrated in the figure.

6. Discussions

The proposed methodology focuses primarily on defining the ICT-related Key Performance Indicators matrix for Afro-Smart Sustainable Cities, with limited consideration given to the other KPIs of cities. However, in terms of adaptability and interoperability, the KPI matrix can be adapted to identify the impact of ICT on the KPIs of other sectors, such as energy and agriculture. This allows for comparisons between African cities with similar characteristics.

A significant challenge in the African context is the limited involvement of citizens in the design of key performance indicators (KPIs), which may reduce the possibility of integrating a highly human-centered element in the design and evaluation phases. It is recommended that community-driven indicators, such as satisfaction with advanced technology services, accessibility of services, quality of service, and community inclusion, be included to ensure that smart city projects in Africa achieve urban performance.

The proposed methodology for developing and evaluating smart city initiatives will help different urban administrators and policymakers in Africa, each in his or her field, understand how ICT-based initiatives' performance can be reached and measured. It aims to involve them as active co-producers of the ICT impact assessment approach. This will guarantee that the resulting KPIs are adapted to the Afro-Smart City projects and that the urban contexts in which they are implemented are relevant to the stakeholders concerned.

Various methods exist for engaging with city planners and stakeholders. Academia and local governments frequently have job positions responsible for overseeing specific datasets related to buildings, ICT, energy, and mobility. However, locating the appropriate individuals and initiating them to share data can be challenging, particularly if they are not directly involved in Smart City projects. Acquiring data from private sector partners is even more complex. While larger corporations are subject to more comprehensive data collection and reporting processes, tech startups and small businesses may not be able to manage or deliver data for their Smart City projects or applications.

7. Conclusions and Way Forward

This research paper discusses two distinct approaches to the concept of a Smart City (SC): The techno-centric approach and the social- and human-oriented one. Distinct components for developing urban "Smartness" are suggested by each of the two approaches, and each one gives a distinct interpretation of the Smart City notion. This conception establishing urban intelligence is exactly the most divisive issue in the Smart City review. The suggested approaches arrange the existing definitions of the Smart City concept and highlight the need for further research and empirical analysis to understand and evaluate the implementation of Smart City initiatives. The two distinguished approaches are as follows:

1. Techno-centric Approach: This approach considers modern technologies as the driving force behind urban development. It focuses on utilizing technological advancements to improve city functioning and enhance the quality of life for residents.
2. Human and social-centric approach: The human- and social-oriented approach views a Smart City as a multifaceted development involving all stakeholders in the city and their active engagement in the processes of decision-making. It emphasizes the inclusion of various social and humanistic factors in the development of smart cities as well as the implementation of good governance principles in Smart City initiatives by prioritizing the engagement, participation, and well-being of the citizens.

Despite their differences, all approaches aim to improve urban residents' quality of life and improve city functionality. This is accomplished by several activities encompassing three dimensions of a Smart City: technology, economy/society, and governance. The paper also acknowledges that the definition-based discussion refers to the conceptual framework of Smart City and not to real-life cases and that the study conducted includes the use of only three databases (Scopus, Web of Science, and Google Scholar) and the exclusion of other sources, which might have provided additional insights and definitions. Therefore, additional future research that could expand the categories and include more examples and subcategories is required.

Furthermore, we suggest that the presented conceptualizations can serve as an initial basis for future studies into developing a complete Smart Urban Planning Model. By combining methods and techniques from each approach, a comprehensive framework for managing smart cities can be developed. In this sense, our research study corroborates that contextual research in Africa was required, which was presented and described in Section 4 . In this section, several key elements to an African smart city approach were defined considering that the Smart City in Africa is a new and fast-growing research area aiming to respond to the challenges posed by these exponential population growths.

The notion of the Afro-Smart City has gained worldwide recognition, but it often lacks a standardized assessment methodology. A thorough review of the literature has unveiled a research gap in the African Smart City paradigm concerning the need for integrated approaches to evaluate Smart Sustainable Cities. This research paper aims to fill the gap by developing a Key Performance Indicators (KPIs) Matrix to assess the performance of an Afro-Smart City. The KPI Matrix has been proposed in terms of evaluating the contributions of adaptive ICT-based solutions in making African cities smarter and more sustainable.

The Key Performance Indicator (KPI) matrix may be useful for city planners and policymakers to understand the progress of African Smart Sustainable Cities initiatives and evaluate the impact of information and communication technology (ICT).

The paper concludes by proposing that the KPI matrix could be used as criteria for creating Afro-Smart City assessment. Instead of comparing African Smart Cities in a fragmented manner, a ranking based on a shared approach to the Smart City concept implementation could provide more meaningful and insightful comparisons between cities. Such rankings could shed light on which cities perform better in specific dimensions or approaches based on the developed KPIs. The identified Assessing Afro-Smart City methodology can serve as a basis for developing a holistic African Smart City framework assessment model applied to a city case study, and it can also be used as criteria for creating Afro-Smart City rankings, providing meaningful comparisons between African cities.

Author Contributions: Conceptualization, N.M.; methodology, N.M.; validation, H.R., K.M.N. and H.J.O.; formal analysis, N.M.; investigation, N.M.; resources, N.M.; data curation, N.M.; writing—original draft preparation, N.M.; writing—review and editing, N.M., K.M.N., H.R. and H.J.O.; visualization, N.M.; supervision, H.R. and K.M.N.; funding acquisition, H.R. All authors have read and agreed to the published version of the manuscript.

Funding: This research received no external funding.

Conflicts of Interest: The authors declare no conflict of interest.

References

1. Bhushan, B.; Khamparia, A.; Sagayam, K.M.; Sharma, S.K.; Ahad, M.A.; Debnath, N.C. Blockchain for smart cities: A review of architectures, integration trends and future research directions. *Sustain. Cities Soc.* **2020**, *61*, 102360. [CrossRef]
2. Gaur, L.; Agarwal, V.; Chatterjee, P. *Intelligent Decision Support Systems for Smart City Applications*; Scrivener Publishing LLC: Beverly, MA, USA, 2023.
3. Attaran, H.; Kheibari, N.; Bahrepour, D. Toward integrated smart city: A new model for implementation and design challenges. *GeoJournal* **2022**, *87*, 511–526. [CrossRef]
4. Myeong, S.; Park, J.; Lee, M. Research Models and Methodologies on the Smart City: A Systematic Literature Review. *Sustainability* **2022**, *14*, 1687. [CrossRef]
5. Pieroni, A.; Scarpato, N.; Di Nunzio, L.; Fallucchi, F.; Raso, M. Smarter City: Smart energy grid based on Blockchain technology. *Int. J. Adv. Sci. Eng. Inf. Technol.* **2018**, *8*, 298–306. [CrossRef]

6. Orecchini, F.; Santiangeli, A.; Zuccari, F.; Pieroni, A.; Suppa, T. Blockchain Technology in Smart City: A New Opportunity for Smart Environment and Smart Mobility. In *Advances in Intelligent Systems and Computing*; Springer: Cham, Switzerland, 2019.
7. Wu, M.; Yan, B.; Huang, Y.; Sarker, M.N.I. Big Data-Driven Urban Management: Potential for Urban Sustainability. *Land* **2022**, *11*, 680. [CrossRef]
8. Marvi, L.T.; Behzadfar, M.; Shemirani, S.M.M. International Journal of Human Capital in Urban Management Defining the social-sustainable framework for smart cities. *Int. J. Hum. Cap. Urban Manag.* **2023**, *8*, 95–110. [CrossRef]
9. Lai, C.S.; Jia, Y.; Dong, Z.; Wang, D.; Tao, Y.; Lai, Q.H.; Wong, R.T.K.; Zobaa, A.F.; Wu, R.; Lai, L.L. A Review of Technical Standards for Smart Cities. *Clean Technol.* **2020**, *2*, 290–310. [CrossRef]
10. Schiavo, F.T.; Magalhães, C.F.d. Smart Sustainable Cities: The Essentials for Managers' and Leaders' Initiatives within the Complex Context of Differing Definitions and Assessments. *Smart Cities* **2022**, *5*, 994–1024. [CrossRef]
11. Paskaleva, K.; Evans, J.; Watson, K. Co-producing smart cities: A Quadruple Helix approach to assessment. *Eur. Urban Reg. Stud.* **2021**, *28*, 395–412. [CrossRef]
12. Singh, T.; Solanki, A.; Sharma, S.K.; Nayyar, A.; Paul, A. A Decade Review on Smart Cities: Paradigms, Challenges and Opportunities. *IEEE Access* **2022**, *10*, 68319–68364. [CrossRef]
13. Shtebunaev, S.; Gullino, S.; Larkham, P.J. Planning the Smart City with Young People: Teenagers' Perceptions, Values and Visions of Smartness. *Urban Plan.* **2023**, *8*, 57–69. [CrossRef]
14. Myeong, S.; Jung, Y.; Lee, E. A Study on Determinant Factors in Smart City Development: An Analytic Hierarchy Process Analysis. *Sustainability* **2018**, *10*, 2606. [CrossRef]
15. Batty, M.; Axhausen, K.W.; Giannotti, F.; Pozdnoukhov, A.; Bazzani, A.; Wachowicz, M.; Ouzounis, G.; Portugali, Y. Smart cities of the future. *Eur. Phys. J. Spec. Top.* **2012**, *214*, 481–518. [CrossRef]
16. Allam, Z.; Newman, P. Redefining the smart city: Culture, metabolism and governance. *Smart Cities* **2018**, *1*, 4–25. [CrossRef]
17. Slavova, M.; Okwechime, E. African Smart Cities Strategies for Agenda 2063. *Afr. J. Manag.* **2016**, *2*, 210–229. [CrossRef]
18. Allam, Z.; Dhunny, Z.A. On big data, artificial intelligence and smart cities. *Cities* **2019**, *89*, 80–91. [CrossRef]
19. Ismagilova, E.; Hughes, L.; Rana, N.P.; Dwivedi, Y.K. Security, Privacy and Risks within Smart Cities: Literature Review and Development of a Smart City Interaction Framework. *Inf. Syst. Front.* **2020**, *24*, 393–414. [CrossRef]
20. Lazaroiu, C.; Roscia, M. Smart Resilient City and IoT Towards Sustainability of Africa. In Proceedings of the 2018 7th International Conference on Renewable Energy Research and Applications (ICRERA), Paris, France, 14–17 October 2018; pp. 1292–1298.
21. Zhan, J.; Dong, S.; Hu, W. IoE-supported smart logistics network communication with optimization and security. *Sustain. Energy Technol. Assess.* **2022**, *52*, 102052. [CrossRef]
22. Mildon, P. The importance of data privacy in smart cities. *Open Access Government*. February 2023. Available online: https://www.openaccessgovernment.org/importance-data-privacy-smart-cities/152918/ (accessed on 1 December 2023).
23. Fabrègue, B.F.G.; Bogoni, A. Privacy and Security Concerns in the Smart City. *Smart Cities* **2023**, *6*, 586–613. [CrossRef]
24. Kuang, Z.; Chen, C. Research on smart city data encryption and communication efficiency improvement under federated learning framework. *Egypt. Inform. J.* **2023**, *24*, 217–227. [CrossRef]
25. Ma, C. Smart city and cyber-security; technologies used, leading challenges and future recommendations. *Energy Rep.* **2021**, *7*, 7999–8012. [CrossRef]
26. Sookhak, M.; Tang, H.; He, Y.; Yu, F.R. Security and Privacy of Smart Cities: A Survey, Research Issues and Challenges. *IEEE Commun. Surv. Tutor.* **2019**, *21*, 1718–1743. [CrossRef]
27. Romani, G.F.; Pinochet, L.H.C.; Pardim, V.I.; de Souza, C.A. Security as a key factor for the smart city, citizens' trust, and the use of technologies. *Rev. Adm. Publica* **2023**, *57*, e2022-0145. [CrossRef]
28. Rejeb, A.; Rejeb, K.; Abdollahi, A.; Keogh, J.G.; Zailani, S.; Iranmanesh, M. Smart city research: A bibliometric and main path analysis. *J. Data Inf. Manag.* **2022**, *4*, 343–370. [CrossRef]
29. Lau, B.P.L.; Marakkalage, S.H.; Zhou, Y.; Hassan, N.U.; Yuen, C.; Zhang, M.; Tan, U.X. A survey of data fusion in smart city applications. *Inf. Fusion* **2019**, *52*, 357–374. [CrossRef]
30. Barro, P.; Degila, J.; Zennaro, M.; Wamba, S. Towards Smart and Sustainable Future Cities Based on Internet of Things for Developing Countries: What Approach for Africa? *EAI Endorsed Trans. Internet Things* **2018**, *4*, 155481. [CrossRef]
31. Campbell, T. *Beyond Smart Cities: How Cities Network, Learn and Innovate*; Routledge: London, UK, 2013.
32. Harrison, C.; Eckman, B.; Hamilton, R.; Hartswick, P.; Kalagnanam, J.; Paraszczak, J.; Williams, P. Foundations for Smarter Cities. *IBM J. Res. Dev.* **2010**, *54*, 1–16. [CrossRef]
33. Kourtit, K.; Nijkamp, P.; Arribas, D. Smart cities in perspective—A comparative European study by means of self-organizing maps. *Innov. Eur. J. Soc. Sci. Res.* **2012**, *25*, 229–246. [CrossRef]
34. Leitheiser, S.; Follmann, A. The social innovation–(re)politicisation nexus: Unlocking the political in actually existing smart city campaigns? The case of SmartCity Cologne, Germany. *Urban Stud.* **2020**, *57*, 894–915. [CrossRef]
35. Guo, Y.M.; Huang, Z.L.; Guo, J.; Li, H.; Guo, X.R.; Nkeli, M.J. Bibliometric analysis on smart cities research. *Sustainability* **2019**, *11*, 3606. [CrossRef]
36. Ziosi, M.; Hewitt, B.; Juneja, P.; Taddeo, M.; Floridi, L. Smart cities: Reviewing the debate about their ethical implications. *AI Soc.* **2022**. [CrossRef]
37. Visvizi, A.; Lytras, M.D. Smart cities research and debate: What is in there? In *Smart Cities: Issues and Challenges Mapping Political, Social and Economic Risks and Threats*; Elsevier: Amsterdam, The Netherlands, 2019; pp. 1–14.

38. Han, H.; Hawken, S. Introduction: Innovation and identity in next-generation smart cities. *City Cult. Soc.* **2018**, *12*, 1–4. [CrossRef]
39. Giffinger, R.; Fertner, C.; Kramar, H.; Meijers, E. City-ranking of European medium-sized cities. *Cent. Reg. Sci. Vienna UT* **2007**, *9*, 1–12.
40. Caragliu, A.; Del Bo, C.; Nijkamp, P. Smart Cities in Europe. *J. Urban Technol.* **2011**, *18*, 65–82. [CrossRef]
41. Hollands, R.G. Will the real smart city please stand up? Intelligent, progressive or entrepreneurial? *City* **2008**, *12*, 303–320. [CrossRef]
42. Mora, L.; Deakin, M. *Untangling Smart Cities: From Utopian Dreams to Innovation Systems for a Technology-Enabled Urban Sustainability*; Elsevier: Amsterdam, The Netherlands, 2019.
43. Anthopoulos, L.G. Understanding the Smart City Domain: A Literature Review. In *Public Administration and Information Technology*; Springer: Berlin/Heidelberg, Germany, 2015; pp. 9–21.
44. Gracias, J.S.; Parnell, G.S.; Specking, E.; Pohl, E.A.; Buchanan, R. Smart Cities—A Structured Literature Review. *Smart Cities* **2023**, *6*, 1719–1743. [CrossRef]
45. Angelidou, M. Smart cities: A conjuncture of four forces. *Cities* **2015**, *47*, 95–106. [CrossRef]
46. Yigitcanlar, T.; Kamruzzaman, M.; Foth, M.; Sabatini-Marques, J.; da Costa, E.; Ioppolo, G. Can cities become smart without being sustainable? A systematic review of the literature. *Sustain. Cities Soc.* **2019**, *45*, 348–365. [CrossRef]
47. Nam, T.; Pardo, T.A. Conceptualizing smart city with dimensions of technology, people, and institutions. In Proceedings of the ACM International Conference Proceeding Series, College Park, MD, USA, 12–15 June 2011; pp. 282–291.
48. Komninos, N.; Pallot, M.; Schaffers, H. Special Issue on Smart Cities and the Future Internet in Europe. *J. Knowl. Econ.* **2013**, *4*, 119–134. [CrossRef]
49. Duan, W.; Nasiri, R.; Karamizadeh, S. Smart city concepts and dimensions. In Proceedings of the ACM International Conference Proceeding Series, Shanghai, China, 20–23 December 2019; Association for Computing Machinery: New York, NY, USA, 2019; pp. 488–492.
50. Esfandiari, M.; Mousakhani, M. Designing a knowledge management model to observing urban information in smart cities. *Strateg. Manag. Stud. Natl. Def. Stud.* **2022**, *12*, 31–60.
51. Giffinger, R.; Gudrun, H. Smart cities ranking: An effective instrument for the positioning of the cities? *ACE Archit. City Environ.* **2010**, *4*, 7–26. [CrossRef]
52. Pourahmad, A.; Ziari, K.; Hataminejad, H.; Parsa, S. Explanation of Concept and Features of a Smart City. *Bagh-e Nazar* **2018**, *15*, 5–26.
53. Hamamurad, Q.H.; Mat Jusoh, N.; Ujang, U. Concept and Evaluating of Smart City. *J. Adv. Geospat. Sci. Technol.* **2022**, *1*, 92–111.
54. Andreev, D. The "Smart City" concept and its implementation prospects. *E3S Web Conf.* **2023**, *389*, 06014. [CrossRef]
55. Carrillo, F.J. *Knowledge Cities: Approaches, Experiences, and Perspectives*; Routledge: London, UK, 2006; pp. 1–290. [CrossRef]
56. Ergazakis, K.; Metaxiotis, K.; Psarras, J. An emerging pattern of successful knowledge cities' main features. In *Knowledge Cities: Approaches, Experiences, and Perspectives*; Taylor and Francis Inc.: Abingdon, UK, 2006; pp. 3–15.
57. Nordin, R. Creating Knowledge-Based Clusters through Urban Development: A Study of Cyberjaya. Master's Thesis, Rheinische Friedrich-Wilhelms-Universität Bonn, Bonn, Germany, 2012.
58. Smith, A.; Stirling, A.; Berkhout, F. The Governance Of Sustainable Socio-Technical Transitions. *Res. Policy* **2005**, *34*, 1491–1510. [CrossRef]
59. Sassen, S. *Brutality and Complexity in the Global Economy*; Harvard University Press: Cambridge, MA, USA, 2014.
60. Flores, R. Rebel Cities: From the Right to the City to the Urban Revolution. *Community Dev. J.* **2013**, *48*, 339–341. [CrossRef]
61. Zanella, A.; Bui, N.; Castellani, A.; Vangelista, L.; Zorzi, M. Internet of things for smart cities. *IEEE Internet Things J.* **2014**, *1*, 22–32. [CrossRef]
62. Chourabi, H.; Nam, T.; Walker, S.; Gil-Garcia, J.R.; Mellouli, S.; Nahon, K.; Pardo, T.A.; Scholl, H.J. Understanding smart cities: An integrative framework. In Proceedings of the Annual Hawaii International Conference on System Sciences, Maui, HI, USA, 4–7 January 2012; IEEE Computer Society: Washington, DC, USA, 2012; pp. 2289–2297.
63. Khan, M.A.; Siddiqui, M.S.; Rahmani, M.K.I.; Husain, S. Investigation of Big Data Analytics for Sustainable Smart City Development: An Emerging Country. *IEEE Access* **2022**, *10*, 16028–16036. [CrossRef]
64. Angel, S.; Parent, J.; Civco, D.; Blei, A.; Potere, D. The Dimensions of Global Urban Expansion: Estimates and Projections for All Countries, 2000–2050. *Prog. Plan.* **2011**, *75*, 53–107. [CrossRef]
65. Kitchin, R. Thinking critically about and researching algorithms. *Inf. Commun. Soc.* **2017**, *20*, 14–29. [CrossRef]
66. Baraniewicz-Kotasińska, S. Smart city. Four approaches to the concept of understanding. *Urban Res. Pract.* **2020**. [CrossRef]
67. Safiullin, A.; Krasnyuk, L.; Kapelyuk, Z. Integration of Industry 4.0 technologies for "smart cities" development. In *IOP Conference Series: Materials Science and Engineering, Saint-Petersburg, Russia, 21–22 November 2018*; Institute of Physics Publishing: Bristol, UK, 2019.
68. Albino, V.; Berardi, U.; Dangelico, R.M. Smart cities: Definitions, dimensions, performance, and initiatives. *J. Urban Technol.* **2015**, *22*, 3–21. [CrossRef]
69. Deakin, M.; Alwaer, H. From intelligent to smart cities. *Intell. Build. Int.* **2011**, *3*, 140–152. [CrossRef]
70. Pandiyan, P.; Saravanan, S.; Usha, K.; Kannadasan, R.; Alsharif, M.H.; Kim, M.-K. Technological advancements toward smart energy management in smart cities. *Energy Rep.* **2023**, *10*, 648–677. [CrossRef]

71. Linders, D. From E-Government to We-Government: Defining a Typology for Citizen Coproduction in the Age of Social Media. *Gov. Inf. Q.* **2012**, *29*, 446–454. [CrossRef]
72. Silva, B.N.; Khan, M.; Han, K. Towards sustainable smart cities: A review of trends, architectures, components, and open challenges in smart cities. *Sustain. Cities Soc.* **2018**, *38*, 697–713. [CrossRef]
73. Cretu, L.G. Smart Cities Design using Event-driven Paradigm and Semantic Web. *Inform. Econ.* **2012**, *16*, 57–67.
74. Barrionuevo, J.M.; Berrone, P.; Ricart Costa, J.E. Smart Cities, Sustainable Progress: Opportunities for Urban Development. *IESE Insight* **2012**, *14*, 50–57. [CrossRef]
75. Washburn, D.; Sindhu, U. Helping CIOs Understand "Smart City" Initiatives. *Growth* **2009**, *17*, 1–17.
76. Bibri, S.E. The IoT for smart sustainable cities of the future: An analytical framework for sensor-based big data applications for environmental sustainability. *Sustain. Cities Soc.* **2018**, *38*, 230–253. [CrossRef]
77. Sikora- Fernandez, D.; Stawasz, D. The Concept of Smart City in the Theory and Practice of Urban Development Management. *Rom. J. Reg. Sci.* **2016**, *10*, 86–99.
78. Cohen, B. The top 10 smart cities on the planet. *Fast Co.* **2012**, *11*, 181.
79. Joss, S.; Sengers, F.; Schraven, D.; Caprotti, F.; Dayot, Y. The Smart City as Global Discourse: Storylines and Critical Junctures across 27 Cities. *J. Urban Technol.* **2019**, *26*, 3–34. [CrossRef]
80. Colombo, M.; Hurle, S.; Portmann, E.; Schafer, E. A Framework for a Crowdsourced Creation of Smart City Wheels. In Proceedings of the 2020 7th International Conference on eDemocracy and eGovernment, ICEDEG 2020, Buenos Aires, Argentina, 22–24 April 2020; Institute of Electrical and Electronics Engineers Inc.: New York, NY, USA, 2020; pp. 305–308.
81. Lara, A.P.; Da Costa, E.M.; Furlani, T.Z.; Yigitcanlar, T. Smartness that matters: Towards a comprehensive and human-centred characterisation of smart cities. *J. Open Innov. Technol. Mark. Complex.* **2016**, *2*, 1–13. [CrossRef]
82. Bibri, S.E.; Krogstie, J. Generating a Vision for Smart Sustainable Cities of the Future: A Scholarly Backcasting Approach. *Eur. J. Futures Res.* **2019**, *7*. [CrossRef]
83. Caragliu, A.; Del Bo, C.F. Smart innovative cities: The impact of Smart City policies on urban innovation. *Technol. Forecast. Soc. Chang.* **2019**, *142*, 373–383. [CrossRef]
84. Wilhelm, R.; Ruhlandt, S. The governance of smart cities: A systematic literature review. *Cities* **2018**, *81*, 1–23. [CrossRef]
85. Garg, S.; Mittal, S.K.; Sharma, S. ScienceDirect Role of E-Trainings in Building Smart Cities Role of E-Trainings in Building Smart Cities. *Procedia Comput. Sci.* **2017**, *111*, 24–30. [CrossRef]
86. Maciej Błaszak, M.B.; Artur Fojud, A.F. Trzy wymiary użytecznego miasta. *Człowiek I Społeczeństwo* **2016**, 219–231. [CrossRef]
87. Manville, C.; Europe, R.; Millard, J.; Technological Institute, D.; Liebe, A.; Massink, R. Directorate General for Internal Policies Policy Department A: Economic and Scientific Policy. In *Mapping Smart Cities in the EU Study*; European Parliamentary Research Service: Brussels, Belgium, 2014.
88. Giffinger, R. *Smart Cities Ranking of European Medium-Sized Cities*; Centre of Regional Science: Vienna, Austria, 2007.
89. Bibri, S.E.; Krogstie, J. The emerging data–driven Smart City and its innovative applied solutions for sustainability: The cases of London and Barcelona. *Energy Inform.* **2020**, *3*, 5. [CrossRef]
90. Yigitcanlar, T.; Kamruzzaman, M.; Buys, L.; Ioppolo, G.; Sabatini-Marques, J.; da Costa, E.M.; Yun, J.H.J. Understanding 'smart cities': Intertwining development drivers with desired outcomes in a multidimensional framework. *Cities* **2018**, *81*, 145–160. [CrossRef]
91. Yigitcanlar, T.; Cugurullo, F. The sustainability of artificial intelligence: An urbanistic viewpoint from the lens of smart and sustainable cities. *Sustainability* **2020**, *12*, 8548. [CrossRef]
92. Gomez, C.; Chessa, S.; Fleury, A.; Roussos, G.; Preuveneers, D. Internet of Things for enabling smart environments: A technology-centric perspective. *J. Ambient Intell. Smart Environ.* **2019**, *11*, 23–43. [CrossRef]
93. Voordijk, H.; Dorrestijn, S. Smart city technologies and figures of technical mediation. *Urban Res. Pract.* **2021**, *14*, 1–26. [CrossRef]
94. O'Dwyer, E.; Pan, I.; Acha, S.; Shah, N. Smart energy systems for sustainable smart cities: Current developments, trends and future directions. *Appl. Energy* **2019**, *237*, 581–597. [CrossRef]
95. Serrano, W. Digital systems in smart city and infrastructure: Digital as a service. *Smart Cities* **2018**, *1*, 134–154. [CrossRef]
96. Bibri, S.E.; Alexandre, A.; Sharifi, A.; Krogstie, J. Environmentally sustainable smart cities and their converging AI, IoT, and big data technologies and solutions: An integrated approach to an extensive literature review. *Energy Inform.* **2023**, *6*, 9. [CrossRef]
97. Kumar, H.; Singh, M.K.; Gupta, M.P.; Madaan, J. Moving towards smart cities: Solutions that lead to the Smart City Transformation Framework. *Technol. Forecast. Soc. Chang.* **2020**, *153*, 119281. [CrossRef]
98. Sugandha; Freestone, R.; Favaro, P. The social sustainability of smart cities: A conceptual framework. *City Cult. Soc.* **2022**, *29*, 100460. [CrossRef]
99. Mgwebi, S. Africa, Be Smart about New Cities. Available online: https://itweb.africa/content/5yONP7ErJPWMXWrb (accessed on 20 December 2023).
100. Rwanda Information Society Authority—RISA: Africa Smart Cities Investment Summit Kicked Off with the Launch of Smart City Hub. Available online: https://www.risa.gov.rw/news-detail/africa-smart-cities-investment-summit-kicked-off-with-the-launch-of-smart-city-hub (accessed on 23 December 2023).
101. Echendu, A.J.; Okafor, P.C.C. Smart city technology: A potential solution to Africa's growing population and rapid urbanization? *Dev. Stud. Res.* **2021**, *8*, 82–93. [CrossRef]

102. Bernard Binagwaho Smart City: Towards More Sustainable Cities in Africa ?—Tactis. Available online: https://www.tactis.fr/smart-city-towards-more-sustainable-cities-in-africa/?lang=en (accessed on 23 December 2023).
103. Mbassi, J.P.E. Smart Cities—The African story. *New Mag. UCLG-Afr.* **2016**, *5*.
104. IMD Smart City Index Report 2023. Available online: https://imd.cld.bz/IMD-Smart-City-Index-Report-2023/109/ (accessed on 23 December 2023).
105. Aurigi, A.; Odendaal, N. From "Smart in the Box" to "Smart in the City": Rethinking the Socially Sustainable Smart City in Context. *J. Urban Technol.* **2021**, *28*, 55–70. [CrossRef]
106. Sharifi, A. A typology of smart city assessment tools and indicator sets. *Sustain. Cities Soc.* **2020**, *53*, 101936. [CrossRef]
107. Neumann, H.-M.; Iglár, B.; Huovila, A.; Airaksinen, M. Recommendations for a smart city index. *CITYkeys-Smart City Perform. Meas. Framew.* **2016**. [CrossRef]
108. Caird, S.P.; Hallett, S.H. Towards evaluation design for smart city development. *J. Urban Des.* **2019**, *24*, 188–209. [CrossRef]
109. Airaksinen, M.; Huovila, A. *CITYkeys Indicators for Smart City Projects and Smart Cities*; The European Commission: Helsinki, Finland, 2017. [CrossRef]
110. Ericsson Ltd. *Networked Society City Index 2014*; Ericsson Ltd.: Stockholm, Sweden, 2012.
111. Shirazi, M.R.; Keivani, R. The triad of social sustainability: Defining and measuring social sustainability of urban neighbourhoods. *Urban Res. Pract.* **2019**, *12*, 448–471. [CrossRef]
112. International Telecommunication Union. *ITU-T FG-SSC Telecommunication Standardization Sector of ITU, Key Performance Indicators Definitions for Smart Sustainable Cities*; Focus Group Technical Report; ITU-T: Geneva, Switzerland, 2015.

applied sciences

MDPI

Article

Structural Health Monitoring for Prefabricated Building Envelope under Stress Tests

Laura Vandi [1], Maria Teresa Calcagni [2], Francesco Belletti [1], Giuseppe Pandarese [2], Milena Martarelli [2], Gian Marco Revel [2], Vincent Docter [3] and Alessandro Pracucci [1,4,*]

[1] Focchi S.p.A., 47824 Poggio Torriana, Italy; l.vandi@focchi.it (L.V.); f.belletti@focchi.it (F.B.)
[2] Department of Industrial Engineering and Mathematical Sciences, Università Politecnica delle Marche, 60131 Ancona, Italy; m.t.calcagni@pm.univpm.it (M.T.C.); g.pandarese@staff.univpm.it (G.P.); m.martarelli@staff.univpm.it (M.M.); gm.revel@staff.univpm.it (G.M.R.)
[3] PhotonFirst International, 1812 SC Alkmaar, The Netherlands; vincent.docter@photonfirst.com
[4] Levery S.r.l. Società Benefit, 47814 Bellaria Igea Marina, Italy
* Correspondence: a.pracucci@focchi.it or a.pracucci@levery.it

Abstract: This paper details the comprehensive activities conducted in a laboratory setting to assess the structural health monitoring (SHM) of prefabricated building envelopes. Integrating sensors into building components like curtain wall facades poses challenges but offers opportunities for monitoring structural health, requiring compliance with regulatory standards. The research investigates the possibility of defining a kit of conventional and multi-parameter sensors integrated within the building envelope to monitor its behavior during the performance test conducted. The kit of sensors also includes Fiber Optic Sensors for effectively monitoring building envelope behavior and structural integrity. In this context, the European project InComEss (H2020-GA862597) aims to define a stand-alone solution for SHM using Piezoelectric Energy Harvesting Systems (PE-EHS) for façade monitoring through FBG/FOS system. After analyzing the main façade structural stress, a series of FBGs, accelerometers, and force washers were integrated within a 1:1 scale façade prototype and tested in a laboratory following the test sequence parameters required by the curtain wall standard EN 13830. The data collected were analyzed with the aim of monitoring the façade behavior before and after the tests. The results show that the façade's performance passed the assessing test criteria without reporting any damages. In addition, the outcomes demonstrated the effectiveness of the defined kit of multi-parameter sensors for the building envelope's SHM.

Keywords: building envelope; façade; structural health monitoring; fiber optic sensor; fiber brag grating; IoT

1. Introduction

The building environment's increasing reliance on smart technologies opens up opportunities for a deeper understanding and consequent service operations of building performance within the so-called Intelligent Building [1,2]. The relevance of data-driven approaches within the smart built environment is consolidating dynamic control for energy optimization and occupant well-being within building automation [3,4].

This field involves the continuous and real-time assessment of a structure's structural integrity, allowing for the early detection of potential issues such as deterioration, damage, or unforeseen structural changes. In the dynamic and ever-changing urban environment, where buildings are subjected to diverse loads, environmental conditions, and unforeseen events, structural health monitoring (SHM) serves as a proactive measure to mitigate risks and enhance the resilience of structures [5]. By leveraging advanced sensing technologies, data analytics, and monitoring systems, SHM empowers engineers and stakeholders with valuable insights, enabling timely interventions, maintenance, and, ultimately, the optimization of overall performance and safety.

In recent years, both civil construction and the building sector have explored the utilization of sensing technologies for SHM embedded in their structure [6,7]; in particular, focus has been given to the SHM of bridges [8,9]. However, the integration of sensors into non-load-bearing structures, such as a curtain wall façade, represents an increasing topic that needs to be further investigated. Indeed, this component faces the risk of sudden breakage, vulnerability to severe weather conditions, and unexpected collisions. These variables present significant obstacles to maintaining the longevity and dependability of these architectural elements. Recent research focused on the SHM of glass curtain walls of a super-high-rise building by integrating thermocouples in a real building to monitor the temperature distribution [10] with the aim of avoiding thermal shock or monitoring the wind effect on a single-layer cable net with modern curtain walls [11]. SHM sensors in no-load-bearing building components are also useful for analyzing the actual behavior of buildings and the relationship between the building response and the damage caused by earthquakes [12]. Moreover, other solutions for façade SHM could be investigated without directly integrating sensors within the façade; indeed, the study conducted by Brell-Cokcan and Co. investigated the possibility of developing an automated system that scans the façade surface through a projector light with the aim of detecting defections and alerting the automized components' replacing system positioned at the façade perimeters [13].

Investigating sensing solutions embedded in prefabricated façades becomes a perspective opportunity, as demonstrated by the exploration of smart applications for other building applications [14,15]. Indeed, as the interface between the internal and external environments, building envelopes serve as invaluable indicators of a building's structural health. For these reasons, compliance with regulatory standards, such as EN 13830 [16] and CWCT's 'Standard for Systemized Building Envelopes'-Part 8 Testing-Section 8.12.2 [17], is compulsory. Therefore, performing a mock-up validation is required to ensure the façade's functional efficacy and safety. The validation process, outlined in detail in Table A1 and reported in Appendix A of these standards, provides a systematic sequence of steps for conducting the validation phase. Each parameter critical to the façade's performance and safety undergoes evaluation within this framework. Structural integrity and functioning are scrutinized to guarantee compliance with established benchmarks. Indeed, by adhering to these standards, the industry demonstrates a commitment to upholding safety and operational standards in curtain wall façade construction ensuring safety of curtain wall façades, mitigating risks, and safeguarding the well-being of occupants. The standards also defined assessing criteria for each test paraments for verified the façade behavior during each test. These criteria, reported in Table A1 of Appendix A, determine the test pass or failure limit values.

Within this scientific and technological context focused on façade applications, this research investigates the possibility of defining a kit of conventional and multi-parameter sensors integrated within the building envelope to monitor its behavior during performance tests. During the research, the kit is validated with the aim of demonstrating its effectiveness and replicability for structural health monitoring of building envelopes. Therefore, an analysis of conventional sensors was conducted to determine the most suitable ones. Fiber Optic Sensors (FOS) featuring Fiber Bragg Gratings (FBGs) [18,19] are examined with the goal of demonstrating potentialities and limitations for their application in building envelopes. Together with FBG sensors, the possibility of integrating conventional devices was investigated; therefore, vibration sensors and force washers have been explored to monitor structural conditions [20]. Numerous research works have not only considered the use of accelerometers to detect the impact of weathering and external forces on buildings for structural analysis, to monitor the behavior in infrastructure [21] and historical building behavior [22], or to quantify the seismic damage in building structure [23], but also for the identification and verification of possible structural damage because of the stresses imposed on the structure under investigation by performing a modal analysis before and after the events [24]. In fact, modal parameters can be used to evaluate the safety condition of a structure and assess structural health based on modal parameters [25]. Deng et al. [9]

demonstrated that temperature is an important factor that affects the natural frequency of the bridge, and it is not negligible. Therefore, the temperature must be controlled in order to not consider a variation in frequency only due to temperature change as a damage indicator. Strain gauge-based force washers have also been considered as commercial sensors to monitor bolt irregularities [26]. This research focuses on monitoring the structure behavior during the performance of the standard compliance tests for wall curtain façades; the temperature variation results in a few degrees between days since they were accelerated tests, and thus they were performed in short duration even if the temperature is monitored during the performance of the test.

In line with these research and technological scenarios, the EU-funded InComEss project [27] aims to define a stand-alone solution for SHM using Piezoelectric Energy Harvesting Systems (PE-EHS) for façade monitoring [28], and this paper presents the structural health monitoring sensing solutions within the InComEss architectural system [29]. The architectural design of the InComEss system is geared towards harnessing the potential of Piezoelectric Energy Harvesting Systems (PE-EHS) to supply energy to a fiber optic sensor, MonadGator. This interrogator reads data from FBG and communicates with a Power Conditioner Circuit (PCC) seamlessly integrated with a Printed Circuit Board (PCB) powered by a locally implemented Supercapacitor (SC). The connection between the PCC/PCB, an Internet of Things (IoT) gateway, and a cloud-based IoT monitoring platform facilitates data collection and analysis. The overarching objective is to showcase the practicality of Wireless Sensor Networks (WSNs) for structural health monitoring (SHM). Based on the InComEss system architecture conceptualization, the PE-EHS and FBG sensors are designed to demonstrate their integrability within a prefabricated façade. Figure 1 shows the InComEss architecture underlining the components embedded within the façade and the digital components.

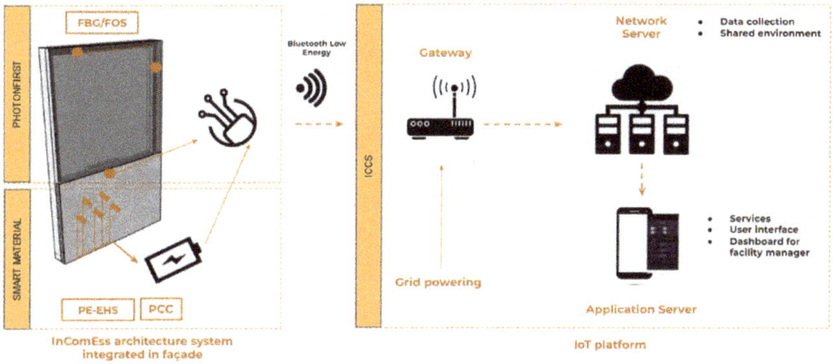

Figure 1. Scheme of InComEss system architecture with components integrated into façade and communication to IoT platform.

The first considerations presented are related to physical components such as the FBG sensors. This product represents an application for the façade demo case, and during the InComEss project, it will demonstrate the potential of its applicability for SHM in the building envelope market. An optical fiber consists of a core and cladding. The cladding of the fiber has a typical diameter of 125 μm while the core has a diameter of typically ~10 μm. The core has a higher refractive index than the cladding, such that the light is captured in the core. A Fiber Bragg Grating is a periodic modulation of the refractive index in the fiber core. A representation can be seen in Figure 2.

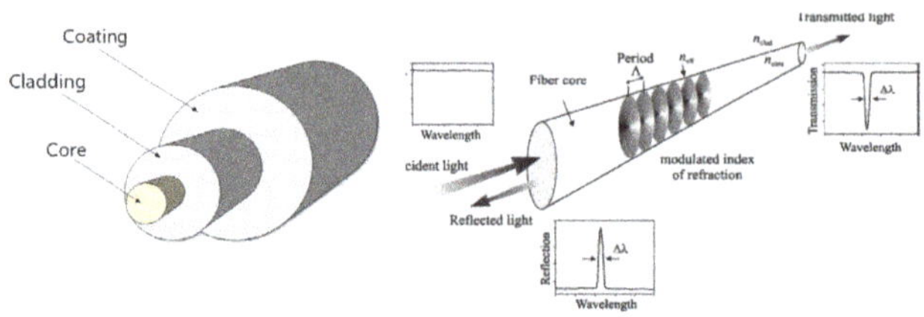

Figure 2. Schematic representation of a fiber and Fiber Bragg Grating inscribed into the core of an optical fiber.

The period of the refractive index modulation determines the wavelength of light that is reflected. Due to temperature or strain, the period of the modulation, and thus, its reflected wavelength, will change. Due to this principle, FBG sensors serve dual roles as both strain and temperature detectors. When mounted within a hollow capillary, an FBG sensor exclusively detects temperature changes, disregarding strain. The sensor's wavelength indicates temperature variations through a fixed, linear conversion based on glass material properties. The MonadGator interrogator is the measurement device developed to measure the shift in Bragg wavelength of the FBG sensors. The system uses a light source that emits a broadband wavelength spectrum through the fiber toward the FBG sensors. The reflected wavelength is measured by the MonadGator and converted to a strain or temperature. This MonadGator is designed to operate with low power consumption. Figure 3 reports the InComEss architecture, showing how the energy generated powered the MonadGator for the FBG sensors.

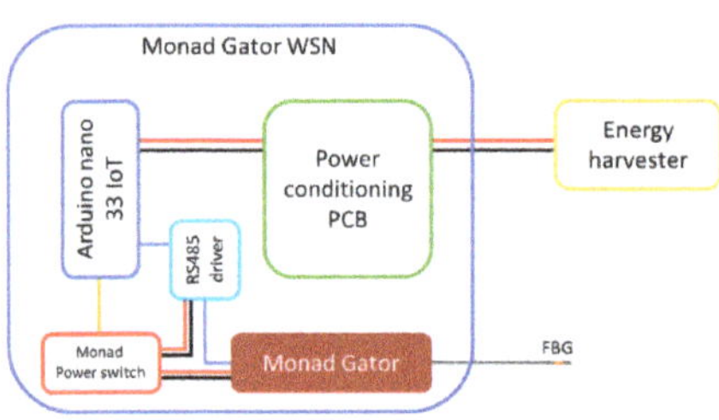

Figure 3. Scheme of InComEss architecture.

An Arduino nano is used to rapidly start up the MonadGator and execute a measurement of the FBG wavelength. The broadband light source is turned on for a period of 70 ms. In that time span, the FBG sensor is read, and the data from the photodiodes are acquired by the Arduino. These data are transferred using a Bluetooth Low Energy (BLE) connection to the IOT gateway. The Arduino is powered by a power conditioning circuit, which is charged with the energy that is harvested from the energy harvester connected to this circuit. A total measurement cycle, including wireless data transfer, takes less than 150 ms and uses only 0.08 J of energy per measurement.

In this scenario, the research aims to investigate the possibility of integrating a kit of sensors for structural health monitoring (SHM) into the curtain wall facade to monitor its behavior post-installation. Therefore, this paper outlines the results achieved by the SHM sensors during a performance test, wherein a full-scale facade prototype was subjected to various weather and mechanical conditions in a controlled laboratory environment. The objective was to monitor the facade's behavior under stress from weather conditions and accidental impacts. The paper's objectives are to investigate and provide original contributions in the scientific field to support analyses useful for establishing monitoring solutions integrated into building envelopes using low-consumption sensing technology. Additionally, it aims to explore potential applications within facade components to define the stress application of Fiber Bragg Grating (FBG) and Fiber Optic Sensors (FOS).

Furthermore, the study seeks to validate the facade's integrability and the effectiveness of the multi-parameter sensor kit for facade SHM, along with the measurement method employed.

This paper is structured as follows: Section 2 presents the methods used for the identification of SHM's sensors to be integrated into the façade, along with the tests conducted for their validation and the design and integration activities in the façade; Section 3 presents the outcomes of the laboratory test activities, displaying collected data and graphical representations for SHM sensing technologies; Section 4 highlights the relevant insights that SHM integrated into the façade can provide and identifies gaps for its application; Section 5 summarizes the main achievements related to the paper's goal of analyzing SHM opportunities within stand-alone solutions as the one proposed by InComEss architecture and suggests further investigations needed for market adoption.

2. Materials and Methods

This section offers an overview of the developed methodology, including the stages involved, and delineates the materials utilized for conducting the research activities presented in this paper, specifically focusing on the selection of SHM sensors and the testing set-up.

2.1. Method

The methods are focused on the stages deployed for the SHM's sensing solutions adopted within the InComEss system architecture with the following development stages.

2.1.1. Analysis of Façade Structural Stress

The analysis of structuring components to be monitored is investigated with the objective of examining the stress conditions and the range of operations, consequently deriving the sensing solution needed. The comprehensive evaluation of these components to be monitored addresses the final structural health monitoring sensing technologies to be included in an SHM configuration.

Given their exposure to various environmental stressors, such as wind, temperature fluctuations, moisture, accidental impact, and seismic activity, façades are susceptible to degradation and structural issues over time. Consequently, defining the components and parameters relevant to façade monitoring becomes important for effective maintenance, timely detection of defects, solution replicability, and proactive mitigation of risks. Indeed, during the research, a list of possible façade defects to be monitored and relative parameters based on structural analysis and an internal key stakeholders' interview was defined. Table 1 presents the façade parameters, their description, and the range of value acceptable for the façade structural health together with the selected monitoring sensors.

Therefore, Table 1 summarizes the most relevant parameters in terms of façade safety, scheduled for monitoring, accompanied by a selection of the suitability of employing FBG technology and conventional sensors for each defined objective. Alongside FBG, conventional sensors will also be applied in the façade system with the aim of defining a complete kit of sensors for building envelope SHM. Therefore, the kit of sensors selected for the façade monitoring, based on possible façade defects, are described as follows:

- FBGs—Previous applications demonstrated the effectiveness of these technologies in the building and infrastructure sectors, such as bridges [29–31], concrete, wood [32], and steel structures, where strain and temperature have so far been the dominating measurands of interest. Utilizing conventional FBG sensors for SHM in building façades offers significant advantages over electrical strain gauges. Unlike electrical strain gauges, which require multiple wires per sensor point, leading to scalability issues, FBG sensors allow for multiplexing multiple sensors into a single fiber. This feature makes FBG sensors highly suitable for environments requiring numerous sensor points, such as the building envelope sector. The main advantages are their lightweight characteristics, single-ended connections, water and corrosion resistance, and absence of electric current in the measurement array, making them suitable for embedding within or attaching to a structure [33], which make them suitable for embedding within or attaching to a structure. In addition, research on integrating FBG cables into the façade system involved examining critical factors to determine the optimal configuration. Key considerations included the FBG cable's bending radius, path within façade components towards the monitoring system for efficiency and aesthetics, sensor position, cable dimensions, lengths, thickness, and material properties for fixing method (e.g., glue, silicone). In this research, the use of FBG for temperature monitoring is integrated into the InComEss architecture. Given that the MonadGator can monitor only one FBG at a time, it was connected to the FBG for temperature, which requires punctual monitoring. The integration of FGB for temperature in the glazed façade design should prevent thermal shock episodes in vision and spandrel parts caused by solar radiation and light converting to IR. Installing temperature sensors can prevent damage and provide data for future thermal shock designs. Conversely, conventional FBGs for strain were selected to investigate further applications and to compare their results to those of conventional sensors such as accelerometers. In this case, FBGs for strain are connected to a conventional switchgator, allowing for measuring multiple FBGs and powered by commercial energy.

- Accelerometers—Vibrational sensors have been considered for the monitoring of a structure combined with IoT systems [20]. The vibration signals encompass parameters such as displacement, velocity, and acceleration. Accelerometers prove to be efficient instruments for detecting vibrations. Several studies have been conducted with accelerometers, starting from the structure analysis to identify the weathering and excitation agents affecting the buildings [21–23,34,35] Positioned on curtain wall façades, they detect dynamic forces such as wind or seismic activity, providing insights into structural health by converting vibrations into electrical signals for frequency, amplitude, and duration analysis [35]. Indeed, in this research, accelerometers were used with the following two aims: The first was to evaluate the structural status of the curtain wall façade, performing a modal analysis—one before the tests of compliance and the other one at the end of the tests. The second was the monitoring of the structure during the compliance test to understand the phenomenology of the event affecting the structure. For the façade integration, the sensors were selected based on their dimensions, water-tightness characteristics, and functionalities. Accelerometers are connected to commercial data collectors.
- Force washers—The curtain wall façade relies on precisely torqued screws within a bracket system, which is essential for accurate positioning and optimal performance [36,37]. Therefore, an issue that could affect the life cycle of a structure could be bolt anomalies. Dominika Ziaja et al. analyzed and proposed a procedure for fault detection, as well as for the determination of their location and type, using IoT [26]. This is why, for our project, a non-destructive system, the strain gauge-based force washers—useful for the measurements of the bolt's load—were installed and strategically placed within these brackets to facilitate accurate measurements, providing insights into façade response under diverse conditions. These washers record structural deformations, aiding in ongoing assessment of façade integrity and performance, enhancing structural resilience and efficiency.

In this research, the information exploited by the force transducers was combined with the information from the accelerometers.

Table 1. List of defined most relevant parameters to be monitored in terms of safety.

Label	Description	Range	Selected Sensor
SHM—thermal shock—temperature	Glazed façade needs to be designed to prevent episodes of thermal shock in vision and spandrel parts linked to the thermal variation due to solar radiation and light conversion into IR on glass and other façade components. The installation of a temperature sensor allows for collecting a set of data to support further thermal shock design.	Temperature of service with a daily excursion between 5 °C and 120 °C Temperature of service with a daily excursion between −5 °C and 80 °C	FBG
SHM—mechanical stress—Strain and vibration due to accidental impacts	Façade could be stressed by extraordinary events as accidental and not possible to foreseen impacts, which could compromise the structural façade's behavior.	Accidental impact—range between 6 J (1.224 mm height) and 343 J (700 mm height)	FBG Accelerometer Force washers
SHM—mechanical stress—Strain and vibration due to dynamic pressure	Façade is stressed by ordinary (wind load) and extraordinary (seismic load) loads and its mechanical behavior needs to be monitored to guarantee the structural integrity	Pressure ranges from −3000 Pa to +2625 Pa	FBG Accelerometer Force washers
SHM—mechanical stress—torque screw	Façade hangs on a brackets system fixed to the load-bearing structure through the utilization of screws specifically torqued to guarantee the right placement and performance of the façade. This torque needs to be preserved during building service to avoid a loss of façade performance and safety issues. The monitoring of torque (e.g., brackets on steelwork) can directly intervene to preserve the façade serviceability and detect defects as distance façade/slab and façade/edge (±5 mm), stack joint distance (±5 mm).	60–120 N	Force washers

2.1.2. Sensing Technologies Integration in Façade

Structural sensing technologies are designed for integration in façade, studying their applicability during off-site façade manufacturing, as well as during installation.

The integration of these sensors underwent a thorough examination involving an analysis of several key parameters. Dimensions were examined to determine compatibility and correctness for its application. Further consideration was given to the sensors' positioning, factoring in orientation and functionality to optimize data collection effectiveness. This included assessing the efficacy of data obtained, co-locating commercial sensors near Fiber Bragg Grating (FBG) sensors for direct comparative analysis, and situating sensors to monitor critical components of the facade, such as the center of transom width, the center of mullion height, and the center of the ventilated cavity. Additionally, the routing and length of cables connecting the sensors were analyzed to ensure efficient signal transmission and practical installation. The fixing system methodology was examined to guarantee stability and reliability in data acquisition. This comprehensive analysis contributes to an informed understanding of the integrated sensor system's performance within the context of the monitored facade. In Section 2.3, the final sensor integration designed for the testing activities is reported.

2.1.3. Testing Activities and Outcomes Analysis

The laboratory tests for structural health monitoring sensing technologies were analyzed based on multiple testing conditions to obtain a complete comprehension of their behavior under weather conditions. The FBG/FOS and conventional sensors were analyzed to wrap up potentialities and limitations within real environment applications.

2.2. Materials

The materials adopted for the research activities were:

- InComEss components—the research was based on component development within the InComEss project [27] concerning the FOS MonadGator. The FOS MonadGator is based on a low-energy consumption solution for data gathering and transmission of the InComEss system architecture. The FOS MonadGator needs 3.3 V to be powered.
- On-market sensing solutions for SHM:
    a. FBG for FOS to monitor stress conditions of façade. The selection of FBG is based on well-established sensors on the market. The InComEss project did not expect to investigate and develop FBG solutions. The FBG used for this research was a fiber type SM1250B, with a length of 9 mm, a reflectivity of 45%, and 3 dB bandwidth of 0.16 nm. The FBG coating is a fiber polyamide.
- Sensors in the field of structural health monitoring to monitor stress conditions of façade to collect data not collectible from FBG, such as the following:
    a. Accelerometers—On the external side of the façade prototypes, a triaxial accelerometer PCB 354C03 is integrated with IP66 characteristics and a compact dimension of $27 \times 21 \times 11$ mm. It is feasible to be integrated into the façade ventilated cavity. On the internal side, the selected accelerometers are monoaxial sensors model PCB 352C33 with no IP66 characteristics and dimensions of $18 \times 11 \times 17$ mm;
    b. Strain gauge-based force washers—the selected model was the K-KMR+200K-01M5-Q with connectors D-SUB HD 15 polis, and the monitoring system was MX840, 8 channels. This washer has a nominal force of 200 kN based on the standard force applied to façade stresses. Force washers are connected to commercial data collectors.
- Monitoring systems for data collection:
    a. FBG interrogator MonadGator, in which the wavelength range is 1575–1582 nm; noise level: $\sigma < 1$ pm; sampling speed: 2 kHz; FBG's per channel: 1; and the number of channels available is 1;
    b. FBG interrogator switchgator, in which wavelength range is 1516–1583 nm; noise level: $\sigma < 1$ pm; sampling speed: 19.23 kHz; FBG per channel is 8; and the available number of channels is 8;
    c. NI 9234 for accelerometer monitoring;
    d. For the strain gauge-based force, it was the QUANTUM X MX840B monitoring system.
- Prefabricated façade—a unitized façade system for multifactional façades is selected to improve solutions in the same product development.
- Method statement for testing activities conducted in a laboratory environment to validate façade system modules based on EN 13830:2015 [16] and EN 14019:2016 [38] for curtain walling—impact resistance—performance requirements, as referenced in Appendix A. These tests involved the use of a fan positioned at 600 mm and rain-simulating nozzles targeting façade joints, delivering a continuous flow rate of 2 L/min·sqm at 400 mm from the façade. To replicate rain in a controlled laboratory setting, a system employing nozzles at typical joint locations on the façade was used. Two types of rain tests were conducted:
    a. Dynamic Rain Test: this test simulates wind gusts, generating pulsating pressure variations every 3 s, fluctuating between 750 Pa and 250 Pa to mimic the dynamic nature of wind-induced pressure changes.
    b. Static Rain Test: this test aimed to assess behavior under constant rain conditions, maintaining a constant pressure of 600 Pa throughout the test duration.

Table A1 in Appendix A also includes the fail/pass criteria for each parameter.

2.3. Experimental Design

This chapter describes the design of the testing activities, focusing on the sensor's integration within the façade prototype and the test sequence. The façade prototype is composed of 6 prefabricated modules with dimensions of 1400 × 3450 mm. Indeed, the realized 1:1 scale façade mock-up total dimension is 6.90 × 4.50 mt, and the technology used is aluminum frames and double skin glaze. The façade modules were assembled off-site while the sensors were integrated on-site once the façade was installed in the laboratory. The sensor integration was deeply investigated, considering the sensor characteristics and façade components or orientations. In particular, the position of the sensor within the façade was defined according to the following:

- The test standards (EN 13830 [16]), which defined the façade positions and parameters to assess the façade behavior. Indeed, the positions correspond to the most stressful façade position in the central part of aluminum profiles and glazed panels.
- Façade axis. Both FBG and accelerometers were positioned with different orientations with the aim of testing the façade behavior in all directions. As a reference for sensor monitoring, the different axes were considered, such as *X*—left, right; *Y*—up, down; and *Z*—inside, outside.
- Avoiding impact position test—the standard EN 14019 and EN 12600 [39] determine several façade positions where to conduct the impact test; the sensors were not positioned there to avoid damage.

Table 2 shows the list of sensors integrated within the façade prototype for the testing activities.

Table 2. Summit of integration sensor within the façade and key information.

Quantity	Type of Sensor	Measured Physical Quantity	Data Collection System	Application on the Façade	Axis	Sensor Model	Sample Rate [Hz]	Acquisition System [bit]	Sensitivity
6	Monoaxial accelerometers	Acceleration	NI 9234 [40]	Internal position	X, Y, Z	PCB 352C33	5000	24	100 mV/g
1	Triaxial accelerometers	Acceleration	NI 9234	External position	X, Y, Z	PCB 354C03	5000	24	1000 mV/g
8	Forced washers	Bolt Tightness	QUANTUM X MX840B [41]	Façade bolt brackets		K-KMR+200K	30	24	2 mV/V Fnom = 200 kN
6	FBGs (channel 2)	Strain	Switchgator	Internal on aluminum profile	X	0.15 nm FBG	1000	18	-
3	FBGs (channel 3)	Strain	Switchgator	External on glass surface	Z	0.15 nm FBG	1000–19,230 (for impact tests)	18	-
1	FBGs (channel 4)	Temperature	Monadgator	External on glass surface	-	0.15 nm FBG	<1	14	-
6	FBGs (channel 5)	Strain	Switchgator	Internal on aluminum profile	Z	0.15 nm FBG	1000–10,230 (for impact tests)	18	-

The FBG sensors are period refractive index modulation in fiber and are completely solid state. The sensitivity is, therefore, not limited by the FBG sensor but by the measurement system connected to it. At the maximum sample rate of 19.2 kHz, the switchgator interrogator can measure strain changes of 1 microstrain or, when measuring temperature, a temperature change of 0.1 °C. By resampling to lower speeds, this sensitivity is improved by the square root of the resampling factor.

2.3.1. FBG Integration within the Façade

According to the previous outcomes, the fiber optic cables were applied to the glass surface and the aluminum profiles in specific positions. The FBG sensors monitor the direction of its application; thus, two fiber cables were installed to monitor the x and y direction, while one monitored the *z* direction. Figure 4 shows the cable path and the FBG positions within the façade.

- Channel 2—central glazed panel for temperature;
- Channel 3—slightly central glazed panel for strain (to avoid the impact test);
- Channel 5—aluminum central profiles (x direction);
- Channel 6—aluminum central profiles (z direction).

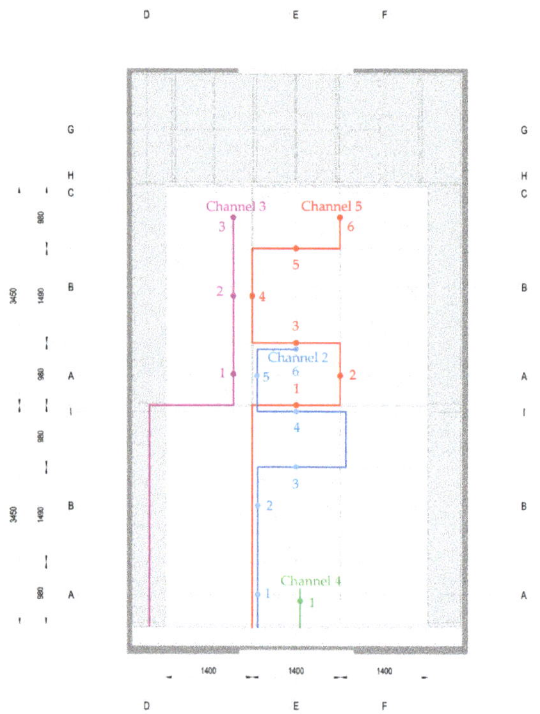

Figure 4. Scheme of FBGs channel position.

2.3.2. Accelerometers Integration within the Façade

Figure 5 shows the accelerometer positions, including both external (colored in blue) and internal ones (colored in green). The internal monoaxial accelerometers were located on the aluminum profile corresponding to the FBGs sensors. Indeed, 6 accelerometers were placed beside 6 FBG sensors (the sequence follows channel 5, FBG positioning). Both sensors were positioned in the central part of aluminum mullions and transoms. Considering the middle upper façade modules, there were 6 positions to monitor (upper mullions and transom, openable vent mullions and transom, and lower mullion and transom). For each position, 1 accelerometer and 1 FBG were placed.

2.3.3. Force Washers' Integration within the Façade

Figure 6 shows the No. 8 force washer positions located in the façade brackets bolt to monitor the behavior of each bracket position.

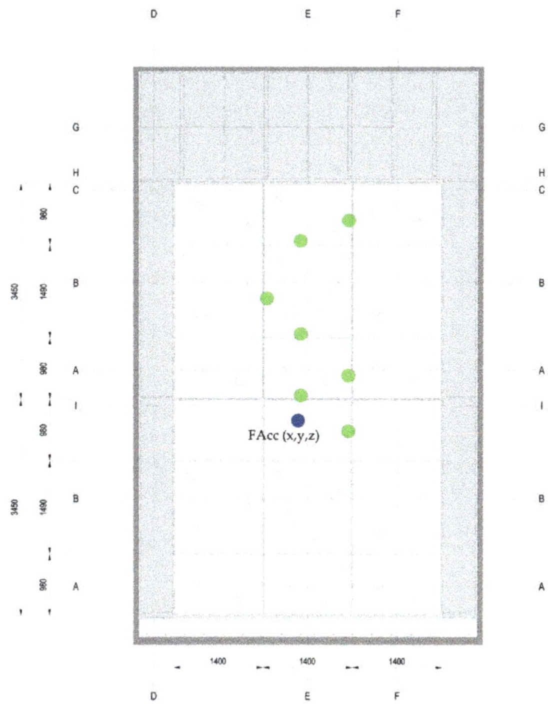

Figure 5. Accelerometer positions—façade external view.

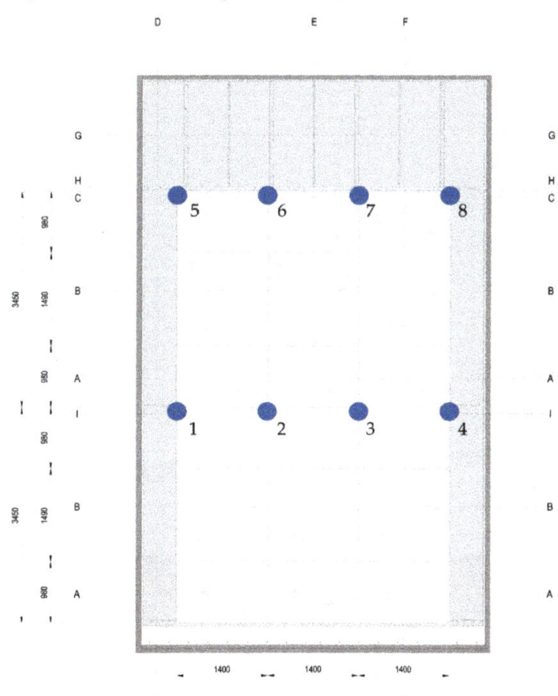

Figure 6. Force washers' positions.

2.4. Testing Methods

The test aims to monitor the façade behavior and validate the integration of sensors, both regarding the InComEss architecture and the defined kit of conventional sensors. Therefore, the outcomes of previous phases were validated in a controlled lab environment by collecting data from integrated sensors for structural health monitoring (SHM) analysis. The defined set of SHM sensors integrated into a façade prototype was tested under controlled weather conditions following the curtain wall standard EN 13830. During the test, a defined tests sequence was conducted to stress the façade's behavior. The series included air infiltration and exfiltration (+600/−600 Pa), wind pressure and depressure (+1750/2000 Pa for serviceability and +2625/−3000 Pa for safety), and rain in static and dynamic regimes (+600 Pa static test and dynamic water penetration test with a fan pulsing every 3 s from 750 Pa to 250 Pa). The tests were conducted with a fan at 600 mm and with a water source simulating rain in correspondence of façade joints with a flow rate of 2 L/min·sqm and at 400 mm from the façade.

Additionally, mechanical tests such as building movement (horizontal and vertical) and impacts at different times and forces were performed. The full test lasted three days and was performed following a determinate sequence of parameters, as reported in Appendix A. During each test, the façade was constantly monitored and checked in real-time while the sensor's data were collected, considering the test duration and parameters, to facilitate the analysis of the results. The full test sequence is reported in Appendix A (Table A1).

2.5. Sample Preparation

The sample preparation involved manufacturing and installation of façade modules and then integrated sensors within the façade. Figure 7 shows the off-site process of façade manufacturing, and Figure 8 shows the on-site installation.

Figure 7. Façade manufacturing process.

Figure 8. Façade installation process.

For FBG, 4 different cables were installed within the façade: 2 in the glass panel and 2 on the aluminum profiles. Indeed, two cables were fixed on the glass surface through epoxy resin DP125 for bonding on glass (Figure 9), while DP410 was used for bonding on aluminum profiles (Figure 10).

Figure 9. FBGs sensors applied on the glass surface for temperature monitoring.

Figure 10. FBGs application on the aluminum profile for strain monitoring.

Accelerometers were installed on the curtain wall façade to acquire vibrational phenomena. A triaxial accelerometer was integrated in the external façade side, while 6 monoaxial accelerometers were positioned in the internal façade's side along the central module of the façade as shown in Figures 11 and 12. The triaxial sensor, positioned in the ventilated cavity (thick 67 mm), is fully integrated and not visible from the outside due to its compact dimension (27 × 21 × 11 mm), while the internal monoaxial sensors are visible and not completely integrated into the façade aluminum profiles due to their dimensions.

Figure 11. Accelerometer applied on aluminum transom.

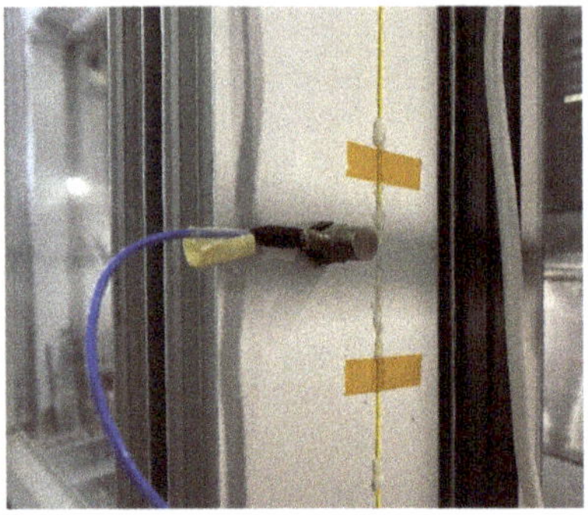

Figure 12. Accelerometer applied on aluminum mullion.

Strain gauge-based force washers were installed on the façade bolts. There were 8 sensors installed, one on each façade bracket. The integration of force washers is shown in Figures 13 and 14.

Figure 13. Zoom on the force washer installed on the bracket's bolt.

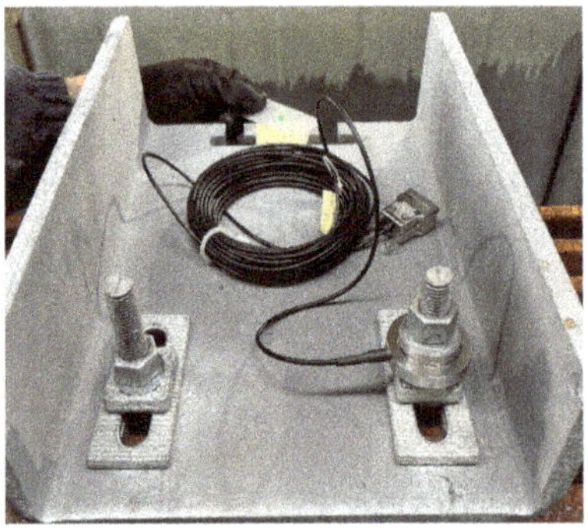

Figure 14. Façade bracket with force washer installation.

3. Results and Test Analysis

In the following paragraphs, the results obtained during the test and the relative analysis are reported.

3.1. Modal Analysis (Pre/Post)

Two modal analyses were conducted to assess any changes in structure that might have occurred due to the conformance tests explained in Appendix A. The procedure for the evaluation of the structure was applied to the central module of the wall façade, and the geometry was configured based on the dimension of the module of interest: 54 nodes were created, as shown in Figure 15.

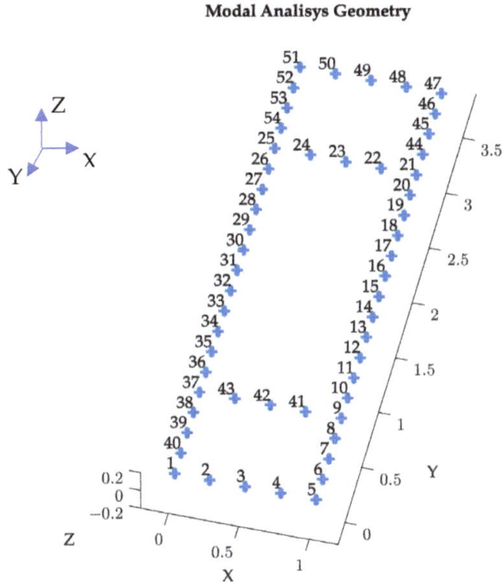

Figure 15. Modal analysis geometry consists of 54 nodes representing the modulus of the facade used for performing modal analysis.

An instrumental hammer was used for the excitation of the surface, and the dynamic response at the nodes reported in Figure 15 was measured with monoaxial accelerometers installed for measuring in Z, moved all over the nodes. The force exerted by the hammer and the acceleration were collected by means of a SIEMENS acquisition system (SCADAS). At each node, the frequency response function was calculated as a response over force ratio. The sum of all the FRFs acquired is reported in Figure 16.

A modal analysis was performed on the FRF dataset collected to estimate the dynamic parameters of the structure under test (natural frequencies, damping loss factors, and mode shapes). A comparison between those parameters was estimated in the modal tests performed before and after the stress tests were performed. To compare mode shapes, the Modal Assurance Criterion (MAC) was used; to compare the variation in the natural frequencies, the Natural Frequency Difference (NFD) was estimated. The MAC and the SFD are reported in Figures 17 and 18, respectively.

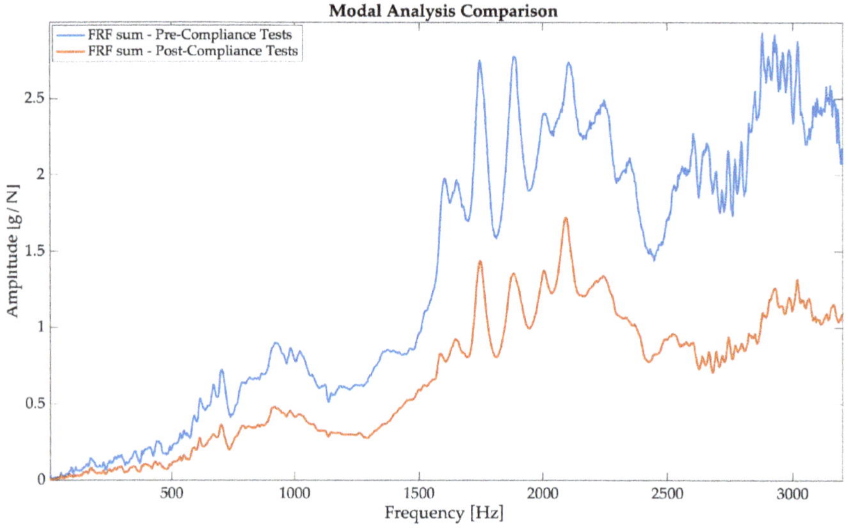

Figure 16. FRF sum comparison between pre- and post-compliance tests.

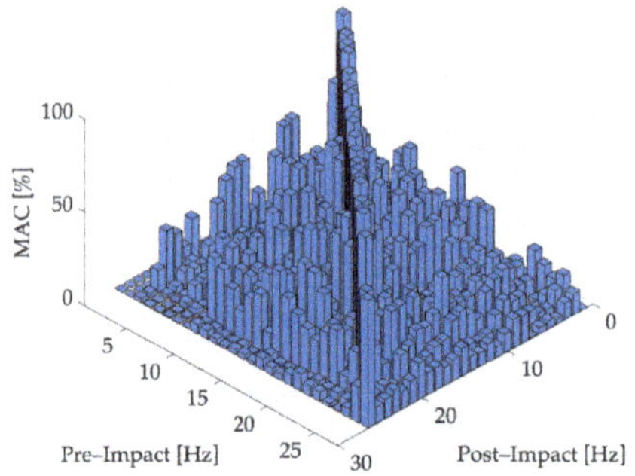

Figure 17. MAC (Modal Assurance Criterion) between the two performed modal analysis.

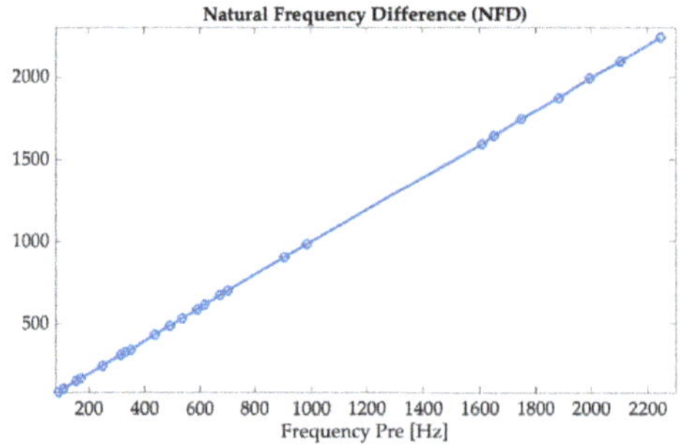

Figure 18. NFD (Natural Frequency Difference) between the two performed modal analysis.

Two modes were selected to have an example of the structure movements at selected frequencies, as reported in Figure 19.

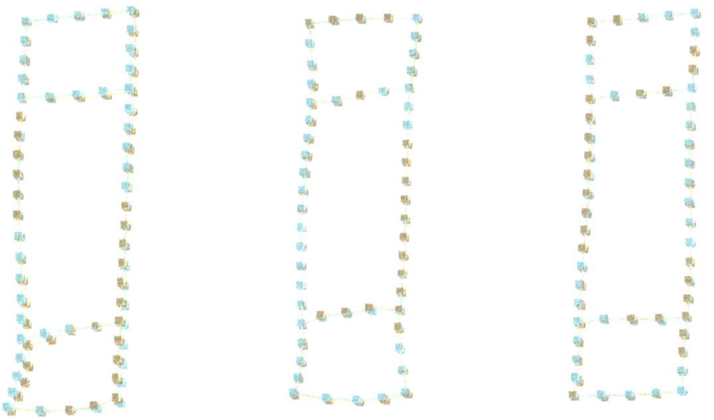

Figure 19. Three modal shapes comparison between pre-compliance test and post-compliance test: Yellow markers represent the mode shape estimated before the compliance test and the cyan ones represent the mode shape estimated after the compliance test.

The natural frequencies of all the mode shapes identified with the modal analysis performed on the FRF dataset acquired on the structure before and after the compliance tests are reported in Table 3.

Table 3. Modal frequencies pre-compliance test (on the left), post-compliance test (on the right).

0–400 Hz		400–1000 Hz		1000–3000 Hz	
Pre-Test	**Post-Test**	**Pre-Test**	**Post-Test**	**Pre-Test**	**Post-Test**
46.9	45.6	436.24	435.11	1608.31	1593.50
90.2	82.5	489.64	489.37	1650.3	1644.79
106.6	103.7	534.23	533.5	1746.9	1746.79
154.2	152.14	589.13	587.5	1882.86	1874.07
169.3	168.14	615.8	614.87	1993.2	1995.06
313.9	312.19	672.64	672.34	2105.16	2096.57
329.5	328.9	701.09	700.95	2246.17	2241.69
351.22	343.04	984.54	982.11	-	-

3.2. Accelerometers

In this section, the most relevant results are reported. As mentioned above, the accelerometer results could be comparable, with no significant differences among them. Therefore, in this article, the reported results consist of the data collected and analyzed from the external accelerometer's 'z' direction called FAcc—Z.

The data analysis reveals that relevant and significant results for the façade were obtained. The SHM can be performed considering the fixed installation of accelerometers on a wall curtain façade, as explained in Section 2.5.

At first, the analysis of the signal over time was performed to understand the global response of the accelerometers subjected to the compliance tests. In fact, a Root Mean Square (RMS) analysis of the signal over time was performed for each test day, as reported in Figure 19. In the following figure (Figure 20), the error bar is shown on RMS data for each day. This value was calculated by applying the accelerometer sensitivity value of $\pm 10\%$ to the accelerometer signals. The error band is outlined by a dashed line surrounding the accelerometer signal curve.

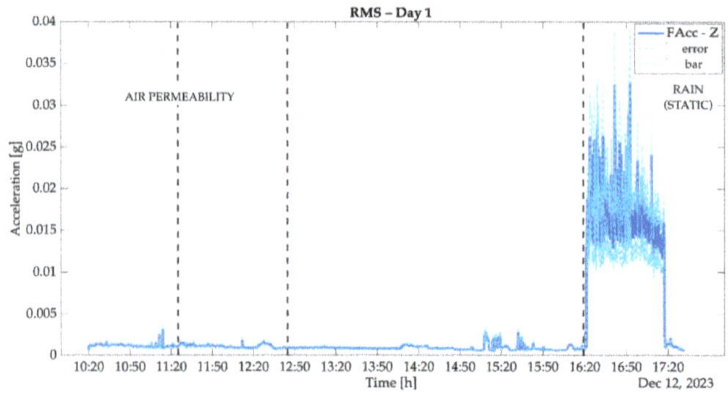

(a)

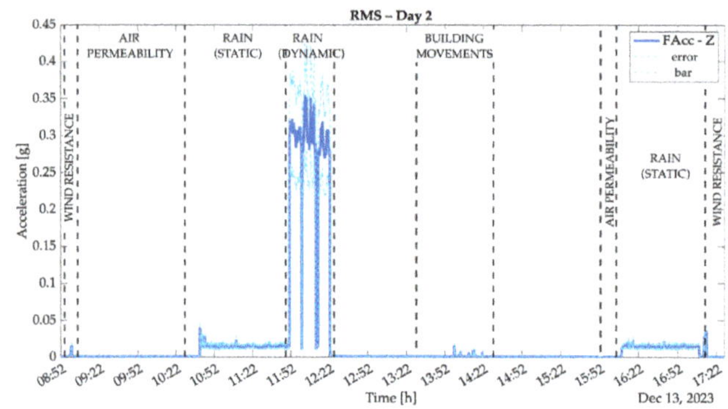

(b)

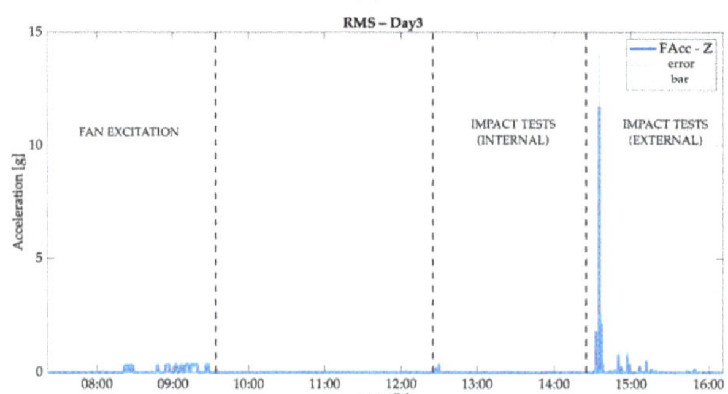

(c)

Figure 20. Signal RMS analysis. In sequence, top figure: RMS of day 1 (**a**); middle figure: RMS of day 2 (**b**); bottom figure: RMS of day 3 (**c**).

The structure excitation was evaluated for each day; the graph corresponding to each excitation event is reported in time on the left and in frequency on the right.

The tests performed during day 1 were acquired and are shown in Figure 21. During this day, air permeability and rain were performed. The power spectrum estimation function was used to reduce the signal noise and better evaluate information. Each 10 s time history was divided into 0.8 s chunks with a 50% overlap to obtain 25 averages.

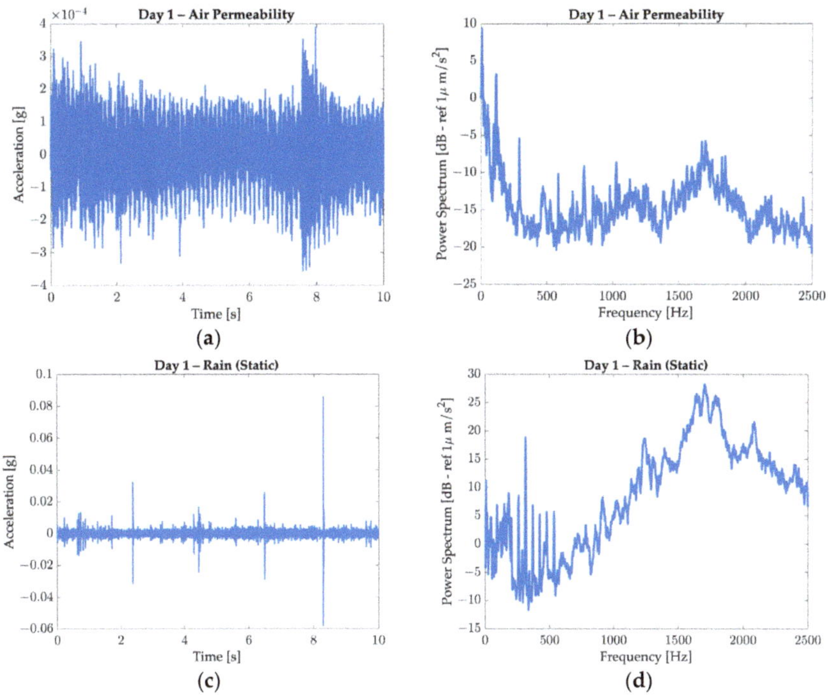

Figure 21. Day 1 tests: respectively, story time and power spectrum for each test: air permeability acceleration/time (**a**); air permeability power spectrum /frequency (**b**); rain (static) acceleration/time (**c**); rain (static) power spectrum /frequency (**d**).

The tests performed during day 2 were acquired and are shown in Figure 22, are presented following the chronological order of the appearance of events. During this day, wind (repeated), air permeability (repeated), rain (repeated), and building movements were performed. To reduce the signal–noise ratio, the power spectra of the signals were calculated using a frequency-domain averaging procedure. The 10 s time histories were divided into 0.8 s chunks with a 50% overlap to obtain 25 averages.

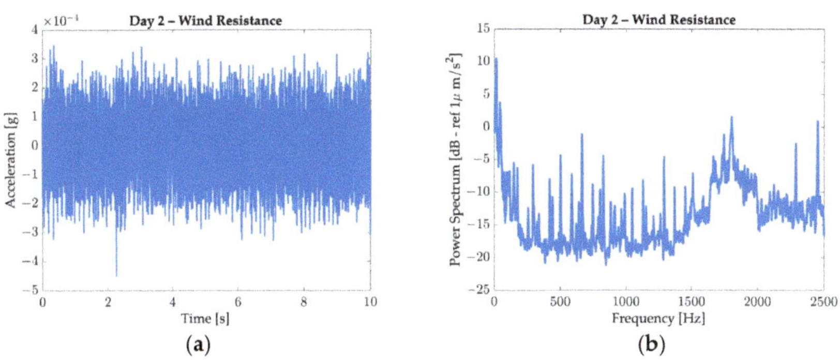

Figure 22. *Cont.*

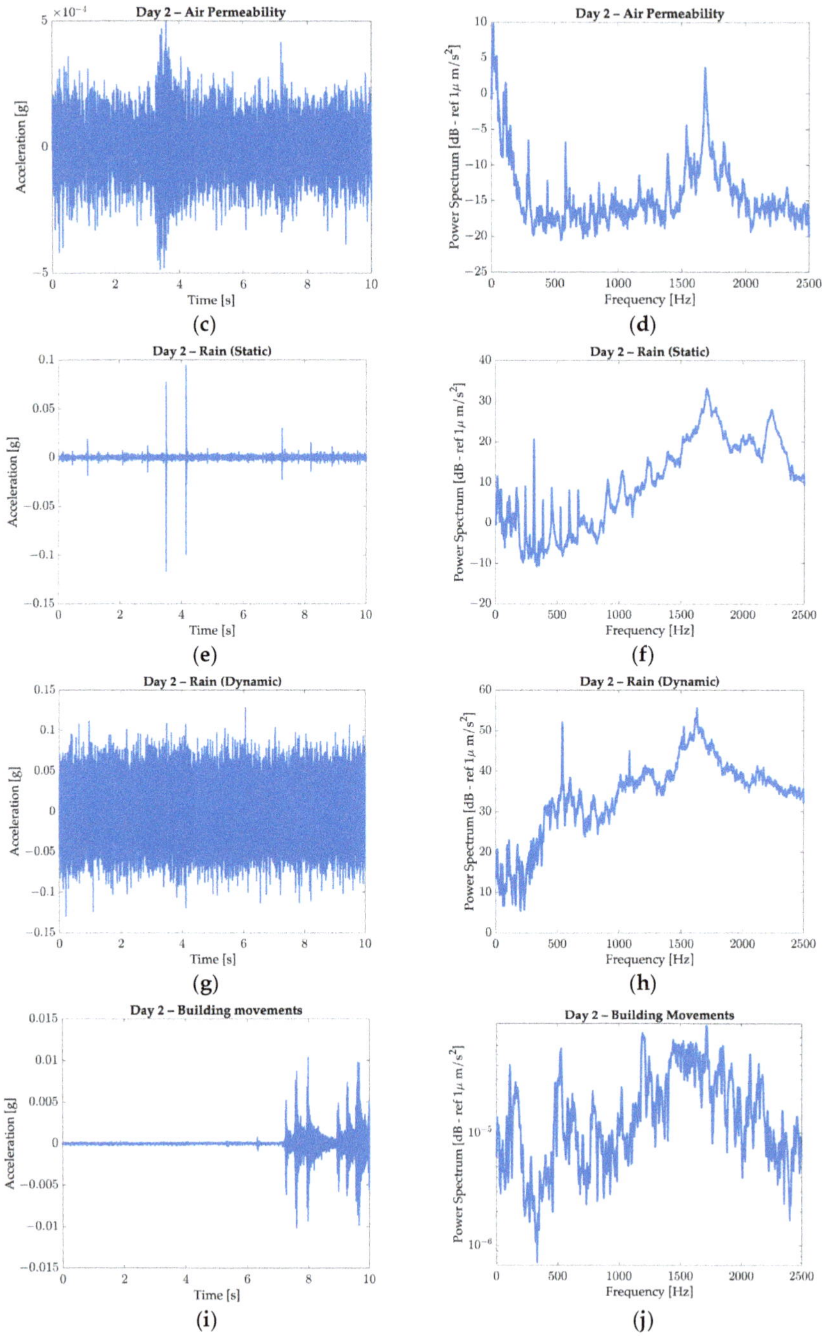

Figure 22. *Cont.*

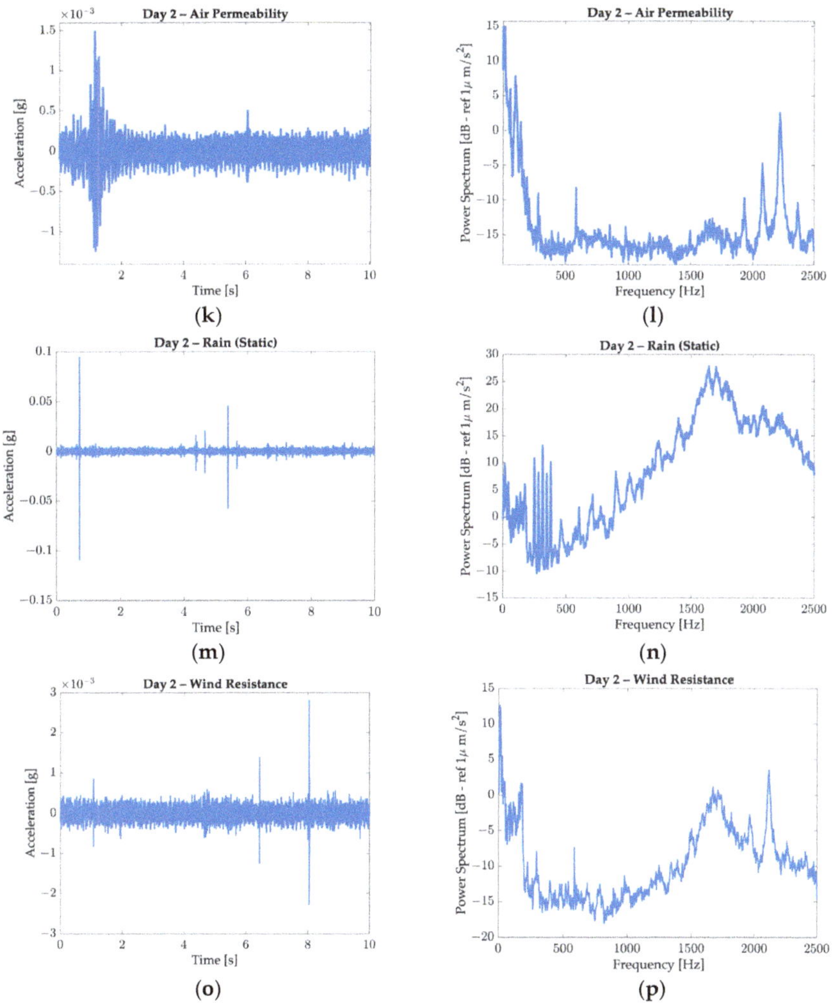

Figure 22. Day 2 tests: respectively, story time and power spectrum for each test: wind resistance acceleration/time (**a**); wind resistance power spectrum/frequency (**b**); air permeability acceleration/time (**c**); air permeability power spectral power spectrum/frequency (**d**); rain (static) acceleration/time (**e**); rain (static) power spectrum/frequency (**f**); rain (dynamic) acceleration/time (**g**); rain (dynamic) power spectrum/frequency (**h**); building movements acceleration/time (**i**); building movements power spectrum/frequency (**j**); air permeability acceleration/time (**k**); air permeability power spectrum/frequency (**l**); rain (static) acceleration/time (**m**); rain (static) power spectrum/frequency (**n**); wind resistance acceleration/time (**o**); wind resistance power spectrum/frequency (**p**).

The time history registered during the building movement test Figure 22i is evidenced by several transients. The power spectrum of each transient signal was calculated and then averaged to improve the signal-to-noise ratio.

The tests performed during day 3 were acquired and are shown in Figure 23 and are presented following the chronological order of the appearance of events. During this day, fan excitation and impact tests were performed. The power spectrum estimation function was used to reduce the signal noise and better evaluate information. Each 10 s time history was divided into 0.8 s chunks with a 50% overlap to obtain 25 averages.

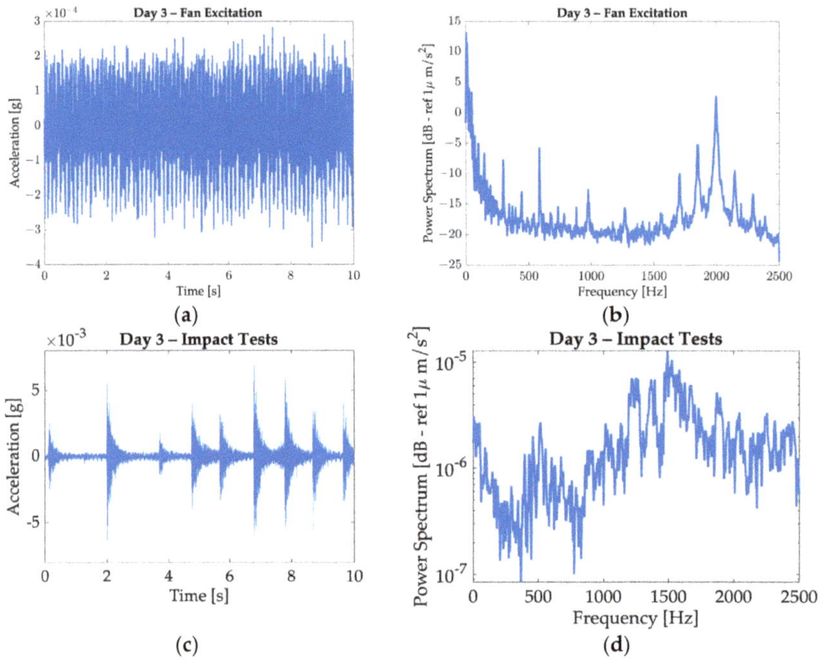

Figure 23. Day 3 tests: respectively, story time and power spectrum for each test: fan excitation acceleration/time (**a**); fan excitation power spectrum/frequency (**b**); impact test acceleration/time (**c**); impact test power spectrum/frequency (**d**).

The time histories registered during the impact tests show transient signals typical of impact excitation. The power spectra of these time histories were computed by selecting only three transient signals and averaging their power spectra to improve the signal-to-noise ratio.

The physical phenomenon is demonstrated with the example shown in Figure 24, in which two comparable tests—rain (static)—between two different days (day 1 and day 2) are presented. By comparing the power spectrum acquired during the first rain test (blue line) and the one acquired during the second rain test (red line), it is evident that the resonances excited by the simulated environmental dynamic load (represented by the impacting rain) are shifted towards lower values. This allows the inference that the structure underwent a stiffness decrease, which can be due to structural yielding.

After conducting the tests, it became evident that the structure is stressed during dynamic tests, as monitored by the RMS level of the accelerometers. These dynamic tests allow for the extraction of a frequency response function, revealing various modes of vibration and any changes in the structural response during testing. The responses of the accelerometers can be correlated with those of the other examined sensors.

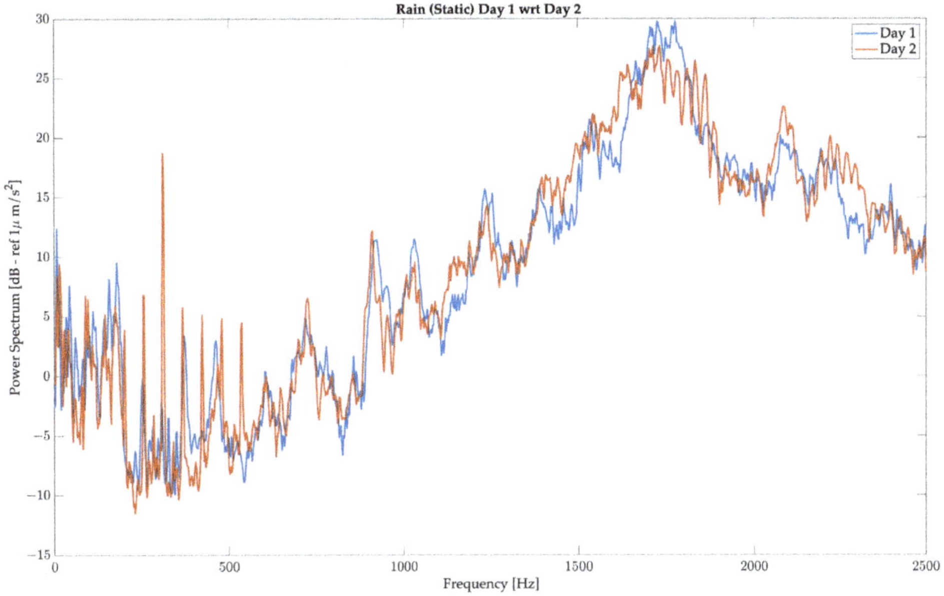

Figure 24. Power spectrum comparison between rain (static) test of day 1 and test of day 2.

3.3. Force Washers

This paragraph reports structure behavior during the stress test analysis of the forces. The results obtained are evident in the graphs below (Figure 25). The error bar is delineated by a light blue dashed line surrounding the force washer's signal curves. This band was computed considering the uncertainty related to the sensitivity deviation declared by the washer constructor, which is 1.5%.

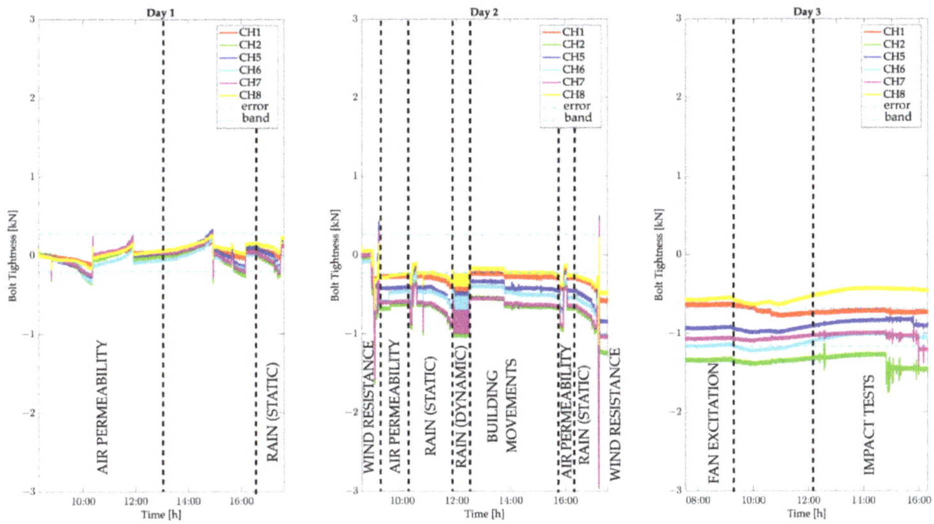

(a) (b) (c)

Figure 25. Strain gauge-based force washer data collected during the 3 days: day 1 (**a**); day 2 (**b**); day 3 (**c**).

By observing the trend of the bolt's tightness during day 1, it is possible to notice that the tightness at the end of the test is equal to the tightness measured before starting the test; this means that the structure did not experience any modification. On the contrary, during the second day, a negative trend of the bolt's tightness is evident. Table 4 reports the value of the tightness measured by all the washers at the beginning and at the end of the tests performed during day 2 and their difference. It is possible to see that all the washers experience a decrease in tightness, particularly those connected to channels 2 and 7 (see Figure 4) for the force channels position. An important drop in the bolt tightness is visible immediately after the first test (wind resistance). As for day 1 and day 3, the tightness of the bolts remains almost constant.

Table 4. Load values of the load cells during day 2.

Channel	Tightness at the Beginning of the Test on Day 2	Tightness at the End of the Test on Day 2	Delta
1	0.4	−0.6	0.64
2	−0.06	−1.3	1.2
5	0.02	−0.8	0.82
6	−0.08	−1.02	0.94
7	−0.004	−1.03	1.026
8	0.04	−0.4	0.44

The force washers allow for the visualization of trends in the load curves of individual tests, enabling the identification of instances when the grab brackets experience stress in shear or compression, as well as any loosening of the anchor bolts. Continuous monitoring of the clamping forces will be highly beneficial for the health monitoring of the entire structure and for triggering alerts when the clamping forces fall below the specified limits.

3.4. FBG Sensors

FBGs for temperature—Figure 26 displays the results obtained by the FBG sensor for temperature (Channel 4) associated with the InComEss architecture. A noticeable increase in temperature was recorded during the first day of testing, indicating the correct procedure for data collection. The markable timestamps are as follows: 11.50 h—temperature decrease recorded during the air exfiltration test (−600 Pa) (i); 12.10 h—temperature increase of 0.5 degrees can be observed during the test since a heat lamp was placed near the temperature sensor (ii); 16.30 h—rapid temperature increases from 16 to 18 degrees due to the rain test likely due to the water temperature (iii).

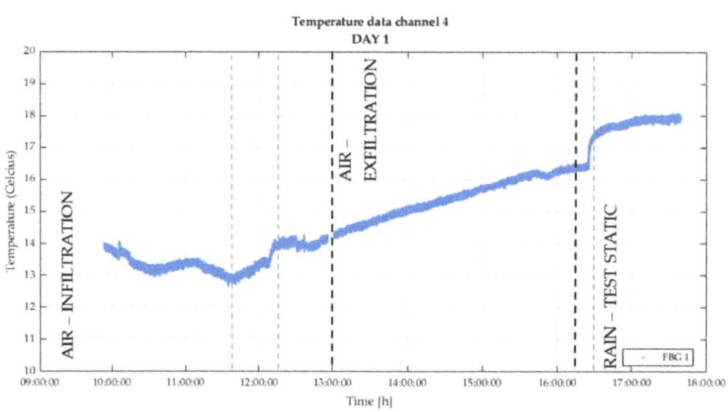

Figure 26. Day 1 FBG for temperature—Channel 4.

The temperature on this test day was much more stable throughout the second day, as recorded by the temperature sensor reported in Figure 27.

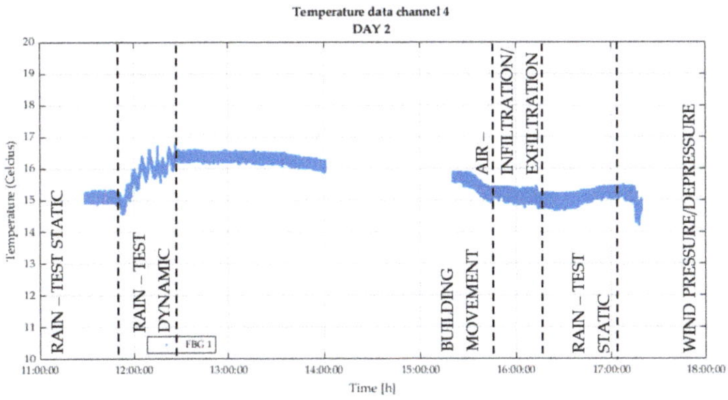

Figure 27. Day 2 FBG for temperature—Channel 4.

In this paragraph, the FBG for strain's test results are reported. In Figure 28, the strain results obtained by the cable applied on the glass panel during the first day can be observed. The test started with air infiltration followed by exfiltration, repeated twice. The glass surface reached negative values of strain during the increasing pressure and positive ones during the negative pressure. Indeed, the FBG sensors in the first case were compressed, while in the second, they were expanded.

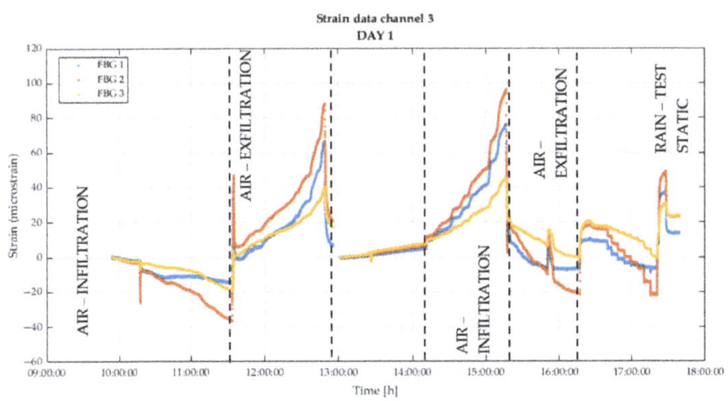

Figure 28. Day 1 FBG for strain—Channel 3.

The strain levels on channel 5 (the cable on the aluminum profile) are minor because the sensors are applied on the aluminum profile where the façade strain due to deformation is contained (Figure 29).

The highest strain levels were measured at the end of the second day in channel 3 (z) when the wind pressure and depressure were executed; indeed, in less than 5″, the façade was stressed from 0 to +2625 Pa, and from 0 to −3000 Pa (Figures 30 and 31).

Figure 31 displays the result obtained for channel 5 (x, y) during day 2.

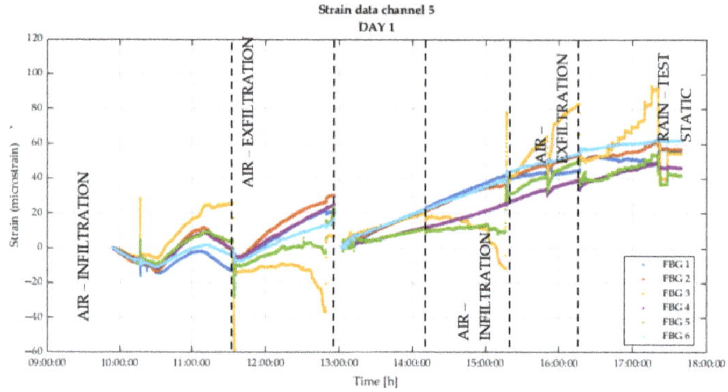

Figure 29. Day 1 FBG for strain—Channel 5.

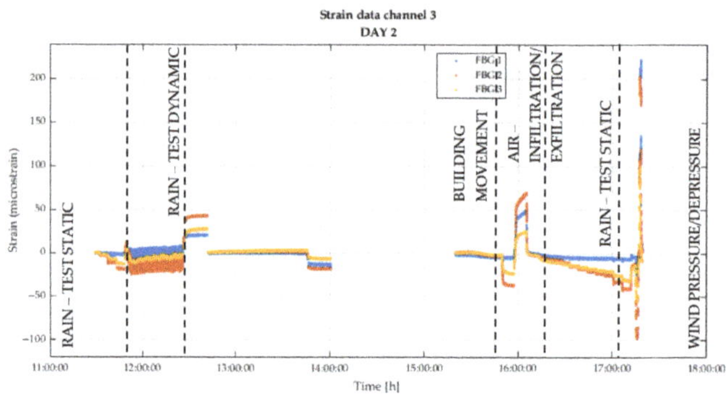

Figure 30. Day 2 FBG for strain—Channel 3.

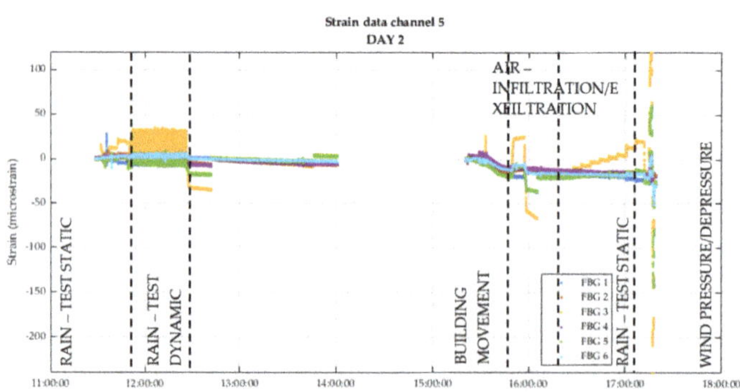

Figure 31. Day 2 FBG for strain—Channel 5.

Figure 32 reports the result obtained for channel 2 during day 2. Strain results obtained from channel 2 (z direction monitoring) are higher than channel 5 (x, y). This can clearly be observed in the rain—dynamic test, which executes from 11:50 h to 12:20 h, and the rain—test static, which executes from 16:20 h to 17:20 h, where the strain level is raised in steps following the pressure curve increasing(+600 Pa).

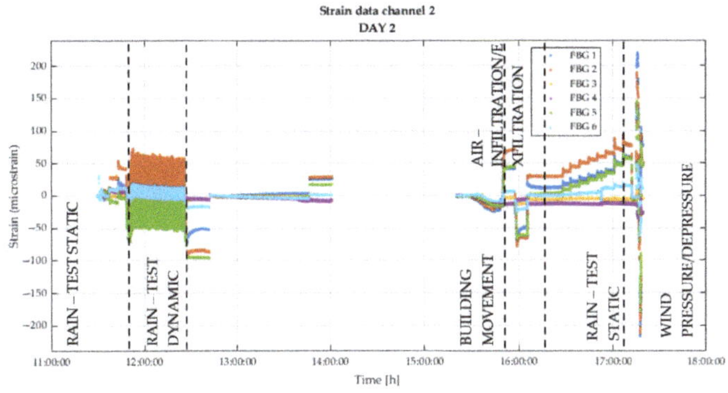

Figure 32. Day 2 FBG for strain—Channel 2.

4. Discussion

The results demonstrate the efficacy of the developed technique, which is capable of monitoring various parameters. However, it is important to mention that during the test, the façade passed all the assessing criteria defined by the standard. Even though the façade did not undergo damage or deformation, the installed kit demonstrated its potential to be applied in SHM. Indeed, it is sensitive to minor structural variations, suggesting its potential for early damage detection.

Several key findings regarding the performance of different sensor types for building envelope health monitoring under various loading conditions were highlighted by the activities to validate sensing technologies:

- Accelerometers monitor dynamic evidence and impulsive effects acting on the structure and can highlight any critical effects and whether the structure has, therefore, been compromised. Structural variation can be highlighted by performing a comparison between different modal analyses or spectral response variations. From the accelerometers, several considerations can be made from the data collected. From the modal analyses performed before and after the monitoring of the phenomena acting on the glass façade, it was possible to notice a decrease in the modal frequencies of the structure after the exemption of the tests on the wall. This is evidenced by the results obtained in Section 3.2. Regarding the analysis of the structural vibration monitored with the accelerometers attached to the structure, two types of observations can be made: The first one is based on the RMS level of the acceleration that allowed for sensing a significant signal increase when the building was subjected to rain (dynamic), impacts, wind, building movements, and fan excitation. Also, a static rain and air permeability test was sensed by the accelerometers thanks to their high sensitivity (100.5 mV/g), even though the dynamic effects of those loads are limited due to their static nature.
- The second observation is the monitoring of the dynamic behavior of the building, which was monitored by the power spectra estimated from the time histories measured by the accelerometers. The spectra show different signatures depending on the type of test that the building is subjected to. For example, in the fan excitation, the typical harmonic pattern is visible, consisting of the fan blade passing frequency and its high-order harmonics. Another piece of evidence is the deviation detected between the power spectra acquired on the first day of static rain and the same test realized on the second day, see Figure 24. Also, in this case, a decrease in the resonance frequencies of the building was observed, which confirms what is evidenced by the modal analysis.
- Conversely, force washers effectively monitored structural health under static and dynamic loads. They accurately measured applied loads, providing valuable insights into the façade's stress response. Interestingly, washers with embedded sensors revealed normal behavior under typical service loads (600 Pa), regaining initial torque

after stress as demonstrated at the end of tests in the 3 days. However, wind pressure tests (2625 Pa positive, −3000 Pa negative) showed permanent torque changes, indicating high-stress conditions. These sensors detected changes in the washer tightening torque, signifying critical stress levels requiring additional maintenance. This demonstrates the potential of integrating such sensors into façades for continuous monitoring of wind-vulnerable areas, confirming the force loads' suitability for SHM applications. This type of sensor made it possible to visualize the changes during all three days of the tests, and it enabled the capture of static phenomena more prominently. On the second day, it was possible to see that the tests tended to degrade the tension on the screws of the structure from the initial state. Also, considering the uncertainty band of the sensors, it is possible to evidence that some events produced an important tension variation on the screw, at least on some of them and particularly during the tests performed on the second day.

- FBG for temperature—Since the temperature increased all day, the FBG sensors in the other fibers responded to the temperature. The noise contributions in this temperature measurement originate mostly from the FBG interrogator. This causes a noise band up to ~0.3 °C. Resampling of the data to a lower frequency could reduce the noise band. Other noise contributions to the temperature sensor are negligible since the temperature sensor is mounted without strain. FBG sensors react to temperature and strain, so drift in the strain measurements can originate from temperature changes. However, is it possible to correct this when the temperature is known or when a second FBG is used as a temperature sensor for more accurate results? The obtained results validate the feasibility of integrating FBGs into the building envelope to measure temperature, aiming to prevent thermal shock damage. In comparison to conventional sensors such as thermocouples, FBGs offer a valuable alternative due to their compact dimensions, both for the sensor and connectors, and the ability to incorporate multiple tens of sensors in a single fiber. Similar considerations could be applied to the following:
- FBGs for strain and vibration—The obtained results could be compared to those obtained from the conventional force washers. Indeed, the outcomes show the strain and vibration registered during the test provoked by accidental impacts or dynamic pressure. The FBG integration within the façade represents a potential solution due to their characteristic with a particular focus on their dimensions compared to conventional force washers, which have larger dimensions. FBG sensors are solid-state sensors of glass and have no moving or active components. Therefore, the lifetime of the sensors is not critical. Furthermore, no effects are expected on the structural integrity since the sensors are passive and the fiber is less than 1 mm in diameter.
- With respect to the state of the art, this paper presents a methodology based on the use of different types of sensors for measuring a wide number of physical parameters to make identifying possible damages in the SHM field more robust.

5. Conclusions

The data-driven approach presented in this paper unlocks a deep understanding of how building envelopes respond to various stresses, including wind, seismic activity, material strain, temperature changes, and other dynamic loads. The research conducted demonstrated the potential of various sensor types for SHM in building envelopes. Considering the nature of different environmental events, a multivariant approach is the best solution for SHM. The facade, in this case, overcame all standard tests, but the different sensors detected minimal structural variation. Modal analysis is useful even if a critical event has been registered and allows for printing the status of the structure in a specific time frame. The final considerations on the obtained results are as follows:

- From the modal analysis point of view, the maximum deviation of the natural frequencies experienced is 1.2% between the structure status at the beginning and at the end of the certification tests.

- This was also confirmed by the monitoring accelerometers and observing the power spectra of the time histories registered during rain (static) between day 1 and day 2 (Figure 16, where the shift towards lower natural frequency is evident)
- Force washers evidenced that between day 1 and day 2, the bolts undergo loosening, especially concerning the bolt registered by the sensor installed on channel 2 (a decrease in the tension of that bolt of about 8% was registered).
- Integration of FBGs into building envelopes for temperature monitoring prevents thermal shock damage, offering compact size and multiplexing capabilities. FBGs also excel in strain, ensuring longevity and structural integrity, with potential applications in façades to improve safety during the entire life of the structure.

The achieved results demonstrate that sensing façade for SHM can open new approaches in the following: preventive maintenance schedules within the building envelope, identifying potential issues before they become critical, thereby reducing downtime and costs and minimizing environmental impact, reducing unnecessary repairs and interventions by prioritizing targeted maintenance based on data-driven insights; prioritizing retrofitting requirements directing resources towards façade areas most in need of improvement for enhanced safety; guiding design improvements integrating real-world performance data into future building designs for increased resilience and sustainability. For future works, a threshold that identifies and detects anomalies will be studied for each variable. In addition, it is also necessary to optimize sensor placement, with a particular focus on integration within the façade, finding a balance between aesthetic harmony, sensor monitoring, and maintenance. Sensor integration should be conducted during manufacturing activities, with the aim of inserting them into the aluminum profile cavity and isolated chamber within double-glazed panels. Other applications could be investigated, such as embedding Fiber Bragg Gratings (FBGs) for temperature monitoring in glass PVB components during the glass lamination phase, aiming to prevent thermal shock glass defects and guarantee aesthetic harmony in the building envelope. Additionally, integrating FBGs for strain monitoring in structural silicone, which are responsible for supporting hundreds of kilograms of glass in minimal space, can ensure the façade's safety and serviceability.

Author Contributions: Conceptualization, A.P. and L.V.; methodology A.P. and L.V.; software, M.T.C. and G.P.; validation, A.P. and L.V.; formal analysis, A.P., L.V. and M.M.; investigation, A.P., L.V. and G.M.R.; resources, A.P., L.V., F.B., V.D., M.T.C. and G.P.; data curation, A.P., L.V., V.D., M.T.C. and G.P.; writing—original draft preparation, A.P. and L.V.; writing—review and editing, A.P., L.V., F.B., M.M., G.M.R., V.D., M.T.C. and G.P.; visualization, L.V., F.B., V.D., M.T.C. and G.P.; supervision, A.P.; project administration, A.P.; funding acquisition, A.P. All authors have read and agreed to the published version of the manuscript.

Funding: This research was funded by the European Union's Horizon 2020 Research and Innovation Program, grant number 862597.

Institutional Review Board Statement: Not applicable.

Informed Consent Statement: Not applicable.

Data Availability Statement: The original contributions presented in the study are included in the article, further inquiries can be directed to the corresponding author.

Conflicts of Interest: Authors Laura Vandi, Francesco Belletti and Alessandro Pracucci was employed by the company Focchi S.p.A. Author Vincent Docter was employed by the company PhotonFirst International. Author Alessandro Pracucci was employed by the company Levery S.r.l. Società Benefit. Author Maria Teresa Calcagni, affiliated with Università Politecnica delle Marche, was engaged with a service contract by the company Focchi S.p.A. The remaining authors declare that the research was conducted in the absence of any commercial or financial relationships that could be construed as a potential conflict of interest.

Appendix A

Table A1. Laboratory environment's tests method statement.

Sequence	Test	Activity—Range Values	Pass/Fail Criteria =	EU Reference	Test Time
1A	Air—infiltration	Test Pressure: +600 Pa (Class A4)	Air leakage $\leq$ 1.5 m^3/hm^2	EN 12152 [42], EN 12153 [43]	45′
1B	Air—exfiltration	Test Pressure: −600 Pa (Class A4)	Air exfiltration rate $\leq$1.5 m^3/hm^2 at test pressures up to 100 Pa		1 h 15′
2	Rain—test static	Test pressure: 600 Pa (Class R7)	No leakage at 600 Pa	EN 12154 [44], EN 12155 [45]	55′
3A	Wind—pression	Test pressure: +1750 Pa	Mullion deflection limit: 3435/300 + 5 = 16.45 mm (ABC) Residual deformation: 005 *Max measured deformation or 1 mm)	EN 12179 [46]	8′
3B	Wind—depression	Test pressure: −2000 Pa			8′
4A	Air—infiltration	Test Pressure: +600 Pa (Class A4)	Air leakage shall not exceed that measured at point 1B by more than 0.3 m^3/hm^2	EN 12152, EN 12153	8′
4B	Air—exfiltration	Test Pressure: −600 Pa (Class A4)	Air leakage shall not exceed that measured at point 2 by more than 0.3 m^3/hm^2		8′
5	Rain—test static	Test pressure: 600 Pa (Class R7)	No leakage at 600 Pa	EN 12154, EN 12155	1 h 5′
6	Rain—test dynamic	Dynamic water penetration test with fan with a pulsing every 3 s from 750 Pa to 250 Pa	No leakage	CWCT 'Standard Method for building envelope' part 8. clause 8.7.2.1	36′
7A	Building movement—vertical	1. Vertical offset of the intermediate unit: uz = ±7 [mm]—2 cycles	-	CWCT 'Standard Method for building envelope' part 17	n.a.
7B	Building movement—horizontal	2. Horizontal offset of the intermediate beam: uz = ±7 [mm]—2 cycles	-	CWCT 'Standard Method for building envelope' part 17	n.a.
8A	Air—infiltration	Test Pressure: +600 Pa (Class A4)	Air leakage shall not exceed that measured at point 1B by more than 0.3 m^3/hm^2	EN 12152, EN 12153	7′ 30′′
8B	Air—exfiltration	Test Pressure: −600 Pa (Class A4)	Air leakage shall not exceed that measured at point 2 by more than 0.3 m^3/hm^2		7′ 30′′
9	Rain—test static	Test pressure: 600 Pa (Class R7)	No leakage at 600 Pa	EN 12154, EN 12155	1 h
10A	Wind—pression	Test pressure: 2625 Pa	Integrity: Residual deformation = 6.90 mm (3435 mm/500)	EN 12179	2′
10B	Wind—depression	Test pressure: −3000 Pa			2′
11	Fan excitation	Dynamic test	-	-	1 h 35′
12A	Impacts test- hard body	6 J (1.224 mm height with 0.5 kg steel ball) 10 J (1.020 mm height with 1.0 kg steel ball)	Negligible risk (TN76)	CWCT TN 76	n.a.
12B	Impact test—soft body	120 J (245 mm height) 500 J (1020 mm height)	Negligible risk (TN76)		n.a.
12C	Impact test—double tyre	343 J (700 mm height)	No part exceeding the mass of 50 g shall fall. No holing shall occur permitting a test block E2 according to EN 1630 (ellipse) to be passed through it;	EN 14019 and BS 12600	n.a.

References

1. Wang, G.; Ke, J. Literature Review on the Structural Health Monitoring (SHM) of Sustainable Civil Infrastructure: An Analysis of Influencing Factors in the Implementation. *Buildings* **2024**, *14*, 402. [CrossRef]
2. Intelligent Buildings—An Overview | ScienceDirect Topics. Available online: https://www.sciencedirect.com/topics/engineering/intelligent-buildings (accessed on 29 January 2024).
3. Blum, D. *Data-Driven Smart Buildings: State-of-the-Art Review*; Energy in Building and Communities Programme; CSIRO: Newcastle, NSW, Australia, 2023; p. 103.
4. Kaboli, A.; Shirowzhan, S. *Advances and Technologies in Building Construction and Structural Analysis*; IntechOpen: London, UK, 2021; ISBN 978-1-83881-141-9.
5. Faridi, A.; Roy, K.; Singhal, V. *Damage Quantification in Beam-Type Structures Using Modal Curvature Ratio*; Innovative Infrastructure Solutions; Springer International Publishing: Berlin/Heidelberg, Germany, 2024.
6. Valinejadshoubi, M.; Bagchi, A.; Moselhi, O. Structural Health Monitoring of Buildings and Infrastructure. *Int. J. Civ. Environ. Eng.* **2016**, *10*, 9.
7. Ferreira, P.M.; Machado, M.A.; Carvalho, M.S.; Vidal, C. Embedded Sensors for Structural Health Monitoring: Methodologies and Applications Review. *Sensors* **2022**, *22*, 8320. [CrossRef] [PubMed]

8. Comisu, C.-C.; Taranu, N.; Boaca, G.; Scutaru, M.-C. Structural Health Monitoring System of Bridges. *Procedia Eng.* **2017**, *199*, 2054–2059. [CrossRef]
9. Deng, Z.; Huang, M.; Wan, N.; Zhang, J. The Current Development of Structural Health Monitoring for Bridges: A Review. *Buildings* **2023**, *13*, 1360. [CrossRef]
10. Li, S.; Chen, S. Field Monitoring and Prediction on Temperature Distribution of Glass Curtain Walls of a Super High-Rise Building. *Eng. Struct.* **2022**, *250*, 113405. [CrossRef]
11. Yang, B.; Zhu, H.; Wüchner, R.; Zhang, Q. Monitoring of Wind Effects on a Wind-Sensitive Hybrid Structure with Single-Layer Cable-Net Curtain Walls under Typhoon Muifa. *J. Build. Eng.* **2021**, *44*, 102960. [CrossRef]
12. Ogasawara, S.; Kanda, K.; Suzuki, Y. Damage Investigation of Non-Structural Components in Buildings with SHM System in the 2018 Osaka Earthquake. In Proceedings of the 17th World Conference on Earthquake Engineering, Sendai, Japan, 13–18 September 2020.
13. Brell-Cokcan, S.; Adams, T.J.; Kerber, E.; Dai, R.; Lee, H.J.; Kirner, L.T.; von Hilchen, M.; Adams, T.J. *Construction & Robotics: Research Driven Project, Volume 1: Research Paper/SS 2021*; RWTH Aachen University: Aachen, Germany, 2022; 179p.
14. Bouzan, G.B. Building Facade Inspection: A System Based on Automated Data Acquisition, Machine Learing, and Deep Learning Image Classification Methods. *ARPN J. Eng. Appl. Sci.* **2021**, *16*, 1516.
15. Giovanardi, M.; Baietta, A.; Belletti, F.; Magnani, S.; Casadei, O.; Pracucci, A. Exploiting the Value of Active and Multifunctional Façade Technology through the IoT and AI. *Appl. Sci.* **2024**, *14*, 1145. [CrossRef]
16. *EN13830-2022*; Curtain Walling. CEN—Comitè Europeen de Normalisation: Bruxelles, Belgium, 2022.
17. Armstrong, S.; Auty, M.; Aylward, T.; Campbell, J.; Claridge, P.; De Bleecker, H. *Standard Test Methods for Building Envelopes*; Centre of Window and Cladding Technology: Bath, UK, 2005; ISBN 1-874003-39-4.
18. Majumder, M.; Gangopadhyay, T.K.; Chakraborty, A.K.; Dasgupta, K.; Bhattacharya, D.K. Fibre Bragg Gratings in Structural Health Monitoring—Present Status and Applications. *Sens. Actuators Phys.* **2008**, *147*, 150–164. [CrossRef]
19. Bremer, K.; Wollweber, M.; Weigand, F.; Rahlves, M.; Kuhne, M.; Helbig, R.; Roth, B. Fibre Optic Sensors for the Structural Health Monitoring of Building Structures. *Procedia Technol.* **2016**, *26*, 524–529. [CrossRef]
20. Ur Rehman, S.; Usman, M.; Younus Toor, M.H.; Hussaini, Q. Advancing Structural Health Monitoring: A Vibration-Based IoT Approach for Remote Real-Time Systems. *Sens. Actuators Phys.* **2024**, *365*, 114863. [CrossRef]
21. Mazzei, M.; Lellis, A.M.D. Capacitive Accelerometers at Low Frequency for Infrastructure Monitoring. *Procedia Struct. Integr.* **2023**, *44*, 1212–1219. [CrossRef]
22. Arezzo, D.; Quarchioni, S.; Nicoletti, V.; Carbonari, S.; Gara, F.; Leonardo, C.; Leoni, G. SHM of Historical Buildings: The Case Study of Santa Maria in Via Church in Camerino (Italy). *Procedia Struct. Integr.* **2023**, *44*, 2098–2105. [CrossRef]
23. Lobianco, A.L.; Zoppo, M.D.; Ludovico, M.D. Correlation of Local and Global Structural Damage State for SHM. *Procedia Struct. Integr.* **2023**, *44*, 910–917. [CrossRef]
24. Li, J.; Hao, H. Substructure Damage Identification Based on Wavelet-Domain Response Reconstruction. *Struct. Health Monit.* **2014**, *13*, 389–405. [CrossRef]
25. Li, J.; Hao, H.; Lo, J. Structural Damage Identification with Power Spectral Density Transmissibility: Numerical and Experimental Studies. *Smart Struct. Syst.* **2015**, *15*, 15–40. [CrossRef]
26. Ziaja, D.; Nazarko, P. SHM System for Anomaly Detection of Bolted Joints in Engineering Structures. *Structures* **2021**, *33*, 3877–3884. [CrossRef]
27. InComEss. Available online: https://www.incomess-project.com (accessed on 29 January 2024).
28. Pracucci, A.; Vandi, L.; Belletti, F.; Melo, A.R.A.; Vlachos, M.; Amditis, A.; Calcagni, M.T.; Esteves, D.S. Integration of Piezoelectric Energy Harvesting Systems in Building Envelope for Structural Health Monitoring with Fiber Optic Sensing Technology. *Preprints* **2024**, 2024031168. [CrossRef]
29. Tommasino, D.; Moro, F.; De Pablo Corona, E.; Vandi, L.; Baietta, A.; Pracucci, A.; Doria, A. Optimization of a Piezoelectric Wind-Excited Cantilever for Energy Harvesting from Facades. In *Advances in Italian Mechanism Science*; Niola, V., Gasparetto, A., Quaglia, G., Carbone, G., Eds.; Mechanisms and Machine Science; Springer International Publishing: Cham, Switzerland, 2022; Volume 122, pp. 848–856. ISBN 978-3-031-10775-7.
30. Ye, X.-W.; Su, Y.-H.; Xi, P.-S. Statistical Analysis of Stress Signals from Bridge Monitoring by FBG System. *Sensors* **2018**, *18*, 491. [CrossRef]
31. Zhou, Z.; Ou, J. Development of FBG Sensors for Structural Health Monitoring in Civil Infrastructures. In *Sensing Issues in Civil Structural Health Monitoring*; Ansari, F., Ed.; Springer: Dordrecht, The Netherlands, 2005; pp. 197–207.
32. Čápová, K.; Velebil, L.; Včelák, J. Laboratory and In-Situ Testing of Integrated FBG Sensors for SHM for Concrete and Timber Structures. *Sensors* **2020**, *20*, 1661. [CrossRef] [PubMed]
33. Mieloszyk, M.; Majewska, K.; Zywica, G.; Kaczmarczyk, T.Z.; Jurek, M.; Ostachowicz, W. Fibre Bragg Grating Sensors as a Measurement Tool for an Organic Rankine Cycle Micro-Turbogenerator. *Measurement* **2020**, *157*, 107666. [CrossRef]
34. Gomasa, R.; Talakokula, V.; Kalyana Rama Jyosyula, S.; Bansal, T. A Review on Health Monitoring of Concrete Structures Using Embedded Piezoelectric Sensor. *Constr. Build. Mater.* **2023**, *405*, 133179. [CrossRef]
35. Lei, Y.-J.; Li, R.-J.; Zhang, L.-S.; Hu, P.-H.; Huang, Q.-X. Optical Accelerometers for Detecting Low-Frequency Micro-Vibrations. *Appl. Sci.* **2022**, *12*, 3994. [CrossRef]

36. Lee, H.; Oh, M.; Seo, J.; Kim, W. Seismic and Energy Performance Evaluation of Large-Scale Curtain Walls Subjected to Displacement Control Fasteners. *Appl. Sci.* **2021**, *11*, 6725. [CrossRef]
37. Aiello, C. In-Plane Seismic Response of a Glazed Curtain Wall: Full-Scale Laboratory Test and Non-Linear Modelling. In Proceedings of the 7th International Conference on Computational Methods in Structural Dynamics and Earthquake Engineering (COMPDYN 2019), Crete, Greece, 24–26 June 2019; Institute of Structural Analysis and Antiseismic Research School of Civil Engineering National Technical University of Athens (NTUA) Greece: Crete, Greece, 2019; pp. 269–281.
38. *EN 14019:2016*; Curtain Walling—Impact Resistance—Performance Requirements. Slovenski Inštitut za Standardizacijo: Ljubljana, Slovenia, 2016. Available online: https://standards.iteh.ai/catalog/standards/cen/1bfe4fa0-6bf4-4ae7-b2a2-0a87ced89e0d/en-14019-2016 (accessed on 12 February 2024).
39. La Norma Uni En 12600: Prova Del Pendolo–Metodo Della Prova Di Impatto E Classificazione Per Il Vetro Piano. Available online: https://www.ingenio-web.it/files/guida-alle-norme-tecniche-del-vetro-per-edilizia.pdf (accessed on 9 February 2024).
40. NI-9234 Specifications—NI. Available online: https://www.ni.com/docs/en-US/bundle/ni-9234-specs/page/specs.html (accessed on 13 March 2024).
41. HBK (Hottinger, Brüel & Kjær). QuantumX MX840B Universal Amplifier. Available online: https://www.hbm.com/en/2129/quantumx-mx840b-8-channel-universal-amplifier/?product_type_no=QuantumX%20MX840B/MX440B:%20Universal%20Data%20Acquisition%20Module (accessed on 9 February 2024).
42. *UNI BS EN 12152_2002*; Facciate Continue Permeabilità All'aria Requisiti Prestazionali e Classificazione. 2003.
43. *UNI EN 12153:2002*; Facciate continue-Permeabilità all'aria-Metodo Di Prova. 2002.
44. *UNI BS EN 12154_1999*; Curtain Walling—Watertighness-Performance Requirments and Classification. 1999.
45. *UNI EN 12155:2002*; Facciate Continue-Tenuta All'acqua-Prova Di Laboratorio Sotto Pressione Statica. 2002.
46. *UNI EN 12179:2002*; Facciate Continue-Resistenza al Carico Del Vento-Metodo Di Prova. 2002.

sensors

MDPI

Article

Portable Arduino-Based Multi-Sensor Device (SBEDAD): Measuring the Built Environment in Street Cycling Spaces

Chuanwen Luo [1,*], Linyuan Hui [1], Zikun Shang [1], Chenlong Wang [1], Mingyu Jin [1], Xiaobo Wang [1] and Ning Li [2,*]

1 Department of Architecture, School of Architecture and Art, North China University of China, Jinyuanzhuang Road 5, Shijingshan District, Beijing 100144, China; huilinyuan@mail.ncut.edu.cn (L.H.); shangzikun@mail.ncut.edu.cn (Z.S.); rain_wcl@163.com (C.W.); jinmingyu@ncut.edu.cn (M.J.); wangxiaobo@ncut.edu.cn (X.W.)

2 Beijing Historical Building Protection Engineering Technology Research Center, Beijing University of Technology, Beijing 100124, China

* Correspondence: witchwen@ncut.edu.cn (C.L.); ning_li11@bjut.edu.cn (N.L.)

Abstract: The built environment's impact on human activities has been a hot issue in urban research. Compared to motorized spaces, the built environment of pedestrian and cycling street spaces dramatically influences people's travel experience and travel mode choice. The streets' built environment data play a vital role in urban design and management. However, the multi-source, heterogeneous, and massive data acquisition methods and tools for the built environment have become obstacles for urban design and management. To better realize the data acquisition and for deeper understanding of the urban built environment, this study develops a new portable, low-cost Arduino-based multi-sensor array integrated into a single portable unit for built environment measurements of street cycling spaces. The system consists of five sensors and an Arduino Mega board, aimed at measuring the characteristics of the street cycling space. It takes air quality, human sensation, road quality, and greenery as the detection objects. An integrated particulate matter laser sensor, a light intensity sensor, a temperature and humidity sensor, noise sensors, and an 8K panoramic camera are used for multi-source data acquisition in the street. The device has a mobile power supply display and a secure digital card to improve its portability. The study took Beijing as a sample case. A total of 127.97 G of video data and 4794 Kb of txt records were acquired in 36 working hours using the street built environment data acquisition device. The efficiency rose to 8474.21% compared to last year. As an alternative to conventional hardware used for this similar purpose, the device avoids the need to carry multiple types and models of sensing devices, making it possible to target multi-sensor data-based street built environment research. Second, the device's power and storage capabilities make it portable, independent, and scalable, accelerating self-motivated development. Third, it dramatically reduces the cost. The device provides a methodological and technological basis for conceptualizing new research scenarios and potential applications.

Keywords: built environment measurement; sensor integration; cycling infrastructure; smart city technology

1. Introduction

Urban and architectural design processes are increasingly using built environmental data acquisition to support and optimize design outcomes. The design of fast-response, autonomous, and affordable mobile urban sensing technologies plays a significant role in acquiring high-spatial-resolution environmental data for urban planning [1]. By integrating spatial data and innovative technologies (e.g., geographic information systems, remote sensing, and the Internet of Things (IoT)), urban planners can gather valuable insights to design environmentally friendly indoor and outdoor spaces that are responsive to the needs of urban development [2–4]. Whether it is monitoring environmental parameters, assessing urban vitality, evaluating the function of urban furniture, or analyzing the impact of smart city planning, data collection plays a crucial role in urban design [5,6].

The term *built environment* refers to the human-made surroundings that provide the setting for human activity. It comprises urban design, land use, and the transportation system and encompasses patterns of human activity within the physical environment [7,8]. Data related to the built environment commonly used in planning and urban design are often macro or meso scale data, such as topography and geomorphology, weather, and points of interest (POI). These data are authoritative in macro-recognition. However, in mesoscale urban design, landscape design, and other urban open space designs, it is not possible to integrate multiple sources of data to achieve a deeper understanding of the city and provide a database for renewal construction. Therefore, there is an urgent need to supplement the fine granularity of data at the street level to promote a better understanding of the city by the designers and managers. The interplay between the built environment and group behavior can be fundamentally recognized based on a deep understanding of the microenvironment. Fine-grained urban open space data can help designers think about the spatial layout and connection of different functional areas in a neighborhood. Designers can then better consider the effect of urban open space on the shading of the surrounding buildings, wind, and the comfort and safety of pedestrians, which will help them improve the comfort and convenience of urban open spaces, and create a pleasant urban open space.

By combining advanced technologies such as multi-source data, 3D modeling, indoor positioning, and image recognition, designers can assess outdoor comfort, air quality, and microclimate to create more sustainable and livable environments. Support is also provided for environmental pollution, emergency response, flood risk management, and urban landscape design [9–11]. In addition, integrating data-driven approaches such as machine learning and big data analytics can provide valuable insights for optimizing urban planning and design processes. Using data from various sources and applying computational techniques, urban designers can evaluate design parameters, assess ventilation potential, and improve the economic and social benefits of urban spaces [12–15]. Currently, digital research related to urban design is mainly carried out by applying computer vision (CV) [16,17] and street view images (SVI) combined with machine learning, e.g., by analyzing SVI to understand landscape features such as buildings, streets, and parks in the city [18], which can help urban planners better understand the layout and structure of the city and provide the necessary data support for urban planning. Transportation planning can be optimized by analyzing traffic flow and pedestrian activity to improve efficiency and enhance pedestrian safety [19]. The city's degree of greening and environmental quality can be assessed by reviewing green spaces and tree cover to provide data support for sustainable urban development [20]. Machine learning can be used to predict urban growth patterns and help formulate sustainable urban development strategies [21]. In the direction of research related to hardware sensing, Zhang collected data in real-time by deploying noise sensors in the city to provide an essential reference for improving the urban noise environment [22]. Based on a single intelligent sensor, the researchers developed a research methodology for data collection, analysis, feature extraction, machine learning prediction, and optimization decision-making, which brought an important impetus to sustainable urban development.

With the rapid development of artificial intelligence, machine learning, and deep learning technologies, the ability to process and analyze massive multi-source image and sensor data has also been greatly improved [23,24]. Data are the premise and foundation for machine learning and deep learning. At present, temperature, humidity, light, particulate matter (PM), noise, greenery, and other data from urban open spaces are seldom fused and collected for application. One reason for this is that there is simply no portable device that can ultimately collect the above data. Although some professional devices are single-sized and portable, there are no multi-source sensor devices designed for street-level urban data platforms. Realizing data collection in the same period will require more than five devices, and it must be ensured that the sensing ends of the devices are in adjacent spaces. If it is a split collection of different devices, it will lose its scientific nature due to different times and environments, and the problem of carrying a large number of devices and power consumption will also make the collection challenging to realize. For example, in understanding and measuring parks or streets in walking or cycling environments, motorized vehicles carrying large, complex multiple sensors cannot be used because they cannot be driven in a particular area; portable equipment is thus required. At the same time, understanding urban open space does not require the same stringent data collection accuracy and error requirements as laboratory research. In addition, cost reduction is a major consideration for small- and medium-sized companies or research departments.

The current Internet of Things (IoT) sensors have been applied to environmental detection and understanding in building and urban design. It can be divided into perception, transmission, and application networks according to their functions. SVI, data acquisition using sensors, and data fusion applications are the keys to realizing street-scale understanding and prediction of the urban built environment. Therefore, there is an urgent need to build a street built environment data platform (Figure 1).

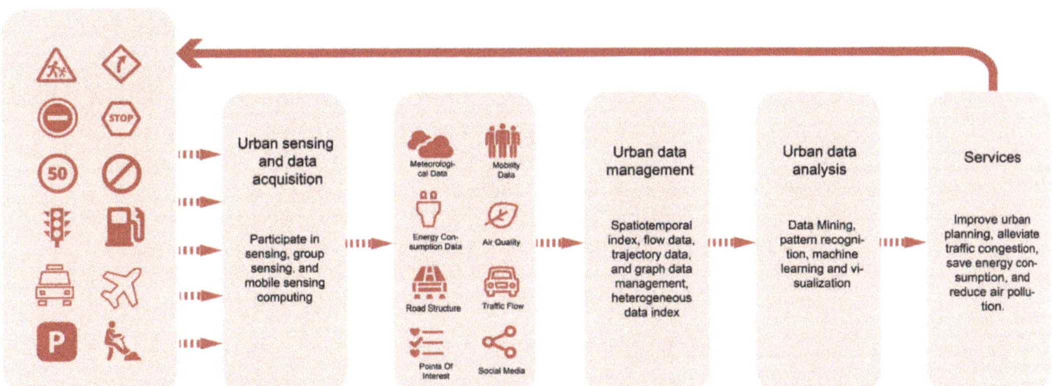

Figure 1. Street built environment data platform based on sensors and IoT technology to realize urban sensing and data acquisition; organize and obtain a variety of data, such as human mobility data, social media data, etc.; and provide services for the healthy development of the city through urban data management and urban data analysis.

This study designs, develops, and makes a low-cost, portable, multi-channel street built environment data acquisition device (SBEDAD) to provide a data platform for city managers and designers for designing, recognizing, and building a smart city.

2. System Requirements and Design

Urban monitoring data collection through a large number of sensors can support innovative applications for a variety of scenarios, such as environmental monitoring, security, health, education, and urban mobility [25]. Multi-sensors can detect urban land cover changes and surface urban heat islands (SUHI) within cities [26–28]. Vibration and deformation sensors can perform structural integrity measurements for historic buildings [29]. However, distributing and maintaining sensors in metropolitan areas is a challenge due to technological limitations [25,30]. The use of mobile sensors can significantly reduce the need to distribute static sensors to cover most urban areas for urban management and monitoring [31] and increase the level of spatial and temporal granularity [32], especially when using a cloud-based platform for crowd sensing in urban scenarios [33]. Detecting load levels and optimizing truck routes using intelligent monitoring sensors reduces the cost of urban waste management [34,35]. Sensors have been placed in vehicles, bicycles, or backpacks of pedestrians for opportunistic surveillance. Meteorological data and built environment data have been collected from mobile monitoring sensors and stationary sites to expand the spatial extent and temporal resolution [20,36–38]. Overall, there are various innovative approaches and technologies for monitoring urban environments efficiently and effectively.

Based on existing research, we realized that urban sensing research has a great need for first-hand, sensor-based mobile data collection. However, current research struggles to obtain urban stationery sensing data, which are not sufficiently dense. Developing or customizing the required equipment is both expensive and time-consuming, as it requires interdisciplinary knowledge. Last but not least, the type of sensing data cannot be changed to suit the needs of the study.

Arduino is a simple and easy-to-use open platform; it includes an open-source hardware and software IDE (version 2.2.1), providing unlimited possibilities for developers. It is developer-friendly and suitable for designers, other non-professional programmers, and non-professional embedded system developers [39]. The research team chose this platform to design and develop a low-cost portable Arduino-based street built environment data acquisition device (SBEDAD). It can not only measure the basic functions needed for this research but also add intelligent sensing components to the device as needed in future research and development processes to prepare for the research and development of urban open space.

2.1. Basis of the System Design

The system is based on the basic core of a microcontroller released by the open hardware Arduino Mega 2560 (Arduino, Turin, Italy). Mega 2560 is a microcontroller application development board printed circuit board (PCB) with an advanced control system (AVR) microcontroller as the core controller, and its ecosystem also provides a large number of application libraries to avoid complex register manipulations. Programming is conducted through programming languages derived from C and C++ to enable different sensors by powering the digital and analog pins on Mega 2560. (See Table 1).

Table 1. Technical parameters of the Arduino Mega 25.60 [40].

Mega 2560 Specification Parameter	
Operating voltage	5 V
Input voltage (recommended)	7–12 V
Input voltage (limit)	6–20 V
Digital I/O pins	54
Analog I/O pins	16
DC I/O pin	40 mA
DC for 3.3 V pin	50 mA
Flash memory	256 KB
SRAM	8 KB
EEPROM	4 KB
Clock frequency	16 MHz

The SBEDAD designed in this research aims to provide a data platform for future urban design and macro urban environmental studies. It integrates feedback on cycling and walking physical sensations, road quality, and greening quality and is fitted with several sensors, including PM concentration, luminous intensity, temperature, humidity, and noise sensors, as well as an 8K panoramic camera. The Mega board has sufficient pins, such as 54 digital input/output pins (of which 15 can be used as PWM outputs), 16 analog inputs, and 4 universal asynchronous receiver/transmitter (UART) hardware serial ports [40]. The GPS and PM sensors, which communicate based on the URAT protocol, are connected to the serial port. The temperature and humidity sensors, which are based on the transistor–transistor logic (TTL) protocol, are connected to the digital pin, while the light intensity sensor and noise sensor are connected to the analog port. The 3.2-inch LCD screen and SD card module, which are based on the serial peripheral interface (SPI) protocol, are connected to the SPI pins. To ensure fast and correct data transfer from sensors using different protocols and ports, we added an expansion board and a module in the script to check the flash availability before recording sensing data. Table 2 summarizes the configuration of the entire device. The circuit system was ultimately mounted in a specially designed enclosure to facilitate assembly and disassembly. The enclosure was made using a Bambu P1P machine, 3D printed in PLA material and secured with drop-in glue and screws. (See Figure 2).

Table 2. Configuration of SBEDAD.

Purpose	Type	Model	Response Time
Printed circuit board	Arduino	MEGA2560	--
Longitude and latitude	GPS and BDS	ATGM336H-5N	0.1 ms
Temperature and humidity	Humidity and temperature	DHT22	10 ms
Air pollution	PM	Plantower G7003S	100 ms
Light	Light sensor	TEMT6000	20 ms
	Light sensor	BH1750	5 ms
Noise	Voice sensor	Croco-Sound	20 ms
Air flow	Air flow sensor	Air flow sensor	10 ms
Status display	LCD	HX8357-B	
Data storage	SD	OV7670	
Video record	Panoramic camera	Pilot One	

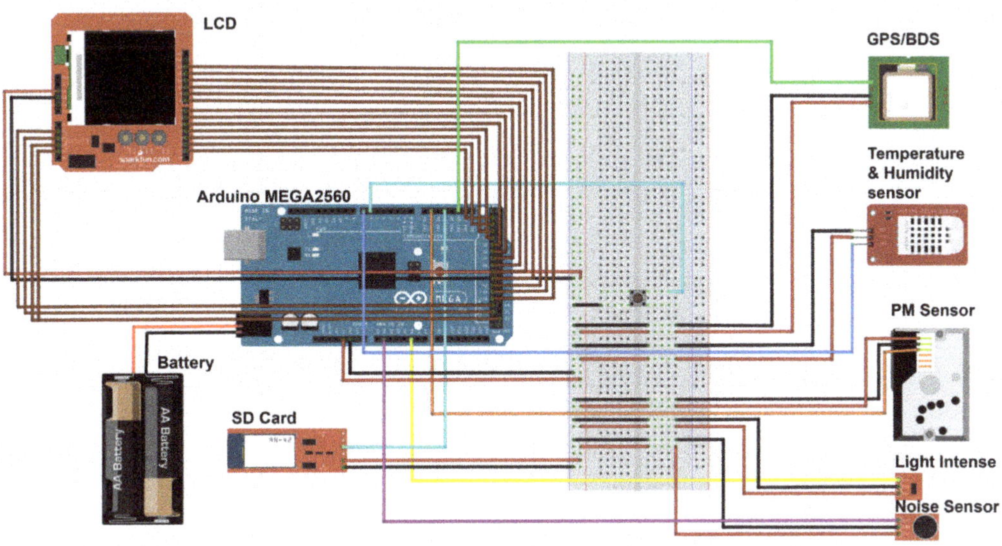

Figure 2. Electronic circuit of the SBEDAD. The Arduino built-in UART enables two devices to communicate at a time, so SBEDAD has the GPS and PM sensors connected to the serial port. The DHT22 data are collected at the digital pin, while the BH1750 and noise sensor data are collected at the analog pins. The availability of each component is confirmed in the overall function realization.

Currently, sample intervals in urban open spaces are usually two to five seconds [20]. Data acquisition for urban open spaces is a unidirectional sensing data collection process rather than an interaction process. The precision of temperature and humidity, air quality, light, noise, and other to-be-measured data in the urban built environment, which are used to study the impact of the urban environment on human behavior, is usually required at the level of seconds. Data with ultra-high precision merely have an impact on the study objectives. It should be emphasized that the ADC of the Arduino board is of 10-bit resolution and measures 0 to 5 V. This 10-bit resolution allows us to achieve an accuracy of

±4.882 mV (beyond the technical requirements). Moreover, to ensure that the user can use the device in daily environments and that the device works properly, the device provides the user with a screen, a secure digital (SD) card for direct storage of data, a power supply, and a reboot switch.

In configuring the parameters for SBEDAD, careful consideration was given to the specific values of the various setup options to meet the characteristics of the target use case and optimize system performance.

Based on the Air Quality Guidelines (AQG) issued by the World Health Organization (WHO) in 2001, standard values were set for a wide range of air pollutants, such as PM2.5 and O_3. The guidelines provide optional milestones for air quality management in countries around the world. The PM2.5 concentrations and PM10 concentration thresholds were set to 15 and 45 $\mu g/m^3$ [41]. These data can help complete an exposure risk assessment of pollutant concentrations.

Temperature and humidity are two important meteorological elements that determine physical comfort. The human body has a certain ability to adapt to changes in external temperature and can maintain balance with the help of thermoregulation. However, the body's regulation has a certain limit, in which the upper limit temperature is 32 °C and the appropriate relative humidity is 30% to 80% [42].

The perceived quality of street lighting affects cyclists and pedestrians' sense of safety and visual comfort, as well as outdoor activities at night. Existing research has revealed that illumination at 5 to 17 lx is a satisfactory level, and people feel safer if the light is warm, even at night [43].

High levels of noise can lead to physical and psychological problems for residents. Noise in the street space has a severe impact, especially on cyclists and pedestrians, and the construction of highly dynamic noise data streams can be used to assess and effectively manage traffic noise pollution [20].

Based on existing research, the data on PM, temperature, humidity, light, and noise in the street space are factors that have a greater impact on the comfort and experience of human beings directly exposed to the street space. Therefore, the data collection scope of SBEDAD is based on these data, and the sensors can be simply replaced to obtain different data according to the needs of different urban designers and managers in the future.

2.2. Low Cost

The manufacturing budget for SBEDAD is broken down into different components in Table 3. The total budget for the obligatory part of the basic version of the project is USD 68.78. The cost of adding other extended functionality parts will increase with the price of the required components, but it will not exceed USD 150.

Table 3. Price and features of SBEDAD.

Model	Nature	Commercial	Price
MEGA2560	Obligatory	YES	35.28
ATGM336H-5N	Obligatory	YES	10.93
DHT22	Obligatory	YES	1.08
Plantower G7003S	Obligatory	YES	12.18
TEMT6000	Obligatory	YES	0.90
BH1750	Obligatory	YES	1.1
Croco Sound	Obligatory	YES	1.37
Air flow sensor	Optional	YES	6.64
HX8357-B	Obligatory	YES	1.38
OV7670	Obligatory	YES	0.48
PTQS1005	Optional	YES	69.16
Heart beat rate sensor	Optional	YES	1.51
3D-printed shell	Optional	No	5
Batteries	Obligatory	YES	2.77
Cable and welding	Obligatory	No	1
Switching button	Optional	YES	1.11
Resistor	Obligatory	YES	0.1

2.3. Open-Source Software

The hardware of the device was programmed using an integrated development platform (IDE) that is compatible with different platforms (Linux, Windows, and Mac) and is available free of charge under the GNU General Public License. The programming language used for the device was derived from C and C++, and the program flowchart is shown below (Figure 3).

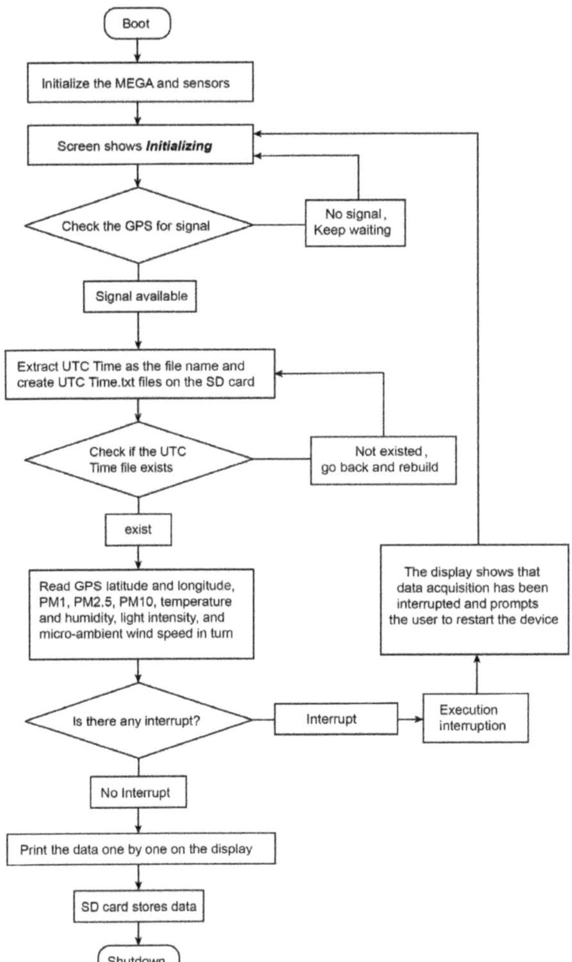

Figure 3. Flow chart of SBEDAD.

To connect multi-source sensors, this study built an initial prototype using an extension board to connect the sensors to the Mega board for multi-channel power supply and receiving data. The extension board allows easy changes in wiring and connections, exchange of components, and software testing until the desired functionality is achieved. After the sensor function tests are completed one by one, the PCB board and all its components are encapsulated in a custom-made 3D-printed thermoplastic case of PLA material. Power is provided by a rechargeable 5–12 V lithium battery connected to the external power socket of the Arduino board. Figure 4 shows the SBEDAD structure.

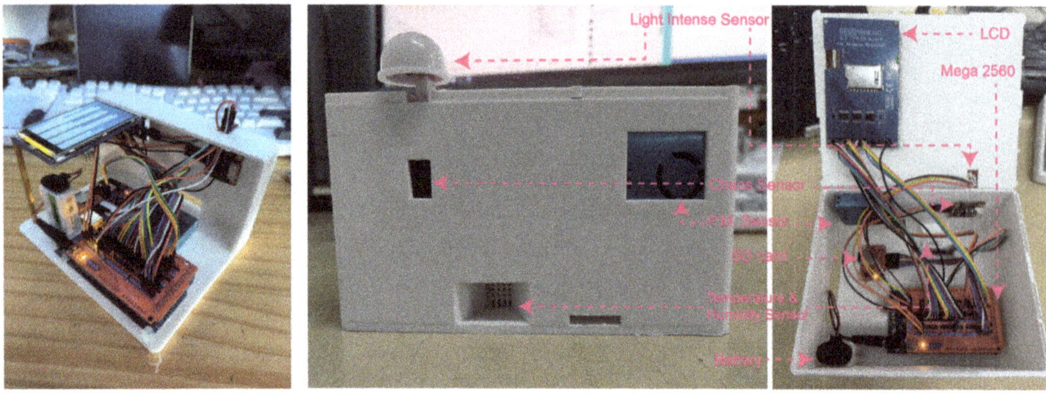

(**a**) (**b**)

Figure 4. (**a**) The 1st version SBEDAD, half-shelled, with TEMT6000 as the light intensity sensor; the screen is printing the sensing data to ensure the users get the running status of the equipment. (**b**) SBEDAD with PLA shell (2nd version) using BH1750 as the light intensity sensor.

3. Functional Validation

We undertook a comprehensive functional validation of the multi-source data acquisition capabilities of the street built environment data acquisition device (SBEDAD), underpinned by environmental sensing theories and methodologies for urban data analysis. Commencing with the environmental protection standards for air pollution monitoring, we integrated models of human thermal comfort, frameworks for evaluating road conditions, and ecological impact theories of urban greening, thereby systematically analyzing and substantiating the validity and reliability of the data procured by SBEDAD. This approach thoroughly examined SBEDAD's performance across multiple dimensions of the urban built environment.

Following the established criteria for evaluating green spaces, we employed image data captured by an 8K panoramic camera to calculate vegetation coverage rates, utilizing an algorithm specifically designed for this purpose. A comparative analysis was conducted against empirical survey data (Table 4), revealing a correlation coefficient of r = 0.87 between the two datasets. This finding attests to a high degree of consistency in SBEDAD's assessment of urban green spaces, thereby validating its efficacy in quantifying and characterizing green infrastructure within urban environments.

Table 4. Accuracy of data acquired by each sensor of SBEDAD.

Model	Range	Error Precision	SBEDAD Accuracy
Temperature	18–25 °C [44,45]	±1 °C	±0.5 °C
Humidity	40–70% [44,45]	>1%	±1%
PM	0–500 $\mu g/m^3$ [46]	±25%	±10–100 $\mu g/m^3$
Light	10,000–18,000 Lux [47]	±20%	>10 lux
Sound	40–120 db [48]	±5 db	0–40 m/s
GPS	--	--	2 m

Through experimentation and analysis, the accuracy and error of each sensor used in SBEDAD were determined, as well as the corrections required for the final readings of the device. It should be noted that the device lost some of its accuracy requirements to achieve portability compared to the high accuracy of industrial equipment. The reason for this is that, for the accuracy of sensory data acquisition in urban open spaces, such as relating to temperature, the perceptual threshold of the body's somatosensory senses in outdoor environments is approximately 1 °C [44,45]. Based on existing research, the required accuracy of SBEDAD was analyzed using empirical values, and finally, the accuracy of the data acquired by each sensor of SBEDAD was determined (Table 4).

According to the existing research results, the temperature data acquisition precision was set to one decimal place, i.e., 0.1 °C ± 10%. Humidity acquisition accuracy was 0.1% ± 10%; light intensity was 100 lux ± 10%. The wind speed accuracy was 5 m/s ± 0.1 m/s. Each value of PM was 2 $\mu g/m^3$ ± 10%.

Two different evaluation calibration activities were performed to confirm the accuracy of different sensors and compare the sensors' readings directly with professional industrial instruments and equipment, including temperature, humidity, light, PM, and noise sensors for acquiring values. These tests represent practical examples of the use of the Arduino-based development of SBEDAD in a well-established technical program. These experiments were carried out in normal temperatures between −10 and 45 °C, with humidity below 80%, to represent the real acquisition environment in urban open spaces instead of considering calibration residuals in the extreme temperature and humidity environment of the laboratory. (See Figure 5).

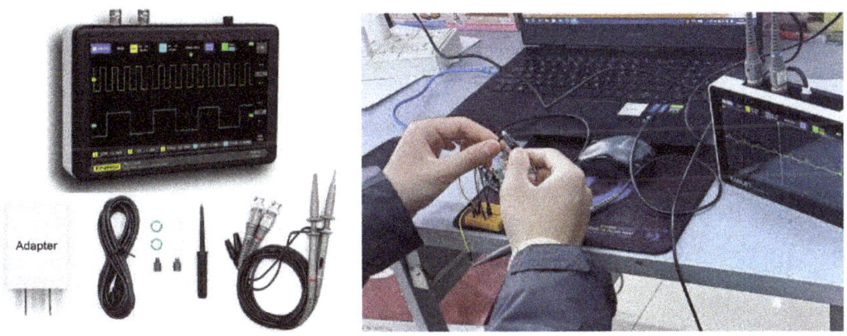

Figure 5. Evaluation calibrations of SBEDAD using the digital oscilloscope FNIRSI-1013D.

3.1. GPS and BDS Longitude and Latitude

The SBEDAD is designed to support urban design and city management, so the data it acquires should have temporal and spatial characteristics. By fusing data on temperature, humidity, air pollution, light, and noise with geographic location and time correspondence, it is possible to obtain mutually reinforcing knowledge from data sources with different attributes and origins and ultimately establish correlations between different data. (See Table 5).

Table 5. GPS and BDS model.

Model	Nature
Manufacturer	Yahboom
Cold-start acquisition sensitivity	−148 dBm
Hot-start acquisition sensitivity	−156 dBm
Recapture sensitivity	−160 dBm
Tracking sensitivity	−162 dBm
Positioning accuracy	2.5 m(CEP50)
Typical power consumption	25 mA @3.3 V
Agreement	NMEA0183
Maximum height	8000 m
Maximum speed	515 m/s
Operating temperature	−40 °C to + 85 °C
Size and weight	10.1 mm × 9.7 mm × 2.4 mm, 0.6 g

The SBEDAD is equipped with two sets of latitude and longitude positioning sensors, namely, the global positioning system (GPS) and the BeiDou system (BDS), for determining the location geographically. At the same time, the coordinated universal time (UTC) is taken as a marker to unify with the data from the panoramic camera and then to fuse and analyze the spatial–temporal characteristics of the urban open space. We programmed the Arduino IDE (Figure 6a) to obtain real-time GPS and BDS information, including UTC, latitude, and longitude, through serial communication. During the research process, we also used a set of Pilot One 8K panoramic video cameras, which have a built-in GPS sensor, to collect video data. Eventually, this study carried out calibrations on the Baidu map by comparing it with GPS (ATGMH-5N) and BDS data.

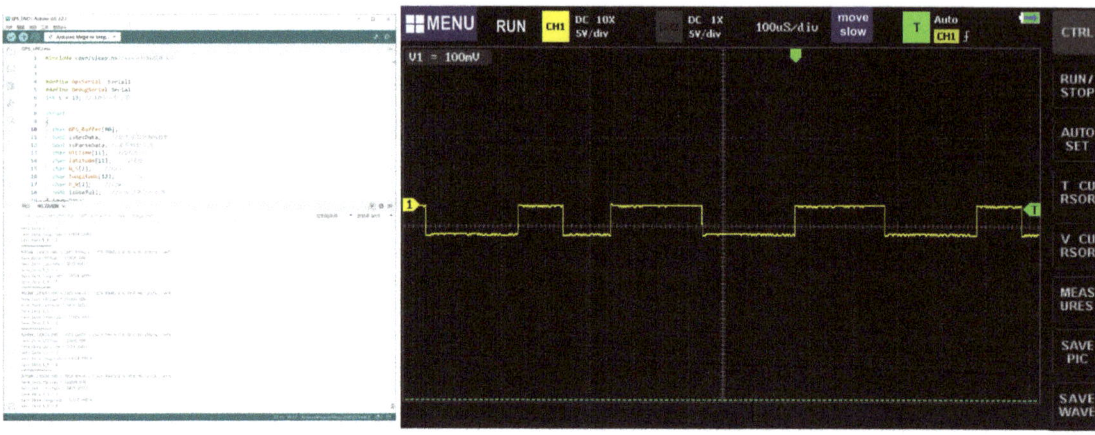

(a) (b)

Figure 6. (**a**) Arduino IDE reads GPS/BDS data via a serial port. (**b**) Measurement of the real frequency of GPS/BDS sensor.

It should be noted that the ATGM336H-5N module is an electrostatic-sensitive device. Constant electrostatic contact can cause unexpected module damage. It should connect the GND pin first and then the signal cable before use. An oscilloscope was also used to read the real frequency of the ATGM336H-5N module, which was 100 milliseconds (Figure 6b), much less than the sampling frequency required by the SBEDAD (1 s, as mentioned earlier).

We conducted several field data acquisition experiments for GPS and BDS sensors and finally decided to use Mega's serial port 3 for ATGM336H-5N data reading. We compared the data acquired by this sensor in Beijing, a city in North China whose temperatures range from −11 to 6 °C with an average temperature of 2 °C, in January 2024, and in Sanya, the southernmost city of China, where the temperatures range from 14 to 29 °C with an average temperature of 22 °C, in January 2024. The accuracy of the GPS module was verified through field tests in different outdoor temperature and humidity environments as well as latitude and longitude geographic locations.

A set of positions were obtained through the built-in sensing elements in Beijing and Sanya. Encrypted location information was then translated and analyzed on the Baidu map (Figure 7), which is compatible with both GPS and BDS. The results showed that the latitude and longitude position deviation was within 2.5 m, which meets the accuracy of information acquisition for urban spatial information.

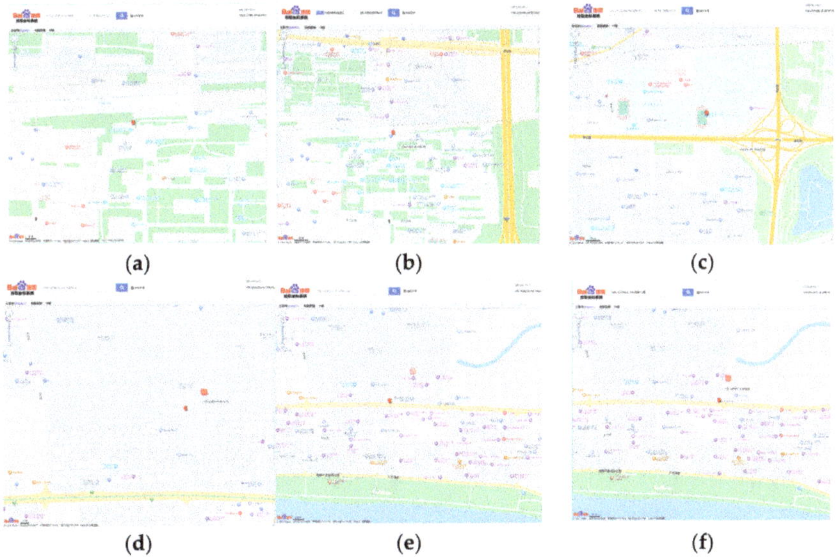

Figure 7. Field test of GPS/BDS sensor in Beijing (**a–c**) and Sanya (**d–f**). By inverting the geographic location by latitude and longitude on the Baidu map (upper left column), the result (the red icons) showed that ATGMH-5N has high accuracy and meets the urban spatial data requirements.

3.2. Environmental Temperature and Relative Humidity

The calibration of the temperature and humidity sensors was performed in a real office environment. The experiment compared the data collected by two sets of sensors (DHT11 and DHT22) and an industrial-grade commercial sensing device (UNI-T UT332+) (Table 6). The data acquired by the sensors was logged, read, and printed on the local side through the serial port. The UT332+ could log 100 sets of data in seconds, then read and print them through the host computer application (Figure 8a). The real frequency of the DHT22 obtained by the oscillograph was 10 milliseconds (Figure 8b).

Table 6. Temperature and humidity model.

	DHT11	DHT22	UNI-T UT332+

Temperature range	0 to 50 °C +/−2 °C	−40 to 80 °C +/−0.5 °C	−20 to 80 °C
Humidity range	20 to 90% +/−5%	0 to 100% +/−2%	0 to 100%
Resolution	Humidity: 1% Temperature: 1 °C	Humidity: 0.1% Temperature: 0.1 °C	Humidity: 0.1% Temperature: 0.1 °C
Operating voltage	3–5.5 V DC	3–6 V DC	3–5 V DC
Current supply	0.5–2.5 mA	1–1.5 mA	0–10 A
Sampling period	1 s	2 s	1 s
Price	USD 1 to 5	USD 4 to 10	USD 60 to 80

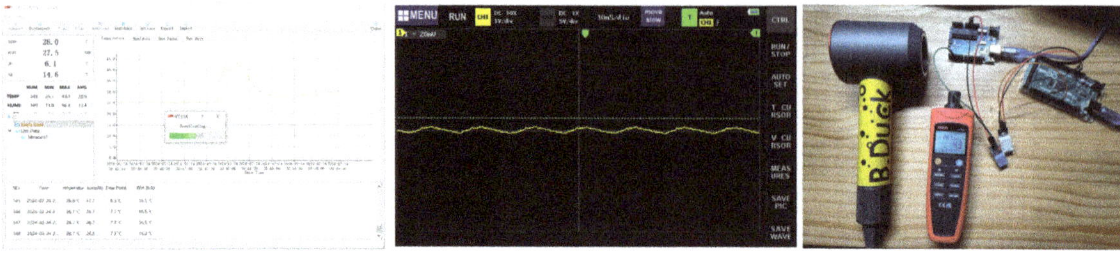

(a) (b) (c)

Figure 8. Lab test of the temperature and humidity sensor. (**a**) Temperature and humidity data for UT332+ through the host computer application. (**b**) Oscillograph of the DHT22. (**c**) Testing the sensor's ability to respond quickly to changes in temperature and humidity using a hairdryer.

To test out the accuracy of the temperature and humidity sensors more efficiently, the experiment used four environmental stimuli to compare the response sensitivity and accuracy (Figure 8c):

- Hairdryer heated blowing mode;
- Hairdryer cool blowing mode;
- Ice-water mode;
- Breathe warm air mode.

As shown in Figure 9, we collected 600 sets of data to compare the temperature and humidity on DHT11, DHT22, and UT332+ at the same time and in the same scenario. In the temperature-sensing performance, a percentage difference below 2% was recorded in 29.18% and 47.87% of cases, respectively. Most data fell within a percentage difference of 5%, with 64.89% and 80.85% of cases, respectively. In the humidity-sensing performance, a percentage difference below 2% was recorded in 38.30% and 58.82% of cases, respectively, and a difference below 5% was recorded in 72.19% and 78.82% of cases (Figure 9). In both temperature and humidity aspects, DHT22 showed more accurate sensing ability and stable environmental sensing ability than DHT11.

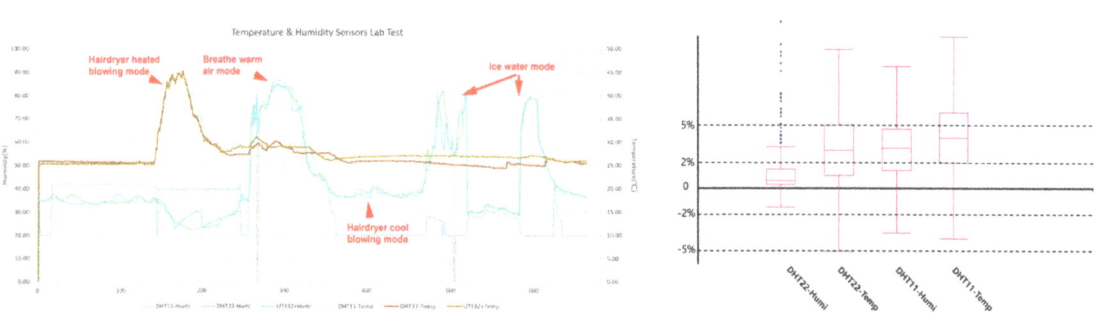

(a) (b)

Figure 9. (**a**) The 600-set data comparison of DHT11, DHT22, and UT332+ in different modes. (**b**) Statistical analyses of the error rate between the sensors and professional tester data for the 600 sets of data.

Through comparisons in the laboratory environment, SBEDAD finally selected the DHT22 as the final option for the temperature and humidity sensor.

3.3. *Air Pollution of PM1, PM2.5, and PM10*

The sensing data for air quality is mainly for particulate matter (PM) values, and sensors for CO, NO, NO_2, SO_2, and other gas detection can also be added in future research. The three most representative suspended particulate matter detections are PM1, PM2.5, and PM10 because they can be suspended in the air for a longer period, and the higher the concentration of their content in the air, the more serious the air pollution.

In this study, the Plantower G7003S laser particle sensor was selected, which uses the laser scattering principle to achieve accurate measurements with an extremely low false alarm rate and a minimum resolved particle size of 0.3 μm. (see Table 7).

Table 7. Plantower G7003S model.

Parameter	Indicator	Unit
Principle	Laser Scattering	---
Measuring range of particles	0.3~1.0; 1.0~2.5; 2.5~10	μm
Effective range of concentration	0~500	$\mu g/m^3$
Maximum mass	≥1000	$\mu g/m^3$
Mass concentration resolution	1	$\mu g/m^3$
DC voltage	4.5–5.5	V
Working current	≤100	mA
Standby current	≤200	μA
operating temperature range	−10~+60	°C
Operating humidity range	0~99%	V

By the oscilloscope test, the real frequency of the Plantower G7003S was 100 milliseconds (Figure 10a), which meets the SBEDAD requirements. Conventional PM measurements performed by gravimetric methods detect the change in mass of PM as it passes through a filter membrane. Plantower G7003S, on the other hand, uses the optical method of measurement, which has a short sampling period and allows for real-time monitoring, providing timely PM values. The technology is relatively simple and does not require complex equipment or operating procedures.

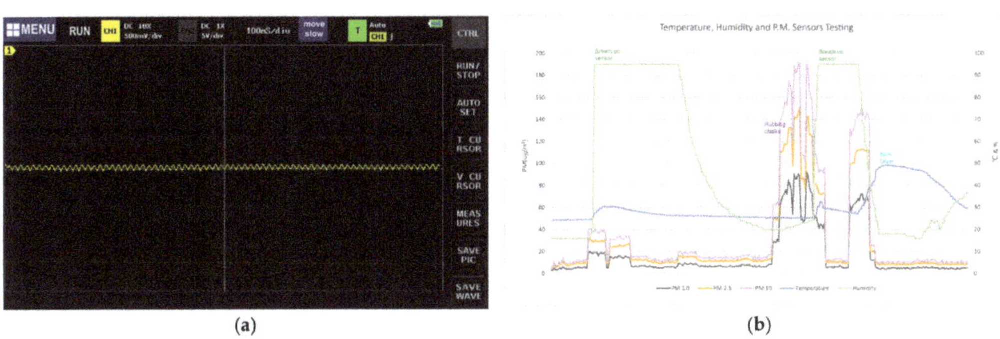

(a) (b)

Figure 10. (**a**) Measurement of the real frequency of Plantower G7003s. (**b**) The lab test with the Plantower G7003S using sandpaper to rub nails, plastics, and chalks.

We used sandpaper to sand different materials, such as nails, plastics, and chalks, and then blew the particles away to generate different particulate matter concentrations in the environment. The accuracy of the Plantower G7003s was verified through experimental analyses. In the end, we chose the G7003S.

3.4. *Illumination*

The illumination on the road has a very strong and obvious impact on the cycling experience. SVIs, which have also been used to analyze shading and illumination, are mostly collected by the data sampling vehicle that traveled on the driveways and are usually more open to the driving view. This kind of SVI does not inherit the ambient light feature on the cycling path. The SBEDAD collects the value change of the illumination on the cycling road, assisted by the panoramic camera and the temperature and humidity sensing data, which can be used to prove the comfort of the cycling experience in multiple dimensions.

The calibration of ambient lighting was performed in two environments: laboratory and outdoor. This experiment compared the data collected by two low-cost sensors (TEMT6000 and BH1750) and an industrial-grade commercial sensing device (the Konica T-10A). (See Table 8).

Table 8. Light intensity model.

	TEMT6000	BH1750	Konica T-10A

Intense range	0–1023	1–100,000 lx	0.01–299,900 lx
Deviation	1	1 lx	0.01 lux
Operating voltage	1.5–6 V DC	1.8–4.5 V DC	1.5 V DC
Operating temp	−10–100 °C	−10–85 °C	−20–55 °C
Current supply	20 mA	0.5–2.5 mA	400 mA
Sampling period	5 ms	20 ms	28 ms
Power dissipation	100 mw	260 mw	244 mw
Price	USD 1	USD 3	USD 2000

Based on the experimental design, the TEMT6000, BH1750, and Konica T-10A were placed in a dimly lit room under a lamp as the light source. The illuminance was measured at a distance of 20 cm from the light source to ensure that the light cone had a large enough diameter and to ensure uniform conditions for the different sensors in the measurement plane.

In addition, the real-frequency measurements of both sensors were conducted with an oscillograph. The data showed 5 and 20 ms, respectively (Figure 11). Through experimental analysis, it was found that the analog value of TEMT6000 stayed around 950 when the light intensity exceeded 1800 lx, which showed great instability and inaccuracy. Therefore, we ruled out this sensor.

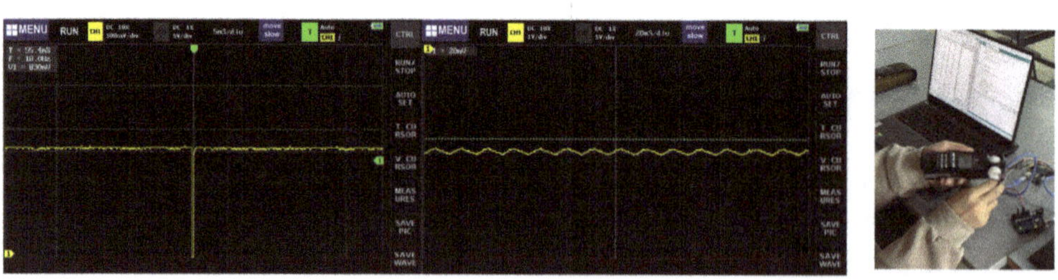

(a) (b) (c)

Figure 11. Measurement of the real frequencies of TEMT6000 (**a**) and BH1750 (**b**). Lab test of TEMT6000, BH1750, and Konica T-10A (**c**).

After determining the portable sensor, its accuracy needs to be checked and calibrated within the field test. We collected data on light in a building's corridor, in the entrance hall, under the canopies, and in the outside open space around buildings (Figure 12b). We collected 100 sets of data in a daily light environment and fit the data from BH1750 to T-10A to perform the data calibration for BH1750.

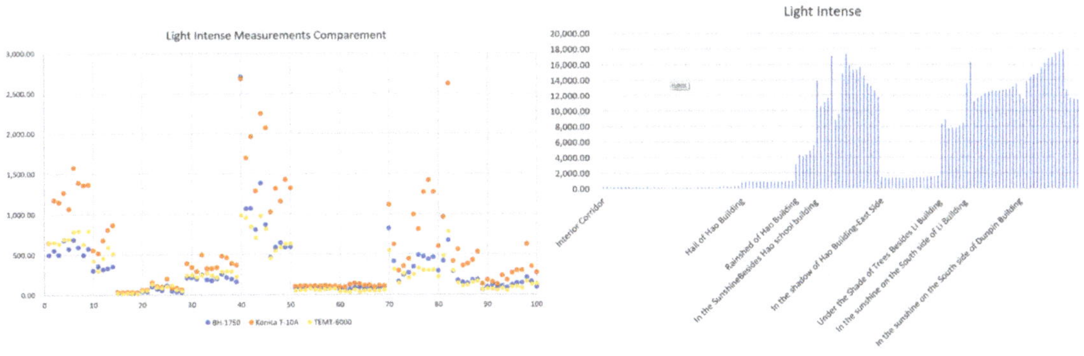

(a) **(b)**

Figure 12. (**a**) Lab test of light intensity with TEMT6000, BH1750, and Konica T-10A. The TEMT6000 shows accuracy in limited light environments (green dots), but in environments with high luminous flux, the results deviate greatly from professional measuring instruments (orange dots). (**b**) Field test of light intensity with BH1750 (**b**).

The BH1750 data were fitted to the Konica T-10A data using both polynomial and exponential functions to calibrate the data taken by the BH1750 (Figure 13). The formulas are as follows:

$$y = 1E - 07x^6 - 4E - 05x^5 + 0.0043x^{\,4} - 0.2085x^3 + 4.733x^2 - 39.174x + 131.47 \tag{1}$$

$$y = 46.375 \times e^{0.0386x} \tag{2}$$

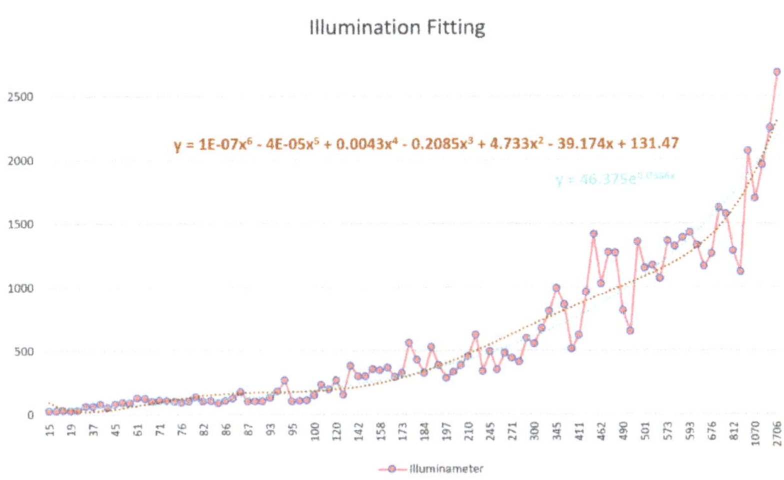

Figure 13. Illumination fitting of BH1750. A total of 100 sets of data in a daily light environment were collected (pink dots), and the data from BH1750 were fit to T-10A to perform data calibration for BH1750 in two ways, the fitting curve in brown is the result of Equation (1), while fitting curve in green is the result of Equation (2).

By contrast, we chose Formula (2) as the prediction and calibration formula. There are two ways to use this equation correction: we can directly input the formula in the Arduino IDE for numerical correction, or we can correct it after obtaining the values.

3.5. Environmental Noise Intensity

Traffic noise occupies more than 60% of the urban acoustic environment [20]. It is an important factor in polluting urban transportation space. The noise made by vehicles has little effect on drivers but has a much more serious impact on cyclists and pedestrians. Therefore, it has been mentioned in several studies [20,49–51] that the noise in the street space is a very important aspect of studying the urban street spatial environment.

We compared sensors with stable performance in different temperature and humidity environments and finally chose Yahboom's professional sound sensors (see Table 9), which are sensitive to ambient noise and have a range suitable for acquiring ambient noise in street spaces. Also, to improve the accuracy of the data, we calibrated the Yahboom data using the sensor inside the cell phone.

Table 9. Environmental noise model.

	Yahboom	Noise App

Intense range	0–1023	0–140 dB
Deviation	±1%	0.1 dB
Operating voltage	3.3–5 V DC	0.8–1 V DC
Operating temperature	−40–100 °C	−10–85 °C
Current supply	20 mA	80 mA
Sampling period	20 ms	1000 ms
Power dissipation	100 mw	18–26 W
Price	USD 0.9	USD 12

We used low-pass and high-pass filters with amplifiers to reduce the noise in this sound level measurement circuit so that the accuracy could be improved. Similarly, we used an oscilloscope to test the real frequency of the sensor, which was 10 ms (Figure 14).

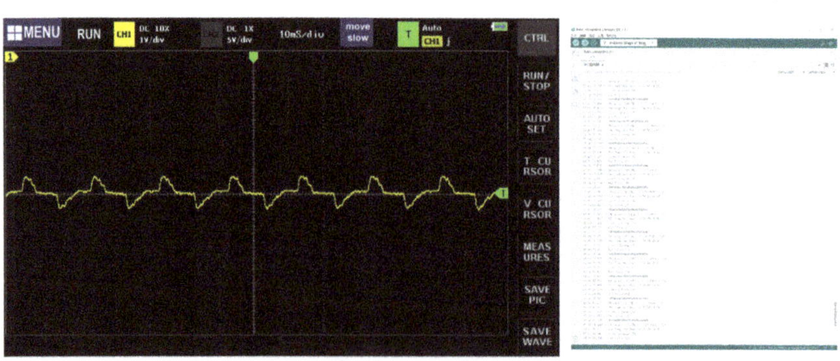

(a) (b)

Figure 14. (**a**) Measurements of the real frequency of the Yahboom sensor. (**b**) We obtained environmental noise in the lab by playing different types of sounds.

The relationship between the decibel value (dB) and the simulated value is not linear, so we need to compare the two values at different intervals. We noted down 100 sets of the ADC value obtained by the Yahboom sensor and the dB value shown on the phone, then used the "linear regression" method to complete the data fitting. This converted the irregular blue scatter to the closest possible curve (red) and generated an equation for that curve. We conducted curve fit, 5th-degree polynomials fit, and 10th-degree polynomials fit using a Python script (Figure 15) and eventually chose the 10th-degree polynomials formula (Formula (3)) to deliver data fitting.

$$dB = -1.179e-21\,x^{10} + 2.569e-18\,x^{9} - 2.38e-15\,x^{8} + 1.237e-12\,x^{7} - 4.013e-10\,x^{6} + 8.554e-08\,x^{5} - 1.216e-05\,x^{4} + 0.001132\,x^{3} - 0.06548\,x^{2} + 2.14\,x + 26.25 \quad (3)$$

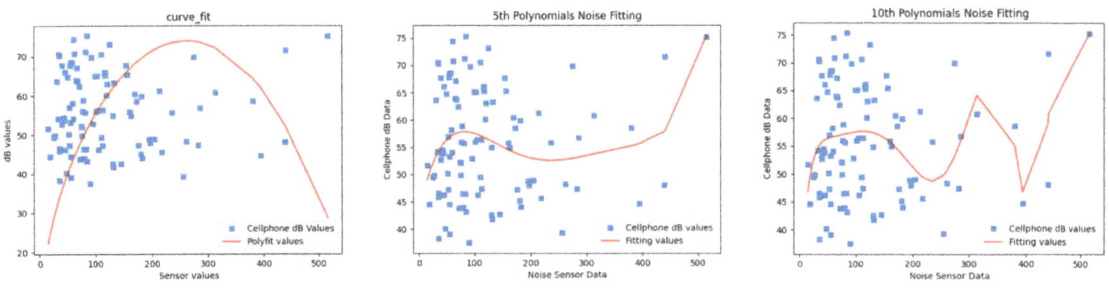

Figure 15. Data fitting with the Yahboom sensor and cellphone sensor. We fit with three modes: (**a**) curves, (**b**) 5th degree polynomials, and (**c**) 10th degree polynomials to the data using a Python script.

4. Field Measurement Results

The experiment finally selected a sample of street cycling space in central Beijing for data collection using SBEDAD, which proved to be efficient and accurate, showing good reliability through the continuity and changing characteristics of the data. The application can be expanded to work with other slow-walking spaces for multi-source data collection, providing a basis for the use of specialized sensors as well as a more critical approach to conceptualizing new scenarios and potential applications.

4.1. Field Data Acquisition Process and Results

The study replicated two sets of equipment in two groups of students for a total of 12 data collection sessions over three days (36 h in total) and collected 127.97 G of video data and 4794 Kb of txt records. Tests were carried out on 15 streets in the center of Beijing to check the operation of the equipment under real-world conditions. Compared to the completion of field data collection of similar streets by students in the researcher's design studio in the same period in the previous year, only 5.97 G of non-panoramic video data and 100 questionnaires were acquired by four research teams in 144 h during 12 days. Moreover, less than 50 questionnaires were valid. The working efficiency using SBEDAD rose to 8474.21% compared to last year. In addition, the streets' GPS location records, temperature, humidity, concentration of PM, light intensity, and noise data were acquired simultaneously in the form of text. An increased variety and density of data provides comprehensive information on the functioning and changes in cities and can lead to an intrinsic change in data-based research. (See Figure 16).

		1st Field Measurement	2nd Field Measurement	3rd Field Measurement
Date		5th March	12th March	19th March
Team A	Street Information	Street 1-1 Fuchenmen Street Dedicated/No 1km	Street 2-1 Xisi South Street Dedicated 1km	Street 3-1 Bingmasi Hutong Dedicated 423m
		Street 1-2 Picai Hutong Dedicated 1.5km	Street 2-2 Xianmen Street Dedicated 700m	Street 3-2 Fengsheng Hutong Dedicated 753m
		Street 1-3 West Huangchenggen Street Dedicated 1.1km	Street 2-3 Zhuanta Hutong Bike Lane 756m	Street 3-3 Shuyin Hutong Dedicated 480m
		Street 1-4 Taipingqiao Street Dedicated 1.1km	Street 2-4 Yangrou Hutong Dedicated 734m	Street 3-4 Shifang Street Dedicated 374m
	Data	Video 22.92G + TxtRecord 486Kb	Video 11.78G + Record 199Kb	Video 8.93G
	Weather	Light rain to cloudy	Sunny	Sunny
Team B	Street Information	Street 1-5 Dongzhimenwai Street Dedicated 2.5km	Street 2-5 Workers Stadium North Stree Dedicated/ 2.5km	Street 3-5 Yaxiu West Road Dedicated 1.5km
		Street 1-6 Sanlitun Road Dedicated 2km	Street 2-6 Xindong Road Dedicated 1.5km	Street 3-6 Yaxiu East Road No Bike Lan 222m
		Street 1-7 South Sanlitun Road Dedicated 2km	Street 2-7 Taiguli North Road No Bike La 434m	Street 3-7 Sanlitun East 3rd Road No Bike Lan 334m
		Street 1-8 Sanlitun East 2nd Road No Bike Lane 525m		
	Data	Video 20.2G + Record 337Kb	Video 8.08G+ Record 102Kb	TxtRecord 1220Kb
	Weather	Cloudy	Sunny	Sunny
Team C	Street Information	Street 1-9 Zhongguancun North Street Dedicated 3.3km	Street 2-8 Shuangqing Road Bike Lane 2km	Street 3-8 4th Ring West Road Dedicated 5.1km
		Street 1-10 Chengfu Road Dedicated 2.1km	Street 2-9 Qinghua East Road Dedicated 2.3km	Street 3-9 Xueqing Road Dedicated 3.2km
		Street 1-11 Zhixin Road Dedicated 2.1km	Street 2-10 Wangzhuang East Road Dedicated 1.6km	Street 3-1 Zhixin West Road Bike Lane 1.6km
	Data	Video 16.39G + TxtRecord 607Kb	Video 18.85G + TxtRecord 413Kb	Video 20.82G + TxtRecord 1430Kb
	Weather	Cloudy	Sunny	Sunny

(a)

(b) (c) (d)

Figure 16. Field research with SBEDAD in three major areas in the Beijing core area. (**a**) Statistical information of field measurements using SBEDAD. (**b–d**) The streets being measured by three teams.

The use of multi-source and multi-sensor data collection tools greatly improved the efficiency of street spatial analysis and research and provided a database and scientific basis for subsequent scientific research.

The research used both the panoramic camera and the SBEDAD (Figure 17). This device is convenient, portable, consumes low energy, and can be placed in the basket of a shared bicycle for data collection.

(a) (b)

Figure 17. Field test with SBEDAD. (**a**) The first version of SBEDAD. (**b**) The measurements used both the panoramic camera and SBEDAD.

Information extraction from SVI obtained by a panoramic camera enables the prediction of urban perception. The development of computer vision (CV) ensures the automatic and efficient processing of large numbers of photographs. We used a Python script to frame-pump the video. Because SBEDAD data were sampled per second, one frame was extracted from the video captured by Pilot One, which was set up to 7 frames per second, and then fused with the SBEDAD data along with the timestamp. For instance, through semantic segmentation, a deep learning technology, we could automatically and efficiently obtain the greenery coverage of different street spaces, combined with the temperature and humidity data, which are directly affected by the greenery rate, to calculate and predict the impact of different built environments on the human thermal comfort sensation.

4.2. Fusing Data to Understand Cities

Using SBEDAD to acquire diverse and massive urban spatial data, i.e., realizing urban sensing and data acquisition, can help to understand the complexity of the city. The problems of data sparsity and missing data can also be solved by SBEDAD.

Based on SBEDAD, users, i.e., students of the researcher's Urban Design Studio, can simply and easily acquire data and use data visualization tools to visualize spatial data (Figure 18) through the forms of colors, symbols, labels, etc., transforming abstract data into intuitive visual forms. Intuitively displaying data changes and trends in the city helps us better understand and analyze the spatial characteristics of the city and supports design, decision-making, and problem-solving in the city. (See Figure 19).

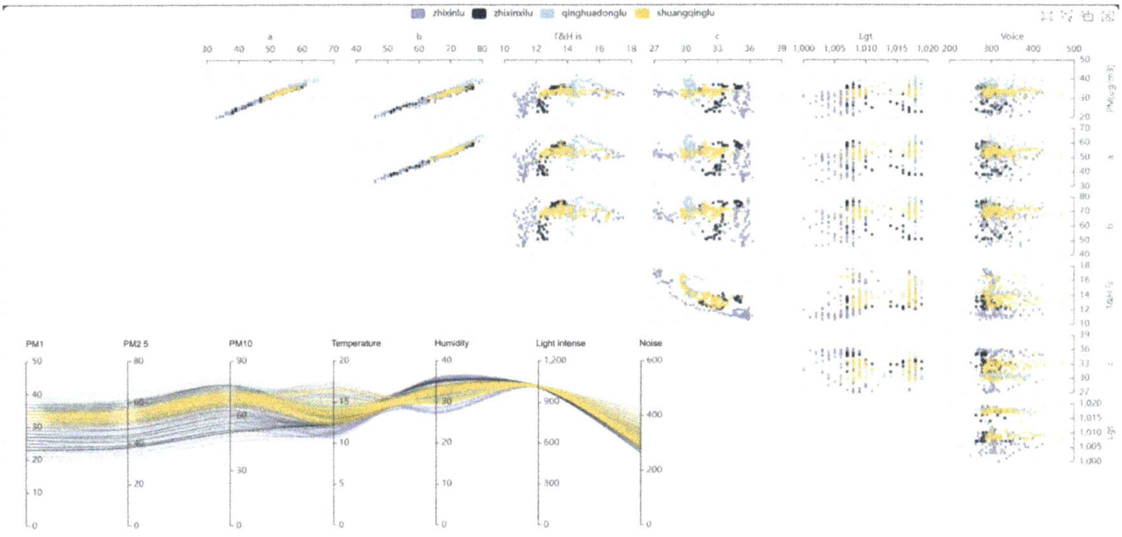

Figure 18. Urban spatial data visualization by Urban Design Studio students using the data collected from SBEDAD in conjunction with the D3.js visualization script to visualize and analyze the spatial–temporal features of urban sensing data to deeply reveal the features of human activities in the urban physical built environment.

Sensing the city for effective planning and governance is critical to the development of the city. Simply interpreting a city through one type of data is probably not comprehensive. Understanding the city requires the integration of knowledge from multiple data sources. Interpreting data with a humanized and sustainable view of the city and objectively judging the state of the city's operation are important. Ultimately, human feelings, consciousness, and social relationships can be improved, and data science can be used to simulate and deduce the effects of data-based urban optimization.

Multiple artificial intelligence algorithms and deep learning methods can be utilized to facilitate cross-domain data fusion in smart cities to capitalize on the intrinsic connections between data streams from different cities. Based on first-hand, multi-source, and cross-domain urban built environment spatial data, the combination of machine learning and deep learning techniques can promote the development of data fusion methods in the direction of urban computing.

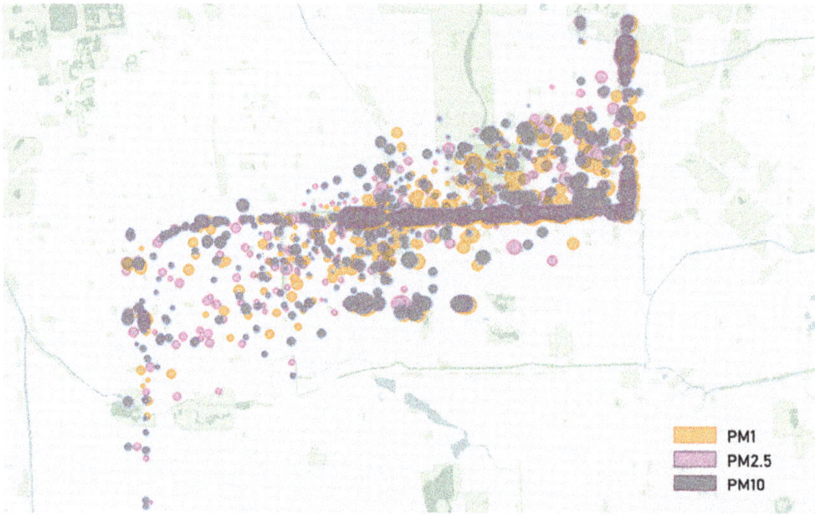

Figure 19. Urban spatial data visualization by Urban Design Studio students using the data collected from SBEDAD in conjunction with the Tableau to visualize and analyze the spatial–temporal features of PM concentration.

5. Conclusions and Future Work

Multi-modal causal learning based on multi-source urban data acquired in this study aims to improve the interpretability of intricate and dynamic urban systems. GPS location, temperature, humidity, and light intensity data can explain the micro-environment of a specific urban space in detail. Data with temporal and spatial characteristics can be combined with artificial intelligence algorithms to learn and predict street spaces' spatial and temporal characteristics.

The device based on the open-source Arduino platform presented in this study has proven to be a good choice for measuring built environment perception data in urban open spaces. The results obtained show that it is feasible to use a low-cost device for the measurement and recording of this type of data. In addition, the simplicity, portability, low energy consumption, room temperature stability, replicability, and low cost of the device compared to other commercial systems, as well as the possibility of scalability, adjustment, and improvement needed for future research, mean a reliable database can be obtained for city managers and program designers. Urban spatial analysis based on multi-source data therefore becomes possible. SBEDAD provides higher spatial resolution and area coverage for urban sensing and effectively addresses data inconsistencies arising from sensors and cameras.

This device has some limitations. Firstly, due to the short stay of the device, SBEDAD data has a high spatial resolution but a low temporal resolution, so machine learning and deep learning processing based on temporal features will be limited. Secondly, to ensure the data collected by SBEDAD had relatively high stability and reliability during data acquisition and transmission, GPS and Plantower G7003S were defined when using Arduino Mega 2560's universal asynchronous receiver/transmitter (UART) ports, which are only equipped with three pairs, which means the extensibility is limited. Although it is possible to transfer data using multiple Mega board communication, it is complex and easy to lose data. Moreover, SBEDAD was initially set up as a monolithic device that reads and stores data locally to ensure the time-resolution granularity of the data, reduce data loss and frame loss, and ensure the security of the data during transmission. SBEDAD is not equipped with any mobile data transfer module or networking module, which means it does not have real-time feedback capabilities. In future research, cloud-based data networking may be attempted, which will lead to some new technical challenges, such as data transfer protocols, data collaboration, edge computing, and so on.

In the original design, we had hoped to provide feedback on the quality of the road surface and add jitter body sensing detection. However, because the ceramic bottom jitter sensor was damaged in the field research and generated frequent error data, the jitter sensor could not effectively provide feedback on the jitter experience while riding due to the time-limited feedback and the limited range and accuracy. It is expected that an IMU system will be added to future experiments to provide pilot feedback on bumps caused by road surface quality in urban environments.

Another future research plan is to apply a Python script to create a customer-friendly interface for SBEDAD data reading, which will synthesize image frame extraction with latitude and longitude data acquisition, sensor data cleaning, and final tabulation of valid data and waveform graph visualization. This is to facilitate its use by urban designers, outdoor built environment designers, and even indoor built environment designers who do not have basic knowledge of programming and to provide designers with new perspectives on the data to improve their attention to the environment. Based on the results of this research, we hope to liberate the shackles of data on urban researchers and managers, ignite more curiosity, and cultivate an enduring passion for cross-disciplinary urban research, thereby realizing accurate urban intelligence.

Author Contributions: Conceptualization, C.L. and N.L.; methodology, C.L.; software, C.L. and L.H.; validation, L.H., Z.S. and C.W.; formal analysis, C.L.; investigation, L.H. and Z.S.; data curation, Z.S.; writing—original draft preparation, C.L.; writing—review and editing, C.L. and N.L.; visualization, C.L.; supervision, M.J.; project administration, N.L.; funding acquisition, X.W. All authors have read and agreed to the published version of the manuscript.

Funding: This research was funded by the National Natural Science Foundation of China, grant number 52278045.

Data Availability Statement: Data unavailable due to privacy restrictions.

Conflicts of Interest: The authors declare no conflicts of interest.

References

1. Llaguno-Munitxa, M.; Bou-Zeid, E. Mobile Urban Sensing Technologies (MUST) for High Spatial Resolution Urban Environmental Mapping. In *Proceedings of the 2020 Digital FUTURES*; Springer: Singapore, 2021; pp. 124–133. [CrossRef]
2. Bibri, S.E.; Krogstie, J. A Novel Model for Data-Driven Smart Sustainable Cities of the Future: A Strategic Roadmap to Transformational Change in the Era of Big Data. *Future Cities Environ.* **2021**, *7*, 2–37. [CrossRef]
3. Ganji, A.; Youssefi, O.; Xu, J.; Mallinen, K.; Lloyd, M.; Wang, A.; Bakhtari, A.; Weichenthal, S.; Hatzopoulou, M. Design, Calibration, and Testing of a Mobile Sensor System for Air Pollution and Built Environment Data Collection: The Urban Scanner Platform. *Environ. Pollut.* **2022**, *317*, 120720. [CrossRef] [PubMed]
4. Noseir, D.; Khalifa, M.; Serag, Y.; El Fayoumi, M.A. Open-Data and Data Acquisition for Smart Cities and Urban Mobility Studies: Potential Approaches and Current Challenges. *IJSBAR* **2023**, *69*, 1–15.
5. Jia, F. Neural Network Model of Urban Landscape Design Based on Multi-Target Detection. *Comput. Intell. Neurosci.* **2022**, *2022*, 9383273. [CrossRef] [PubMed]
6. Maider, M.; Bou-Zeid, E.; Bogosian, B.; Al Tair, A.; Radcliff, D.; Fisher, S.; Ryu, Y. Sensing and Information Technologies for the Environment (SITE): Hardware and Software Innovations in Mobile Sensing Technologies. In Proceedings of the International Building Physics Conference, Syracuse, NY, USA, 23–26 September 2018; Volume 9, p. 24.
7. Portella, A.A. Built Environment. In *Encyclopedia of Quality of Life and Well-Being Research*, 2nd ed.; Michalos, A.C., Ed.; Springer: Dordrecht, The Netherlands, 2014. [CrossRef]
8. Handy, S.L.; Boarnet, M.G.; Ewing, R.; Killingsworth, R.E. How the Built Environment Affects Physical Activity: Views from Urban Planning. *Am. J. Prev. Med.* **2002**, *23* (Suppl. S2), 64–73. [CrossRef] [PubMed]
9. Nouman, A.S.; Chokhachian, A.; Santucci, D.; Auer, T. Prototyping of Environmental Kit for Georeferenced Transient Outdoor Comfort Assessment. *ISPRS Int. J. Geo-Inf.* **2019**, *8*, 76. [CrossRef]
10. Apte, J.S.; Messier, K.P.; Gani, S.; Brauer, M.; Kirchstetter, T.W.; Lunden, M.M.; Marshall, J.D.; Portier, C.J.; Vermeulen, R.C.; Hamburg, S.P. High-Resolution Air Pollution Mapping with Google Street View Cars: Exploiting Big Data. *Environ. Sci. Technol.* **2017**, *51*, 6999–7008. [CrossRef] [PubMed]
11. Tsoka, S.; Tsikaloudaki, K.; Theodosiou, T.; Bikas, D. Urban Warming and Cities' Microclimates: Investigation Methods and Mitigation Strategies—A Review. *Energies* **2020**, *13*, 1414. [CrossRef]

12. Behzadfar, M.; Faizi, M.; Asl, S.R.; Beiklou, B.H. A New Collaborative Pattern between Landscape Architects and Urban Designers in Environmental Design. *Am. J. Environ. Sci.* **2010**, *6*, 344–349. [CrossRef]
13. Nguyen, H.-T.; Lee, Y.; Ahn, J.; Han, T.H.; Park, J.K. A Low-Cost Lightweight Deflectometer with an Arduino-Based Signal Interpretation Kit to Evaluate Soil Modulus. *Sensors* **2023**, *23*, 9710. [CrossRef] [PubMed]
14. Cannistraro, M.; Cao, J.; Ponterio, L. New Technique Monitoring and Transmission Environmental Data with Mobile Systems. *Instrum. Mes. Métrologie* **2018**, *18*, 549–562. [CrossRef]
15. Ottaviano, M.; Beltrán Jaunsarás, M.; Terius-Padron, J. Empowering Citizens through Perceptual Sensing of Urban Environmental and Health Data following a Participative Citizen Science Approach. *Sensors* **2019**, *19*, 2940. [CrossRef] [PubMed]
16. Nagata, S.; Nakaya, T.; Hanibuchi, T.; Amagasa, S.; Kikuchi, H.; Inoue, S. Objective Scoring of Streetscape Walkability Related to Leisure Walking: Statistical Modeling Approach with Semantic Segmentation of Google Street View Images. *Health Place* **2020**, *66*, 102428. [CrossRef] [PubMed]
17. Najafizadeh, L.; Froehlich, J.E. A Feasibility Study of Using Google Street View and Computer Vision to Track the Evolution of Urban Accessibility. In Proceedings of the 20th International ACM SIGACCESS Conference on Computers and Accessibility, Galway, Ireland, 22–24 October 2018; Association for Computing Machinery: New York, NY, USA, 2018; pp. 340–342. [CrossRef]
18. Vinci, B.; Tonacci, A.; Caudai, C.; Rosa, P.; Nencini, L.; Pratali, L. The Senseable Pisa Project: Citizen-Participation in Monitoring Acoustic Climate of Mediterranean City Centers. *Clean-Soil Air Water* **2017**, *45*, 1600137. [CrossRef]
19. Gu, P.; Han, Z.; Cao, Z.; Chen, Y.; Jiang, Y. Using Open Source Data to Measure Street Walkability and Bikeability in China: A Case of Four Cities. *Transp. Res. Rec.* **2018**, *2672*, 63–75. [CrossRef]
20. Ye, Y.; Richards, D.; Lu, Y.; Song, X.; Zhuang, Y.; Zeng, W.; Zhong, T. Measuring Daily Accessed Street Greenery: A Human-Scale Approach for Informing Better Urban Planning Practices. *Landsc. Urban Plan.* **2019**, *191*, 103434. [CrossRef]
21. Ito, K.; Biljecki, F. Assessing Bikeability with Street View Imagery and Computer Vision. *Transp. Res. Part C* **2021**, *132*, 103371. [CrossRef]
22. Zhang, Y.; Zhao, H.; Li, Y.; Long, Y.; Liang, W. Predicting Highly Dynamic Traffic Noise Using Rotating Mobile Monitoring and Machine Learning Method. *Environ. Res.* **2023**, *229*, 115896. [CrossRef]
23. Anguelov, D.; Dulong, C.; Filip, D.; Frueh, C.; Lafon, S.; Lyon, R.; Ogale, A.; Vincent, L.; Weaver, J. Google Street View: Capturing the World at Street Level. *Computer* **2010**, *43*, 32–38. [CrossRef]
24. Li, X.; Ratti, C. Using Google Street View for Street-Level Urban Form Analysis, a Case Study in Cambridge, Massachusetts. In *The Mathematics of Urban Morphology*; D'Acci, L., Ed.; Springer International Publishing: Cham, Switzerland, 2019; pp. 457–470. [CrossRef]
25. Costa, R.; Resende, R.; Fernandes, S. Urban Data Collectors: A Pragmatic Approach to Leveraging Urban Sensing. In Proceedings of the 2015 IFIP/IEEE International Symposium on Integrated Network Management (IM), Ottawa, ON, Canada, 11–15 May 2015; IEEE: Piscataway, NJ, USA, 2015; pp. 886–889.
26. Kabisch, N.; Selsam, P.; Kirsten, T.; Lausch, A.; Bumberger, J. A Multi-sensor and Multi-temporal Remote Sensing Approach to Detect Land Cover Change Dynamics in Heterogeneous Urban Landscapes. *Ecol. Indic.* **2018**, *99*, 273–282. [CrossRef]
27. Campbell, A.T.; Eisenman, S.B.; Lane, N.D.; Miluzzo, E.; Peterson, R.A. People-Centric Urban Sensing. In Proceedings of the 2nd Annual International Workshop on Wireless Internet, Boston, MA, USA, 2 August 2006. [CrossRef]
28. Zhou, D.; Xiao, J.; Bonafoni, S.; Berger, C.; Deilami, K.; Zhou, Y.; Frolking, S.; Yao, R.; Qiao, Z.; Sobrino, J.A. Satellite Remote Sensing of Surface Urban Heat Islands: Progress, Challenges, and Perspectives. *Remote Sens.* **2019**, *11*, 48. [CrossRef]
29. Lynch, J.P.; Kenneth, J.L. A Summary Review of Wireless Sensors and Sensor Networks for Structural Health Monitoring. *Shock. Vib. Dig.* **2006**, *38*, 91–130. [CrossRef]
30. Zanella, A.; Bui, N.; Castellani, A.; Vangelista, L.; Zorzi, M. Internet of Things for Smart Cities. *IEEE Internet Things J.* **2014**, *1*, 22–32. [CrossRef]
31. Cardone, G.; Bellavista, P.; Corradi, A.; Foschini, L. Effective Collaborative Monitoring in Smart Cities: Converging MANET and WSN for Fast Data Collection. In Proceedings of the Kaleidoscope 2011: K-2011, Proceedings of ITU, Cape Town, South Africa, 12–14 December 2011; pp. 1–8.
32. Sideris, N.; Bardis, G.; Voulodimos, A.; Miaoulis, G.; Ghazanfarpour, D. Enhancing Urban Data Analysis: Leveraging Graph-Based Convolutional Neural Networks for a Visual Semantic Decision Support System. *Sensors* **2024**, *24*, 1335. [CrossRef] [PubMed]
33. Longo, A.; Zappatore, M.; Bochicchio, M.; Navathe, S.B. Crowd-Sourced Data Collection for Urban Monitoring via Mobile Sensors. *ACM Trans. Internet Technol.* **2017**, *18*, 1–21. [CrossRef]
34. Nuortio, T.; Kytöjoki, J.; Niska, H.; Bräysy, O. Improved Route Planning and Scheduling of Waste Collection and Transport. *Expert Syst. Appl.* **2006**, *30*, 223–232. [CrossRef]
35. Chen, Y.; Nakazawa, J.; Yonezawa, T.; Tokuda, H. Cruisers: An Automotive Sensing Platform for Smart Cities Using Door-to-door Garbage Collecting Trucks. *Ad Hoc Netw.* **2019**, *85*, 32–45. [CrossRef]
36. O'keeffe, K.; Santi, P.; Wang, B.; Ratti, C. Urban Sensing as a Random Search Process. *Phys. A Stat. Mech. Its Appl.* **2020**, *562*, 125307. [CrossRef]
37. Quintero, G.; Balastegui, A.; Romeu, J. Traffic Noise Assessment Based on Mobile Measurements. *Environ. Impact Assess. Rev.* **2021**, *86*, 106488. [CrossRef]
38. Guillaume, G.; Aumond, P.; Chobeau, P.; Can, A. Statistical Study of the Relationships between Mobile and Fixed Stations Measurements in Urban Environment. *Build. Environ.* **2019**, *149*, 404–414. [CrossRef]

39. Kang, Y.; Zhang, F.; Gao, S.; Lin, H.; Liu, Y. A Review of Urban Physical Environment Sensing Using Street View Imagery in Public Health Studies. *Spat. Sci.* **2020**, *26*, 261–275. [CrossRef]
40. The Official Website of Arduino. Available online: https://www.arduino.cc (accessed on 15 April 2024).
41. World Health Organization. *WHO Global Air Quality Guidelines: Particulate Matter (PM2.5 and PM10), Ozone, Nitrogen Dioxide, Sulfur Dioxide and Carbon Monoxide*; World Health Organization: Geneva, Switzerland, 2021.
42. Giannopoulou, K.; Livada, I. The influence of air temperature and humidity on human thermal comfort over the greater Athens area. *Sustain. Cities Soc.* **2021**, *10*, 184–194. [CrossRef]
43. Liu, M.; Zhang, B.; Luo, T.; Liu, Y.; Portnov, B.A.; Trop, T.; Jiao, W.; Liu, H.; Li, Y.; Liu, Q. Evaluating Street Lighting Quality in Residential Areas by Combining. Remote Sensing Tools and a Survey on Pedestrians' Perceptions of Safety and Visual Comfort. *Remote Sens.* **2022**, *14*, 826. [CrossRef]
44. Sengupta, P.; Garrity, P. Sensing Temperature. *Curr. Biol.* **2013**, *23*, 304–307. [CrossRef]
45. Luger, T.; Amagai, M.; Dreno, B.; Dagnelie, M.-A.; Liao, W.; Kabashima, K.; Schikowski, T.; Proksch, E.; Elias, P.M.; Simon, M.; et al. Atopic Dermatitis: Role of the Skin Barrier, Environment, Microbiome, and Therapeutic Agents. *J. Dermatol. Sci.* **2021**, *102*, 142–157. [CrossRef] [PubMed]
46. Fitoussi, R.; Faure, M.-O.; Beauchef, G.; Achard, S. Human Skin Responses to Environmental Pollutants: A Review of Current Scientific Models. *Environ. Pollut.* **2022**, *306*, 119316. [CrossRef] [PubMed]
47. Bahashwan, E. Awareness and Knowledge of Sun Exposure and Use of Sunscreen Among Adults in Aseer Region, Saudi Arabia. *Saudi Pharm. J.* **2024**, *32*, 102019. [CrossRef]
48. Thakre, C.; Laxmi, V.; Vijay, R.; Killedar, D.J.; Kumar, R. Traffic Noise Prediction Model of an Indian Road: An Increased Scenario of Vehicles and Honking. *Environ. Sci. Pollut. Res.* **2020**, *27*, 38311–38320. [CrossRef] [PubMed]
49. Mioduszewski, P.; Ejsmont, J.A.; Grabowski, J.; Karpiński, D. Noise Map Validation by Continuous Noise Monitoring. *Appl. Acoust.* **2011**, *72*, 582–589. [CrossRef]
50. Zambon, G.; Roman, H.E.; Smiraglia, M.; Benocci, R. Monitoring and Prediction of Traffic Noise in Large Urban Areas. *Appl. Sci.* **2018**, *8*, 251. [CrossRef]
51. Zou, X.; Yan, Y.; Hao, X.; Hu, Y.; Wen, H.; Liu, E.; Zhang, J.; Li, Y.; Li, T.; Zheng, Y.; et al. Deep Learning for Cross-Domain Data Fusion in Urban Computing: Taxonomy, Advances, and Outlook. Computer Science, Environmental Science, Engineering. *arXiv* **2024**, arXiv:2402.19348.

smart cities

MDPI

Article

Optimization of Geothermal Heat Pump Systems for Sustainable Urban Development in Southeast Asia

Thiti Chanchayanon [1], Susit Chaiprakaikeow [2,*], Apiniti Jotisankasa [2] and Shinya Inazumi [3,*]

1 Graduate School of Engineering and Science, Shibaura Institute of Technology, Tokyo 135-8548, Japan; na23108@shibaura-it.ac.jp
2 Department of Civil Engineering, Kasetsart University, Bangkok 10900, Thailand; fengatj@ku.ac.th
3 College of Engineering, Shibaura Institute of Technology, Tokyo 135-8548, Japan
* Correspondence: fengssck@ku.ac.th (S.C.); inazumi@shibaura-it.ac.jp (S.I.)

Abstract: This study examines the optimization of ground source heat pump (GSHP) systems and energy piles for sustainable urban development, focusing on Southeast Asia. GSHPs, which utilize geothermal energy for indoor HVAC needs, offer a sustainable alternative to traditional systems by utilizing consistent subsurface temperatures for heating and cooling. The study highlights the importance of understanding thermal movement within the soil, especially in soft marine clays prevalent in Southeast Asia, to improve GSHP system efficiency. Using a one-dimensional finite difference model, the study examines the effects of soil thermal conductivity and density on system performance. The results show that GSHP systems, especially when integrated with energy piles, significantly reduce electricity consumption and greenhouse gas emissions, underscoring their potential to mitigate the urban heat island effect in densely populated areas. Despite challenges posed by the region's hot and humid climate, which could affect long-term effectiveness, the study highlights the need for further study, including field experiments and advanced modeling techniques, to optimize GSHP configurations and fully exploit geothermal energy in urban environments. The study's insights into soil thermal dynamics and system design optimization contribute to advancing sustainable urban infrastructure development.

Keywords: energy pile; finite difference method; ground source heat pump system; soil thermal conductivity; sustainable urban development; urban heat island effect

1. Introduction

In the construction of smart cities, implementing energy piles and assessing heat transfer within the ground is paramount for efficient and sustainable urban development. The use of shallow geothermal systems, also known as ground source heat pump (GSHP) systems, utilizes the geothermal energy available beneath the surface to provide either heating or cooling resources. This technology meets the thermal requirements for indoor Heating, Ventilation, and Air-Conditioning (HVAC) systems, as described by Sarbu and Sebarchievici [1]. GSHP systems can be configured in several forms, including vertical GSHP systems integrated into the pile foundations, which are essential for energy piles in smart cities, horizontal GSHP systems, and GSHP systems that use a reservoir for cooling. For smaller residential buildings, a closed-loop system is often recommended due to its ease of maintenance and operation, as shown in Figure 1. Shallow GSHP systems, defined by Eugster and Sanner [2] as operating at a depth of 3 to 100 m, use the soil mass as a thermal reservoir. This approach provides a consistent temperature source throughout the year, in contrast to the variable air temperatures at the surface [3,4]. Shallow geothermal systems are primarily used for heating and cooling by exploiting subsurface temperatures below 25 °C, making them suitable for small-scale and residential applications in various locations. Conversely, deep geothermal systems target temperatures above 25 °C and can reach up to 200 °C, serving different energy needs [2].

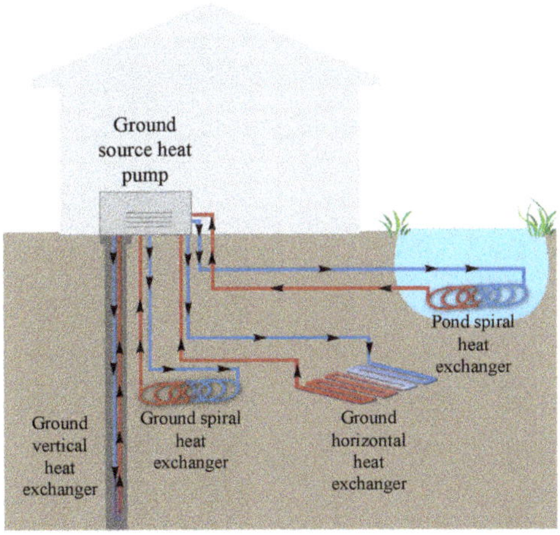

Figure 1. General overview of shallow ground source heat pump (GSHP) installations.

This study explores the use of GSHP systems, with a focus on shallow geothermal systems, for smart city development. GSHPs use the constant temperature below the earth's surface for heating or cooling, efficiently meeting indoor HVAC requirements. The study emphasizes the integration of energy piles—structures that both support buildings and function as part of a GSHP system—and highlights their potential in sustainable urban construction. By evaluating thermal movement within the soil, particularly in soft marine clays common in Southeast Asia, the study advances the understanding of how GSHP systems can be optimized for tropical climates where the urban heat island effect is a significant challenge. A notable aspect of this study is the use of a one-dimensional finite difference model to predict the temperature distribution around energy piles over time, allowing for a simplified yet effective analysis. This approach facilitates the evaluation of the effect of soil thermal conductivity and density on GSHP system efficiency. The results indicate that soil density affects the thermal performance of the system. Denser soils improve thermal conductivity. By focusing on the practical application of GSHP systems in smart cities, especially in regions such as Bangkok with specific geotechnical challenges, this study provides valuable insights into the deployment of sustainable energy solutions. It highlights the need for further study to optimize GSHP configurations and operating parameters to realize the full potential of geothermal energy in urban development. The study's implications for smart city construction include the reduction of electricity consumption, mitigation of greenhouse gas emissions, and improvement of HVAC system efficiency, demonstrating a promising path towards sustainable urban living environments.

While existing studies focus primarily on the technical efficiency and economic aspects of GSHP systems in temperate climates, this study extends existing studies to tropical regions, addressing a significant knowledge gap. Unlike previous studies that rarely evaluate the interaction between soil thermal properties and GSHP performance in hot and humid climates, this study uses a one-dimensional finite difference model to analyze the effects of soil thermal conductivity and density on system efficiency. This approach is novel in its application to the soft marine clays prevalent in Southeast Asia, which differ significantly from the more commonly studied temperate zone soils. By integrating energy piles (which are typically used as part of a building's foundation) into the GSHP system, this study utilizes a dual-function infrastructure that enhances both structural support and energy efficiency. The results illustrate how variations in soil properties can affect system performance, which is a critical consideration for urban planners and engineers seeking to deploy these systems in similar urban settings. The originality of this study lies in its methodological innovation and its focus on a region where high ambient temperatures and urban heat island effects pose unique challenges. The integration of field experiments and advanced modeling provides new insights into the optimization of GSHP configurations, which is critical to realizing the full potential of geothermal energy in urban sustainability initiatives. This study not only fills an important research gap but also sets the stage for future innovations in geothermal energy use in tropical urban environments.

2. Evaluation and Prospects of GSHP Systems

2.1. GSHP Systems and Thermal Movement Evaluation

Especially in the integration of energy piles, the evaluation of thermal movement in the ground becomes crucial. This evaluation is essential for effectively operating GSHP systems in cooling modes. The core architecture of the GSHP system consists of three essential elements: the heat source, the heat sink, and the heat exchanger. Specifically, the system's heat source is derived from the operation of the HVAC system that regulates the building's indoor environment. Conversely, the ground acts as a heat sink, absorbing excess heat. The heat exchanger transfers heat from the source to the sink. The heat transfer fluid, typically a water-antifreeze solution, is circulated by a pump, facilitating heat transfer by conduction and convection. Heat exchangers can be divided into two types: open-loop and closed-loop systems. An open-loop system uses groundwater drawn from wells to cool and returns the heated water to the aquifer. In contrast, a closed-loop system uses a network of High-Density Polyethylene (HDPE) pipes in series or parallel configuration, as shown in Figure 2 [5]. The most basic design within this system is the U-shaped exchanger, which is vertically embedded in boreholes or integrated into the building's foundation piles, providing a path for the circulating medium. To improve heat dissipation, various configurations are used, such as the double U-pipe in parallel or cross arrangement or the spiral/coil structure, to maximize the contact surface area for efficient heat exchange.

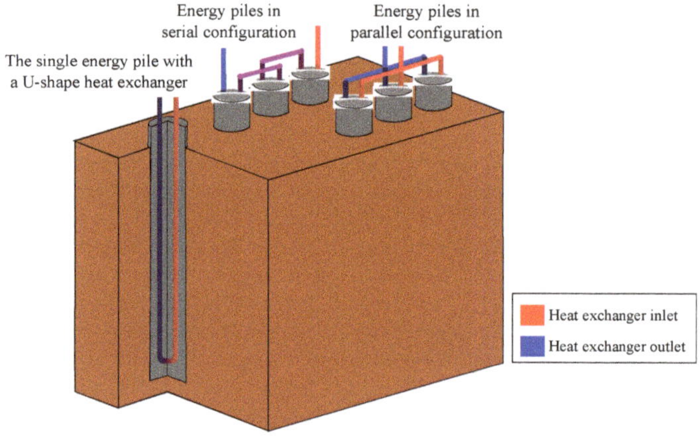

Figure 2. Schematic mechanism of a heat exchanger in an energy pile.

Understanding and evaluating thermal movement in the ground is paramount in implementing energy piles. This understanding hinges on the concept of thermal conductivity, a material property that indicates how well heat can move through a substance. Clay soil, with thermal conductivity values ranging from approximately 1 to 2.5 W/(m·K), has a significantly higher capacity for heat transfer than air, which has a thermal conductivity of only 0.026 W/(m·K), demonstrating the superior efficiency of the ground in conducting heat by a factor of nearly 40 [6]. This characteristic underscores the GSHP system's environmental benefits, which significantly reduces electricity consumption and greenhouse gas emissions compared to traditional air source heat pump systems. The versatility of GSHPs makes them an optimal choice for various settings, including residential, commercial, and industrial buildings, as highlighted by Tissen et al. [7]. Specifically, GSHP systems are distinguished by their quieter operation, longer lifespan, and lower maintenance requirements than their air source counterparts. Although the initial investment for GSHP installations is generally higher than for air source systems, the long-term energy savings achieved typically offset the upfront costs. From 1995 to 2020, there was a remarkable increase in the adoption of ground source heat pumps, indicating a growing degree of recognition of their benefits and efficiency. According to the most recent data, the leading countries in terms of installed GSHP capacity include China, the United States, Sweden, Germany, and Finland, with capacities of 26,450, 20,230, 6680, 4400, and 2300 MWt, respectively [8,9]. This global trend underscores the growing appreciation of GSHP systems as a sustainable and efficient solution in the development of smart cities and underscores the critical role of accurately assessing the potential of geothermal energy in urban planning and construction.

2.2. Regional Focus on Europe and Southeast Asia

The origin of GSHP technology can be traced to the European region, where GSHPs are mostly operated in heating mode to provide heat to the heating infrastructure. Sayegh et al. (2019) [10] discussed how the increasing popularity of district heating in Europe had a positive impact on the development and contributed to the advancement of heat pump technology, of which GSHP is a subset. The integration of heat pump technology into district heating networks led to an increase in efficiency and flexibility, resulting in a lower level of carbon emissions. The four basic heat pump integration scenarios are as follows: (i) heat pump integration into the existing network without major changes, (ii) heat pump integration into an extended network, (iii) deep renovation of the existing district heating system, and (iv) the design of a new district heating system supplied by a heat pump. In the current situation, heat pumps are considered a renewable energy technology in Europe, where they account for 5% to 20% of the European renewable energy targets. The success of this policy implementation will help Europe to achieve its sustainability goals faster, to the benefit of its inhabitants.

The viable heat source for GSHP systems used in residential and commercial buildings in Europe mainly comes from the shallow soil layer. Brys et al. (2019) [11] concluded the results of a thermal analysis of over 10 years (August 2007 to July 2017). They collected the measurements of net solar radiation (the difference between incoming solar radiation and outgoing reflected solar radiation) data of positive heat fluxes ($G > 0$) and negative heat fluxes ($G < 0$) in bare and grassy soil in Wroclaw, Poland. Heat fluxes (a measure of how much heat is transferred per unit area per unit time) varied seasonally and annually, with bare soil showing higher heat flux values than grass-covered soil. The ten-year average sums for negative soil heat fluxes, which are suitable for GSHP operation in heating mode, reached 91.8 kWh/m^2 for the bare soil and 47.4 kWh/m^2 for the turfed area. The negative heat flux reaches its minimum value in May and its maximum value in October. This confirms that the shallow soil layer has the potential to be an efficient and environmentally friendly heat source for GSHP operation throughout the year.

Focusing on the Southeast Asia region, an area that has recently begun to attract attention for exploring the implementation of GSHP systems, the effects of urbanization come to the fore. The region's rapid urban expansion, coupled with an influx of population into densely populated urban centers, contributes to the urban heat island effect. This phenomenon results in urban areas experiencing significantly higher temperatures than their rural counterparts due to the heat-trapping properties of buildings, roads, and other infrastructure. Such conditions not only reduce the efficiency of conventional air source HVAC systems, but also increase their energy consumption.

Due to the hot and humid climate, most buildings in Southeast Asia are equipped with air-conditioning systems. In urban areas, the density of these units is particularly high, leading to a concentration of noise. Poor maintenance of air conditioners can result in increased noise due to worn components, loose parts, or inefficient operation, combined with inadequate noise regulations and poorly enforced codes. This allows building owners and operators to neglect noise mitigation measures, exacerbating the problem. The outdoor installation of the air conditioner's heat exchanger, cooling tower, and compressor is located in confined spaces and close to residential areas, often on street-facing walls or in narrow alleys. This amplifies the noise and directs it into public spaces. This means that noise from AC units can easily reach residential areas, disturbing residents and affecting their quality of life.

2.3. *Performance Metrics and Future Prospects*

2.3.1. Performance Metrics for GSHP

To measure the efficiency of the GSHP compared to the traditional air source heat pump, it would be necessary to use the index numbers. The Coefficient of Performance (COP) is the ratio of energy output (watts unit) to energy input (watts unit). A higher COP means the air conditioner can provide more cooling while using less energy at a given point in time under ideal conditions. A higher COP means the heat pump is producing more heat output for a given amount of electrical input, indicating higher efficiency, translating to lower electricity consumption, resulting in lower energy bills. On the other hand, the Seasonal Coefficient of Performance (SCOP) is a more comprehensive index that considers a heat pump's efficiency over an entire cooling season that lasts several months. It takes into account varying temperatures and weather conditions. A higher SCOP indicates better overall efficiency throughout the year and helps to reduce the greenhouse gas emissions and overall environmental impact. Another commonly used measurement index is the Energy Efficiency Ratio (EER). This metric is generally used to measure the cooling efficiency of a heat pump or GSHP at a given point in time. It is the ratio of cooling output (BTU unit) to energy input (watts unit). A higher EER indicates that the air conditioner can provide more cooling while using less energy. The Seasonal Energy Efficiency Ratio (SEER), similar to SCOP, is an index that evaluates the cooling efficiency of a heat pump over an entire cooling season, taking into account varying temperatures and conditions. A higher SEER indicates better cooling performance throughout the year.

COP/SCOP and EER/SEER are typically compared between ASHP and GSHP. The GSHP yields the higher score in all COP, SCOP, EER, and SEER indexes compared to traditional ASHP at 2.5 to 4.0 compared to 3.1–5.0 in the COP index, 2.5 to 3.5 compared to 3.5–4.5 in the SCOP index, 8.0 to 12 compared to 14.0–20.0 in the EER index, and 13.0 to 20 compared to 16.0–30.0 in the SEER index. This is because the ground maintains a more constant temperature than the air, resulting in more efficient heat exchange. However, GSHPs typically have higher installation costs due to the need for ground loop systems but come with the benefit of lower operating costs compared to ASHPs, while ASHPs are usually cheaper to install but may have higher operating costs.

2.3.2. Challenges and Prospects in Southeast Asia

The current issue of energy piles and ground source heat pump (GSHP) systems in Southeast Asia is still in its early stages, with minimal research on both practical and simulated models. Given the region's high ambient temperatures, there are concerns about the efficiency and environmental impact of GSHP installations from both a technical and environmental perspective. As a result, available data is scarce. However, there is considerable room for development and these systems may become commercially viable. In the future, I expect the popularity of GSHPs to increase in Southeast Asia. Because of the high initial prices, they may be used in high-rise and large structures at first, but I think they will eventually become ubiquitous in residential areas as part of standard air-conditioning systems.

Faced with these challenges, GSHP systems emerge as a potential solution to mitigate the adverse effects of urban heat islands. However, the tropical, consistently hot, and humid climate of Southeast Asia presents a unique technical hurdle. The relatively low ground-to-air temperature differential characteristic of the region suggests that GSHPs would operate predominantly in cooling mode. As a result, the long-term effectiveness and efficiency of GSHP systems could be compromised as the ground temperature could eventually converge with the ambient air temperature, reducing the cooling potential of the system. Despite these challenges, the volume of field experiments and studies specific to GSHP performance in such climates remains limited. This gap underscores the need for a comprehensive study to evaluate the viability and optimization of GSHP systems under tropical conditions.

2.3.3. Case Studies and Soil Analysis in Southeast Asia

Figure 3 [4] shows the geographical distribution of studies on GSHP systems in the Southeast Asian region, showing a wide range of studies in Thailand [12–16], Indonesia [17], Malaysia [18], Singapore [19], and Vietnam. In particular, the study conducted in Kamphaeng Phet by Yasukawa et al. [16] is recognized as the first investigation of GSHP technology in the region, highlighting the potential benefits of such applications at the study site. These studies were conducted primarily at sites underlain by soft marine clay deposits, with soil samples collected at depths ranging from 5 to 20 m below the surface. Table 1 shows that the clay samples have a similar specific gravity (Gs), liquid limit (LL), plastic limit (PL), and sand-silt-clay percentage content to Bangkok clay. As shown in Table 1 [20–32], a commonality among these samples is their classification as marine deposit clays. Such clays, formed by sedimentation processes in marine contexts such as deltaic and coastal plains, include notable varieties such as Ariake clay from Japan, Singapore marine clay, Hong Kong marine clay, Kuantan clay from Indonesia, and Ho Chi Minh City marine clay. The latter has significant engineering similarities to Bangkok's soft clay, characterized by high plasticity, low permeability, and significant compressibility. These soil characteristics pose substantial challenges to construction efforts, particularly in urban areas where soil conditions play a critical role in ensuring the structural integrity of densely populated building projects. The inclusion of GSHP systems in such environments requires a thorough understanding of local geotechnical conditions to optimize design and functionality, underscoring the critical importance of evaluating thermal movement on the ground as part of smart city infrastructure development. The focus is on the use of energy piles and shallow GSHPs where soft marine clays are present. For example, one location that has started to gain attraction is around Bangkok, where the study of GSHP system implementation has begun. If the result of the study in Bangkok is obtained, it can be used in other places with similar soil. The background temperature of the soil in the Bangkok area was studied in 2011 [15]. The temperature was obtained from 56 temperature depth profiles throughout the central region of Thailand; the measurements were taken at every 2 m interval with an accuracy of 0.01 °C. The temperature at shallow depths from 0 to 50 m below the ground surface is in the range of 27 °C to 30 °C, which is, on average, 5 °C lower than the monthly average maximum atmospheric temperature.

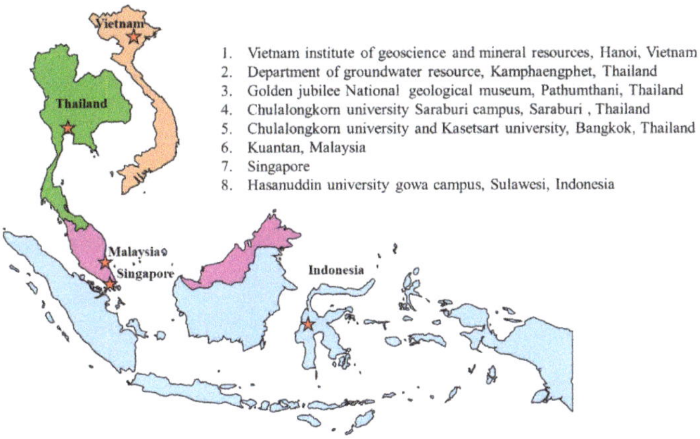

Figure 3. Locations of GSHP experiments within Southeast Asian countries in tropical climates [4].

Table 1. Basic engineering properties of clay sedimentary formations in different countries.

Soil	Unit Weight (kN/m^3)	Water Content (%)	Thickness (m)	Thermal Conductivity (W/m·c)	Specific Gravity	Plastic Limit (%)	Liquid Limit (%)
Ariake clay (Japan)	13.50–14.00	90.0 to 120.0	15.0 to 20.0	0.65 to 0.75	2.65	25.0 to 56.0	61.6 to 88.7
Singapore marine clay	14.23–15.70	80.0 to 95.0	14.0 to 17.0	N/A	2.60 to 2.72	20.0 to 28.0	80.0 to 95.0
Hong Kong marine clay	13.9–18.25	40.8 to 98.6	10.0 to 55.0	N/A	2.62 to 2.67	18.6 to 50.8	36.7 to 87.3
Kuantan clay (Indonesia)	15.00	N/A	N/A	0.45	2.64 to 2.78	18.9 to 25.7	30.5 to 63.5
Bangkok clay	15.20–16.18	60.0 to 80.0	10.0 to 14.0	0.80 to 1.00	2.56	95.0 to 100.0	60.0 to 80.0

For each clay property, Ariake clay refers to [20–22], Singapore marine clay to [23,24], Hong Kong marine clay to [25], Kuantan clay to [26,27], and Bangkok clay to [28–32], respectively.

In 2020, a collaborative study effort between Japanese and Thai academics, highlighted by the work of Shimada et al. [14], was conducted at the facilities of Chulalongkorn University to conduct both experiments and simulations on an operational GSHP system. The investigation included a Thermal Response Test (TRT) to validate the simulation model within a pilot plant setup. This GSHP configuration included two boreholes 6 m apart, each 50 m deep and 0.23 m in diameter. The first borehole was equipped with two sets of U-tubes at 10 and 15 m depths connected in series, while the second borehole had U-tubes installed at the 50 m mark. Notably, this GSHP system was integrated into the air-conditioning infrastructure of a compact building on the Bangkok campus known as "Parot Racha". The results of this study showed that the GSHP system was more efficient than traditional air source heat pump systems, achieving a reduction in electricity consumption of approximately 30% until the heat pump inlet temperature reached 5 °C above ambient. This result underscores the significant potential of GSHP technology to reduce long-term energy consumption in buildings despite Bangkok's elevated daytime temperatures. Further progress in 2020 has been documented by Jotisankasa et al. [13], who conducted a comprehensive energy pile study in a residential building. The study used a 17.5 m-long, 152 mm diameter steel pipe pile equipped with a U-tube heat exchanger in a parallel configuration. The GSHP setup consisted of two different loops, labeled number 1 and number 2, with the former consisting of six interconnected piles and the latter consisting of seven. Preliminary evaluations of the GSHP efficiency, as measured by the thermal conductivity of the system, yielded values of 2.01 W/m-K for loop number 1 and 1.49 W/m-K for loop number 2. These evaluations were conducted with the piles positioned in the stiff clay layers of Bangkok, highlighting the significant impact of groundwater infiltration on the performance of the GSHP system. This study underscores the critical role of assessing thermal dynamics within the soil for the effective implementation of GSHP technologies in smart city developments, particularly in energy pile designs.

Understanding the role of thermal movement in the soil is critical to building smart cities and integrating energy piles. Thermal conductivity, a determinant of how well heat moves through the soil, varies among different soil types, even those with similar engineering properties. The main factors affecting thermal conductivity are moisture content and density. In Southeast Asia, natural soft clays at shallow depths are often completely saturated with water. High moisture content and saturation levels reduce the presence of air, which acts as an insulating layer in the voids between soil particles. In addition, the movement of water through these voids enhances heat convection, resulting in increased thermal conductivity. Similarly, the density of the soil plays an important role; denser soil allows for better contact between particles, which increases heat conduction and, consequently, thermal conductivity. The application of GSHP and energy pile technologies in Southeast Asia is still in its infancy, indicating a significant area for study and exploration. The specific effects of variations in the effective thermal conductivity of GSHP systems, as well as the thermal conductivity and density of soft clays, on temperature modulation and heat dissipation in the soil, remain poorly understood. Thailand, which is at the forefront of the GSHP study and has extensive data on the characteristics of Bangkok soft clay, provides an excellent opportunity for detailed case studies. Such a study could provide valuable insights applicable to other Southeast Asian regions with similar geotechnical characteristics, thereby advancing GSHP feasibility studies, system designs, and performance evaluations. This effort is essential to realize the full potential of GSHP systems in supporting the sustainable infrastructure of smart cities.

3. Fundamental Theory of Heat Transfer in Soil Using Finite Difference Method

Typically, preliminary design heat transfer calculations from the energy foundation to the surrounding soil mass are evaluated based on the magnitude of temperature changes in the soil at various distances from the energy foundation. These calculations are critical for predicting the performance of ground source heat pump (GSHP) systems before the foundation design is finalized.

The basic theories needed to understand the finite difference method for studying soil heat transfer can be summarized as follows: 1. The definition and advantages of the finite difference method; 2. The model simplification and boundary conditions; 3. The model computational domain and computational nodes (meshing); 4. The transformation from differential equation to algebraic equation solution.

The analysis technique using the finite difference method (FDM) offers several advantages, including a relatively simple calculation procedure. It uses algebraic equations instead of the complex heat transfer differential equations typically used in other methods. In addition, this analysis can be performed using a simple spreadsheet, using a mathematical equation derived from a more complex differential equation that can be solved by entering specific parameters. Finite difference analysis allows for the calculation of the time-dependent temperature evolution of the soil around the energy pile, with the scale of the time step adjustable to seconds, minutes, hours, or days, depending on the study requirements. Given the time constraints of this study, the finite difference method is identified as the most appropriate and direct approach.

In this study, the finite difference model of the energy piles is conceptualized as elongated cylindrical shapes within an infinite soil mass, simplifying the three-dimensional calculations to a one-dimensional analysis focused on radial temperature changes from the center of the energy pile outward. The general domain of the model is a vertical plane bisecting the energy pile, containing a heat exchanger that maintains a constant temperature, simulating the heating mode of the GSHP system. Heat transfer is radial, moving outward from the heat exchanger to the infinite groundmass on three different soft marine clay densities. Before starting the calculations, it is necessary to define the reference domain of the model to outline all boundaries and define all required parameters and variables. Figure 4 illustrates the definition of a constant temperature boundary, also known as a "Dirichlet boundary", which assumes that the soil mass around the energy pile or GSHP system acts as a large, finite heat reservoir. The temperature along certain edges or surfaces of the model domain is fixed at a predetermined soil mass temperature, ignoring time-dependent temperature variations. This boundary condition ensures the stability and accuracy of the calculations by preventing instability. The bottom and side boundaries of the soil mass domain (as shown in Figure 4, No. 1) are set to the natural background temperature of the soil type. Conversely, the surface of the energy piles (as shown in Figure 4, No. 2) represents the operating scenario of the GSHP by injecting a constant temperature similar to the temperature of the heat exchanger. The next step is to define the calculation node (as shown in Figure 4, No. 3) as a reference point within the domain, overlaying a grid array. The distance Δr represents the distance between computation nodes, with $\Delta r = 0$ at the center of the energy pile. The time iterations, or time step Δt, indicate the interval between calculation iterations, which is essential for analyzing the time-dependent changes in the soil temperature. The temperature of the neighboring calculation nodes T_0, T_1, T_2, T_3, and T_4 within the mesh facilitates the process description.

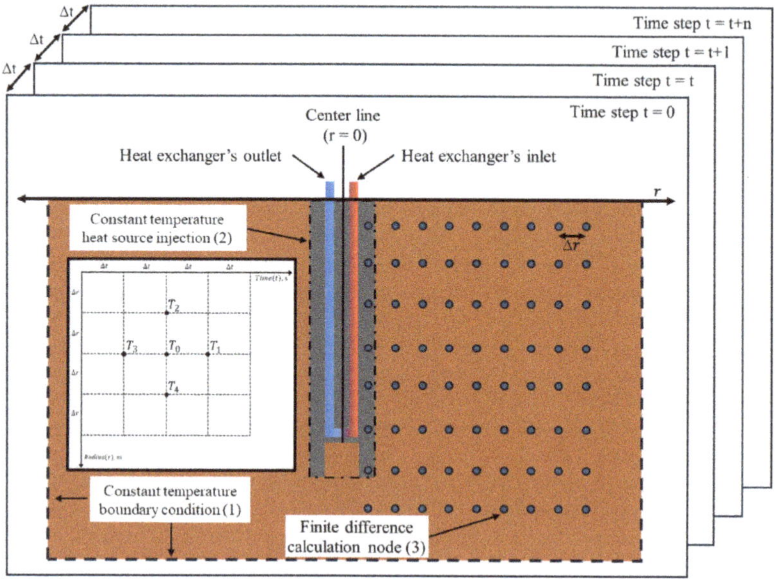

Figure 4. Analysis domain for FDM model with calculation node and boundary conditions.

The temperature rise (T) within any domain element can be calculated using the one-dimensional transient heat balance differential equation, as shown in Equation (1):

$$\frac{\partial^2 T}{\partial r^2} + \frac{1}{r} \cdot \frac{\partial T}{\partial r} = \rho \frac{1}{\alpha_d} \frac{\partial T}{\partial t} \tag{1}$$

where the thermal diffusivity (α_d) is the product of the thermal conductivity (λ) divided by the soil density (ρ) and specific heat capacity (c), t is the time, and r is the radial distance.

Break down the finite differential equation (Equation (1)) into Equations (2)–(4) separately and solve for the explicit form of the functions individually. The named calculation nodes T_1 to T_4 referred to in Figure 4 are used to translate the differential term into an ordinary algebraic equation.

$$\frac{\partial T}{\partial t} = \frac{T_1 - T_0}{\Delta t} \tag{2}$$

$$\frac{\partial T}{\partial r} = \frac{1}{2} \cdot \left(\frac{T_4 - T_0}{\Delta r} + \frac{T_0 - T_2}{\Delta r} \right) = \frac{T_4 - T_0}{2\Delta r} \tag{3}$$

$$\frac{\partial^2 T}{\partial r^2} = \frac{1}{\Delta r} \left(\frac{T_4 - T_0}{\Delta r} - \frac{T_0 - T_2}{\Delta r} \right) = \frac{T_4 + T_2 - 2T_0}{(\Delta r)^2} \tag{4}$$

By substituting Equation (2) to Equation (3) into the heat balance equation, the final explicit solution for the temperature of point T_1 can be derived as given in Equation (5):

$$T_1 = T_0 + \left[\frac{T_4 + T_2 - 2T_0}{(\Delta r)^2} + \frac{1}{r} \cdot \frac{T_4 - T_2}{2\Delta r} \right] \alpha_d \Delta t \tag{5}$$

where the required parameters for the calculation of the changing temperature based on Equation (5) are the temperature of the surrounding points (T_0, T_2, and T_4), the radius of each node from the center line (r), the changing radius (Δr), the thermal diffusivity (α_d), and the thermal time step (Δt).

4. Finite Difference Modeling for Geothermal Temperatures in GSHP Systems

4.1. Methodology

When building smart cities, the integration of energy piles into the urban infrastructure and the evaluation of the resulting heat movement on the ground is of paramount importance. These considerations are critical for optimizing the efficiency and sustainability of GSHP systems, which play a significant role in the energy framework of smart cities. In this study, the predictive power of a one-dimensional FDM was used to estimate and project the soil temperatures around an energy pile over various distances and time periods. The input parameters, which include the detailed physical and thermal characteristics of the materials and soil, were carefully collected from an extensive literature review of GSHP systems in Southeast Asia, design manuals, and reputable textbooks [12–16,33]. The model formulation is further enhanced by an in-depth study conducted at a physical GSHP test site and an energy pile installation at Kasetsart University, Thailand [13].

The focus of our analysis will primarily be placed on two critical parameters: the apparent thermal conductivity of the soil and its density. The model assumes that the thermal conductivity increases proportionally with the temperature increase in the heat exchanger connected to the GSHP system. Variations in soil density will reflect the different geographical locations in Southeast Asia where GSHP and energy pile systems could be deployed. This approach allows for a comprehensive examination of how soil density affects temperature rise, with further details on the different soil densities and their underlying rationale to be provided in a future section.

A distinctive feature of the finite difference model is its ability to distill complex real-world heat transfer processes into a manageable mathematical model. This simplification facilitates the analysis of the vertical heat exchanger integrated into the energy pile, underscoring the model's utility in advancing the development of smart city infrastructure.

4.2. Model Definition

As urban areas in Southeast Asian regions expand, there is an increase in the construction of small to medium-sized buildings to meet the growing demand. These structures often require pile foundations to be installed in firmer soil layers to minimize settlement, especially in the soft clayey soil layer (see Table 1). The development of a one-dimensional FDM model requires simplified inputs.

The concept of energy piles and their foundations as applied in FDM modeling is shown in Figure 5. The model assumes that the energy pile is in a soft clay layer with an average thickness of 10 to 15 m below the ground surface. For the integration of a GSHP system into the pile, it is assumed that a suitable pile type is a spun pile with a diameter of 0.3 m and an opening width of 0.15 m along the depth of the pile. After the installation of the GSHP system, the pile is filled with concrete. The length of the pile is determined to be 12 m, with a working depth of 10 m for the heat exchanger, considering the soil plug at the top of the pile. The U-tube heat exchanger, which is essential to the system, has a total length of 20 m, equally divided between the inlet and the outlet. The 10 m heat exchanger's inlet, illustrated in red on the right side of the pile's figure, provides the downward flow path of the high temperature-heat medium. In contrast, the 10 m heat exchanger's outlet, illustrated in blue on the left side of the pile's figure, provides the upward flow path. Thermal input parameters for all components of the GSHP system and energy pile, as referenced in the American Society of Heating, Refrigerating and Air-Conditioning Engineers (ASHRAE) [34] and the Analysis and Design of Energy Geostructures [6], are provided in Table 2 for use in the numerical model.

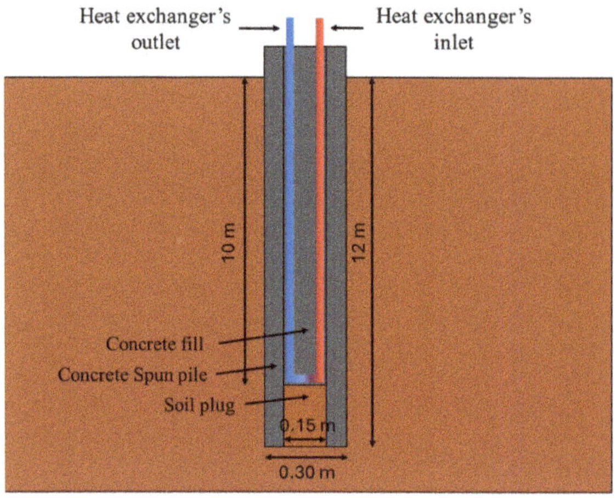

Figure 5. General definition of energy pile in FDM model.

Table 2. Input parameters for FDM analysis.

Parameters	Value	Unit
Global parameters		
Background temperature	29	°C
Inlet U-tube temperature	37	°C
Outlet U-tube temperature	33	°C
Time step (Δt)	7200	s
Changing radius (Δr)	0.15	m
GSHP on/off time	12	h
HDPE U-tube heat exchanger		
Heat exchanger length	20	m
Thickness	0.005	m
Density (ρ)	950	kg/m^3
HDPE Specific heat capacity (c_{HDPE})	2400	J/(kg·K)
HDPE Thermal conductivity (λ_{HDPE})	0.48	W/(m·C)
HDPE Thermal diffusivity (α_{HDPE})	2.11×10^{-7}	m^2/s
Concrete pile		
Diameter (ϕ)	0.30	m
Concrete thickness	0.15	m
Concrete pile Density ($\rho_{concrete}$)	2400	kg/m^3
Concrete pile Specific heat capacity ($c_{concrete}$)	900	J/(kg·K)
Concrete pile Thermal conductivity ($\lambda_{concrete}$)	0.8	W/(m·C)
Concrete pile Thermal diffusivity ($\alpha_{concrete}$)	3.70×10^{-7}	m^2/s
Marine clay soil		
Density (ρ)	1400, 1600, 1800	kg/m^3
Clayey Soil Specific heat capacity (c_{soil})	800	J/(kg·K)
Clayey Soil Thermal conductivity (λ_{soil})	varies	W/(m·C)
Clayey Soil Thermal diffusivity (α_{soil})	varies	m^2/s

The mesh is the grid or mesh of nodes used to compute the numerical solution of the governing equations. It divides the entire domain into small rectangular sections in 2D. The number of nodes can vary across the domain, with finer meshes used in areas where the solution needs to be finer or change rapidly. At each node, the governing equations are approximated by finite difference formulas that are replaced by references to the values of the neighboring computational nodes. The location and setup of the computing nodes has a significant impact on the accuracy and efficiency of the FDM result.

A finer mesh generally results in higher accuracy and resolution, but higher computational cost. Conversely, a coarser mesh reduces computational time but may sacrifice accuracy. Figure 4 provides an explanation of the mesh used in this study; each computation point is placed at the horizontal spacing of 0.15 m, and the spacing distance can be referred to as the increment r. In addition, the vertical spacing between each horizontal roller is set to 1.0 m to accommodate the varying temperature of the heat exchanger at depth.

4.3. Constant-Temperature Boundary Condition

To simulate the initial operating condition of the energy pile system, the temperature of a 20 m-long heat exchanger is set as a constant temperature boundary within the analysis spreadsheet. This approach simplifies the model by ignoring the radial heat convection within the heat exchanger and the fluid medium it contains. It is important to consider the different heat contributions of the medium flowing downward through the 10 m-long inlet and upward through the 10 m-long outlet of the heat exchanger and treat them as independent heat sources.

Previous studies [12,13,16,33] have monitored the average temperatures at the inlet and outlet of the heat exchanger, showing that the inlet water temperature is 37 °C, while the outlet water temperature is measured at 33 °C. Direct interpolation is used to calculate the temperature at each meter along the heat exchanger, which indicates a decrease of 0.2 °C/m along its length, as shown in Table 3. The operation of the GSHP is often tied to the demand of the building's air-conditioning system, which requires 12 h of operation to maintain a constant temperature range of 33 to 37 °C, depending on the depth of the heat exchanger. This is followed by a 12 h shutdown of the GSHP system, during which the temperature at the constant temperature boundary is adjusted to match the ambient ground temperature.

Table 3. The temperature of the HDPE heat exchanger at different depths of energy pile.

Outlet Side		Inlet Side	
Depth (m)	**Temperature (°C)**	**Depth (m)**	**Temperature (°C)**
0 to 1	33.0	0 to 1	37.0
1 to 2	33.2	1 to 2	36.8
2 to 3	33.4	2 to 3	36.6
3 to 4	33.6	3 to 4	36.4
4 to 5	33.8	4 to 5	36.2
5 to 6	34.0	5 to 6	36.0
6 to 7	34.2	6 to 7	35.8
7 to 8	34.4	7 to 8	35.6
8 to 9	34.6	8 to 9	35.4
9 to 10	34.8	9 to 10	35.2

Studies conducted in the Bangkok area [12–14], using Thermal Response Tests (TRT) or operational GSHP air-conditioning systems, have found that the background temperature of the Bangkok clay varies between 29 °C and 31 °C during different seasons. On average, this is 5 °C cooler than the ambient air temperature [15]. For the FDM temperature boundary calculations, the background temperature of Bangkok clay is set to 29 °C.

To help understand the concept of a constant temperature boundary, Figure 6 illustrates the asymmetric temperature bulb surrounding the energy pile in the soil. This is based on direct interpolation of the constant temperature boundary 1 year after the simulation, using a finite element model (FEM) through PLAXIS 2D CONNECT V21.01.00.479 software [35,36], within a 3 m radius of the pile centers.

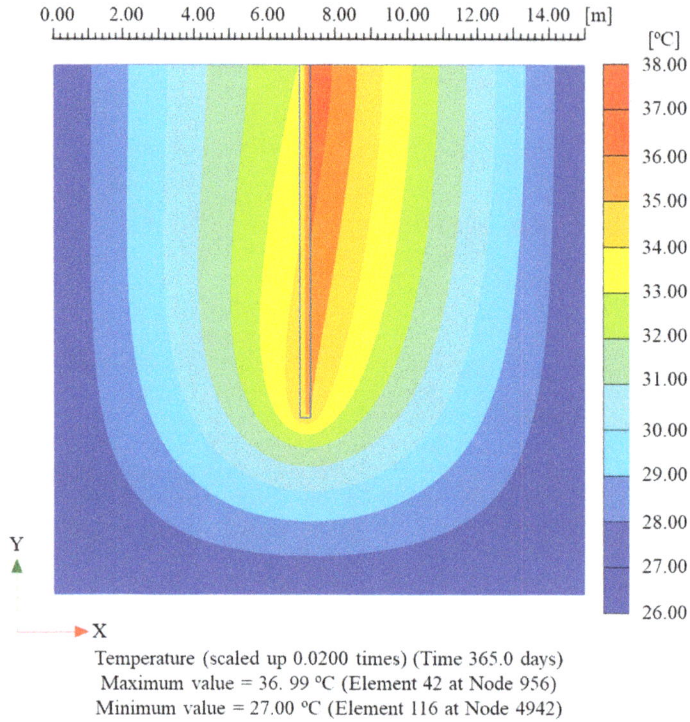

Figure 6. Asymmetrical soil temperature around energy pile after 1 year, simulated with FEM model.

4.4. Variable Thermal Conductivities and Soil Densities

The thermal conductivity of the GSHP system, which is integral to the performance of energy piles, is affected by various physical factors of the system components. These include the mineralogy, dry density, water content, and gradation of the soil materials as identified in previous studies [37,38]. Experimental evidence from lateritic clay samples [39] indicates that thermal conductivity values for samples from the Guilin and Liuzhou areas in the Guangxi region of China are temperature-dependent. In laboratory tests, at temperatures ranging from 5 to 90 °C, clay samples at saturation levels of 4%, 44.90%, and 83.90% exhibited a linear increase in thermal conductivity, with values ranging from 0.2 to 0.4 W/m·°C, 0.6 to 1.6 W/m·°C, and 0.9 to 1.3 W/m·°C, respectively, as shown in Figure 7.

Assuming a linear relationship between temperature and thermal conductivity for clay samples, this relationship is applied in the FDM model. The thermal conductivity of soft clay from different locations is expressed as a linear function of the average inlet-outlet temperature of the GSHP system. Bangkok soft clay, which has been extensively studied, serves as a representative of other soft clays. Its thermal and engineering properties are similar to those of other East Asian soft marine clays, such as Ariake clay, Singapore marine clay, Hong Kong marine clay, Kuantan clay, and Ho Chi Minh City marine clay. They may provide insight into the thermal performance of similar soil types. Formed during the Holocene by sediments from the Chao Phraya River delta, Bangkok clay is characterized by its high moisture content, low strength, and high compressibility, which cause it to pose geotechnical challenges. Notable studies include those by Abuel-Naga et al. [28,29], who used temperature-controlled triaxial equipment at 20 °C to 90 °C under undrained conditions and found thermal conductivities of 0.825, 1.1, and 1.3 W/m·°C. Jotisankasa et al. [13], who implemented a full-scale energy pile system in a residential setting, found thermal conductivities for steel and spun piles of 1.49, 2.01, and 3.3 W/m·°C. Another study in Hanoi, Vietnam [40] recorded heat exchanger temperatures ranging from 37 °C to 40 °C with resulting thermal conductivities of 1.82 and 1.97 W/m·°C.

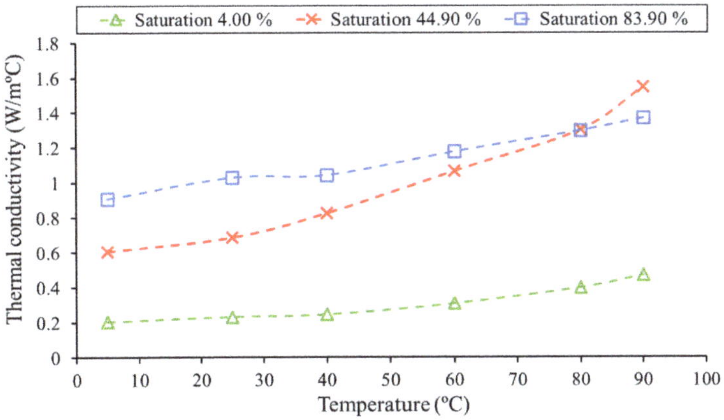

Figure 7. Guilin lateritic clay's temperature-dependent thermal conductivity.

The study derives variable thermal conductivities for the GSHP system from these results, which are summarized in Table 4, which provides the data for constructing four variable thermal conductivity equations. These linear equations correlate the temperature of the heat exchanger to its thermal conductivity, with cases no. 2 and no. 4 based on Jotisankasa et al. [13] and Widiatmojo et al. [40], and cases no. 1 and no. 3 offering adjusted slopes to provide intermediate ranges of thermal conductivity for the model. For clarity, all four thermal conductivity equations are shown in Figure 8, starting at a temperature of 32 °C and an initial thermal conductivity of 1.49 W/m·°C. The upper limit of the average temperature of the heat exchanger will not exceed 40 °C, according to reviews of in situ GSHP experiments.

Table 4. Summary of deriving variable thermal conductivities in GSHP systems for constructing four variable thermal conductivity equations.

Case	Temperature (°C)	Thermal Conductivity (w/m·°C)	Thermal Conductivity Equation	R^2
no. 1	32 to 40	1.49 to 2.99	y = 0.0500 x − 14.5100	-
no. 2 [13]	32, 34, 38	1.49, 2.09, 3.31	y = 0.3046 x − 8.2943	0.9971
no. 3	32 to 40	1.49 to 2.69	y = 0.1500 x − 3.3100	-
no. 4 [40]	32, 37, 40	1.49, 1.82, 1.97	y = 0.0606 x − 0.4422	0.9951

The variation in soil density is another crucial variable that correlates with the geolocation of the GSHP system. This study considers the density of soft clays varying from 1400 kg/m^3 (representing Singapore marine clay and Ariake clay), through 1600 kg/m^3 (Bangkok clay), to 1800 kg/m^3 (Hong Kong clay). Additional variables and their effect on system performance are detailed in Table 5.

Table 5. The difference in density for clay in different countries.

Soil Name	Density (kg/m^3)
Singapore marine clay and Ariake clay	1400
Bangkok clay	1600
Hong Kong clay	1800

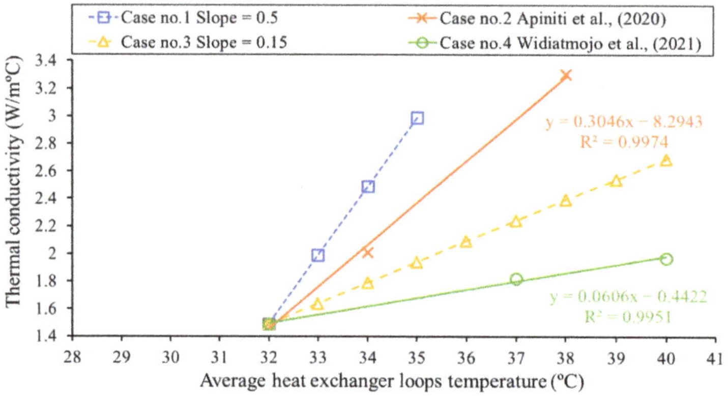

Figure 8. Comparative thermal conductivity in GSHP operation range [13,40].

5. Results and Discussion

5.1. Evaluation of Temperature Distribution around Energy Piles

The operation of GSHP and energy piles is crucial for the efficient performance of smart cities, with typical operating times of up to 12 h per day. This study meticulously examines the temperature dynamics surrounding the soil adjacent to energy piles, using a detailed temporal analysis ranging from short-term (initial 7 days) to long-term (annual overview) perspectives.

During the first week of observation, temperature variations around the energy pile exhibit a cyclical pattern reminiscent of a sine wave, as shown in Figure 9. This cycle is attributed to the alternating operating phases of the GSHP system. Despite the cyclic nature, a gradual upward trend in temperature is observed, indicating an accumulation of heat within the soil mass over time. This phenomenon is primarily governed by the mechanisms of heat conduction and convection, which become more pronounced with increasing distance from the core of the energy pile. Specifically, at distances of 0.15 m, 0.30 m, 0.45 m, and 0.60 m from the pile center, the temperature variations progressively decrease, indicating a decrease in the thermal gradient due to the operating cycles of the energy pile. For example, at a 0.15 m distance, the temperature variations range from 31.25 °C to 30.25 °C, showing a variation of 1 °C. This variation narrows significantly at distances of 0.3 m (0.25 °C fluctuation) and 0.45 m (0.17 °C fluctuation), illustrating thermal diffusion and distribution patterns within the soil.

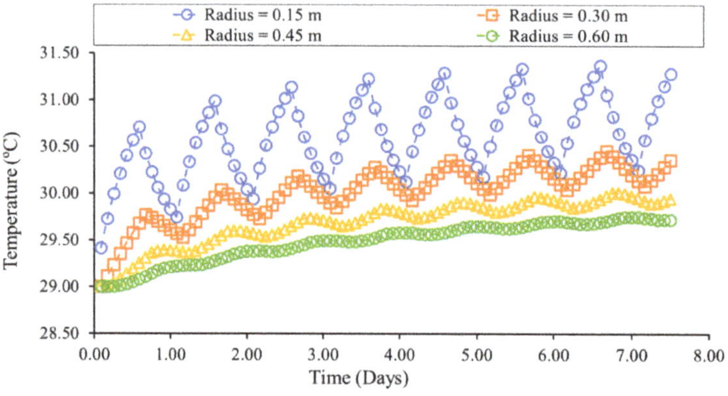

Figure 9. Seven-day temperature cycle at various radii.

In contrast, the annual perspective shows a relatively stable temperature distribution with negligible long-term increases. At the end of the year, the maximum recorded soil temperature near the energy pile (at a radius of 0.15 m) stabilizes around 31.5 °C, as shown in Figure 10. This stability underscores the equilibrium achieved between the heat input from the GSHP system and the natural heat dissipation mechanisms within the soil over an extended period.

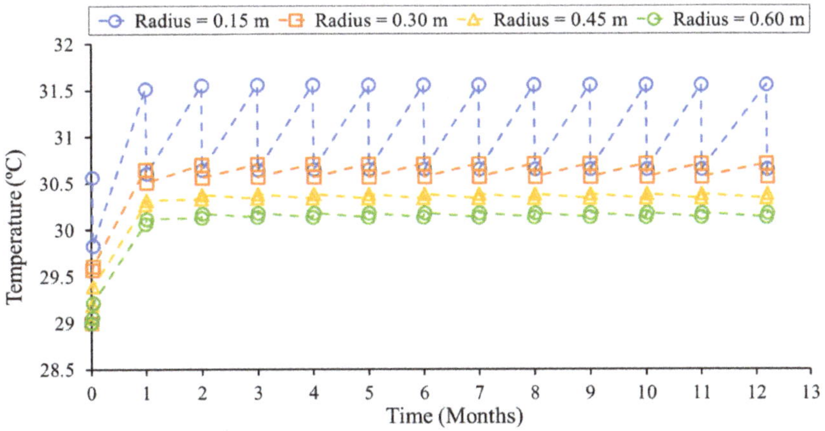

Figure 10. One-year temperature cycle at various radii.

Figure 11 further elaborates on the temperature dynamics by comparing the effects of alternating 12 h operating cycles of the GSHP system over 1 month and 1 year. Within a radius of 1 m from the center of the pile, temperatures are observed to range from 29.5 to 32 °C, with a broader dispersion observed within a radius of 1 to 2 m, where temperatures range from 29.0 to 30 °C. These observations suggest that the effects of thermal consolidation, both primary and secondary, have minimal influence on soil volume change within the temperature range studied (25 to 90 °C), referencing the work of Abuel-Naga et al. [28,29] on thermally induced volume change in soft Bangkok clay.

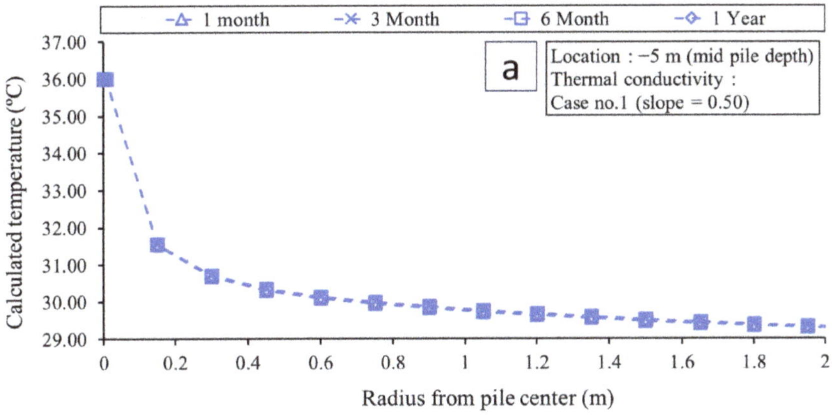

Figure 11. *Cont.*

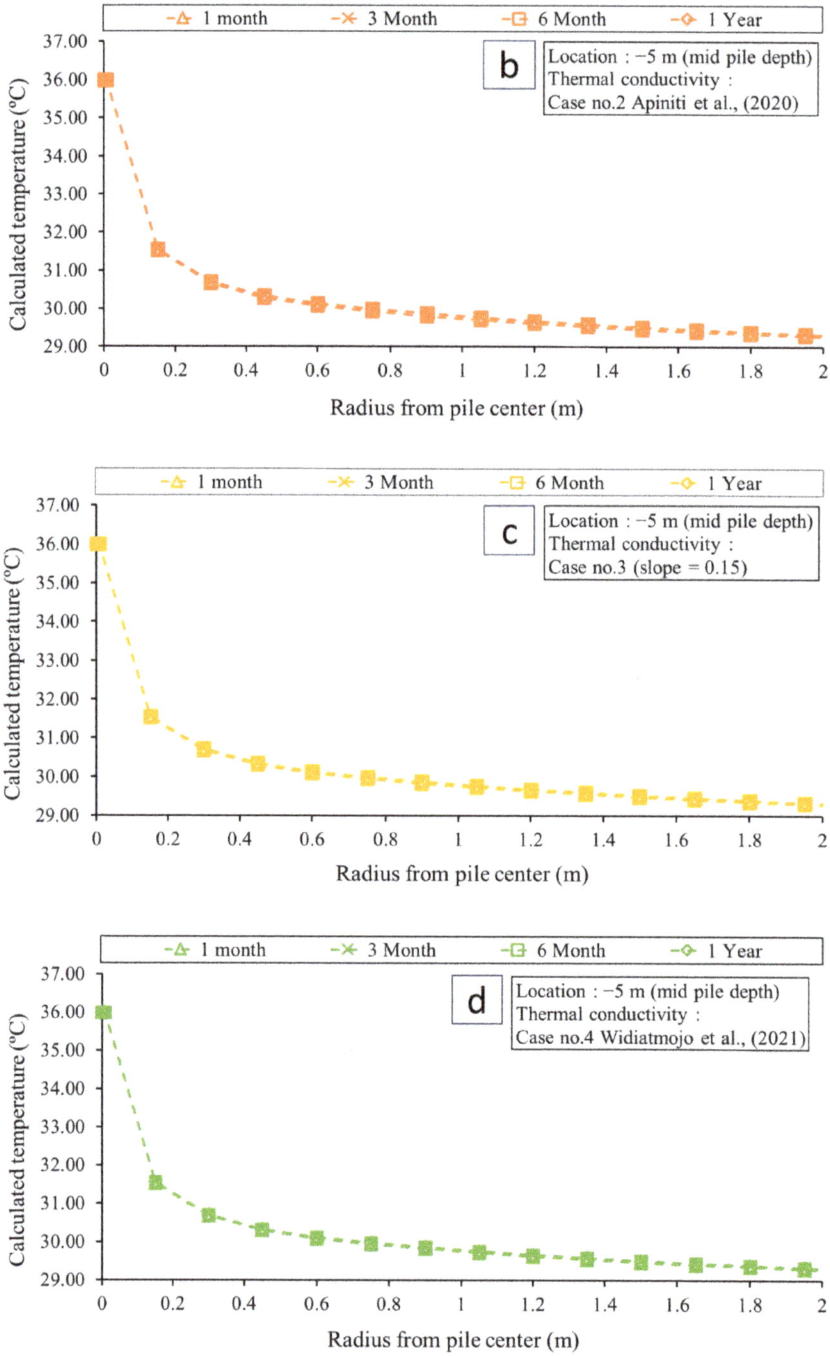

Figure 11. Predicted temperatures at various radii for different cases of linear thermal conductivity functions (cases no. 1, 2, 3, and 4) [13,40].

The methodological framework of this study employs FDM, recognizing its inherent limitations, particularly in the spatial resolution of the temperature nodes, which are placed at 0.15 m intervals from the pile centerline. This limitation ensures model accuracy but limits the granularity of the temperature distribution analysis. The study contrasts the results of the finite difference method with potentially more accurate modeling techniques, such as the FEM, to highlight areas for future study improvements. Specifically, the study highlights the potential benefits of including heat energy as a boundary condition within FEM analyses to capture a more accurate and realistic representation of GSHP system operation.

In addition, this study identifies a significant data gap regarding the temperature-dependent thermal conductivity of clayey soils, particularly for Bangkok clay, which is critical for refining the accuracy of temperature distribution models around energy piles. Future study directions include conducting laboratory experiments to determine specific thermal conductivity values for Bangkok clay and other similar soils to improve the predictability and reliability of temperature distribution models in the context of smart city infrastructure development.

5.2. Evaluation of Thermal Conductivity Variations on Temperature Distribution

Exploring the role of thermal conductivity variations within different time frames, namely 1 month, 3 months, and 1 year, provides crucial insights into the thermal dynamics around energy piles, which are an integral part of smart city infrastructures. The meticulously detailed analysis, summarized in Figure 12, provides a clear visualization of the temperature profiles around energy piles, meticulously examining the potential shifts in temperature distribution attributable to changes in the thermal conductivity of the surrounding soil medium. This analysis is critical because thermal conductivity is a fundamental property affecting heat transfer efficiency within the soil. It directly affects the operational effectiveness of GSHP systems and energy piles, which are critical components in the sustainable energy framework of smart cities. Therefore, understanding the sensitivity of temperature distribution to variations in thermal conductivity is essential for predicting the long-term performance of these systems.

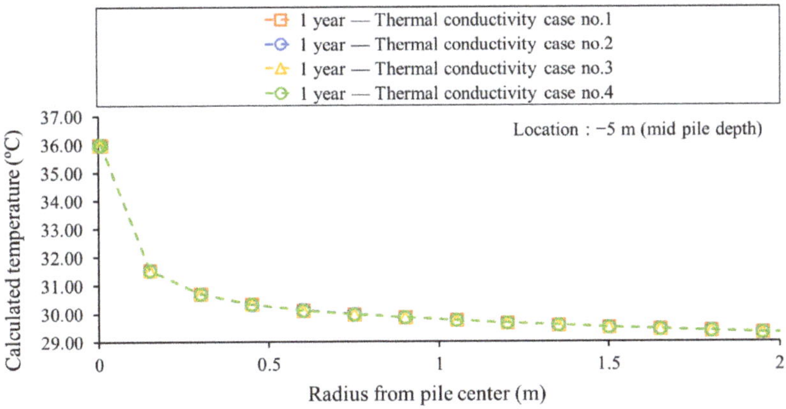

Figure 12. One-year predicted temperatures with different thermal conductivities.

The results of the study are quite revealing, indicating that within the observed range of thermal conductivity variations, there is remarkable stability in the temperature distribution around the energy piles over the time periods studied. This observation suggests that the thermal behavior of the system exhibits a degree of resilience to changes in soil thermal conductivity. This resilience is particularly noteworthy in the context of smart city planning and development, where energy efficiency and sustainability are paramount. Such stability in temperature distribution, despite variations in thermal conductivity, underscores a critical aspect of the thermal dynamics involved in the operation of energy piles. It suggests that the design of these systems can afford a degree of flexibility with respect to the thermal properties of the soil. This finding is invaluable to engineers and urban planners, as it implies that minor variations in soil thermal conductivity may not require adjustments to the design or operation of GSHP systems.

However, it is important to contextualize these findings within the limitations of the study. The analysis, while comprehensive, is based on the range of thermal conductivity variations observed within the scope of the study. This range is inherently related to the specific soil types, moisture content, and other environmental conditions pertinent to the study area. Consequently, extending these findings to different regions or soil conditions requires careful consideration of local soil properties and environmental factors. Furthermore, the implications of this stability go beyond the immediate operational considerations of energy piles. They touch on broader issues of sustainability and resilience in urban planning. In a world where climate variability and environmental change are of increasing concern, the ability to design energy systems that are robust to changes in underlying environmental characteristics is invaluable. The findings of this study contribute to a growing body of knowledge that supports the development of more sustainable and resilient urban environments.

5.3. Impact of Clay Density Variation on Temperature Predictions

This study explores the intricate dynamics of soil density variation and its subsequent impact on the thermal behavior around energy piles, a key component in the thermal management systems of smart cities. Using FDMs, this study categorizes soft clay into three distinct density types: Singapore marine clay and Ariake clay with a density of 1400 kg/m^3, soft Bangkok clay with a density of 1600 kg/m^3, and Hong Kong clay with a density of 2200 kg/m^3, as contextually implied in subsequent discussions. These categorizations are critical to understanding the nuanced impact of soil density on the thermal properties and temperature regulation capabilities of the soil surrounding energy piles. Figure 13 provides a visual representation of how variations in soil density affect temperature predictions over a one-year observation period. Notably, the graph illustrates a convergence of temperature differences over time, suggesting a diminishing effect of density variations on temperature fluctuations with long-term system operation. This phenomenon underscores the complex interplay between soil physical properties and thermal dynamics and highlights the critical role of soil density in determining the thermal efficiency of subsurface energy systems.

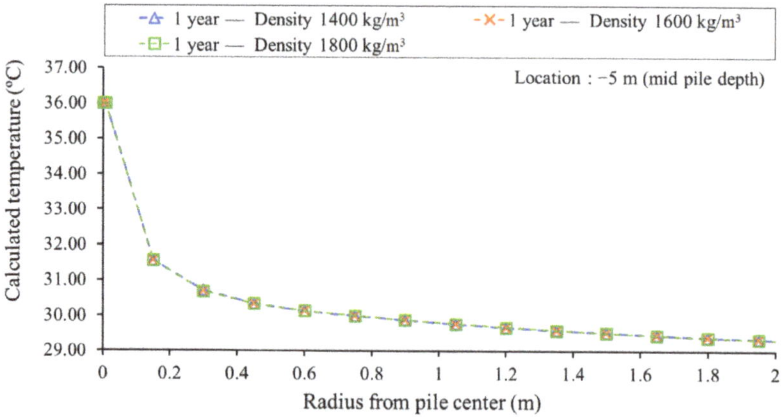

Figure 13. One-year predicted temperatures with different densities of Bangkok clay.

Further analysis, as shown in Figures 14–16, systematically evaluates the temperature response to increasing soil density at different radial distances from the energy pile (0.3 m,

0.6 m, and 1.0 m) and over different time periods (3 months, 6 months, and 1 year). This multidimensional analysis reveals a consistent trend: as soil density increases from 1400 to 1800 kg/m^3, a 28.57% increase, the resulting effect on soil temperature shows a decreasing trajectory. Specifically, temperature decreases of 0.02 °C, 0.01 °C, and 0.0025 °C are observed with each density increase over the specified time periods. This trend is further quantified by noting that for every 10% increase in soil density, there is a corresponding average temperature decrease of 0.007 °C at a 0.3 m radius from the energy pile. The granularity of this analysis is further expanded in Figures 14–16, where the temperature effects of increased soil density are assessed at radial distances of 0.6 m and 1 m from the energy pile. These results provide a nuanced understanding of the thermal dynamics of the soil, revealing a proportional relationship between soil density and temperature modulation. For example, at a 0.6 m radius, each 10% increase in soil density correlates with an average temperature decrease of 0.003 °C, while at a 1 m radius, the average temperature decrease is 0.0009 °C for the same density increase.

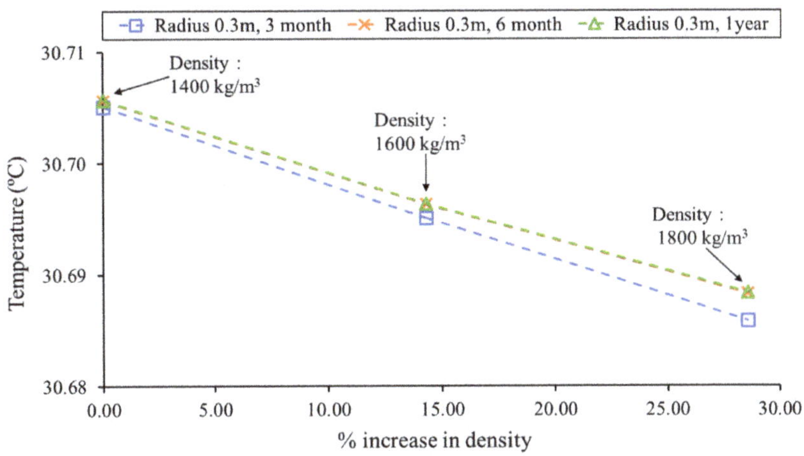

Figure 14. Temperature prediction in 1 month, 6 months, and 1 year and Bangkok clay density correlation over time at 0.3 m radius.

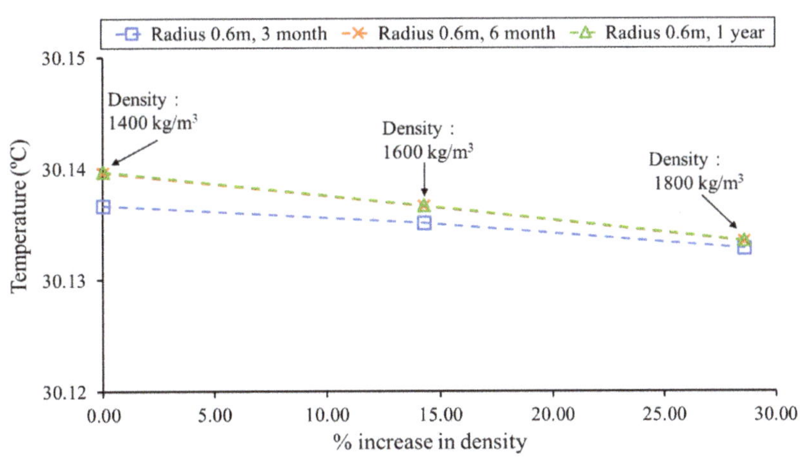

Figure 15. Temperature prediction in 3 months, 6 months, and 1 year and Bangkok clay density correlation over time at 0.6 m radius.

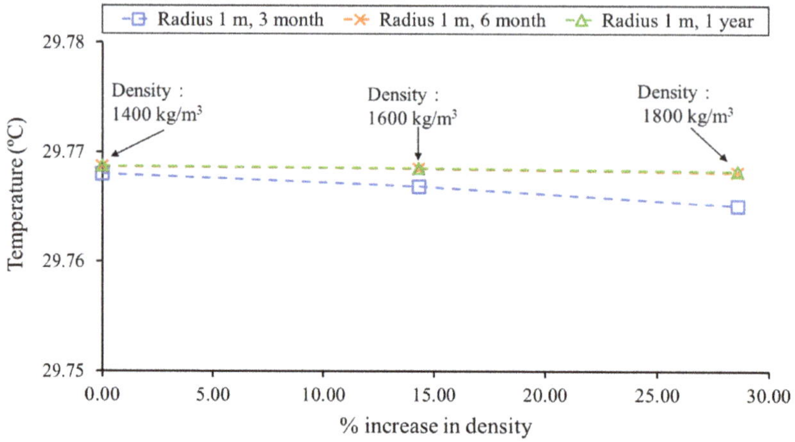

Figure 16. Temperature prediction in 3 months, 6 months, and 1 year and Bangkok clay density correlation over time at 0.3 m radius.

This comprehensive study of the effects of soil density on the thermal behavior around energy piles provides valuable insights for the design and operation of geothermal systems in smart cities. It highlights the importance of the physical properties of soil, particularly density, in influencing thermal interactions within the soil mass, which in turn affects the efficiency and sustainability of energy pile systems. The results suggest that careful consideration of soil density can improve the predictive accuracy of thermal models, thereby facilitating optimized design strategies for subsurface energy solutions. In addition, the results of the study underscore the importance of incorporating detailed soil property analyses into the planning and implementation phases of geothermal projects. Understanding the specific effects of soil density variations on temperature regulation can lead to more effective thermal management practices, contributing to the overall resilience and sustainability of urban infrastructure. This study not only advances the scientific understanding of soil-thermal interactions but also provides a pragmatic guide for engineers and urban planners seeking to integrate efficient geothermal technologies into the fabric of smart cities.

6. Conclusions

The study on the use of GSHP systems and energy piles in smart city development has yielded several significant findings:

(1) GSHP systems, especially when integrated with energy piles, have proven to be efficient and sustainable solutions for heating and cooling in urban environments, significantly reducing electricity consumption and greenhouse gas emissions.
(2) The study emphasized the critical role of understanding thermal movement within the soil, particularly in soft marine clays common in Southeast Asia, for optimal implementation of GSHP systems. Soil thermal conductivity and density were identified as key factors affecting system performance.
(3) Using a finite difference method (FDM), the study successfully modeled temperature distribution around energy piles and evaluated the impact of variations in soil thermal conductivity and density on system efficiency.
(4) GSHP systems offer a promising solution to mitigate the urban heat island effect, especially in densely populated areas where conventional HVAC systems are less efficient.

Based on the results of this study, several avenues for future study were identified:

(1) While FDM provided valuable insights, future studies could explore more sophisticated modeling techniques, such as the finite element method (FEM), to achieve greater accuracy in predicting ground temperature distribution and system performance.
(2) Additional study is needed to determine the temperature-dependent thermal conductivity of different types of clayey soils, particularly Bangkok clay, to refine the accuracy of temperature distribution models around energy piles.
(3) Investigating the effects of rapid urbanization and the subsequent increase in the urban heat island effect on the efficiency of GSHP systems could provide valuable insights for urban planning and smart city development.
(4) Future work should focus on optimizing GSHP system configurations and operating parameters to maximize the use of geothermal energy in urban development.
(5) Further studies could explore the integration of GSHP systems within the broader context of sustainable and resilient urban infrastructure, considering climate variability and environmental change.

This study contributes significantly to the growing body of knowledge on sustainable urban development and highlights the potential of GSHP systems and energy piles in smart cities. By addressing the outlined future study directions, the full potential of geothermal energy in urban environments can be realized, supporting the transition to more sustainable and energy-efficient urban habitats.

This study has several limitations that should be acknowledged. First, the short period of data collection may not adequately represent the long-term performance and seasonal variations of ground source heat pump (GSHP) systems. Second, the lack of temperature data from thermal response tests (TRTs) and current GSHP installations in Southeast Asia hinders comprehensive regional studies and comparisons. Third, the one-dimensional finite difference model, while computationally efficient, simplifies thermal behavior by ignoring vertical temperature changes and possibly omitting localized effects. Finally, the study emphasizes the need to use heat energy as a boundary condition in finite element method (FEM) calculations to improve accuracy. This suggests that the existing model could be improved using more complex methods. These limitations highlight potential areas for future research to improve GSHP modeling and assessment in the region.

Author Contributions: Conceptualization, S.I. and A.J.; methodology, A.J.; software, T.C. and S.C.; validation, A.J. and S.I.; formal analysis, T.C. and S.C.; investigation, T.C. and S.C.; resources, A.J. and S.I.; data curation, T.C. and A.J.; writing—original draft preparation, T.C.; writing—review and editing, S.I.; visualization, S.I.; supervision, A.J., S.C. and S.I.; project administration, S.I.; funding acquisition, S.I. All authors have read and agreed to the published version of the manuscript.

Funding: This research received no external funding.

Data Availability Statement: The data that support the findings of this study are available from the corresponding author, Shinya Inazumi, upon reasonable request.

Conflicts of Interest: The authors declare no conflicts of interest.

References

1. Sarbu, I.; Sebarchievici, C. General review of ground-source heat pump systems for heating and cooling of buildings. *Energy Build.* **2014**, *70*, 441–454. [CrossRef]
2. Eugster, W.; Sanner, B. Technological status of shallow geothermal energy in Europe. In Proceedings of the European Geothermal Congress, Unterhaching, Germany, 30 May–1 June 2007; pp. 1–8.
3. Liu, C.; Shi, B.; Tang, C.; Gao, L. A numerical and field investigation of underground temperatures under Urban Heat Island. *Build. Environ.* **2011**, *46*, 1205–1210. [CrossRef]
4. Ratchawang, S.; Chotpantarat, S.; Chokchai, S.; Takashima, I.; Uchida, Y.; Charusiri, P. A Review of Ground Source Heat Pump Application for Space Cooling in Southeast Asia. *Energies* **2022**, *15*, 4992. [CrossRef]
5. Fadejev, J.; Simson, R.; Kurnitski, J.; Haghighat, F. A review on energy piles design, sizing and modelling. *Energy* **2017**, *122*, 390–407. [CrossRef]
6. Laloui, L.; Loria, A.F.R. *Analysis and Design of Energy Geostructures: Theoretical Essentials and Practical Application*; Academic Press: Cambridge, MA, USA, 2019.

7. Tissen, C.; Menberg, K.; Bayer, P.; Blum, P. Meeting the demand: Geothermal heat supply rates for an urban quarter in Germany. *Geotherm. Energy* **2019**, *7*, 9. [CrossRef]
8. Lund, J.W.; Toth, A.N. Direct utilization of geothermal energy 2020 worldwide review. *Geothermics* **2021**, *90*, 101915. [CrossRef]
9. Shelare, S.; Kumar, R.; Gajbhiye, T.; Kanchan, S. Role of Geothermal Energy in Sustainable Water Desalination—A Review on Current Status, Parameters, and Challenges. *Energies* **2023**, *16*, 2901. [CrossRef]
10. Sayegh, M.A.; Jadwiszczak, P.; Axcell, B.P.; Niemierka, E.; Bryś, K.; Jouhara, H. Heat pump placement, connection and operational modes in European district heating. *Energy Build.* **2018**, *166*, 122–144. [CrossRef]
11. Bryś, K.; Bryś, T.; Sayegh, M.A.; Ojrzyńska, H. Characteristics of heat fluxes in subsurface shallow depth soil layer as a renewable thermal source for ground coupled heat pumps. *Renew. Energy* **2020**, *146*, 1846–1866. [CrossRef]
12. Chokchai, S.; Chotpantarat, S.; Takashima, I.; Uchida, Y.; Widiatmojo, A.; Yasukawa, K.; Charusiri, P. A pilot study on geothermal heat pump (GHP) use for cooling operations, and on ghp site selection in tropical regions based on a case study in Thailand. *Energies* **2018**, *11*, 2356. [CrossRef]
13. Jotisankasa, A.; Chantawarangul, K.; Sittidumrong, J.; Thangkhumwong, K.; Aowphitak, P.; Laohasrisakul, P.; Chanchayanon, T.; Aonkam, S.; Satayanan, S. Investigation of the feasibility of energy piles in Bangkok and their impacts on pile behavior. *E3S Web Conf.* **2020**, *205*, 05018. [CrossRef]
14. Shimada, Y.; Uchida, Y.; Takashima, I.; Chotpantarat, S.; Widiatmojo, A.; Chokchai, S.; Charusiri, P.; Kurishima, H.; Tokimatsu, K. A study on the operational condition of a ground source heat pump in Bangkok based on a field experiment and simulation. *Energies* **2020**, *13*, 274. [CrossRef]
15. Uchida, Y.; Yasukawa, K.; Tenma, N.; Taguchi, Y.; Ishii, T.; Suwanlert, J.; Lorphensri, O. Subsurface temperature survey in Thailand for geothermal heat pump application. *J. Geotherm. Res. Soc. Jpn.* **2011**, *33*, 93–98.
16. Yasukawa, K.; Takashima, I.; Uchida, Y.; Tenma, N.; Lorphensri, O. Geothermal heat pump for space cooling in Kamphaengphet, Thailand. *Bull. Geol. Surv. Jpn.* **2009**, *60*, 491–501. [CrossRef]
17. Jalaluddin; Miyara, A.; Tarakka, R.; Mochtar, A.A.; Anis, I.M. Thermal performance of shallow spiral-tube ground heat exchanger for ground-source cooling system. *IOP Conf. Ser. Mater. Sci. Eng.* **2019**, *619*, 012012. [CrossRef]
18. Roy, D.; Chakraborty, T.; Basu, D.; Bhattacharjee, B. Feasibility and performance of ground source heat pump systems for commercial applications in tropical and subtropical climates. *Renew. Energy* **2020**, *152*, 467–483. [CrossRef]
19. Liu, Y.; Qin, X.S.; Chiew, Y.M. Investigation on potential applicability of subsurface cooling in Singapore. *Appl. Energy* **2013**, *103*, 197–206. [CrossRef]
20. Hanzawa, H.; Fukaya, T.; Suzuki, K. Evaluation of engineering properties for an Ariake clay. *Soils Found.* **1990**, *30*, 11–24. [CrossRef] [PubMed]
21. Moqsud, M.A.; Hayashi, S.; Du, Y.J.; Suetsugu, D. Assessment of thermal properties of mud of the Ariake sea, Japan. *Daffodil Int. Univ. J. Sci. Technol.* **2007**, *2*, 1.
22. Tanaka, H.; Locat, J.; Shibuya, S.; Soon, T.T.; Shiwakoti, D.R. Characterization of Singapore, Bangkok, and Ariake clays. *Can. Geotech. J.* **2001**, *38*, 378–400. [CrossRef]
23. Arulrajah, A.; Bo, M.W. Characteristics of Singapore marine clay at Changi. *Geotech. Geol. Eng.* **2008**, *26*, 431–441. [CrossRef]
24. Bo, M.W.; Arulrajah, A.; Sukmak, P.; Horpibulsuk, S. Mineralogy and geotechnical properties of Singapore marine clay at Changi. *Soils Found.* **2015**, *55*, 600–613. [CrossRef]
25. Song, X.; Du, G.; Wang, Z.; Wang, Z.; Zhu, M.; Bai, W.; Xie, S.; Xia, K.; Zheng, G.; Cheng, X.; et al. Study on engineering properties of deep marine soft clay in Guangdong-Hong Kong-Macao greater bay area. *J. Phys. Conf. Ser.* **2021**, *2021*, 012032. [CrossRef]
26. Fauzi, A.; Fauzi, U.J.; Nazmi, W.M. Engineering quality improvement of Kuantan clay subgrade using recycling and reused materials as stabilizer. *Procedia Eng.* **2013**, *54*, 675–689. [CrossRef]
27. Aljanabi, Q.A.; Chik, Z.A.M.R.I.; Kasa, A.N.U.A.R. Construction of a new highway embankment on the soft clay soil treatment by stone columns in Malaysia. *J. Eng. Sci. Technol.* **2013**, *8*, 448–456.
28. Abuel-Naga, H.M.; Bergado, D.T.; Bouazza, A. Thermally induced volume change and excess pore water pressure of soft Bangkok clay. *Eng. Geol.* **2007**, *89*, 144–154. [CrossRef]
29. Abuel-Naga, H.M.; Bergado, D.T.; Bouazza, A. Thermal conductivity of soft Bangkok clay. In *GeoCongress 2008: Characterization, Monitoring, and Modeling of GeoSystems*; ASCE Publications: Reston, VA, USA, 2008; pp. 1061–1068. [CrossRef]
30. Horpibulsuk, S.; Shibuya, S.; Fuenkajorn, K.; Katkan, W. Assessment of engineering properties of Bangkok clay. *Can. Geotech. J.* **2007**, *44*, 173–187. [CrossRef]
31. Mairaing, W.; Amonkul, C. Soft Bangkok Clay Zoning. In *EIT-Japan Symposium on Engineering for Geo-Hazards: Earthquakes and Landslides-Surface and Subsurface Structures*; Imperial Queen's Park Hotel: Bangkok, Thailand, 2010; pp. 1–10.
32. Shibuya, S.; Tamrakar, S.B.; Theramas, N. Geotechnical site characterization on engineering properties of Bangkok clay. *Geotech. Eng.* **2001**, *32*, 139–147.
33. Sittidumrong, J.; Jotisankasa, A.; Chantawarangul, K. Effect of thermal cycles on volumetric behaviour of Bangkok sand. *Geomech. Energy Environ.* **2019**, *20*, 100127. [CrossRef]
34. American Society of Heating, Refrigerating and Air-Conditioning Engineers (ASHRAE). *ASHRAE Handbook—Fundamentals*; ASHRAE: Atlanta, GA, USA, 2009.
35. Korshunov, A.A.; Churkin, S.V.; Nevzorov, A.L. Calibration of PLAXIS frozen/unfrozen soil model according to results of laboratory tests and in-situ monitoring. *Transp. Soil Eng. Cold Reg.* **2020**, *2*, 105–120. [CrossRef]

36. Allani, M.; Lysebetten, G.V.; Huybrechts, N. Experimental and numerical study of the thermo-mechanical behaviour of energy piles for Belgian practice. In *Advances in Laboratory Testing and Modelling of Soils and Shales (ATMSS)*; Springer: Cham, Switzerland, 2017; pp. 405–412. [CrossRef]
37. Alrtimi, A.; Rouainia, M.; Haigh, S. Thermal conductivity of a sandy soil. *Appl. Therm. Eng.* **2016**, *106*, 551–560. [CrossRef]
38. Khan, M.I. Factors affecting the thermal properties of concrete and applicability of its prediction models. *Build. Environ.* **2002**, *37*, 607–614. [CrossRef]
39. Xu, Y.; Sun, D.; Zeng, Z.; Lv, H. Effect of temperature on thermal conductivity of lateritic clays over a wide temperature range. *Int. J. Heat Mass Transf.* **2019**, *138*, 562–570. [CrossRef]
40. Widiatmojo, A.; Uchida, Y.; Fujii, H.; Kosukegawa, H.; Takashima, I.; Shimada, Y.; Chotpantarat, S.; Charusiri, P.; Tran, T.T. Numerical simulations on potential application of ground source heat pumps with vertical ground heat exchangers in Bangkok and Hanoi. *Energy Rep.* **2021**, *7*, 6932–6944. [CrossRef]

buildings

MDPI

Article

Energy Self-Sufficiency of Smaller Rural Centers: Experimental Approaches

Elvira Nicolini

Department of Architecture, University of Palermo, 90128 Palermo, Italy; elvira.nicolini@unipa.it

Abstract: Inland areas have been affected by demographic and economic decline over the past decades. New economic models, which are more focused on a humane quality of life, encourage a revival of these territories as newer, healthier places for living. This paper focuses on minor centers, rethought as energy communities and how these can sustain themselves and become new places of living. The first part of the research critically analyzes current strategies of SECAPs (Sustainable Energy and Climate Action Plans) in smaller historic urban centers. The second part of the paper starts with the typological, morphological, and technological interscalar analysis of two case studies, testing a repeatable expeditious knowledge collection and an intervention method on them. For urban environments, the hypothesized interventions include the management of energy production from renewable sources that are compatible with the presence and value of urban and built heritage; concerning rural territories, an agro-energy park is proposed. The document aims to provide a repeatable method for planning strategic actions within SECAPs in smaller urban centers with a high historical connotation. The case studies show that energy self-sufficiency can be an opportunity to valorize the urban center while favoring environmental sustainability and local development.

Keywords: sustainable energy technologies; climate adaptation plan; urban regeneration; small towns and rural communities

1. Introduction

The leading cause of climate change is the greenhouse effect. The energy sector accounts for about two-thirds of greenhouse gas emissions, as more than 80 percent of the world's energy consumption is still met by fossil fuels [1]. The 2030 EU strategies are mainly aimed at decarbonization, accelerating the transition from traditional fuels to renewables; energy efficiency, with actions for making buildings more efficient and constructing infrastructures, including some for sustainable mobility; and strengthening the energy market, increasing electricity interconnections and uniting the markets between member states and third world countries to foster efficient trade [2].

One of the best-known strategies is EPBD III, EU Directive 2018/844 on the energy performance of buildings, which is part of the legislative measures adopted at the European level known as the Clean Energy Package. It fixes the regulatory framework for achieving the new European 2030 energy and climate-related targets [3]. Its September 2022 update requires member states to adopt energy and climate plans with measures to at least double the annual rate of renewable energy by 2030 and the development of a sustainable, competitive, secure, and decarbonized energy system by 2050 [4].

The decarbonization process can advance at a faster pace with the deployment of renewables, but the synergies and the trade-offs between mitigation goals and the use of renewables must be better understood and integrated into climate policy [5]. The European Union's (EU) long-term climate neutrality targets require that by 2050, at least 16% of the electricity generation has its origins in collective projects and almost half of all European households must be involved in renewable energy generation, 37% of which should be engaged in collective projects [6]. More or less spontaneously in the last few years, energy communities have emerged, i.e., associations of individuals who share the benefits of producing and consuming green energy locally. In Europe, various forms of energy communities have been created according to different needs and opportunities in the area, for example, with Smart Cities' regulation. Smart Cities and communities were already defined as a priority and strategy by the previous European Horizon 2020 program and the 17 Sustainable Development Goals set by the UN and the 2030 Agenda. Energy communities are new configurations within the European electricity system. At the legislative level, they have been defined by the Directive on "Common rules for the internal market for electricity" (IEM) 2019/944 (Art.16), published in June 2019 [7]. They are legal entities based on the voluntary "network" participation of businesses, individuals, or municipal governments, whose goal is to create community-based environmental, economic, or social benefits through collective energy production. In Europe, energy communities are a rooted and widespread trend. Germany, Denmark, the Netherlands, and the UK are the European countries with the largest numbers. In Germany, there are now as many as 1700 entities; in Denmark and the Netherlands, there are over 500 active CERs.

The Covenant of Mayors is undoubtedly one of the most centered EU-level actions on this issue and can significantly counter the depopulation of these areas. This voluntary initiative has been promoted as early as 2008 by the European Commission to actively involve cities and their administrators in fighting climate change and achieving EU goals. Signatory local authorities commit to signing a Sustainable Energy Action Plan (SEAP) [8]. With the 2015 European policy (2030 Climate and Energy Package, Adaptation Strategy, and Energy Strategy), the commitment has been extended to updating the strategies according to the new goals and climate adaptation (Sustainable Energy Action Plan—SEAP) [9]. Currently, in Sicily, Italy, 145 out of 390 municipalities have signed a Covenant of Mayors [10]. Only a few have SEAPs, and a very small percentage of these have upgraded it to SECAP, even though in June 2019, the regional Department of Energy and Public Utilities [11] presented the program for the allocation of resources to Sicily's municipalities for the drafting of SECAPs. In particular, municipalities in inland areas, the farthest from major metropolitan centers, have a deficit. This fact is particularly relevant considering that Italy and Spain are the countries in Europe with the most significant number of XS municipalities (under 10,000 inhabitants) among the signatories to the Covenant of Mayors.

2. Background

The ferment towards the need for energy and climate sustainability is tangible in metropolitan cities. However, it is not yet tangible in small towns, whose size could lead to their involvement in eco-neighborhoods or energy community strategies. The most difficult contexts are inland areas, territories with semi-abandoned architectural and urban heritage, and severe depopulation. These urban centers were often founded according to defensive and territorial-control criteria. Thus, they are located in inaccessible places that are difficult to access, and their livelihood relied on the productive territory. Abandonment has resulted in further marginalization and impoverishment, both rural and urban. These scattered settlements undoubtedly lack services. Still, they are 'humane scale' places where the air is healthier, and the problems tied to humane concentration are reduced; their identity consists of ancient cultures and traditions, leading them to represent a treasure trove of knowledge and memory [12]. Still, at least the architectural and urban heritage identity features have been the object of enforced preservation, keeping the traces and original appearance of ancient historic centers of ancient dates [13]. There is still a wide presence of historical, or at least traditional, buildings and fabrics, which lacked the resources for uncontrolled building modernization.

The green economy and circular economy have restored the importance of balance with natural dynamics and resource provisioning, which were the guiding principles of the rural and survival economies [14]. Some urban centers that have kept visible historical features can be reevaluated as new energy self-sufficient territories and favorably accustomed to climate mitigation since they are often surrounded by and integrated with nature. This aligns with the principles of the Italian National Strategy for Inland Areas and aims to reverse demographic trends, maintain natural and territorial capital, and improve the resilience of these places [15,16].

Some indications for inland areas come from the Italian National Strategy for Smart Specialization (European Cohesion Policy 2014–2020 programming). The development trajectories include restoring productive activities, regenerating sustainable communities, and regaining knowledge of exploiting resources (water, energy, MSW, etc.) [17]. Regarding energy self-sufficiency from renewable sources, some actions are already outlined in the Italian National Plan for the Redevelopment of Small Municipalities referred to in Law 6 October 2017, No. 10, particularly for the energy upgrades to the public and private building stock to achieve EU targets [18].

The contribution aims to center again on the value these places can provide to the global climate resilience movement. Today, high economic commitment and technical effort are generally employed—for example, with greening actions—to restore nature in heavily urbanized contexts; yet, places such as inland areas, which are already immersed in nature, can fulfill community goals and raise the quality of life for those who want to inhabit them.

Rural depopulation would not be a limit, but rather a "potential positive opportunity", for a social reorientation due to its natural green advantages and the scarce pressure on natural resources and the environment [19]. However, declining local communities in Europe have limited capacities for long-term comprehensive policy governance responses because depopulation generates fragile communities that are very vulnerable to environmental risks and have little capacity for effective responses [20].

The construction of a high-quality land development, protection, and support system has become a topic of great interest for all societies, but it still needs to be explored [21]. Especially in Italy, in urban center, the integration of renewable energy sources has struggled to take place due to the impact the technologies would have on the existing landscape. The question still arises, especially in centers with a strong historical connotation.

Sustainability assessment methods for urban centers, in general, use different approaches for assessments including LCA (life cycle cost analysis), productivity analysis, operation and maintenance optimization, indoor air quality, and other approaches; however, the existing methods are in a transition phase and the next generation will encompass all aspects of sustainability [22]. The historic urban landscape requires an additional step: landscape impact assessments. Downstream of the consolidated Historic Urban Landscape UNESCO guideline, several member states have experienced poor enforcement of building regulations by planning authorities regarding prohibited interventions in heritage areas. Amendments to national- and local-level legislation are required to integrate impact assessments for heritage projects and all initiatives in historic urban areas. The 2011 recommendation addresses the need to better integrate and frame urban heritage conservation strategies within a broader objective of sustainable development. Most member states have policies pertaining to renewable energy at the national level, which do not place special emphasis on historic urban areas [23].

The Guidelines of the Italian Ministry of Culture (2015) for the improvement of energy efficiency at cultural heritage sites address the delicate implications of an efficient use of energy in relation to the conservation and protection of historic centers and nuclei and rural architecture for landscape purposes [24]. The guidelines suggest viable solutions that pursue more explicit references to the local building tradition, as long they are within a framework of overall compositional and linguistic sobriety. Landscape constraints must be respected, with the condition that the systems are concealed and totally integrated into the built environment. In the case of using existing buildings, it is preferable to identify those of lesser value and of recent construction, and it is also suggested to identify areas for the installation of photovoltaic or solar systems on the ground, suitably identified in a defiladed position, and to carry out the interventions with the interposition, with respect to the observation points, of vegetation borders such as trees and/or hedges that are possibly accompanied and supported by wooden or metal trellises. Pertinent and service buildings may also lend themselves to accommodating micro wind farms, carefully studying their visibility in the surrounding context at various scales. An interesting proposal of the Ministry is to identify a single place outside the built-up area where the plants can be installed cumulatively by the municipality, while allowing individual users to enjoy pro quota the benefits and facilities provided by law: the so-called 'delocalised on-site exchange' provided, in application of Law 99/09, for municipalities with a population of up to 20,000 residents.

The paper reports two experimental case studies using a strategic planning approach to address the needs for climate resilience and energy self-sufficiency from renewable sources. The study focuses on two small urban centers with strong historical connotations. They are both part of the metropolitan area of Palermo but are more than 40 min away from it. They are both characterized by high and heterogeneous marginalities and disadvantages. Their location in impervious and distant places, increasingly marginalized by the infrastructure network and a lack of resources, has for a long time, together with the lack of employment, led to a progressive depopulation and, in some cases, complete abandonment. At the same time, however, they are places that still retain a high architectural and landscape quality, which are now being reviewed as primitive models of balance. Thinking of repopulating them means going back to living in sustainable places, where slow life and zero-kilometer consumption of resources persist, where the environment is still preserved in its essential features and pollution levels are contained.

This study aims to propose sustainable energy and climate actions for these marginal areas and evaluate their possible integration into historic centers. These actions aim to drive the economy of these devalued places, creating a principle of energy self-sufficiency to meet the needs of the built-up area in a vision of subsidiarity with neighboring municipalities. For these towns, energy self-sufficiency can bring a direct economic return in a short time, proving to be a fast-track system that attracts local authorities and citizens towards active management of their own resources. This is just one tile of an ideal vision, which could be implemented with simple technological designs and thematic planning solutions in cities. In addition to being scarcely invasive, they could be an opportunity for concrete development and achievement of new desirability for these places.

3. Methods

Each urban and spatial context suggests specific and tailored solutions. However, both case studies can be addressed through a needs–performance approach, starting from users' requests to identify design objectives and requirements useful for proposing solutions at various scales. Comparing the pre-identified requirements with the expected performance of design solutions allows the ex-ante assessment of projects' efficiency and effectiveness.

The first phase of the research analyzed the state of the art on sustainable energy and climate processes in Italian so-called XS municipalities (with a population below 10,000) that have joined the latest Covenant of Mayors (CoM) 2050 strategy. The study showed that this action still has a limited scope. In this research, experimenting with smaller urban centers—inland areas of the Sicilian mountainous territory—an additional effort was made concerning this method: broadening the perspective to a broader territorial context in a vision of mutual aid between municipalities. Thus, the choice of design solutions involved reflecting on the current territorial situation, considering the possibility of drawing from existing renewable energy sources (RESs) or the complementarity of potential new plants with others already in the area. Moreover, the impact of all actions on the historic and natural landscape was evaluated.

The first work phase involved interscalar context analysis, from a broader territorial to an urban and, finally, an architectural one. The spatial analysis assessed the conditions of marginality and the potential for relationships between neighboring municipalities. To lay the foundation for constructing a SECAP, the characteristics, state of use, and extent of renewable and non-renewable energy production technology solutions were analyzed.

The urban-scale analysis considered settlement principles, including environmental design choices, the city as a whole, its historic core, its relationship with the land, orography, geological formations, natural and man-made historical infrastructure, and the typology of public spaces (hierarchies, functions, and forms). At this scale, total inhabited energy consumption, produced CO_2, and toe for thermal and electrical use in the residential, tertiary, transportation, industrial, and agricultural sectors were measured for a base year.

The architectural analysis defined the high-value historical heritage and urban spaces to protect in design choices. A survey was conducted on the building units' type, morphology, technology, potential for retrofit interventions, and current performance, including historical and state of preservation analysis. These are useful for configuring suitable spaces to accommodate new design solutions. The three analyses required constantly considering urban planning instruments in force, which are helpful in landscape characterization, geology, constraint definition, and urban center zoning.

Project proposals include a preliminary study of applicable technologies. They are formulated to reduce energy and CO_2 consumption, thus achieving greater environmental sustainability, self-sufficiency, and the possibility of integrating technological solutions into the historic city. The survey of national and European best practices allowed for the identification of strategic territorial and urban actions and interventions such as climate mitigation actions at urban and architectural scales.

The premise is that the solution to energy problems cannot lie in a technological, large-scale, and invasive solution but in a mix of solutions, more tailored and, above all, tuned to the sensitivity of the historic urban and rural landscape context.

The energy and climate actions proposed here, at the architectural, urban, and sub-urban scales, are calibrated in response to the needs of the current and potential resident population. The analyses performed and the proposed actions can be preliminary for implementing a SECAP. Transversally to the technological actions here suggested, institutions are expected to perform sensitization actions with the local community, combined with energy retrofit interventions on lighting and public buildings, and with the development of agreements between public and private institutions to foster eco-sustainable systems.

The method presupposes the development of scenarios and verification of the fulfillment of requirements and constraints which, in the case of the historic urban landscape, are more complex (morphological, typological, material, perceptual constraints, etc.). This required various analyses: analysis of the current urban planning tools; orography of mountain ranges; individuation of watersheds; study of the geological context, with a specific focus on orogenetic episodes; study of the territorial constitutional state; and the landscape constraints at various urban scales. For these analyses, the main source was cartography at various scales: municipality, provincial, and maps in the regional landscape plan. At the territorial scale, authorized renewable energy production plants near the studies were surveyed. At the urban scale, the size of the built environment, the construction typologies, and the building typologies were analyzed, with a specific focus on the architectural works with notable historical-architectural value.

The energy requirements of the technological systems were established through demographic analysis of the current state, analysis of the demographic trend, and analysis of energy consumption. The energy demand has been calculated by the neighborhood and economic sectors: residential, tertiary, transport, industry, and agriculture. Regarding the residential sector, the main sector in the city, the overall data provided by the Energy Department of the Sicily Region were verified for each neighborhood by crossing an estimate by number of inhabitants with one by building size. Toe and KWh consumption, as well as CO_2 production have been calculated both for thermal and electric uses.

Design phases involved programming strategic actions for energy development and urban regeneration and energy plan hypotheses. The solutions resulted from an iterative process that mainly considered two aspects: the characteristics of each urban space, marked by a specific layout and, more importantly, their infrastructural, technological, and organizational supply; and the fulfillment of energy demand.

4. SECAP: Smaller Italian Urban Centers in Line with the CoM 2050

Covenant of Mayors signatories under the latest CoM 2050 Strategy have committed to prioritizing climate action by setting medium- and long-term goals consistent with those in the Paris Agreement, with renewed ambitions for a fairer, climate-neutral Europe.

The number of XS signatories—municipalities under the population threshold of 10,000 residents—is interesting: Italy and Spain alone have 3940 XS municipalities, corresponding to 91% of the total XS signatories. The remaining 376 XS signatories are spread across 28 EU and non-EU countries, with an average percentage of less than 1 percent. In 2023, these municipalities constitute about 63 percent of the total number of signatories to the Covenant, highlighting that the Covenant is a local action supported by "small" municipalities [25]. This number of signatories makes the Covenant of Mayors for Climate and Energy the world's largest movement for local climate and energy action.

The Covenant's official website database supports advanced searches by region of origin, population, and CO_2 emission reduction target. The Italian XS signatories that have submitted a SECAP following the 2050 strategy are currently few and only located in northern Italy: Arcugnano (VI), Caraglio (CN), Colceresa (VI), Ponte di Piave (TV), and Racconigi (CN). Table 1 below clearly presents their key actions and expected CO_2 t/y savings. In bold are the actions that have the greatest impact.

Table 1. Strategic actions of SECAP in line with the 2050 strategy of smaller Italian centers.

Municipality	Date	N. Inhabitants	Strategic Actions	Reduction (CO_2 t/y)
SECAP of Colcerosa (VI) [26]	2022	5968	**Energy efficiency upgrade of residential buildings**	**2643.23**
			Energy efficiency upgrade of tertiary buildings	1109.13
			Energy efficiency upgrade of public buildings	16.90
			Installation of photovoltaic systems	519.70
			Modernization of the vehicle fleet	1713.62
			Energy efficiency upgrade of public lighting	108.18
			Urban greening	1358.6
Total expected savings				40%
SECAP of Ponte di Piave (TV) [27]	2022	8345	Electricity acquisition from renewable sources	623.00
			Energy efficiency upgrade of tertiary buildings	709.18
			Energy efficiency upgrade of residential buildings	**2356.17**
			Energy efficiency upgrade of public lighting	38.72
			Electric mobility promotion	1142.75
			Modernization of the vehicle fleet	1828.21
			Installation of private photovoltaic systems	1712.19
			Energy efficiency upgrade in industrial processes	2060.16
Total expected savings				40%
SECAP of Racconigi (CN) [28]	2022	9661	Energy efficiency upgrade of residential and tertiary buildings	746.00
			Energy efficiency upgrade of public buildings	104.00
			Energy efficiency upgrade of public lighting	80.00
			Installation of photovoltaic systems on public buildings	75.00
			Energy management of public assets	29.00
			Modernization of the vehicle fleet	2.00
			Establishment of Renewable Energy Communities	921.00
			Electric mobility promotion	**2634.00**
			Sharing mobility promotion	649.00
			Increasing sustainable mobility and bicycle and pedestrian infrastructure	107.00
			Enhancement and efficiency upgrade of public transport	284.00
Total expected savings				55.4%

Table 1. *Cont.*

Municipality	Date	N. Inhabitants	Strategic Actions	Reduction (CO_2 t/y)
SECAP of Caraglio (CN) [29]	2022	6818	Installation of photovoltaic systems on public buildings	71.61
			Solar heating installation on public buildings	8.95
			Energy efficiency upgrade of public lighting	35.80
			Energy efficiency upgrade of public buildings	53.70
			Modernization of the public vehicle fleet	8.95
			Energy efficiency upgrade of residential buildings	**3208.95**
			Energy efficiency upgrade of tertiary buildings	1814.16
			Increasing sustainable mobility and bicycle and pedestrian infrastructure	470.70
			Increase in 30 km zones in the historic center	470.70
			Modernization of private vehicle fleet	3765.61
Total expected savings				40.47%
SECAP of Arcugnano (VI) [30]	2023	7722	**Energy efficiency upgrade of residential buildings**	**3643.04**
			Energy efficiency upgrade of tertiary buildings	3479.33
			Energy efficiency upgrade of buildings for industrial use	1397.73
			Energy efficiency upgrade of public buildings	n/a
			Incentives for electric micro-mobility	294.62
			Modernization of the vehicle fleet	949.32
			Increasing sustainable mobility and bicycle and pedestrian infrastructure	458.29
			Increasing proximity shopping	425.56
			Proposals for eco-driving, carpooling, and telecommuting	523.76
			Improvement of agricultural techniques	240.00
Total expected savings				50%

The CO_2 emission reduction target set by 90 percent of the XS signatories is between 20 and 30 percent. Thus, the signatories described can be considered among the most virtuous ones, as they set an average 2030 abatement target of 40 percent of emissions following the new European Strategy. All of the municipalities surveyed aim for citizens' active involvement. The administration expects them to be willing to upgrade their homes or retrofit their vehicles to produce fewer emissions.

In some cases, the presence of photovoltaic panels is observed in the roofs of historical (yet not monumental) buildings. This phenomenon is likely to increase since the examined SECAPs report the target of a substantial rate of emission reduction by installing photovoltaic systems on building roofs. Being smaller municipalities, the center generally coincides with the historic center, which is also characterized by valuable building types and attractive monumental complexes (think of the castle of Racconigi or the Silk Mill of Caraglio). Thus, there will inevitably be an impact on the historic urban landscape. The issue is also encouraged by the Italian "Energy" Decree-Law no. 17/2022, converted into law no. 34/2022, which establishes a liberalization for installing new solar photovoltaic systems by providing the possibility to build those up to 200 kW with lighter regulations [31]. In the presence of the constraints, performing the interventions is allowed after the issuance of a permit by the competent administration according to the code in Legislative Decree No. 42 of 2004, already mentioned. According to Article 136, Paragraph 1 (c) of Legislative Decree No. 42 of 2004, the provisions also apply to "complexes of immovable property with characteristic appearance having aesthetic and traditional value, including historic centers and cores," limitedly to the installation of integrated panels in roofs. They must not be visible from outdoor public spaces and scenic viewpoints, except for roofs whose coverings are made of traditional local materials. The impact must be focused on, and it is necessary to verify that the emission abatement calculations in SECAPs consider urban planning instruments and municipal building codes while maintaining protections and limits of intervention to buildings of historical, cultural, and architectural significance.

Municipalities' actions primarily concern four economic sectors (residential, tertiary, industrial, and transportation). The largest share is related to the behaviors that private citizens should undertake to improve consumption efficiency and consequently reduce atmospheric emissions. This is underpinned by the public administration's essential commitment to raising public awareness of these interventions/issues through meetings, communication campaigns, and incentives. The action of information and communication aimed at citizenship helps lay the foundations for new habits of consumption and behavior. This includes education on the use of electrical and electronic appliances, care of the property's maintenance status by making it energy efficient with several simple active and passive solutions, e.g., internal relamping, automatic shut-off devices, and remote controls; replacement of heat generators and electric pumps; installation of regulation systems, adoption of construction systems capable of reducing heat loss (replacement of window frames, insulation of vertical and horizontal closures); and use of heat pumps connected to renewable energy capture systems (solar thermal panels, geothermal system, etc.). The action may receive a boost from ongoing incentive systems (heat bills, energy efficiency bonds) and tax deductions.

A good practice is the Sportello Energia (Energy Desk), created at the initiative of the Piedmont region. This desk aims to guide local property owners and condominium managers in their decision to proceed with energy upgrades. The desk staff provides information, answers doubts, and supports them in identifying incentive mechanisms, resulting in more building upgrade projects that help to achieve higher energy savings and faster returns on investments.

All the reviewed SECAPs acknowledge that the transportation sector significantly impacts the environment, leading to further sustainable transport solutions that benefit the environment, society, and the economy. The action aims to support the replacement of the car fleet with low-emission vehicles that meet the limits specified in European Regulation 715/2007. Proponents include practices for free or reserved parking spaces to encourage electric vehicle purchases (or rentals). The action provides for the placement of charging points capable of covering the urban and built-up areas. In addition, they encompass the maintenance and implementation of bicycle routes to encourage private citizens to choose slower mobility options.

One element that makes small towns in inland areas healthy is their integration into the natural green landscape. Little or almost nothing is needed to implement tree planting, except in a few parking areas.

The advantage of these areas' lower degree of urbanization and settlement principle also affects safety regarding the ability to control hydraulic hazard conditions. Greenery provides one more contribution through the waste from mowing and pruning, which is recycled along with that from agricultural production. Together with food production, this activity is among the most important in these contexts. These wastes are used in the production of energy from biogas. The latter derives from the anaerobic fermentation of agricultural plant biomass, as for the plants in Ponte di Piave and Arcugnano.

Due to being smaller centers, this action has a small global impact. So, to amplify it, some centers have decided to extend their choices by involving neighboring municipalities, forming networks with leaders, from a more long-term perspective. Indeed, if each city adopted its own SECAP, this could result in repetitive or disconnected actions among neighboring municipalities at the expense of additional resource expenditures. Small municipalities having submitted a joint SECAP with the CoM 2050 strategy include two in Italy, both in Piedmont: Sommariva Perno-Lisio-Montà (CN) and Quincinetto-Borgofranco D'Ivrea-Burolo-Lessolo-Quassolo-Chiaverano-Montalto Dora (TO). In these cases, the administrations have taken steps to redefine the municipal structure for the various activities under the initiative, identifying a responsible figure and organizing a working group capable of managing relations with the European Commission and, in general, the organization and implementation of the various activities. When drafting the SECAP, to achieve the collective goal of reducing emissions, the municipal administrations carried out an energy-environmental analysis of the territory and the activities insisting on it. They reconstructed the energy balance and realized the CO_2 baseline emission inventory (BEI). Moreover, they evaluated and indicated the potential reduction of final energy consumption and the possibility of increasing local energy production from renewable or other low-impact sources, reconstructing the possible evolutionary scenarios of the local energy system. These data are reported for each municipality and then aggregated. The actions, however, are not based on mutual help between neighboring municipalities according to a principle of subsidiarity. Instead, they share the strategy and overall goals by presenting individual and aggregated results, but the actions are specific to each municipality. The aggregation of small municipalities is a relevant first step. Thanks to this, XS municipalities can produce more substantial results within the global movement aimed at CO_2 abatement. The outcomes could be even more satisfactory if the interventions were conceived through interactions between municipalities, including the joint procurement and management of new sustainable energy and climate services (think, for example, of the continuity of mobility routes between municipalities or the possibility of surplus energy divestment between neighboring territories).

5. Case Studies of Sustainable Energy in Small Urban Centers with Strong Historical Features

The experimentation, part of the workshop "Smartness: Minor Centers as Laboratories of New Sustainable Residentiality" held in the Department of Architecture at the University of Palermo, was aimed at urban centers with strong historical connotations but were on the geographical and economic margins. Related to the theme of sustainable energy and climate adaptation, two urban centers were considered: San Mauro Castelverde and Santa Cristina Gela, both in the province of Palermo. The study aims to investigate the feasibility of implementing climate adaptation solutions and using low-impact renewable energy in a vulnerable context, such as a historic urban environment in rural areas. In the first case study, solutions with geothermal probes compatible with the subsoil characteristics, urban spaces' morphology, and architectural heritage have been identified among the many opportunities to supply energy from renewable sources [32]. In the second context, an agro-energy park is proposed, with the integration of wind power plants with micro-rotors for energy production and distribution in public buildings, a system of urban gardens to serve local populations, and a bicycle and pedestrian park, emphasizing the naturalistic values and relevant touristic and sports vocation of the place [33].

The physical resilience of these urban centers has allowed the permanence of traditional buildings, whose energy retrofit will be an improvement of their intrinsic passive-building characteristics. These buildings were constructed with an awareness of the contextual, environmental, and meteorological factors, without the aid of technical systems, which did not exist at the time or were reduced to chimneys and wood stoves. Their construction techniques aimed to protect against cold and heat with natural materials, taking advantage of exposure, sunlight, openings, and shading, wall inertia, passive ventilation, and shading.

San Mauro Castelverde, in the province of Palermo, is a municipality of 114 km^2 belonging to the territory of the Alte Madonie Mountains, in the homonymous park. It is located at 1050 m a.s.l. and about 22 km from the coast. It is on the edge of the mountainous Madonie territory; thus, it suffers from territorial marginalization in geographical and infrastructural terms. Its settlement and cultural characteristics are also partly affected by its proximity to the Nebrodi mountain range. The positional and geomorphological characteristics of San Mauro Castelverde allow for using geothermal energy to lower CO_2 emissions into the environment and achieve autonomy from the oil supply. The focus has been on designing low-enthalpy systems (geothermal fields with temperatures below 100 °C) for air conditioning, heating, and cooling of buildings intended for civil housing and services. Despite its copious initial cost, this choice can guarantee CO_2 emission abatement, high yields, and almost no perceivable landscape impacts.

The choice of the renewable energy source followed the interscalar analysis of the characteristics of the urban center under study and the suburban landscape to determine the area's most widespread renewable energy sources. The use of photovoltaic panels was ruled out since they would impact the landscape of the historic center. Moreover, hydroelectric power would be disadvantageous since the nearest river is torrential, and the absence of reservoirs does not ensure a continuous flow of water. Moreover, the exploitation of biomass is not advisable since there is already a plant a short distance away (in Castelbuono), receiving organic MSW and pruning waste from San Mauro. After analyzing the state of law to identify landscape constraints, wind poles greater than 18 kW were ruled out. Despite being convenient, a micro-wind farm is contraindicated in the historic center due to the noise pollution caused by the rotation of the blades. The area of the historic center, almost coinciding with the entire urban center, is characterized by a multitude of courtyards suitable for geothermal installations. One more aspect is the outcrop of Numidian Flysch, a rock formation of the Sicilian Oligocene composed of argillitic sediments composed of quartzarenites, with an excellent thermal conductivity of 2.3 W/mk and an extractable power of 65–80 W/m in 1800 h [34].

The analyses (Tables 2 and 3, Figure 1) revealed that the historic center has the highest energy consumption, amounting to 2015 kWh in the heating sector alone, with CO_2 emissions to the environment of 1270 kg.

Table 2. Consumption analysis of San Mauro Castelverde (toe) (Source: Department of Energy of the Sicilian Region).

	Residential	Tertiary	Transport	Industry	Agriculture
Thermal use (toe *)	0.31	0.26	-	0.07	0.10
Electricity use (toe)	0.27	0.19	0.02	0.24	0.04
Fuel (toe)			0.71		
Total	2.21 ktoe				

* toe: ton of oil equivalent; For electricity use 1 toe = 5347.59 kWh; for fuel and thermal use 1 toe = 11,628 kWh; source: Sicilian Region Energy Department.

Table 3. Consumption analysis of San Mauro Castelverde (kWh) (Source: E. Nicolini).

	Residential	Tertiary	Transport	Industry	Agriculture
Thermal use (kWh)	1657.8 *1	1390.37	-	374.33	534.76
Electricity use (kWh)	3139.6 *2	2209.32	232.56	2790.72	465.12
Fuel (kWh)			3796.79		
Total	16,591.37 kWh				

*1 0.31 × 5347.59 = 1657.8 kWh; *2 0.27 × 11,628 = 3139.6 kWh; CO_2 = 16,591.37 kWh × 0.255 kg/kWh = 4230.80 CO_2 produced in the whole town.

The tables show the overall consumption, but the work was performed in detail by district and by building use. Calculations determined that to meet these needs, 149 geothermal probes with shallow exchangers of four types are needed: baskets in backyards, vertical probes along streets, loops in residences' private gardens, and vertical probes, again, outside the walls. For clarity and as an example, the calculation for the residential sector is shown below. According to the guidelines laid down by the Italian Ministry for the Environment, it is considered that to produce 1 kWh of electricity, on average, the equivalent of 2.56 kWh in the form of fossil fuels are burnt, and that approximately 0.53 kg of carbon dioxide is emitted into the air 2.56 kWh × 0.255 kg/kWh. Therefore, 0.31 × 5347.59 = 1657.8 kWh, 0.27 × 11,628 = 3139.6 kWh, giving a total of 4797.4 kWh × 0.255 kWh = 1225 kg of CO_2.

The installation of conical basket-shaped probes (Figures 2 and 3) made of polyethylene pipes attached to iron or plastic reinforcement has been hypothesized in courtyards. Their depth is set at 1.5 m for a height of 1.2 m, with a diameter of 2.4 m, at a distance of 5 m. In gardens, the choice of geothermal loops is meant to preserve trees. They are lined with plastic tubes with internal diameters between 16 and 22 mm; their embedment depth is 80 cm. The rings develop on three mutually parallel planes placed in ditch excavations. Vertical probes would be realized with foundation piles and U-shaped or spiral pipes, have a depth between 15 and 20 m, and be located in the street or outside the walls. The probes were placed throughout the neighborhoods according to the following plan.

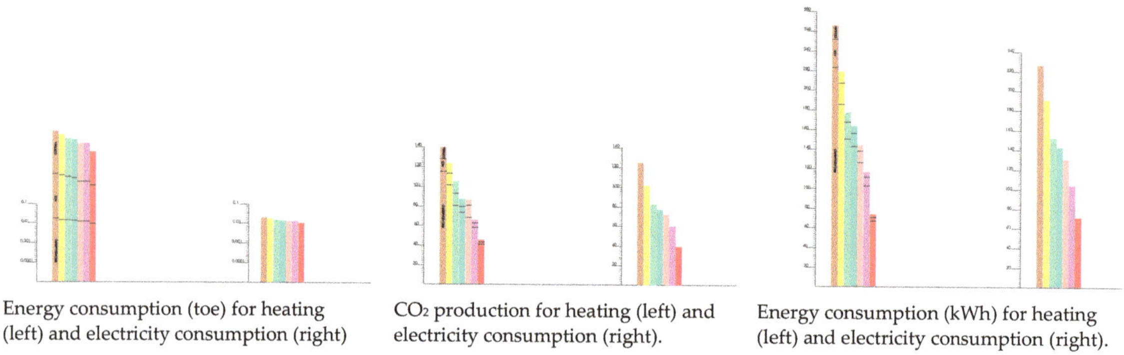

Figure 1. Analysis of neighborhood consumption. The color corresponds to each neighbourhood shown in Figure 2. Images by: Arch. Maria Rosaria Cimilluca.

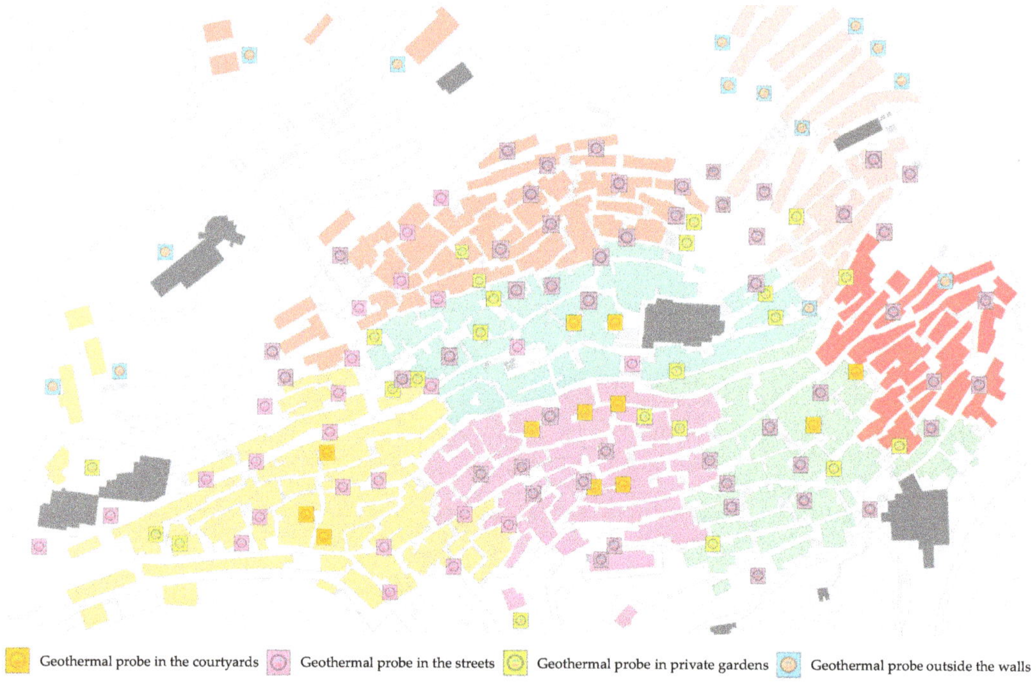

Figure 2. Geothermal probe placement hypothesis in the urban center. Images by: Arch. Maria Rosaria Cimilluca.

Serra district	**245 inhabitants**	Color yellow indicated on the plan of Figure 2						
Building type	**Ground floor use**	**Construction technique**	**Floors**		**Probe classification**	**Kwh produced for heating/refreshing**	**n° probe**	**Kwh produced per district**
Terraced housing	Abandoned local	Mixed-structure buildings	2–3 elevations		Ring probes	15 kWh	5 × 15	75 kWh
Palaces and buildings	Abandoned local	Plastered stone masonry	2–4 elevations		Basket-shaped probes	15 kWh	3 × 15	45 kWh
New buildings	Garages	Reinforced concrete	2–3 elevations		Vertical probes	15 kWh	3 × 15	45 kWh
Terraced housing	Garages	Mixed-structure buildings	2–4 elevations		Vertical probes	15 kWh	14 × 15	210 kWh

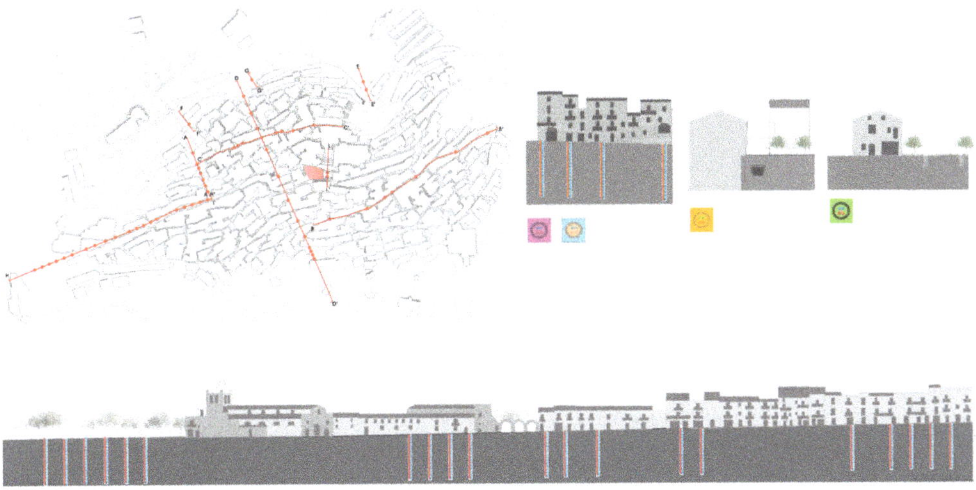

Figure 3. Probe type selection according to building type and calculation of KWh produced. Images by: Arch. Maria Rosaria Cimilluca.

Neighborhood consumption analysis was performed regarding thermal and electric uses for residential, agriculture, service, transportation, and industry sectors. The total consumption is 1.97 toe (1 toe = 11,628 KWh for fuels and 1 toe = 5247.59 KWh for electricity consumption). Considering the standard emission factor, the amount of CO_2 released into the environment expressed in kilograms is 1270.53 kgCO_2 in the heating sector alone. Based on the analysis, the sector that consumes the most is the residential one, especially for heating. Building profiles of some streets with associated geothermal probes were drawn to highlight the relationship between excavation and the built environment. This was followed by a study of the various shallow geometric probes available in the market, with their types, applications, materials, power, and depth of deployment. Ground floors or basement floors, such as cellars or current technical rooms, were identified as places for system components, such as manifolds, with at least a minimum volume of 30 m^3.

After identifying the probe network's placement in the historic center's urban spaces, energy efficiency upgrades to the city hall building were explored (Figure 4). The hypothesis of one more low-enthalpy geothermal system was formulated, involving the installation of probes in the square before the building and a geothermal heat pump inside the building to power the fan coils.

The facility has a traditional thermal system with an oil-fired power supply and thermal output of 110 kW; it was proposed to replace it with a system with 16 vertical probes 12 m deep to meet the energy consumption and needs for heating and cooling. The hypothesized air-conditioning system mainly consists of one to several water-to-water heat pump units and hydraulic circuits with vertical heat-exchange probes. A room in the basement was chosen to place the heat pump and various collectors; a total of 12 vertical recessed fan coils are to be placed as terminals in the various rooms on the aboveground levels.

Sicily is characterized by a scorching summer climate with an asymmetry of solar irradiance and temperatures from March to October. Thus, the extensive use of heat pumps with low-enthalpy geothermal energy as a thermal source can help drastically reduce greenhouse gas emissions produced by air conditioning systems in building structures. Soil is a great energy resource, as it keeps higher average temperatures than average air temperatures in winter while doing the opposite in the summer. Under these conditions and heat pumps being cooled and heated by water instead of air, the resulting efficiency can provide an economically and environmentally valid alternative to combustion-based heating and air conditioning systems.

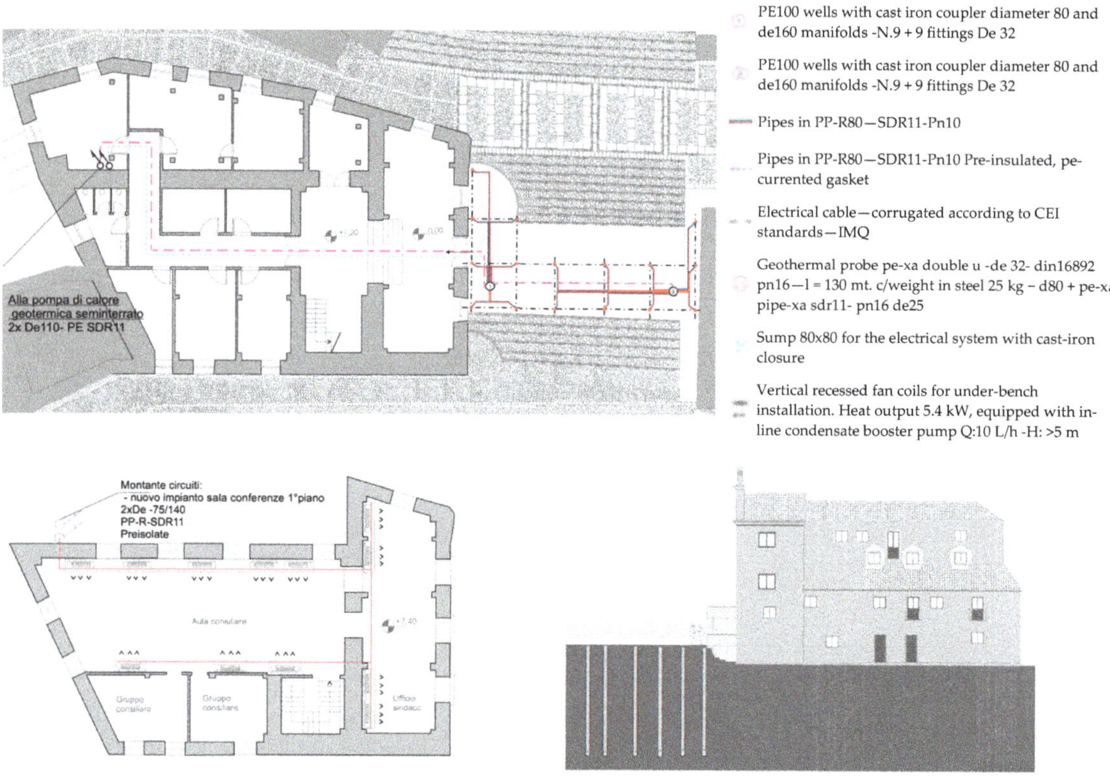

Figure 4. Example of the installation of a geothermal system in a historic building, the Municipal Palace of San Mauro Castelverde. Ground floor plan, first floor plan, and elevation. Images made by: Arch. Maria Rosaria Cimilluca.

Santa Cristina Gela is a small municipality founded in the late 17th century. It is located 25 km from Palermo and now covers 38 km^2. Located in the hilly inland areas behind Palermo, it is among the inland areas that have suffered a gradual abandonment of the countryside toward a major attractor. Demographic movements in the municipality mark a steady demographic decline, especially during the 30 years between 1951 and 1981. It seems the trend has been the opposite in the last few years. The historic core of the township has a valuable eighteenth-century build heritage, which would require an overall redevelopment and valorization project with a focus on cultural tourism. Indeed, urban planning tools identify it as an area to be protected. In addition, the surrounding area is characterized by several geographic peculiarities. Indeed, the municipality belongs to both the Oreto Valley and the protected area of Lake Piana degli Albanesi. They are both notable watersheds devoted to enhancing agricultural resources and with high scenic value. The strategic plan hypothesis of energy self-sufficiency (Figure 5) results from a field analysis showing that Santa Cristina Gela is a windy area with predominant winds from the northwest and southwest with an average speed of 8.20 m/s. The idea involves a new smart grid infrastructure for managing and distributing electricity and advanced metering systems to constantly monitor energy consumption. The hypothesized interventions include upgrading the energy efficiency of the public lighting systems by replacing lighting fixtures with latest-generation LED devices, providing network optimization systems with smart on/off controls, and reconnecting the urban and agricultural areas through urban gardens. The urban garden can also trigger other circular resource reuse actions, such as composting organic waste and reusing rainwater for irrigation.

Figure 5. Energy development and urban redevelopment: strategic action planning. Images by: Arch. Angelo Alabiso.

To meet the energy needs of the entire municipality, estimated to be around 1650 MWh now, an "agro-energy" urban park was hypothesized (Figure 6). It houses a mini-wind farm consisting of 100 vertical-axis blades 8 m high with a power rating of 6 kW and, therefore, a total instantaneous power of 600 kW capable of producing 1140 MWh per year. Its internal park is divided into areas dedicated to typical local crops (almond, citrus, and olive groves). Despite their high performance and constant blade rotation, horizontal-axis wind turbines (mega-wind turbines with a capacity of about 3 MWh, mini-wind turbines with a capacity of 2 kW to 20 kW, and micro-wind turbines with a capacity of 0.5 KW to 2 KW) can impact the landscape both visually and sonically. Therefore, their installation away from urban centers is recommended. Therefore, this type of turbine was ruled out due to the excessive size of the machines (heights between 20 m and 100 m and shovel diameters from 10 m to 50 m) and the risks involved in their inclusion in a natural setting like the case study, which is particularly rich in protected bird species. Vertical-axis wind turbines (mini-wind turbines with power from 2 KW to 10 KW, micro-wind turbines with power from 0.5 KW to 2 KW) do not need to be oriented toward the wind and have low noise levels. Moreover, energy is produced where it is consumed, and thanks to their small dimensions (height from 4 m to 8 m and diameter from 2 to 3 m), they could be integrated into the natural landscape.

In Italy, oil-fired thermal power plants emit an average of 0.65–0.85 $kgCO_2$ into the atmosphere to produce one electric kWh. The emission reduction associated with using wind power can be estimated as follows: CO_2 (t) = 0.3 × A × h × 860/1000. Where 0.3 is a constant that depends on the intermittent nature of wind, availability of wind turbines, and energy losses; h is the number of hours of operation per year (order of magnitude: 7000–8000 h); and A is the estimated theoretical power output (in MWh). Thus, for example, a 1 kW turbine, operating for 8000 h/year, reduces CO_2 production by 20 t/year. The park itself could meet almost all the needs of the municipality if combined with resource-friendly consumption and an energy-efficient regime and, at the same time, contribute to a decrease of 12,000 tons of CO_2/year. In addition, it could complement the ongoing work of a nearby hydroelectric power plant (Agghiastro locality in the municipality of Santa Cristina Gela) without interference. The plant exploits a net energy jump of about 240 m with a maximum flow rate of 1100 L per second, thus producing 2.5 MWh. It already powers several grid municipalities.

In conclusion, the project envisions upgrading and integrating the existing axis by connecting the town center, the new agro-energy park, and the lake through pedestrian–cycling tracks and redesigning the route of a former railway line integrated into the park also as a pedestrian–cycling track.

Figure 6. Consumption analysis, energy vision, and planned energy balance. Agro-energy park project, sections. Images made by Arch. Angelo Alabiso.

6. Considerations and Conclusions

This paper provides a preliminary analysis of actions to draft a SECAP in two historic urban settings characterized by territorial marginality.

One of the most emerging and least-addressed issues in the literature in these urban contexts is the risk of implementing single, unconnected punctual actions, benefiting only some points in the urbanized area without an overall macro-scale reasoning. The contribution recalls this aspect and is meant to support architects, planners, and local government administrations in proposing technological design solutions involving the whole housing community, also in agreement with the surrounding areas, precisely in the logic of sustainable energy communities, aiming at a SECAP. This paper can be useful to inform administrations, especially in inland areas, so that they can rethink their territories as energy communities and achieve self-sufficiency. The limited size of these centers favors the sense of community; with little effort, the active management of energy resources can be a promising solution to fight energy poverty while co-creating sustainable systems suitable for the community's needs. Thus, adjacent municipalities could develop by constructing an energy network to sustain two or more cities through a system of technologies integrated into periurban areas, which are often not valorized. For example, the "Agro-Energy Park" described here can be an opportunity for the regeneration of an area in a state of abandonment; at the same time, it turns into a place for sustainable production. It preserves and extends cultivated areas and realizes an integrated system of mini/micro-wind energy production, sized to produce most of the energy demand of the park and the municipality itself.

The advantages of these actions in small municipalities are, first and foremost, the enhancement of an original vocation for sustainability; the implementation of design solutions with an overall vision, as these territories are less vast and complex than metropolitan contexts; and the possibility of achieving effectiveness and benefits quickly. In addition, tactical actions aimed at energy efficiency, such as those designed here, could be opportunities for concrete development and the conferring of new desirability, as well as having a trailing effect on other initiatives.

Spatial proximity is essential and reveals the need for place-based approaches to developing energy communities. This aspect is underscored by the National Plan for Recovery and Resilience (NRRP): an instrument that outlines the objectives, reforms, and investments that Italy aims at, thanks to Next Generation EU funds to mitigate the economic and social impact of the recent pandemic. Within the plan, the investment dedicated to Energy Communities includes a clear reference to supporting projects focused on areas where the greatest socio-territorial impact is expected, specifying, in particular, municipalities under 5000 inhabitants as targets [35].

Implementing renewable and low-impact energy is the basic principle of a vision to counteract CO_2 emissions to the environment and achieve energy self-sufficiency in these communities. The proposed interventions also converge from the perspective of urban regeneration since historic urban settings, particularly those already in a state of disrepair, such as the cases studied herein, are particularly vulnerable to various climatic threats that jeopardize their preservation.

Due to their scale, energy supply interventions are potentially the most impactful on the landscape, especially in vulnerable-build contexts, among sustainable energy and climate adaptation interventions possibly included in the plan. The challenge is to reconcile the need for climate adaptation with technological elements to strengthen the quality of the built environment. This is why the paper focuses on technologies for energy supply from renewable sources and shows how they can be integrated into the historic fabric with a few targeted interventions, suggesting design choices that arise from the context study and analysis phase. During this study, it has been kept in mind that the cases studied are strongly characterized by evident physical characteristics, and the settlements are strongly influenced by their orography, producing their landscape appeal and infrastructural difficulties. It is understood that the actions presented have a high cost and that this may discourage administrations from implementing them, but because they are modular systems, implementation can be carried out in phases. The technologies that are most suitable in historical contexts and that are on the market still have lower efficiencies and higher costs than traditional ones, and being new, the necessary maintenance and useful life of these elements is still uncertain. Therefore, the calculation of payback time may be improper. In fact, however, these are elements could simultaneously respect the overall perception of the historic urban landscape and drive the energy transition towards the use of RES. Therefore, it is worth investigating. Wherever it is possible to install it, the geothermal plant is a good compromise; micro-wind power can be impactful on the landscape if not integrated in a park, as presented here.

If technologies for energy self-sufficiency from renewable sources were associated with a building renovation aimed at exploiting and enhancing the thermal energy conservation attitudes that traditional architecture generally preserves, fewer active technologies for energy production would be needed; therefore, the use of less-expensive solutions could also be realized. On the other hand, conservation is a strategy for sustainability, and understanding contemporary values to revitalize and rethink future practices in historic cities can help settlements respond to new needs and priorities, such as climate change [36]. I would conclude by reflecting that it is the built heritage itself that gives us the answer; design choices have been defined after careful analysis that highlight various morphological, typological, technical, material, and historical characteristics that suggest how to improve the energy behavior of the built environment.

Funding: This research received no external funding.

Data Availability Statement: The original contributions presented in the study are included in the article, further inquiries can be directed to the corresponding author.

Acknowledgments: Cimilluca, M.R. and Alabiso, D.A. have conducted work under the supervision of Mamì A., Schillaci F., at the DARCH of the University of Palermo, who is acknowledged for his precious contribution. The author Nicolini, E. and Mormino, L., have also served as supervisors. Cimilluca, S. and Pagano, A. have collaborated Scientific Consultants.

Conflicts of Interest: The author declares no conflicts of interest.

References

1. Hennessy, K.; Lawrence, J.; Mackey, B. *IPCC Sixth Assessment Report (AR6): Climate Change 2022-Impacts, Adaptation, and Vulnerability: Regional Factsheet Australasia*; IPCC: Geneva, Switzerland, 2022.
2. European Commission. *Communication from the Commission to the European Parliament, the Council, the European Economic and Social Committee and the Committee of the Regions United in delivering the Energy Union and Climate Action—Setting the Foundations for a Successful Clean Energy Transition*; 18.6.2019, COM(2019) 285; European Union: Brussels, Switzerland, 2019.
3. European Commission. *Directive (EU) 2018/844 of the European Parliament and Council of 30 May 2018 Amending Directive 2010/31/EU on the Energy Performance of Buildings and Directive 2012/27/EU on Energy Efficiency*; European Union: Brussels, Switzerland, 2021.
4. European Commission. *Proposal for a DIRECTIVE OF THE EUROPEAN PARLIAMENT AND OF THE COUNCIL Amending Directive (EU) 2018/2001 on the Promotion of the Use of Energy from Renewable Sources, Directive 2010/31/EU on the Energy Performance of Buildings and DIRECTIVE 2012/27/EU on Energy Efficiency*; European Union: Brussels, Switzerland, 2022.
5. Lesk, C.; Csala, D.; Hasse, R.; Sgouridis, S.; Levesque, A.; Mach, K.J.; Horton, R.M. Mitigation and adaptation emissions embedded in the broader climate transition. *Proc. Natl. Acad. Sci. USA* **2022**, *119*, e2123486119. [CrossRef]
6. Reis, I.F.; Gonçalves, I.; Lopes, M.A.; Antunes, C.H. Business models for energy communities: A review of key issues and trends. *Renew. Sustain. Energy Rev.* **2021**, *144*, 111013. [CrossRef]
7. European Commission. *Directive 2019/944 on Common Rules for the Internal Market for Electricity and Amending Directive 2012/27/EU, and Regulation (EU) 2022/869*; European Union: Brussels, Switzerland, 2019.
8. European Commission. Covenant of Mayors Plans & Actions. Available online: https://www.covenantofmayors.eu/plans-and-actions/action-plans.html (accessed on 17 June 2024).
9. Bertoldi, P. *Guidebook How to Develop a Sustainable Energy and Climate Action Plan (SECAP)*; Publication Office of the European Union: Luxembourg, 2018.
10. Sicily Region. SECAP Platform. Available online: https://www.paesc-sicilia.enea.it/ (accessed on 17 June 2024).
11. Sicily Region, Regional Division of Energy and Public Utility Services, Regional Department of Energy. *"Promuovere la Sostenibilità Energetico-Ambientale nei Comuni Siciliani Attraverso il Patto dei Sindaci" Programma di Ripartizione di Risorse ai Comuni della Sicilia per la Redazione del Piano di Azione per l'Energia Sostenibile e il Clima (PAESC), D.D.G. 1033*; Gazzetta Ufficiale della Regione Siciliana, 2019. Available online: https://pti.regione.sicilia.it/portal/page/portal/PIR_PORTALE/PIR_LaStrutturaRegionale/PIR_AssEnergia/PIR_DipEnergia/PIR_Struttura/PIR_Organizzazioneecompetenze/PIR_CompetAttivita/PIR_CompetenzeAreeServizi/PIR_Serv1Pianifprogrenerg/PIR_pattodeisindaci2/PIR_ModulisticaPAESC/programma%20di%20ripartizione%20PAESC.pdf (accessed on 17 June 2024).
12. Fiore, P.; D'Andria, E. Small Towns... from problem to resource. In *Sustainable Strategies for the Valorization of Building, Landscape and Cultural Heritage in Inland Areas*; FrancoAngeli: Milan, Italy, 2019.

13. Mamì, A. Centri storici e Smart Town: I centri minori come laboratori di nuova residenzialità sostenibile. In *Recupero Valorizzazione Manutenzione nei Centri Storici*; Castagneto, F., Fiore, V., Eds.; LetteraVentidue: Syracuse, Italy, 2013; pp. 250–253.
14. Mamì, A.; Nicolini, E. Riabitare il Patrimonio Urbano ed Edilizio dei Territori Interni: Spazio Digitale per Servizi Sanitari Efficienti. *BDC Boll. Cent. Calza Bini* **2020**, *20*, 317–335.
15. Pinzello, I.; Schilleci, F. (Eds.) Città e Campagna. In *Le Aree di Transizione Come Patrimonio Comune: Le Aree di Transizione Come Patrimonio Comune*; FrancoAngeli: Milan, Italy, 2015.
16. Vitale Brovarone, E.; Cotella, G. *La Strategia Nazionale per le Aree Interne: Una Svolta Place-Based per le Politiche Regionali in Italia*; FrancoAngeli: Milan, Italy, 2020; pp. 22–46.
17. Lucatelli, S. *La Strategia Nazionale, il Riconoscimento delle Aree Interne*; Territorio, (2015/74); FrancoAngeli: Milan, Italy, 2015.
18. Italian Republic. *Decreto del Presidente del Consiglio dei Ministri 16 Maggio 2022, Predisposizione del Piano Nazionale per la Riqualificazione dei Piccoli Comuni*; Official Gazette of the Italian Republic, 2022; General Series n.167, 19-07-2022. Available online: https://www.gazzettaufficiale.it/eli/id/2022/07/19/22A04101/sg (accessed on 17 June 2024).
19. ESPON. *Shrinking Rural Regions in Europe. Towards Smart and Innovative Approaches to Regional Development Challenges in Depopulating Rural Regions*; Policy Brief; European Union: Luxemburg, 2017; pp. 1–14.
20. Paniagua, A. Climate Change and Depopulated Rural Areas in the Global North: Geographical Socio-political Processes and Resistances. In *Handbook of Climate Change Management: Research, Leadership, Transformation*; Springer: Berlin/Heidelberg, Germany, 2020; pp. 1–17.
21. Jin, G.; Peng, J.; Zhang, L.; Zhang, Z. Understanding land for high-quality development. *J. Geogr. Sci.* **2023**, *33*, 217–221. [CrossRef]
22. Jamoussi, B.; Abu-Rizaiza, A.; Al-Haij, A. Sustainable building standards, codes and certification systems: The status quo and future directions in Saudi Arabia. *Sustainability* **2022**, *14*, 10314. [CrossRef]
23. UNESCO. *Urban Heritage for Resilience Consolidated Results of the Implementation of the 2011 Recommendation on the Historic Urban Landscape*; United Nations Educational, Scientific and Cultural Organization: Paris, France, 2023.
24. Italian Republic, Italian Ministry of Culture. Guidelines for the improvement of energy efficiency in cultural heritage architecture and historical urban centers. *J. Technol. Archit. Environ.* **2015**, *12*. Available online: https://www.beniculturali.it/comunicato/linee-di-indirizzo-per-il-miglioramento-dell-efficienza-energetica-nel-patrimonio-culturale-architettura-centri-e-nuclei-storici-ed-urbani (accessed on 17 June 2024).
25. European Commission. Covenant of Mayors—Europe, Local Plans and Actions. Available online: https://eu-mayors.ec.europa.eu/en/action_plan_list (accessed on 17 June 2024).
26. Municipality of Colceresa (VI). *PAESC—Piano d'Azione per l'Energia Sostenibile e il Clima*; Prot. N. 11348 del 31/10/2022; 2022.
27. Municipality of Ponte di Piave (TV). *PAESC—Piano d'Azione per l'Energia Sostenibile e il Clima*; drafted in collaboration with the Province of Treviso, RTI Rekeep/Apleona/Antase, and Divisione Energia srl; 2022.
28. Municipality of Racconigi (CN). *PAESC—Piano d'Azione per l'Energia Sostenibile e il Clima*; drafted in collaboration with Environment Park S.p.A.; 2023.
29. Municipality of Caraglio (CN). *PAESC—Piano d'Azione per l'Energia Sostenibile e il Clima*; drafted in collaboration with Tecnolusso Ingegneria e Architettura; 2022.
30. Municipality of Arcugnano (VI). *PAESC—Piano d'Azione per l'Energia Sostenibile e il Clima con integrazione Programma Qualità dell'Aria (PQA)*; drafted by ADAPTEV—IUAV Spinoff; 2023.
31. Italian Republic, Presidency of the Council of Ministries. Conversione in Legge, con Modificazioni, del Decreto-Legge 1° Marzo 2022, n. 17, Recante Misure Urgenti per il Contenimento dei Costi dell'Energia Elettrica e del gas Naturale, per lo Sviluppo delle Energie Rinnovabili e per il Rilancio delle politiche Industriali. (22G00048). 2022. Available online: https://www.normattiva.it/uri-res/N2Ls?urn:nir:stato:legge:2022-04-27;34 (accessed on 17 June 2024).
32. Cimilluca, M.R. San Mauro Castelverde, come Smart Town? Una Vision Energetica sulla Riqualificazione Urbana Sostenibile. Master's Thesis, University of Palermo, Palermo, Italy, 2014.
33. Alabiso, D.A. Santa Cristina Gela Smart Village? l'Energia come Occasione di Riqualificazione Sostenibile. Master's Thesis, University of Palermo, Palermo, Italy, 2014.
34. Grasso, M.; Lentini, F.; Vezzani, L. *Lineamenti Stratigrafico-Strutturali delle MADONIE (Sicilia Centro-Settentrionale)*; "Flysch Numidico", Carta geologica f.t. Roma; Istituto di Scienze della Terra, Università di Catania: Catania, Italy, 1978.
35. Italian Republic, Ministry of the Environment and Energy Security. Investimento 1.2—Promozione Rinnovabili per le Comunità Energetiche e l'Auto-Consumo. 2022. Available online: https://www.mase.gov.it/pagina/investimento-1-2-promozione-rinnovabili-le-comunita-energetiche-e-l-auto-consumo (accessed on 17 June 2024).
36. UNESCO and Ministry of Education, Culture and Science of The Netherlands. World Heritage City Lab Historic Cities, Climate Change, Water, and Energy. In Proceedings of the 10th Anniversary of the Recommendation on the Historic Urban Landscape, Online, 16–17 December 2021.

MDPI AG
Grosspeteranlage 5
4052 Basel
Switzerland
Tel.: +41 61 683 77 34

MDPI Books Editorial Office
E-mail: books@mdpi.com
www.mdpi.com/books

www.ingramcontent.com/pod-product-compliance
Lightning Source LLC
Chambersburg PA
CBHW061602220926
51850CB00005B/399